Supergravity '81

Supergravity '81

Proceedings of the 1st School on Supergravity held on 22 April - 6 May, 1981, at the International Centre for Theoretical Physics, Trieste, Italy.

Edited by S. Ferrara and J.G. Taylor

CAMBRIDGE UNIVERSITY PRESS
Cambridge
London New York New Rochelle
Melbourne Sydney

Published by the Press Syndicate of the University of Cambridge
The Pitt Building, Trumpington Street, Cambridge CB2 1RP
32 East 57th Street, New York, NY 10022, USA
296 Beaconsfield Parade, Middle Park, Melbourne 3206, Australia

First published 1982

Printed in Great Britain at the University Press, Cambridge

Library of Congress catalogue card number: 82-1204

British Library cataloguing in publication data

Supergravity '81
1. Supergravity
I. Ferrara S. II. Taylor J.G.
531'.5 QC173.59.S/

ISBN 0 521 24738 1

CONTENTS

CONTRIBUTORS

E. CREMMER: Lab. de Physique Théorique, Ecole Normale Supérieure, 24 rue Lhomond, F-75231 Paris CEDEX 05, France

R. D'AURIA: Istituto di Fisica Teorica, University of Turin, Corso M. d'Azeglio 46, I-10125 Turin, Italy

B. DE WIT: NIKHEF, Postbus 41882, 1009 DB Amsterdam, The Netherlands

M. DUFF: Department of Theoretical Physics, Imperial College, South Kensington, London SW7 2AZ

P. FRE: Istituto di Fisica Teorica, University of Turin, Corso M. d'Azeglio 46, I-10125 Turin, Italy

S. FERRARA: CERN, CH 1211 Geneva 23, Switzerland

R. GRIMM: Institut für Theoretische Physik, Universität Karlsruhe, Physikhochhaus, Karlsruhe 1, D-7500 Germany

M.T. GRISARU: Department of Physics, Brandeis University, Waltham, MA 02254, U.S.A.

R. KALLOSH: Lebedev Physical Institute, Leninsky Prospekt 53, 117 924 Moscow, U.S.S.R.

T. REGGE: Istituto di Fisica Teorica, University of Turin, Corso M. d'Azeglio 46, I-10125 Turin, Italy

A. SALAM: International Centre for Theoretical Physics, P.O. Box 586, Miramare, Trieste, I-34100 Italy

C.A. SAVOY: CERN, CH 1211 Geneva 23, Switzerland

J. STRATHDEE: International Centre for Theoretical Physics, P.O. Box 586, Miramare, Trieste, I-34100 Italy

J.G. TAYLOR: Department of Mathematics, King's College London, Strand, London, WC2R 2LS

P. VAN NIEUWENHUIZEN: Department of Physics, SUNY at Stony Brook, Stony Brook, NY 11794, U.S.A.

P.C. WEST: Department of Mathematics, King's College London, Strand, London, WC2R 2LS

EDITORS' PREFACE

S. Ferrara
CERN, Geneva, Switzerland

J. G. Taylor
Dept. of Mathematics,
King's College, London

We are delighted to have been editors of the Proceedings of the first school on supergravity, and thereby enable the elegance and beauty of the theory of supergravity to be accessible to a wider audience than purely those who have worked somewhat extensively in it. The School itself was held in the stimulating atmosphere of the International Centre for Theoretical Physics at Trieste with the active participation of its Director, Professor Salam. The lecturers themselves are among the leading experts in the world in the subject, and the resulting lectures present a very broad view of supergravity and supersymmetry, and also give a tantalising hint of the future developments which promise to unify gravity with the other known forces of nature. We trust that this volume will allow workers more easily to enter the field who are stimulated by this program.

After the School was concluded a topical Workshop on supergravity and supersymmetry was held for 3 days. This had contributions from 30 speakers on the latest developments in these subjects and on attempts at super-unification. In addition to the lecturers at the School, speakers at the Workshop were: P. Aichelburg, C. Aragone, P. Breitenlohner, R. D'Auria, S. Deser, J. Ellis, P. Fre, D.Z. Freedman, S.J. Gates, L. Girardello, W. Lang, J. Kukierski, S. Macdowell, B. Morel, P. Nath, H. Nicolai, C. Orzalesi, A. Rogers, K. Sibold, M. Sohnius, E. Sokatchev, K. Stelle, P. Townsend and A. Van Proeyen. These contributions were all at the highest level and showed the breadth and speed of development in these areas.

We would like to thank the lecturers who made the School so successful and also made this volume possible. We also thank the staff of the Centre at Trieste and the local organising committee, Professors Iengo and Akeampong, who worked so hard to run the School and Workshop so efficiently. Finally we thank the Director of the Centre at Trieste, Abdus Salam, for his enthusiastic support of the School.

INTRODUCTION

A. SALAM
Director, International Centre for Theoretical Physics, Trieste

Supersymmetry - the symmetry between bosons and fermions - is an incredible symmetry. It could have been discovered at any time subsequent to 1935 after the canons of quantum theory of fields had been established. However, even in 1971 when it was first conceived of in the USSR, the existence of such a symmetry went unnoticed, and its significance missed. This situation persisted till the symmetry was rediscovered in 1973. And even thereafter, though one had early recognized its elegance, the freedom of supersymmetric Lagrangians from field-theoretic infinities, and the remarkable positivity of supersymmetric Hamiltonians, the fact that there appeared no direct evidence of its existence meant that supersymmetry still remained a minority interest.

The developments subsequent to 1973 are well known. Two of the early ones concerned firstly the recognition of the existence of extended supersymmetries (N = 2, 3, ...) where internal symmetries are married to simple supersymmetry (N = 1), plus the concomitant phenomenon of central charges and secondly the superspace and superfield method of exhibiting N = 1 simple supersymmetry in field theory.

In 1976, the prize problem of supersymmetrising gravity was solved and the existence of spin-$\frac{3}{2}$ gravitons surmised. With extended supergravities (N = 2, 3, ..., 8) was born the hope that particles of spin two and spin one, mediating all the four fundamental forces (including gravity), plus Higgs, plus "source" matter of half-integer spins may form one single supermultiplet of an extended supergravity theory, uniting the "marble" of gravity with the "base wood" of matter - as Einstein dreamt.

But it was recognized fairly early on that the spectrum of particles in even the maximally extended supergravity theory (N = 8) was inadequate to accommodate presently known quarks, leptons and gauge particles. It was also recognized that after gauging, any such theory must contain two coupling constants - Newtonian and an unacceptably large cosmological constant in lieu of the fine structure constant. These circumstances have led to the belief that possibly the basic supermultiplet of N = 8 supergravity should be interpreted as describing preons, of which quarks, leptons and gauge particles are made, rather than these particles themselves.

On the technical side, it has also come to be recognized that the geometrisation of physics, which started with the Einsteinian revolution of associating gravity with the curvature of space-time would not be complete till one has described supergravities in terms of geometrical entities (super-curvature and super-torsion) in extended space-times - extended bosonically as for compactified Kaluza-Klein type of integer-spin theories, and extended fermionically for curved superspaces. These two problems, the physical relevance of supergravities and the geometrisation of dynamics to be achieved through a super-space formulation are still with us.

The Spring College at Trieste, held during 22 April up to 3 May 1981, whose lectures comprise the present volume is perhaps the first systematically complete didactic attempt to give an exposition of the present status of supersymmetry and supergravity theories. After an introduction to global supersymmetry by J. A. Strathdee, J. G. Taylor introduces extended supersymmetries, auxiliary fields and superfield Lagrangians, including Lagrangians for linearised N = 1, 2 supergravities. This is followed by S. Ferrara's analysis of the multiplet spectrum in extended supersymmetries and M. T. Grisaru's lectures on superfield perturbation techniques. P. C. West then discusses the local field representations of N = 1 supersymmetry giving the minimal version of N = 1 supergravity. Supergravities proper, without using superfields, are the subject of P. van Nieuwenhuizen's lectures who describes how finite S-matrix elements arise at the lowest non-trivial order for supergravity. This is followed by M. Duff's analysis of one-loop counterterms for extended gauged supergravities and the demonstration that for $N > 4$, by a suitable choice of scalar field representations, the appropriate β-function can be shown to vanish.

The construction of the Poincaré supergravity Lagrangians by reduction from the super-conformal N = 2 and 4 supergravities is exhibited by B. de Wit, while E. Cremmer exhibits this construction for N = 8, as well as the hidden symmetries of the resulting Lagrangian, by reducing an N = 1 simple supergravity from eleven dimensional space-time to the physical space. The superfield approach to supergravities, including the investigation of necessary constraints and the investigation of higher loop infinities are the subject of the lectures by R. Grimm and R. Kallosh. And finally T. Regge develops the alternative group-manifold approach to constructing supergravity Lagrangians which he and his collaborators have favoured.

This bald description of the contents of the lectures does no justice to the riches contained in each lecture series in this explosively developing field, with diverse approaches jostling and vying with each other at present. The problems being addressed are well known: (1) the construction of extended off-shell super-gravitational Lagrangian - off-shell in the sense that the super-algebra closes without the use of equations of motion: (2) the marriage of geometry and dynamics, which one hopes may be achievable in a super-space formulation: (3) the elucidation of the significance of central charges: (4) the conjecture that $N > 4$, or at the least the $N = 8$ supergravity S-matrix may be infinity-free: (5) and finally the physical relevance of supergravity theories and in particular $N = 8$ theory for super-unification of all fundamental forces and particles.

At the humbler level there has recently been a revival of interest in supersymmetric grand unifying theories (without incorporating gravity), with the recognition that the only "natural" theory of low-mass Higgs particles is supersymmetric, with Higgs and chiral fundamental fermions in the same multiplet. Such theories offer the hope that super-symmetry may possibly become manifest at relatively low energies $\approx$ 100 GeV in contrast to the distant view of supergravity theories which surmise that the symmetry may possibly not make itself felt till Planck energies are reached. With this new hope, the timing of the Supergravity College and the publication of the record of its lectures appears opportune. Our appreciation goes to the lecturers as well as to the Directors of the College - Professors J. G. Taylor, S. Ferrara and R. Iengo.

INTRODUCTION TO SUPERSYMMETRY

J. Strathdee
International Centre for Theoretical Physics, Trieste, Italy.

Abstract. General features of global supersymmetry are discussed. The algebra is defined and its unitary irreducible representations are summarized. Superfields and superspace are introduced and their roles in realizing the symmetry are examined. Some examples of Lagrangians are included which illustrate how global and local internal symmetries are combined with supersymmetry. Finally, some aspects of the spontaneous breaking of supersymmetry are discussed.

1 INTRODUCTION

Supersymmetry represents an extension of the symmetry of spacetime. It differs from the familiar internal symmetries in that it combines states of different angular momentum in a single irreducible multiplet. More particularly, it combines fermions with bosons. Whether or not this symmetry has any basis in reality, it remains an intriguing mathematical structure and one which certainly deserves further study.

The formalism is somewhat complicated but the underlying idea is a simple one. Ordinary, flat Minkowskian spacetime, whose established symmetries are characterized by the Poincaré group, may have further, as yet hidden, symmetry. According to this hypothesis the Poincaré group would be only a subgroup of the set of spacetime symmetries. Of course there are other symmetries in nature, such as the electromagnetic U(1), which may yet be found to have their origin in the structure of spacetime, but here we are concerned with symmetry operations which act non-trivially on the spacetime co-ordinates themselves. The generators of supersymmetry are of this type in that they do not commute with the Poincaré generators. To be precise, the new generators carry angular momentum. It is because of this property that they are able to relate particles with different spins.

Since supersymmetry involves the adjoining of new generators to those of the Poincaré group, the result is sometimes called the "extended" Poincaré group. One can take the view that this symmetry should be expressed most naturally by its action on a suitably "extended" spacetime. This idea is emphasized in the following where the notion of superspace is introduced. The points of superspace are labelled by eight co-ordinates (x_μ, θ_α) where the new quantities θ_α are taken to be anti-commuting. With respect to ordinary Lorentz transformations they comprise a 4-component Majorana spinor. Fields are defined over superspace and it will be shown how to decompose them into ordinary Fermi- and Bose-type components. The techniques for constructing invariant action functionals are described and, finally, some features of the spontaneous violation of supersymmetry are discussed. One of the most remarkable properties of supersymmetric Lagrangian theories, the cancellation of ultraviolet divergences, is mentioned only in passing. Space does not allow us to enter into this matter. A detailed treatment is available in the review of Fayet and Ferrara (1977). More detail concerning the superspace approach is contained in a review by Salam and Strathdee (1978). New insights into the nature of spontaneous supersymmetry violation are provided by some recent work of Witten (1981).

The algebra of supersymmetry in its simplest form is generated by fourteen hermitian operators. These include the ten generators, P_μ and $J_{\mu\nu}$, of the Poincaré group and four anticommuting generators, Q_α, of what may be called "supertranslations". The Poincaré generators obey the familiar commutation rules, as usual. The new elements, Q_α, obey the following

$$[Q_\alpha, P_\mu] = 0 \quad , \tag{1.1a}$$

$$[Q_\alpha, J_{\mu\nu}] = \frac{1}{2} (\sigma_{\mu\nu})_\alpha{}^\beta \, Q_\beta \quad , \tag{1.1b}$$

$$\{Q_\alpha, Q_\beta\} = -(\gamma_\mu C)_{\alpha\beta} \, P_\mu \quad . \tag{1.1c}$$

Of these rules, the first means that Q_α is unaffected by spacetime translations, the second means that it transforms as a Dirac spinor under homogeneous Lorentz transformations. The anticommutator (1.1c) implies that the supertranslations are not independent, and it is the main ingredient of the symmetry. The generators Q_α are subject to the reality condition,

$$Q_\alpha = C_{\alpha\beta} \, \bar{Q}^\beta \quad . \tag{1.2}$$

They are components of a Majorana spinor. (The charge conjugation matrix C is antisymmetric and has the property that $C^{-1}\gamma_\mu C = -\gamma_\mu^T$. It follows that $\gamma_\mu C$ is symmetric.)

Extensions of supersymmetry are generated by sets of Majorana spinors $Q_{\alpha j}$, $j = 1,2,\ldots,N$. For these extensions the rule (1.1c) is replaced by

$$\{Q_{\alpha i}, Q_{\beta j}\} = -\delta_{ij}(\gamma_\mu C)_{\alpha\beta} P_\mu + C_{\alpha\beta} Z_{ij} + (\gamma_5 C)_{\alpha\beta} Z'_{ij}, \quad (1.1c')$$

where the generators Z_{ij} and Z'_{ij} are antisymmetric in i,j. They are central charges in that they commute with all generators of the system. We shall be concerned only with the simple ($N = 1$) supersymmetry in these lectures.

The structure of the unitary irreducible representations is easily arrived at by applying the methods of Wigner. Firstly, for the massive case one can generate the basis vectors $|p\rangle$ by acting with Lorentz boosts on a set of rest states. The set of states with a fixed value of the 4-momentum p_μ is invariant under the action of Q_α owing to (1.1a). On these states the right-hand side of (1.1c) becomes a c-number matrix, $-\not{p}C$. This means that the four components Q_α generate a Clifford algebra and so can be represented by 4×4 matrices. The manifold of states with fixed 4-momentum is reduced by the action of supertranslations into 4-dimensional invariant subspaces. On the other hand, the rest states are reduced into $(2j+1)$-dimensional subspaces by the action of 3-rotations. Hence the irreducible representations of the extended symmetry are spanned by $4(2j+1)$ rest states. These can be boosted by Wigner's method to give the complete basis for an irreducible representation. A typical representation is characterized by two invariants, $p_\mu^2 = M^2 > 0$ and a spin j. The spin content of the rest states is

$$j \oplus (j - \tfrac{1}{2}) \oplus (j + \tfrac{1}{2}) \oplus j \quad . \quad (1.3)$$

(If $j = 0$ then the representation $j - \frac{1}{2}$ is, of course, absent.)

For the massless case one can again apply the Wigner method but starting from a set of states with fixed lightlike 4-momentum, instead of rest states. This time the right-hand side of (1.1c) becomes a numerical matrix of rank two. From this it follows that the manifold of states with fixed lightlike 4-momentum is reduced by the action of supertranslations into 2-dimensional invariant subspaces. A typical lightlike representation of the extended Poincaré algebra, in addition to having

$p_\mu^2 = 0$, is characterized by an helicity, λ. It contains two irreducible representations of the Poincaré group with helicities

$$\lambda \oplus \lambda + \frac{1}{2} \quad . \tag{1.4}$$

The unitary representations (1.3) and (1.4) define the basic 1-particle states with which supersymmetric theories are outfitted. It is possible to construct a Lagrangian dynamics for these theories by introducing local fields to interpolate the states. Although the Lagrangian can be expressed in terms of the usual Fermi and Bose type fields on 4-dimensional spacetime, such a description does not do full justice to the symmetry. In these lectures I shall introduce the notion of "superfields" which are defined on an "extended" spacetime and which exhibit the supersymmetry in a manifest form.

Just as ordinary spacetime can be thought of as a quotient space,

$$\text{spacetime} = \frac{\text{Poincaré}}{\text{Lorentz}} \quad ,$$

it is possible to define a superspace extension as a new kind of quotient space,

$$\text{extended spacetime} = \frac{\text{extended Poincaré}}{\text{Lorentz}} \quad .$$

Since the extended Poincaré group has fourteen parameters, it follows that superspace is parametrized by eight co-ordinates. For example, one could take

$$e^{ix_\mu P_\mu} \, e^{i\bar{\theta}^\alpha Q_\alpha} \quad ,$$

where x_μ is the usual Minkowski co-ordinate and θ_α is a Majorana spinor. From the algebra (1.1) it is possible to deduce the behaviour of x and θ under the action of supertranslations. The result is,

$$\begin{aligned} x_\mu &\rightarrow x_\mu + \frac{i}{2}\,\bar{\varepsilon}\,\gamma_\mu\,\theta \\ \theta &\rightarrow \theta + \varepsilon \quad , \end{aligned} \tag{1.5}$$

where ε is a Majorana spinor which characterizes the supertranslation. The superfield, $\Phi(x,\theta)$, is simply a function (operator valued in the quantized theory) defined on the extended spacetime, or superspace. Some of the features of these fields are discussed in the following section.

2 SUPERFIELDS

It is natural to require that superfields should, in some sense, be local. By this we shall mean

$$[\Phi(x,\theta),\ \Phi(x',\theta')] = 0 \text{ for spacelike } x-x' \quad . \tag{2.1}$$

To see what this implies, consider the first few terms of an expansion in powers of θ,

$$\Phi(x,\theta) = A(x) + \bar{\theta}\psi(x) + \ldots \tag{2.2}$$

If A(x) is a scalar then it is clear that $\psi(x)$ must be a spinor. Therefore, in order to maintain the usual association between spin and statistics, A(x) should be a commuting (bosonic) field, while $\psi(x)$ should be an anticommuting (fermionic) field. On the other hand, from (2.1) it follows that the combination $\bar{\theta}\psi$ should be bosonic like A. We conclude that θ must itself be an anticommuting quantity,

$$\{\theta_\alpha,\theta_\beta\} = 0 \quad . \tag{2.3}$$

This is important because it implies that the expansion (2.2) must terminate after a finite number of terms. Indeed, since θ_α has only four components (and $\bar{\theta}$ is not independent) the series must end with the fourth order term. It is not possible to construct the necessarily antisymmetric tensors, $\theta_{\alpha_1}\theta_{\alpha_2},\ldots$ of rank greater than four. Hence the expansion (2.2) is really a polynomial of the fourth order. A convenient representation is given by

$$\Phi(x,\theta) = A(x) + \bar{\theta}\psi(x) + \frac{1}{4}\ \bar{\theta}\theta F(x) + \frac{1}{2}\bar{\theta}\gamma_5\theta G(x) +$$

$$+ \frac{1}{4}\bar{\theta}i\gamma_\nu\gamma_5\theta V_\nu(x) + \frac{1}{4}\bar{\theta}\theta\ \bar{\theta}\chi(x) + \frac{1}{32}(\bar{\theta}\theta)^2\ D(x), \tag{2.4}$$

where A, F, G, D are scalars, V_ν is a vector, and ψ ,χ are spinors. All of these component fields may be complex (more generally, they would all belong to the same representation of any symmetry group which commutes with the extended spacetime group) or they could be real, with the spinors ψ, χ being real in the Majorana sense (1.2). It is possible to have $\Phi(x,\theta)$ transforming as a spinor or tensor rather than a scalar. In such cases all the "component" fields A, ψ, F,... must carry an appropriate spinor or tensor index.

It is a simple matter to extract transformation rules for the component fields in (2.4) corresponding to a supertranslation (1.5). For infinitesimal ε one obtains

$$\delta A = \bar{\varepsilon}\psi$$

$$\delta\psi = \frac{1}{2}(F + \gamma_5 G + i\gamma_\nu\gamma_5 V_\nu - i\partial\!\!\!/ A)\varepsilon$$

$$\delta F = \frac{1}{2}\bar{\varepsilon}\psi - \frac{1}{2}\bar{\varepsilon}i\partial\!\!\!/\psi,$$

etc. However, it turns out that these transformations are not irreducible. The simplest way to see this is by constructing a set of differential operators, D_α , which anticommute with the supertranslations.

To define differential operators in superspace it is necessary to adopt a convention. We propose to define the derivative $\partial/\partial\bar{\theta}^\alpha$ by the formula

$$\delta\Phi = \delta\bar{\theta} \frac{\partial\Phi}{\partial\bar{\theta}} = \frac{\partial\Phi}{\partial\theta}\delta\theta \quad , \tag{2.5}$$

where the infinitesimal variation $\delta\bar{\theta}$ stands to the left and $\delta\theta$ stands to the right. If Φ is a real scalar then $\partial\Phi/\partial\bar{\theta}$ is a Majorana spinor. The variation $\delta\Phi$ due to an infinitesimal supertranslation (1.5) can be expressed succinctly,

$$\delta\Phi = -\bar{\varepsilon}\left(\frac{\partial}{\partial\bar{\theta}} + \frac{i}{2}\gamma_\mu\theta\frac{\partial}{\partial x_\mu}\right)\Phi \quad . \tag{2.6}$$

It is easy to verify that the differential operator (2.6) which generates supertranslations anticommutes with the operator D defined by

$$D = \frac{\partial}{\partial\bar{\theta}} - \frac{i}{2}\gamma_\mu\theta\frac{\partial}{\partial x_\mu} \quad . \tag{2.7}$$

The differential operators, D_α , can be shown to satisfy the anticommutation rules

$$\{D_\alpha, D_\beta\} = -i(\gamma_\mu C)_{\alpha\beta}\frac{\partial}{\partial x_\mu} \quad ,$$

from which it follows that their chiral projections

$$D_- = \frac{1 - i\gamma_5}{2} D \tag{2.8}$$

have vanishing anticommutator. This means that the conditions

$$D_-\Phi = 0 \tag{2.9}$$

are mutually compatible as well as being covariant with respect to the full supersymmetry. They are solved by superfields of the form

$$\Phi_+(x,\theta) = \exp\left(-\frac{1}{4}\bar{\theta}\partial\!\!\!/\gamma_5\theta\right)\left(A_+(x) + \bar{\theta}_-\psi_+(x) + \frac{1}{2}\bar{\theta}_-\theta_+ F_+(x)\right) \quad , \tag{2.10}$$

where $\theta_\pm$ denote the chiral projections of θ . The field Φ_+ is known as a chiral superfield and its independent components transform under supertranslations according to

$$\delta A_+ = \bar{\varepsilon}\psi_+$$

$$\delta\psi_+ = F_+\varepsilon_+ - i\partial\!\!\!/ A_+\varepsilon_-$$

$$\delta F_+ = -\bar{\varepsilon}_+ i\partial\!\!\!/\psi_+ \quad . \tag{2.11}$$

The scalar components A_+, F_+ are necessarily complex while the spinor component ψ_+ has positive chirality. (It is possible to define another chiral superfield Φ_- which transforms like the complex conjugate, Φ_+^*, inequivalent to Φ_+.)

There is yet another piece in the scalar superfield. This is the "transverse vector" superfield, Φ_1, defined by

$$\Phi_1(x,\theta) = A_1(x) + \bar{\theta}\psi_1(x) + \frac{1}{4}\bar{\theta}i\gamma_\nu\gamma_5\theta\, V_{1\nu}(x) + $$
$$+ \frac{1}{4}\bar{\theta}\theta\, \bar{\theta}i\partial\!\!\!/\psi_1(x) + \frac{1}{32}(\bar{\theta}\theta)^2\, \partial^2 A_1(x) \qquad (2.12)$$

defined by the second order condition $\bar{D}_+ D_- \Phi_1 = 0$. Its component fields transform according to

$$\delta A_1 = \bar{\varepsilon}\psi_1$$
$$\delta\psi_1 = \frac{1}{2}(i\gamma_\nu\gamma_5 V_{1\nu} - i\partial\!\!\!/ A_1)\,\varepsilon$$
$$\delta V_{1\nu} = -\,\bar{\varepsilon}\,\sigma_{\mu\nu}\,\gamma_5 i\partial_\nu\psi_1 \quad . \qquad (2.13)$$

The components here can be real and the vector component $V_{1\nu}$ is necessarily transverse.

Because the defining condition (2.1) is linear, it follows that the product of two chiral scalars, Φ_+, is also chiral. This is not true of products involving both Φ_+ and Φ_+^* or Φ_1. The latter products make reducible superfields.

Finally, in order to set up an action principle for superfields it is necessary to be able to make integral invariants out of them. Suppose that the Lagrangian density consists of sums of products of $\Phi_+(x,\theta)$ and $\Phi_+^*(x,\theta)$. It must then take the form of a scalar superfield, $\mathcal{L}(x,\theta)$, and so can be represented as a polynomial in θ . Under supertranslations the components of $\mathcal{L}$ will transform in the manner described above. In particular, its D-component, $\mathcal{L}_D(x)$, will change by a term which is a 4-divergence. Hence, in effect, the integral

$$\int d^4x\ \mathcal{L}_D(x)$$

is invariant. Now, the D-component of a superfield can be extracted by differentiating four times with respect to θ. Alternatively, this can be looked upon as <u>integration</u>. That is, paradoxical as it may seem, one is free to define the integral over an anticommuting variable, say the co-ordinate θ_1, as the derivative

$$\int d\theta_1\ f(\theta_1) = \frac{\partial}{\partial\theta_1}\ f(\theta_1) \ . \qquad (2.14)$$

Since $f(\theta_1)$ is necessarily linear in θ_1 this merely defines the integral as the coefficient of θ_1 in $f(\theta_1)$. It has the virtue that integration by parts is always permitted, i.e.

$$\int d\theta_1 \frac{\partial}{\partial\theta_1} f(\theta_1) = 0 \quad . \tag{2.15}$$

Thus, we shall write for the invariant made from the D-component of $\mathcal{L}(x,\theta)$, the expression

$$\int d^4x\, d^4\theta\ \mathcal{L}(x,\theta) \tag{2.16}$$

with the understanding that integration over θ is to be understood in the sense (2.14). This does not exhaust the possibilities, however. It may happen that $\mathcal{L}(x,\theta)$ contains a chiral piece $\mathcal{L}_+(x,\theta)$. Such a term would make no contribution to (2.16) because its D-component is a 4-divergence as can be seen from the representation (2.10). In this case, as can be seen from (2.11), an infinitesimal supertranslation causes the F-component to change by a 4-divergence. The spacetime integral of $\mathcal{L}_{+F}(x)$ is therefore an invariant. Being the coefficient of $\bar{\theta}_-\theta_+ \sim \theta_+\theta_+$, the F-component can be extracted by two differentiations with respect to θ_+ or, alternatively, by a double integration. Hence, the integral

$$\int d^4x\, d^2\theta_+\ \mathcal{L}_+(x,\theta) \tag{2.17}$$

is an invariant. It is necessarily complex but its real part can be used in the action. This exhausts the possible invariants in a local theory of superfields.

3 LAGRANGIANS

For a set of chiral superfields,

$$\Phi_+ = \begin{pmatrix} \Phi_+^1 \\ \Phi_+^2 \\ \vdots \\ \Phi_+^n \end{pmatrix} \tag{3.1}$$

a suitable action functional is easily obtained. One writes simply

$$S = \int d^4x\, d^4\theta\ \Phi_+^\dagger\Phi_+ + \left(\int d^2x\, d^2\theta_+\ W(\Phi_+) + \text{h.c.} \right) \quad , \tag{3.2}$$

where $W(\Phi_+)$ is a polynomial. (More generally one need only require that the complex valued function $W(A_+)$ be analytic.) The suitability of (3.2) can be established by reducing it to component form. On substituting the expansion (2.10) and performing the θ-integration one finds (discarding the suffix + from the scalars A_+ and F_+),

$$S = \int d^4x \Big[\partial_\mu A^\dagger \partial_\mu A + \bar{\psi}_+ i \partial\!\!\!/ \psi_+ + F^\dagger F + \Big\{ \frac{\partial W(A)}{\partial A} F + \psi_+^T C^{-1} \frac{\partial^2 W(A)}{\partial A^T \partial A} \psi_+ + \text{h.c.} \Big\} \Big] \quad . \tag{3.3}$$

The scalars F(x) are clearly redundant variables and can be eliminated. The action then takes the form

$$S = \int d^4x \Big[\partial_\mu A^\dagger \partial_\mu A + \bar{\psi}_+ i \partial\!\!\!/ \psi_+ - \frac{\partial W^*}{\partial A^\dagger} \frac{\partial W}{\partial A} + \Big\{ \psi_+^T C^{-1} \frac{\partial^2 W}{\partial A^T \partial A} \psi_+ + \text{h.c.} \Big\} \Big] \tag{3.4}$$

which describes a set of n complex scalar fields, A(x), interacting with n chiral spinors, $\psi_+(x)$. The potential is given by

$$V(A, A^\dagger) = \frac{\partial W^*}{\partial A^\dagger} \frac{\partial W}{\partial A} \tag{3.5}$$

and is manifestly non-negative. This fact hints already at an intrinsic stability of supersymmetric theories. One sees from (3.4) that a necessary condition for the renormalizability of the theory is that W should be a polynomial of order three

$$W(A) = a_i A^i + \frac{1}{2} b_{ij} A^i A^j + \frac{1}{6} c_{ijk} A^i A^j A^k \quad . \tag{3.6}$$

It can be shown by detailed investigation that this form is also sufficient to ensure renormalizability. One of the most remarkable facts to emerge from such an investigation is the result that the coefficients a, b, c in (3.6) receive no radiative corrections, not even finite ones. The kinetic terms in (3.4), however, acquire logarithmically divergent corrections and require renormalization.

It is simple to devise models which incorporate a global symmetry. If the Φ_+^i belong to a representation of some compact group ($\subseteq U(n)$) then (3.4) will be invariant provided W(A) is invariant up to a phase. This implies appropriate restrictions on the coefficients in (3.6).

To formulate theories with local symmetries is more complicated. Firstly, to be consistent with supersymmetry, transformations which are local in x should also be local in θ. If such a transformation group is to act on chiral fields, Φ_+, then it must also be chiral,

$$\Phi_+ \rightarrow e^{i\Lambda_+} \Phi_+ \quad , \tag{3.7}$$

where $\Lambda_+ = \Lambda_+^a(x,\theta)\, Q_a$ belongs to the algebra of the symmetry group. To make the kinetic terms in (3.2) invariant, make the substitution

$$\int d^4x\, d^4\theta\, \Phi_+^\dagger \Phi_+ \rightarrow \int d^4x\, d^4\theta\, \Phi_+^\dagger\, e^{2V}\, \Phi_+ \quad ,$$

where $V(x,\theta)$ transforms such that

$$e^{2V} \rightarrow e^{i\Lambda_+^\dagger}\, e^{2V}\, e^{-i\Lambda_+} \quad . \tag{3.8}$$

The superfield $V(x,\theta)$ belongs to the algebra of the symmetry group and it is real, $V = V^\dagger$. This field is sometimes called the "superpotential". It represents a generalization of the familiar vector potentials of Yang-Mills theory. It is important to realize that V is not a chiral superfield. This much is clear from the transformation rule (3.8) which involves Λ_+^* as well as Λ_+ . On the other hand,the freedom to make gauge transformations can be exercised to remove some of the components (equal in number to the independent component fields in Λ_+) from V. By this means one can reduce it to the form

$$V(x,\theta) = \frac{1}{4}\,\bar\theta i\gamma_\nu\gamma_5\theta\, V_\nu(x) + \frac{1}{2\sqrt{2}}\,\bar\theta\theta\, \bar\theta\gamma_5\lambda(x) + \frac{1}{16}\,(\bar\theta\theta)^2\, D(x) \ , \tag{3.9}$$

where V_ν, λ and D belong to the symmetry algebra and are real. The vector $V_\nu(x)$ will turn out to be the usual Yang-Mills potential, while $\lambda(x)$ and $D(x)$ are its supersymmetry partners.

The form (3.9) corresponds to a special choice of gauge (due to Wess and Zumino) which is revealing in that it minimizes the number of component fields. But it is not manifestly supersymmetric. If a supertranslation acts on (3.9) then an accompanying gauge transformation must be applied in order to restore the form (3.9). The components are then found to transform according to

$$\begin{aligned}
\delta V_\nu &= -\frac{i}{\sqrt{2}}\,\bar\varepsilon\gamma_\nu\lambda \\
\delta\lambda &= \frac{i}{\sqrt{2}}\,(i\gamma_5 D + \frac{1}{2}\,\sigma_{\mu\nu}\, V_{\mu\nu})\,\varepsilon \\
\delta D &= -\frac{i}{\sqrt{2}}\,\bar\varepsilon\nabla\!\!\!/\,\gamma_5\lambda \quad ,
\end{aligned} \tag{3.10}$$

where the Yang-Mills covariant derivatives appear,

$$\begin{aligned}
V_{\mu\nu} &= \partial_\mu V_\nu - \partial_\nu V_\mu - i[V_\mu, V_\nu] \\
\nabla_\mu\lambda &= \partial_\mu\lambda - i[V_\mu,\lambda] \quad .
\end{aligned} \tag{3.11}$$

Of course, the components of the matter fields Φ_+ are subject to compatible transformations under supertranslations in this gauge,

$$\begin{aligned} \delta A &= \bar{\varepsilon}\psi_+ \\ \delta\psi_+ &= F\varepsilon_+ - i\not{\nabla}A\varepsilon_- \\ \delta F &= \bar{\varepsilon}(-i\not{\nabla}\psi_+ + i\sqrt{2}\,\lambda_- A) \quad , \end{aligned} \tag{3.12}$$

where λ_- denotes the negative chirality projection of λ and the covariant derivatives are given by

$$\begin{aligned} \nabla_\mu A &= \partial_\mu A - i\,V_\mu A \\ \nabla_\mu \psi_+ &= \partial_\mu \psi_+ - iV_\mu \psi_+ \quad . \end{aligned} \tag{3.13}$$

The covariant derivatives (3.11) of the component fields of the superpotential, V, are contained in the chiral spinor superfield $W_{\alpha+}$ defined by

$$W_{\alpha+} = -\frac{i}{2\sqrt{2}}\,\bar{D}_+ D_-\,(e^{-2V}\,D_{\alpha+}\,e^{2V}) \quad , \tag{3.14}$$

which can be shown - on the basis of (3.8) - to transform according to

$$W_{\alpha+} \rightarrow e^{i\Lambda_+}\,W_{\alpha+}\,e^{-i\Lambda_+} \quad . \tag{3.15}$$

This superfield is chiral with respect to both its external spinor index and its θ-dependence. In the Wess-Zumino gauge (3.9) it takes the form

$$W_{\alpha+} = \exp\left(-\frac{1}{4}\,\bar{\theta}\not{\partial}\gamma_5\theta\right)\left[\lambda_+(x) + \frac{i}{\sqrt{2}}\left\{D + \frac{1}{2}\,\sigma_{\mu\nu}\,V_{\mu\nu}\right\} + \right.$$
$$\left. + \frac{1}{2}\,\bar{\theta}_-\theta_+\left(\frac{1}{i}\,\not{\nabla}\lambda_-\right)\right] \quad . \tag{3.16}$$

Given that W is chiral and transforms as in (3.15) one can make an invariant

$$\begin{aligned} S(V) &= \int d^4x\, d^2\theta_+\, \frac{1}{4}\,\mathrm{Tr}(W_+^T\, C^{-1}\, W_+) + \text{h.c.} \\ &= \int d^4x\,\mathrm{Tr}\left(-\frac{1}{4}\,V_{\mu\nu}\,V_{\mu\nu} + \bar{\lambda}_- i\not{\nabla}\lambda_- + \frac{1}{2}\,D^2\right) \quad , \end{aligned} \tag{3.17}$$

where the component form given here is valid only in the Wess-Zumino gauge. This expression makes it clear that $V_\mu(x)$ is a Yang-Mills potential, while $\lambda(x)$ represents a multiplet of fermions associated with it.

The coupling of V to the matter fields Φ_+, together with any self-couplings among the matter fields which are compatible with the local symmetry, are governed by the invariant

$$S(\Phi_+,V) = \int d^4x\, d^4\theta\, \Phi_+^\dagger e^{2V} \Phi_+ + \int d^4x\, d^2\theta_+\, W(\Phi_+) + \text{h.c.}$$

$$= \int d^4x \Big[\nabla_\mu A^\dagger \nabla_\mu A + \bar\psi_+ i \not\nabla \psi_+ + F^\dagger F + \left(i\sqrt{2}\, A^\dagger \bar\lambda_- \psi_+ + \text{h.c.} \right) + A^\dagger D A + \left(\frac{\partial W}{\partial A} F + \psi_+^T C^{-1} \frac{\partial^2 W}{\partial A^T \partial A} \psi_+ + \text{h.c.} \right) \Big] . \quad (3.18)$$

Again, the component form applies only in the gauge (3.9). If the local symmetry contains a U(1) factor then there is another possible invariant,

$$S_\xi(V) = \int d^4x\, d^4\theta\, \xi\, \mathrm{Tr} V = \int d^4x\, \xi\, \mathrm{Tr}\, D \quad , \quad (3.19)$$

where ξ is a parameter with the dimensions of $(\text{mass})^2$. This term is important in models where the supersymmetry breaks spontaneously.

Taken together, the three expressions (3.17), (3.18) and (3.19) comprise a gauge invariant and supersymmetric action functional for matter and gauge fields. The auxiliary fields F and D can be eliminated to leave A, ψ_+, λ_- and V_ν as the active fields. In this way one can see the basic degrees of freedom revealed clearly. However, for practical computations it is usually advantageous to employ a supersymmetric gauge and write Ferynman rules for the superfields themselves, rather than for the component fields.

We conclude this section with the remark that it is possible to associate a fermionic quantum number with the phase transformations,

$$\theta \to e^{\alpha\gamma_5}\, \theta \quad . \quad (3.20)$$

Invariance is achieved by requiring $V(x,\theta)$ to transform as a scalar, while the matter fields transform according to

$$\Phi_+(x,\theta) \to e^{i\alpha f}\, \Phi_+\left(x, e^{-\alpha\gamma_5}\theta\right) \quad , \quad (3.21)$$

where the hermitian matrix f is required to commute with all symmetries and W(A) is restricted such that

$$W(e^{i\alpha f} A) = e^{2i\alpha}\, W(A) \quad . \quad (3.22)$$

This quantum number is unique in that it distinguishes among members of a single supermultiplet. The gauge symmetries discussed above make no such distinction. All members of a supermultiplet necessarily belong to the same representation of the gauge symmetries (which commute with the supertranslations). In the case of extended supersymmetries ($N > 1$) the phase symmetry (3.20) generalizes to U(N).

4 SPONTANEOUS BREAKDOWN OF SUPERSYMMETRY

Unbroken supersymmetry can certainly not be held to provide a realistic description of the world as we see it. At the presently accessible energies there is almost no sign of this symmetry. Therefore, if one wishes to impose it at some hypothetical deeper level, one must make provision for its spontaneous violation. The purpose of this section is to discuss some general aspects of the symmetry breaking phenomenon and to exhibit a simple model (due to O'Raifeartaigh) where supersymmetry breaks in the tree approximation.

A unique and important property of supersymmetric theories is that the energy must vanish in the ground state if the symmetry is unbroken and, conversely, if the ground state is not invariant its energy density is positive. This follows directly from the algebra. Thus, from (1.1c) one can write

$$P_0 = \{Q, Q^\dagger\} \tag{4.1}$$

Introducing a complete set of states and taking the vacuum expectation value,

$$\langle 0|P_0|0\rangle = \frac{1}{2}\sum_n \sum_\alpha |\langle 0|Q_\alpha|n\rangle|^2 . \tag{4.2}$$

If Q_α annihilates the vacuum then $\langle P_0 \rangle$ vanishes; otherwise it is positive.

Consider, for example, the system of chiral scalars Φ_+ governed by the action (3.4). In the tree approximation the ground state energy density is given by the potential (3.5), evaluated at its minimum. The minimum value is clearly not negative. If the supersymmetry is to be unbroken then this minimum value must equal zero. Hence, for unbroken supersymmetry we must satisfy the equations

$$\frac{\partial W(A)}{\partial A} = 0 . \tag{4.3}$$

If these equations cannot be satisfied then supersymmetry is spontaneously violated at the tree level. A simple model that illustrates the latter possibility was constructed by O'Raifeartaigh. It uses three superfields, Φ_{0+}, Φ_{1+} and Φ_{2+}. The function W takes the form

$$W = A_0(s + \frac{h}{2} A_1^2) + m\, A_1 A_2$$

with arbitrary coefficients s, h and m. It is clear that the derivatives

$$\frac{\partial W}{\partial A_0} = s + \frac{h}{2} A_1^2$$

$$\frac{\partial W}{\partial A_2} = m\, A_1$$

cannot vanish simultaneously. Hence the vacuum energy density is positive and supersymmetry is broken.

The spontaneous violation of supersymmetry gives rise to a Goldstone particle. In this respect supersymmetry is like any other global symmetry. Indeed, one can demonstrate the effect by an argument which involves only minor modifications of the standard one used for the ordinary global symmetries. It goes as follows. Since the action is invariant there exists a conserved Noether current $J_{\mu\alpha}(x)$ which can be integrated to give the generators

$$Q_\alpha = \int d^3x\, J_{0\alpha}(x) \quad . \tag{4.4}$$

The variation $\delta\psi_+$ caused by an infinitesimal supertranslation can be written

$$\delta\psi_+(0) = i \int d^3x\, [\psi_+(0), \bar{\varepsilon}\, J_0(x)] \quad . \tag{4.5}$$

But, according to (2.11), $\delta\psi_+ = F\, \varepsilon_+ - i\partial\!\!\!/ A\varepsilon_-$, so that, in the vacuum, we must have

$$\langle F(0) \rangle\, \varepsilon_+ = i \int d^3x \quad \langle 0|[\psi_+(0), \bar{\varepsilon}\, J_0(x)]|0\rangle$$

$$= \frac{1}{i} \int d^4x\, \partial_\mu \langle 0|T(\psi_+(0)\, \bar{\varepsilon}\, J_\mu(x))|0\rangle \quad . \tag{4.6}$$

On the other hand, the auxiliary field F is obtained by varying the action (3.3) with respect to $F^\dagger$. The result is,

$$F = - \frac{\partial W^*}{\partial A^\dagger} \quad . \tag{4.7}$$

But this is precisely the quantity which fails to vanish in the vacuum when supersymmetry breaks spontaneously. In such cases, therefore, the right-hand side of (4.6) is non-vanishing and this implies that the 2-point function $\langle 0|T(\psi_+(0)\, \bar{\varepsilon}J_\mu(x))|0\rangle$ has a zero-mass singularity. Thus, one concludes that the system includes a massless spin-$\frac{1}{2}$ particle, the Goldstone fermion.

The explicit form of the supercurrent $J_{\mu\alpha}$ corresponding to the action (3.3) is

$$J_{\mu-} = F^{\dagger}\gamma_{\mu}\psi_{+} + \gamma_{\nu}\gamma_{\mu}C\bar{\psi}_{+}^{T}\, i\partial_{\nu}A$$
$$= \langle F^{\dagger}\rangle\, \gamma_{\mu}\psi_{+} + \ldots \qquad (4.8)$$

where, in the latter form, only the linear part is retained. This serves to define the combination of fields which interpolates the Goldstone fermion,

$$\nu_{+} = \langle F^{\dagger}\rangle\, \psi_{+} \quad . \qquad (4.9)$$

One of the peculiar features of supersymmetric theories is the fact, established in perturbation series, that this symmetry will not break in any finite order if it does not break at the tree level. This does not preclude the possibility that it might break dynamically. The question has been considered recently in an interesting paper by Witten (1981) who establishes some criteria whereby certain models can be excluded. In such models, that is, the supersymmetry must persist in the exact solution.

Abdus Salam and Strathdee, J. (1978). Fortschritte der Physik 26, 57.

Fayet, P. and Ferrara, S. (1977). Phys. Rep. 32C, 249.

Witten, E. (1981). "Dynamical breaking of supersymmetry", Princeton preprint.

REPRESENTATIONS OF EXTENDED SUPERSYMMETRIES AND LINEARISED SUPERGRAVITIES

J.G. Taylor
Department of Mathematics, King's College, London.

I INTRODUCTION: SUPERSYMMETRY AND SUPERFIELDS

1.1 What is supersymmetry?

Supersymmetry is the symmetry which interchanges bosons and fermions. It arose historically from the dual scattering model (Ramond 1971) but was extracted from it and presented as a model-independent symmetry by Wess and Zumino in 1974 (though it had been considered in a neglected earlier paper by Golfand and Likhtman (1971)). Various interacting field theories were constructed which satisfied supersymmetry and were found to possess amelioration of their ultra-violet divergences when quantised (Wess & Zumino 1974 and Illiopolous & Zumino 1974). This divergence reduction was also found to occur (Grisaru, van Nieuwenhuizen & Vermaseren 1976) at the lowest non-trivial level in supergravity (Freedman, van Nieuwenhuizen & Ferrara 1976, Deser & Zumino, 1976), the supersymmetric version of Einstein's gravitational theory. A possible avenue to quantize gravity in a sensible fashion had thus been opened up. The incorporation of matter fields representing leptons, quarks and gluons into supergravity was then found (Grisaru, van Nieuwenhuizen & Vermaseren 1976, van Nieuwenhuizen & Vermaseren 1977, Deser, Kay & Stelle 1977) to be restricted, by the criterion of ultra-violet divergence cancellation, to what are termed extended supergravity theories. It is these which are presently the centre of a great deal of research activity, some of which will be described here and by other lecturers at the School.

The purpose of my talks is to introduce you to extended supersymmetry and especially to describe its particle content and how that may be used as painlessly as possible. There are a very large number of component fields in the maximally extended supersymmetric theories, and it is necessary to reduce this considerably.

To see what the difficulties are let me introduce supersymmetry as relating a bose field represented by a real scalar field A and a fermi field represented by a Majorana (real) spinor ψ. Since A and ψ have canonical dimensions 1 and $^3/_2$ respectively we can define the infinitesimal supersymmetry variation δA of A and $\delta\psi$ of ψ in terms of the constant Majorana parameter ε of dimension $\frac{1}{2}$ as

$$\delta A = \bar{\varepsilon}\psi, \quad \delta\psi = \frac{1}{2}(\not{p}\varepsilon) A \tag{1}$$

These transformations can be generated by the Majorana generator S_α as

$$\delta A = i[\bar{\varepsilon}S,A]_-, \; \delta\psi = i[\bar{\varepsilon}S,\psi]_- \tag{2}$$

We can perform 2 successive variations δ_1 and δ_2 with parameters ε_1 and ε_2, and can calculate from (1)

$$[\delta_2,\delta_1]_- A = \delta_2 \bar{\varepsilon}_1 \psi - \delta_1 \bar{\varepsilon}_2 \psi = \frac{1}{2}(\bar{\varepsilon}_1 \not{p} \varepsilon_2 - \bar{\varepsilon}_2 \not{p} \varepsilon_1) A$$

$$= - \bar{\varepsilon}_2 \not{p} \varepsilon_1 A = - [\bar{\varepsilon}_2 S, \bar{\varepsilon}_1 S] A \tag{3}$$

We conclude from (3) that, for anti-commuting fermion parameters ε_1 and ε_2,

$$[S_\alpha,S_\beta]_+ = - (\not{p} C^{-1})_{\alpha\beta} \tag{4}$$

where C is the usual Dirac conjugation matrix. This is the basic anti-commutation relation of the super-symmetry generators S_α, and the pair (A,ψ) forms the simplest supermultiplet of spins $(0,\frac{1}{2})$.

We may extend S_α to include an internal symmetry label i, with $1 \leqslant i \leqslant N$ so have $S_{\alpha i}$ with anticommutator taken as a trivial extension of (4) to be

$$[S_{\alpha i},S_{\beta j}]_+ = - (\not{p} C^{-1})_{\alpha\beta} \delta_{ij} \tag{5}$$

This is the basic relation of the N-extended supersymmetry algebra $\mathcal{S}_N$. It possesses an internal SU(N) symmetry which we will discuss shortly.

1.2 Superfields

A natural extension of the notion of a field over space-time as a representation space for momentum and angular momentum generators P_μ

and $J_{\mu\nu}$ is a superfield (Salam & Strathdee 1974, 1975, Volkov & Soroka 1973) defined over superspace as a representation space for the supersymmetry generators S_α. It is easy to check that if we introduce anti-commuting variables θ_α, with $[\theta_\alpha, \theta_\beta]_+ = 0$ and $\bar{\theta}^\alpha = \eta^{\alpha\beta}\theta_\beta$ with $\eta = -C^{-1}$ and take

$$S_\alpha = \left(\frac{\partial}{\partial\bar{\theta}^\alpha} + \frac{i}{2}(\not{\partial}\theta)_\alpha\right) \tag{6}$$

as defined on superfields $\Phi(x,\theta)$ then (4) is satisfied. Similarly we may attach the internal symmetry label i to α in (6) to have a representation of the extended supersymmetry relations (5).

To discover the component field content of a superfield $\Phi(x,\theta_{\alpha i})$ we may expand it in powers of θ; the coefficients will be the component fields. Due to the anticommutation of the θ's there can be at most 2^{4N} terms in such an expansion. Whilst this number is only 16 for N = 1 and 256 for N = 2 it increases exponentially in N, being 65,536 for N = 4 and 4,294,967,296 for N = 8! We will see later that 8 is the largest value of N for which we expect a satisfactory theory of extended supergravity. Clearly we cannot handle such a large number of degrees of freedom without a suitable technology. That is what the representation theory of extended supersymmetries on extended superfields should provide for us.

It might be possible to dispense with superfields altogether, and indeed great progress has been made at the component level in supergravity. However superfields appear indispensible in the maximal utilisation of the ultra-violet divergence cancellation mechanism inherent in supersymmetry (Capper & Leibrandt 1975, Delbourgo 1975; Fujikawa & Lang 1975). Moreover there are elegant geometric formulations of N = 1 supergravity in superspace (Ogievetsky & Sokatchev 1975, Siegel & Gates 1979, Bedding, Downes-Martin & Taylor 1979; Schwarz 1980) and attempts are being made to extend these to higher N (Sokatchev 1981, Gates 1981; Rivelles & Taylor 1981a; Gayduk, Romanov & Schwarz 1981). For these reasons a superfield formulation of N = 8 maximally extended supergravity is the ultimate goal.

II SUPERALGEBRAE: CLASSIFICATION AND REPRESENTATION THEORY

2.1 Extended supersymmetry and superalgebrae

The supersymmetry algebra S_1 involves the supersymmetry generators S_α and by closure the generators of translations P_μ and Lorentz rotations $J_{\mu\nu}$ with commutators and anti-commutators:

$$[S_\alpha, S_\beta]_+ = (\not{P}\eta)_{\alpha\beta},\ [J_{\mu\nu}, S_\alpha]_- = -\tfrac{1}{2}(\sigma_{\mu\nu} S)_\alpha,$$

$$[P_\mu, P_\nu]_- = [P_\mu, S_\alpha]_- = 0,\ [J_{\mu\nu}, P_\lambda]_- = i(\eta_{\nu\lambda} P_\mu - \eta_{\mu\lambda} P_\nu), \quad (7)$$

$$[J_{\mu\nu}, J_{\lambda\sigma}]_- = i(\eta_{\mu\sigma} J_{\nu\lambda} + \ldots)$$

with $\eta^{\mu\nu} = \mathrm{diag}(1, -1, -1, -1)$, $[\gamma^\mu, \gamma^\nu]_+ = 2\eta^{\mu\nu}$, $\sigma_{\mu\nu} = \frac{1}{4i}[\gamma_\mu, \gamma_\nu]_-$. We can combine P_μ and $J_{\nu\lambda}$ into the algebra $L_o = io(3,1)$, the inhomogeneous Lorentz algebra, and denote the set of S_α by L_1, so that $S_1 = L_o + L_1$. Then we have the inclusion relations

$$[L_o, L_o]_- \subset L_o,\ [L_o, L_1]_- \subset L_1, [L_1, L_1]_+ \subset L_o \quad (8)$$

We can forget the details of L_o and L_1 to define a super-algebra L which is a vector space over R or C with a bracket operation [,] of $L \times L \to L$ and a function $g(x)$ on L with values 0 or 1 (called a Z_2-grading) so that

$$[x,y] = -(-1)^{g(x)g(y)}[y,x] \quad \text{(symmetry)}$$

$$[x,[y,z]] = [[x,y],z] + (-1)^{g(x)g(y)}[y,[x,z]] \quad \text{(Jacobi identity)}$$

Then $L_o = (x\colon g(x) = 0)$ is called the even part of L and $L_1 = (x\colon g(x) = 1)$ the odd part. If L is finite dimensional then there will be bases Q_m and R_α for L_o and L_1 with

$$[Q_m, Q_n]_- = f_{mn}{}^p Q_p,\ [Q_m, R_\alpha]_- = F_{m\alpha}{}^\beta R_\beta,\ [R_\alpha, R_\beta]_+ = A^m_{\alpha\beta} Q_m \quad (9)$$

together with quadratic identities arising from the Jacobi identity. Thus from (9) L_o is a Lie algebra and L_1 a representation of L_o.

2.2 Classification of superalgebrae

A useful concept in classification theory is that of an ideal I of L, which is a subset for which $[I,L] \subset I$. L is called simple if there are no non-trivial ideals. A simple superalgebra L for which L_1 is a completely reducible representation of L_o is called a classical Lie superalgebra, whilst if L_1 is not so reducible L is termed a Cartan type superalgebra.

We consider the matrix superalgebra $\begin{pmatrix} A & B \\ C & D \end{pmatrix}$ where A, B, C, D are m × m, m × n, n × m, and n × n matrices respectively, L_o will be composed of matrices of the form $\begin{pmatrix} A & 0 \\ 0 & D \end{pmatrix}$ and L_1 of those of the form $\begin{pmatrix} 0 & B \\ C & 0 \end{pmatrix}$. The classical Lie superalgebrae are of 4 sorts (Kac 1977, Rittenberg 1977, Freund & Kaplansky 1975):

(1) $spl(m,n)$: $\mathrm{tr}\,A = \mathrm{tr}\,D$, or $\mathrm{str}\begin{pmatrix} A & B \\ C & D \end{pmatrix} = \mathrm{tr}\,A - \mathrm{tr}\,D = 0$,

$n \neq m, \quad m,n \geq 1.$

$L_o = sl(m) + sl(n) + gl(1)$; a real form is $su(m,n)$

(2) $osp(m,2p)$: $D^TG + GD = 0$, $A^T = -A$, $B = C^TG$, $G = \begin{pmatrix} 0 & 1_p \\ -1_p & 0 \end{pmatrix}$, $m \geq 1$.

$L_o = o(m) + sp(2p)$

(3) $P(m)$: $m = n$, $A^T + D = 0$, $B = B^T$, $C = -C^T$, $\mathrm{tr}\,A = 0$, $m \geq 3$

$L_o = sl(m)$

The Killing form metric $g^{(R)}_{\mu\nu} = \mathrm{str}(X^R_\mu X^R_\nu) \equiv 0$ for any representation R, where $X_\mu = \begin{pmatrix} A_\mu & B_\mu \\ C_\mu & D_\mu \end{pmatrix}$ form a basis for the (n+m)x(n+m) matrices.

(4) $Q(m)$: $m = n$, $A = D$, $B = C$, $\mathrm{tr}\,B = 0$

$L_o = sl(m)$; $g^{(R)}_{\mu\nu} \equiv 0$ for any representation R.

There are also exceptional cases F_4, G_3, $osp(4,2;\alpha)$ which are described in the literature (Kac 1977, Rittenberg 1977, Freund & Kaplansky 1975). There is also a complete classification of the Cartan-type superalgebrae, into the types W(n), S(n), S(n), $\bar{S}(n)$, H(n); for more details I will also refer the reader to the literature (Kac 1977, Rittenberg 1977, Freund & Kaplansky 1975).

There are several definitions of semi-simplicity which are

equivalent for Lie algebrae but not for superalgebrae:

(i) det $g^{adj}_{\mu\nu} \neq 0$, so L = direct sum of simple algebrae (adj is the adjoint representation)

(ii) all finite dimensional representations are completely reducible, and then L = direct sum of simple Lie algebrae and osp(1,n) superalgebrae

(iii) If I is a maximal solvable ideal ($I^n = 0$, some n) of a Lie superalgebra $\bar{L}$ then $L = \bar{L}/I$ is semi-simple. Then L can be expressed as a (possibly complicated) function of simple superalgebrae.

2.3 The nature of extended supersymmetry algebrae

We wish to fit S_1, and more generally S_N (the N-extended algebra with $\alpha \to \alpha i$ in (7)) into the above analysis. We start with the superalgebra $osp(1,4) = \begin{pmatrix} 0 & S_\alpha \\ \bar{S}_\alpha & sp(4) \end{pmatrix}$ where $sp(4) \sim so(3,2)$ is a 10-parameter group. We can regain iso(3,1) by group contraction as follows. If we denote by J_{ab}, $1 \leqslant a, b \leqslant 5$ the generators of so(3,2) we define $P_\mu = R^{-1} J_{5\mu}$, so$[P_\mu, P_\nu]_- = -iR^{-2} J_{\mu\nu} \to 0$ as $R \to \infty$. The remaining commutator brackets give the other relations for $[J,P]_-$ and [J,J] in io(3,1). More generally $osp(N,4) = \begin{pmatrix} o(N) & S_{\alpha i} \\ \bar{S}_{\alpha i} & sp(4) \end{pmatrix}$ may be contracted to S_N in a similar manner.

What about S_1 or S_N after contraction: are they simple or semi-simple? We define a Lie super-algebra L to be the semi-direct sum $S_1 \oplus_s S_2$ if $[S_1, L] \subset S_1$, $[S_2, S_2] \subset S_2$ and $L = S_1 + S_2$ as a vector space. Then $S_1 = (P_\mu, S_\alpha) \oplus_s (J_{\mu\nu})$, $S_N = (P_\mu, S_{\alpha i}) \oplus_s (J_{\mu\nu}, o(N))$, and in both cases S_1 is a maximal solvable ideal of L with $S_1{}^3 = 0$. Thus S_N is not even semi-simple by the previous definitions, but it has a 'simple' structure. In particular we will see that the representations on superfields can be analysed completely.

2.4 Representations of superalgebrae

Lie algebrae have the properties (a) any finite dimensional representation is completely reducible (b) any irreducible representation (irrep) is equivalent to a hermitian irrep (c) irreps are completely labelled by their highest weights (the eigenvalues of the maximal set of mutually commuting generators) (d) irreps are completely labelled by the eigenvalues of Casimir operators. For Lie superalgebrae (L.S.A.) (a) is only valid for osp(1,n), (c) for all simple L.S.A.'s, (d) is not always valid e.g., not for spl(2,1), and (b) needs the use of the superadjoint $\begin{pmatrix} A^+ & -C^+ \\ B^+ & D^+ \end{pmatrix}$.

For semi-direct sums of algebrae it is possible to get representations by the Frobenius method of induction from irreps of subalgebrae; by that method we may not get all irreps or only irreps. We will consider this in details for S_N in the next section.

III IRREPS OF S_N

3.1 Massive Irreps

On a sub-manifold with given $P_\mu = p_\mu$, $p^2 > 0$, the relation (5) states that $S_{\alpha i}$ belongs to the Clifford algebra with 2^{4N} elements. There is a unique irrep of this algebra by $2^{2N} \times 2^{2N}$ matrices, so that irreps of S_N can be built on 2^{2N} dimensional representation spaces of this Clifford algebra. These may be combined with (2J + 1)-dimensional irreps of the little group so(3) of p_μ to give 2^{2N} (2J + 1)-dim. irreps of S_N by induction.

We can analyse in more detail the 2^{2N}-dim space of the Clifford algebra using the representation of the Dirac matrices with diagonal γ_o:

$$\gamma_o = \begin{pmatrix} 1 & 0 \\ 0 & -1 \end{pmatrix}, \quad \gamma_i = \begin{pmatrix} 0 & \sigma_i \\ -\sigma_i & 0 \end{pmatrix}, \quad \eta = i\gamma_2\gamma_o = \begin{pmatrix} 0 & i\sigma_2 \\ i\sigma_2 & 0 \end{pmatrix},$$

$$\gamma_o\eta = \begin{pmatrix} 0 & i\sigma_2 \\ -i\sigma_2 & 0 \end{pmatrix}.$$

Since S_α is Majorana then $S_1^* = -S_4$, $S_2^* = S_3$, so that in the rest frame $p_\mu = (m, \underline{0})$, (5) becomes

$$[S_{\alpha i}, S^*_{\beta j}]_+ = m\delta_{\alpha\beta}\,\delta_{ij}, \quad [S_{\alpha i}, S_{\beta j}]_+ = [S^*_{\alpha i}, S^*_{\beta j}]_+ = 0$$

$$(\alpha,\beta) = (1,2). \qquad (10)$$

Thus $S^*_{\alpha i}$, $S_{\alpha i}$ act as creation and annihilation operators respectively. We may take a vacuum state $|o\rangle$ with $S_{\alpha i}|o\rangle = 0$, for all i and α; the states of the Clifford representation will be $(\Pi S^*_{\alpha i})|o\rangle$. If $|o\rangle$ has so(3) spin j then the latter set of states will have a range of spins which we now analyse in detail successively for the cases N = 1, 2, 4 and 8.

3.2 N = 1

In this case there are only 2 creation operators S_1^*, S_2^*, so from the vacua $|j,j_3\rangle$ with $S_1|j\ j_3\rangle = S_2|j\ j_3\rangle = 0$ we can form the 4 states $\{|j\ j_3\rangle, S_1^*|j\ j_3\rangle, S_2^*|j\ j_3\rangle, S_1^*\ S_2^*|j\ j_3\rangle\}$. From $[J_{\mu\nu}, S_\alpha^*]_- = -\frac{1}{2}\sigma_{\mu\nu}\ S_\alpha^*$ follows that $S_1^*|j,-j\rangle \propto |j+\frac{1}{2}, -j-\frac{1}{2}\rangle$ and $S_2^*|j\ j\rangle \propto |j+\frac{1}{2}, j+\frac{1}{2}\rangle$. Thus we conclude that $|oo\rangle$, $S_1^*\ S_2^*|oo\rangle$ have spin 0 and $S_1^*|oo\rangle$, $S_2^*|oo\rangle$ have spin $\frac{1}{2}$. We denote this multiplet by $(0^2, \frac{1}{2})$, where j^r denotes a spin j component field with multiplicity r. We can build up higher spins by multiplying $(0^2, \frac{1}{2})$ by a spin j to give the general spin content $((j-\frac{1}{2}), j^2, (j+\frac{1}{2}))$ of any N = 1 multiplet.

3.3 N = 2

The same method as for N = 1 gives the spin assignments to the states $\Pi S_{\alpha i}^*|oo\rangle$:

spin 1: $S_{12}^* S_{11}^*|oo\rangle$, $S_{22}^*\ S_{21}^*|oo\rangle$, $(S_{12}^*\ S_{21}^* + S_{22}^*\ S_{11}^*)|oo\rangle$

spin $\frac{1}{2}$: $S_{1i}^*|oo\rangle$, $S_{2i}^*|oo\rangle$

spin 0: $|oo\rangle$, $S_{11}^*\ S_{21}^*|oo\rangle$, $S_{12}^*\ S_{22}^*|oo\rangle$, $(S_{12}^*\ S_{21}^*$

$- S_{22}^*\ S_{11}^*)|oo\rangle$, $S_{11}^*\ S_{12}^*\ S_{21}^*\ S_{22}^*|oo\rangle$

giving the multiplet $(0^5, \frac{1}{2}^4, 1)$. Higher irreps are obtained as before by multiplication by j to give the spin content $((j\pm 1)^1, (j\pm\frac{1}{2})^4, j^{5+1})$. Multiplication by SO(2) or SU(2) irreps can also be taken; we will turn to the internal symmetry features shortly.

3.4 N = 4 and N = 8

Similar analyses can be performed to give the fundamental multiplet for N = 4 as $(2^1, \frac{3}{2}^8, 1^{27}, \frac{1}{2}^{48}, 0^{42})$, and this can then be multiplied by spin j and any SO(4) or SU(4) irrep to give higher irreps. We see that the multiplicities of various spins in the irreps for N = 1, 2 and 4 are given by the dimensions of the antisymmetric tensor irreps $[N]_r$ of rank r of Sp(2N). Thus for N = 2, $[4]_1 = 4$, $[4]_2 = \frac{1}{2}(4\times 3) - 1 = 5$, (where the -1 is the removal of the trace using the antisymmetric Sp(4) metric) $[4]_3 = 4$, whilst for N = 4, $[8]_1 = 8$, $[8]_2 = \frac{1}{2}(8 \times 7) - 1 = 27$, $[8]_3 = \frac{1}{6}(8 \times 7 \times 6) - 8 = 48$, $[8]_4 = \frac{1}{4!}(8 \times 7 \times 6 \times 5) - 28 = 42$. We would expect a similar situation for N = 8, with the fundamental irrep being $(4^1, \frac{7}{2}^{16}, 3^{119}, \frac{5}{2}^{544}, 2^{1700}, \frac{3}{2}^{3808}, 1^{6188}, \frac{1}{2}^{7072}, 0^{4862})$, and this can be shown explicitly. Higher irreps are again obtained by taking the direct product of this irrep with a Poincaré spin j and an internal symmetry irrep.

3.5 The internal symmetry and USp(2N)

The superalgebra S_N of (7) (with $\alpha \to (\alpha,i)$) is invariant under an internal SO(N) rotation, so we may adjoint to S_N the generators of SO(N). If we take the chiral projections $S_{\alpha\pm i} = \frac{1}{2}[(1 \pm i\gamma_5)S_i]_\alpha$, with $\gamma_5 = \gamma_0\,\gamma_1\,\gamma_2\,\gamma_3$, then

$$[S_{\alpha+i}, S_{\beta+j}]_+ = 0, \quad [S_{\alpha+i}, S_{\beta-j}]_+ = (\not{p}\eta)_{\alpha+\beta-}\,\delta_{ij} \tag{11}$$

Under a U(N) transformation $S_{\alpha+i} \to U_{ij}\,S_{\alpha+j}$, $S_{\alpha-i} \to (U^+)_{ji}\,S_{\alpha-j}$, (11) remains invariant so that SO(N) may be extended to U(N) as the internal symmetry group of S_N. We should therefore be able to classify the components of a given spin in the above irreps of S_N in irreps of U(N). This can be done by recognizing (Ferrara 1981; Rittenberg & Sokatchev 1981; Taylor 1981) that the operators $S_{\alpha+i}$, $S_{\beta-j}$ can again be regarded as creation and annihilation operators in the rest frame in the representation of the γ_μ with γ_5 diagonal. Components of a given spin are then obtained by acting with the products $(\bar{S}_{+i}\,S_{+j})$ (which don't change spin) on the state $\prod_{[i]} S_{\alpha+i}|o\rangle$. This latter is chosen to be a completely antisymmetric representation of SU(N) and so a symmetric representation of spin (acting on α +) with unique spin $\ell/2$ (if there are ℓ factors $S_{\alpha+i}$ acting on $|o\rangle$).

The fact that all of these states (for a given spin) belong to the antisymmetric tensor representation of USp(2N) follows (Ferrara 1981, Rittenberg & Sokatchev 1981, Pickup & Taylor 1981) from the fact that the matrices $b_{ij} = (\varepsilon^{\alpha+\beta+}\,S_{\alpha+i}\,S_{\beta+j})$ and $a_{ij} = (S_{\alpha+i}\,(\sigma_o)^{\alpha+\beta-}\,S_{\beta-j})$ (with $(\sigma_o)^{\alpha+\beta-} = S^{\alpha+\beta-}$) are respectively symmetric and hermitian in i and j. They are the generators of USp(2N), since any 2N × 2N matrix $A = \begin{pmatrix} a & b \\ c & d \end{pmatrix}$ is hermitian and symplectic ($A^TG + GA = 0$ for $G = \begin{pmatrix} 0 & 1 \\ -1 & 0 \end{pmatrix}$) provided

$a^+ = a$, $d^+ = d$, $c^+ = b$, $c = c^T$, $b = b^T$, $a = -d^T$. Thus a and b need to be hermitian and symmetric respectively and then A is determined. This is indeed the case for the a and b expressed in terms of the $S_{\alpha i}$, thus proving the presence of USp(2N). Since a and b, given in terms of $S_{\alpha i}$, do not change the spin of a state then we expect the states of a given spin to be classified by the irreps of USp(2N) (where a_{ij} on the state $\prod_{[i]}^{\ell} S_{\alpha+i}|o\rangle$ may be shown to vanish). Furthermore the anticommutativity of the $S_{\alpha i}$ should lead only to the antisymmetric tensor irreps of USp(2N), as we have found.

The maximal Poincaré spin in the fundamental irrep is N/2. Therefore the largest value of N for which at least one massive irrep of S_N has maximal spin J is N = 4. This implies that off-shell theories with $N \geq 5$ will involve spin $J > 2$, and so have problems with interactions (Berhends, De Wit, van Holten & van Nieuwenhuizen 1980).

3.6 Massless irreps of S_N

For the massless case with $p_\mu = (p, 0, 0, p)$ we take the two component formalism with diagonal γ_5. Then $(\gamma_o - \gamma_3)\eta = i\sigma_2\begin{pmatrix} 0 & 1-\sigma_3 \\ 1+\sigma_3 & 0 \end{pmatrix}$,

so that $[S_{1i}, S^*_{1j}]_+ = 0$, $[S_{2i}, S^*_{2j}]_+ = 2p\delta_{ij}$ where $S_{\alpha i} = \begin{pmatrix} S_{1i} \\ S_{2i} \\ S^*_{1i} \\ S^*_{2i} \end{pmatrix}$.

We can thus ignore S^*_{1i} altogether, so giving a reduction of the number of $S^*_{\alpha i}$ from the value of 2N in the massive case to N. Maximal spins of multiplets will therefore be half that of the massive case, and N can take the value 8 before spins larger than 2 appear.

Since S^*_{2i} lowers helicity by $\frac{1}{2}$ we have that for N = 1 the general irrep has helicity $(\lambda, \lambda - \frac{1}{2})$, for λ half an integer. If we add the CPT-conjugate irrep this becomes $(\pm\lambda, \pm(\lambda - \frac{1}{2}))$. For N > 1, b_{ij} defined in the previous section is now zero, since each term in b_{ij} involves at least one (where we are taking $S_{1+i} = S^*_{1i}$). We have therefore only SU(N) as a classifying group, and the number of states of a given helicity for massless irreps of S_N will have dimension given by those of SU(N). We thus obtain the following helicity values for irreps with maximum helicity 1:

$$N = 1: \quad (0^2, \pm \tfrac{1}{2}); \quad (\pm \tfrac{1}{2}, \pm 1)$$

$$N = 2: \quad (0^4, \pm \tfrac{1}{2}^2); \ (0^2, \pm \tfrac{1}{2}^2, \pm 1)$$

$$N = 3: \quad (0^6, \pm \tfrac{1}{2}^4, \pm 1)$$

$$N = 4: \quad (0^6, \pm \tfrac{1}{2}^4, \pm 1)$$

whilst for maximum helicity 2 we have the irreps (singling out those with a single state of helicity ±2):

$$N = 1: \quad (\pm \tfrac{3}{2}, \pm 2)$$

$$N = 2: \quad (\pm 1, \pm \tfrac{3}{2}^2, \pm 2)$$

$$N = 3: \quad (\pm \tfrac{1}{2}, \pm 1^3, \pm \tfrac{3}{2}^3, \pm 2)$$

$$N = 4: \quad (0^2, \pm \tfrac{1}{2}^4, \pm 1^6, \pm \tfrac{3}{2}^4, \pm 2)$$

$$N = 5: \quad (0^{10}, \pm \tfrac{1}{2}^{11}, \pm 1^{10}, \pm \tfrac{3}{2}^5, \pm 2)$$

$$N = 6: \quad (0^{30}, \pm \tfrac{1}{2}^{26}, \pm 1^{16}, \pm \tfrac{3}{2}^6, \pm 2)$$

$$N = 7: \quad (0^{70}, \pm \tfrac{1}{2}^{56}, \pm 1^{28}, \pm \tfrac{3}{2}^8, \pm 2)$$

$$N = 8: \quad (0^{70}, \pm \tfrac{1}{2}^{56}, \pm 1^{28}, \pm \tfrac{3}{2}^8, \pm 2)$$

(where $\pm\lambda^r$ denotes an r-fold degenerate state of helicity λ).
We note that for N = 1 and 2, on-shell super Yang-Mills theory (with $\lambda_{max} = 1$) can co-exist with matter multiplets whereas for N = 3 and 4 no matter multiplets can be present. Similarly on-shell supergravity can exist with additional matter multiplets up to N = 6 (only up to N = 4 if matter is only allowed to have maximal helicity 1) but no additional matter multiplets can arise for N = 7 or 8.

IV SUPERFIELDS

4.1 Induced representations

We have already mentioned the analogy between io(3,1) and S_N behind the introduction of superfields. In detail we have the similarities:

Poincaré algebra io(3,1) <-> super-Poincaré algebra S_1

$$(P_\mu) \oplus_s (J_{\mu\nu}) \leftrightarrow (P_\mu, S_\alpha) \oplus_s (J_{\mu\nu})$$

Poincaré group IO(3,1) <-> Super-Poincaré group $SIO(3,1)$: $(X^\mu, \Lambda_\mu{}^\nu)$ with $\Lambda^T\eta\Lambda = \eta$ <-> : $(X^\mu, \theta^\alpha, \Lambda_\mu{}^\nu)$ with $\Lambda^T\eta\Lambda = \eta$

$$(X,\Lambda) \, . \, (Y,\Sigma) = (X + \Lambda Y, \Lambda\Sigma) \leftrightarrow (X,\theta,\Lambda) \, . \, (Y,\psi,\Sigma)$$

$$= (X + \Lambda Y + \tfrac{1}{2}\,\bar{\theta}\gamma^\mu\, u(\Lambda)\psi,\; u(\Lambda)\psi + \theta, \Lambda\Sigma)$$

$$IO(3,1) = T_4 \otimes_s SO(3,1) \leftrightarrow SIO(3,1) = (T_4, S_4) \otimes_s SO(3,1)$$

Induced representations of IO(3,1) are obtained by choosing an element $e^{ip^{(o)}t}$ as a complex-valued homomorphism of T_4, so an element of $\hat{T}_4$ (the character group of T_4). The little group $L_{p^{(o)}}$ of $p^{(o)}$ is then defined in SO(3,1) as the set of Λ with $\Lambda p^{(o)} = p^{(o)}$. A unitary irrep U of $L_{p^{(o)}}$ induces an irrep of IO(3,1) on functions ϕ on $IO(3,1)/T_4 \otimes_s L_{p^{(o)}}$. This quotient space can be regarded as the set of points on the orbits of $p^{(o)}$ i.e., as $\{p = \Lambda_p\, p^{(o)}\}$. Then for $g = (t,\Lambda) \in IO(3,1)$ we define the unitary representation U_g on $\phi(p)$ as

$$U_g\phi(p) = e^{ipt}\; U(R_p)\; \phi(\Lambda_p^{-1}\, p)$$

where $R_p = \Lambda_p^{-1}\Lambda\Lambda_{\Lambda^{-1}p} \in L_{p^{(o)}}$. Then P_μ is represented as $i\partial/\partial x^\mu$ and $J_{\mu\nu}$ as $i(x^\mu\, \partial/\partial x^\nu - x^\nu \partial/\partial x^\mu)$ (on a scalar) on the Fourier transform of ϕ. In this way one obtains all irreps with $p^2 > 0$ and O(3) spin j.

By the above analogy one might replace T_4 by the sub-algebra (T_4, S_4) of S_1 in the above construction. However (T_4, S_4) is not Abelian, though since the 4-dimensional representation X_4 of (S_α) is unique for a given $p^{(o)}$ then we may replace $e^{ip^{(o)}t}$ by $e^{ip^{(o)}t}\, X_4$ in the above steps to induce irreps from SO(3) into functions on $SIO(3,1)/(T_4,S_4) \otimes_s L_{p^{(o)}}$. This quotient space is again $\underline{R}^4$, so we find irreps on functions $\Phi_j(x)\, X_4$ which were the irreps discussed in the previous section.

It is necessary to take a different approach to obtain superfields, the simplest of these being simply to extend the space-time variables x to superspace variables (x,θ) as in section 1.2. We are thus considering functions on the quotient of the group manifold by SO(3,1), so the extension from space-time to superspace is

$$(T_4) \otimes_s SO(3,1)/SO(3,1) \to (T_4,S_4) \otimes_s SO(3,1)/SO(3,1).$$

Superfields are thus functions $\Phi_j\,(x,\theta)$, where j denotes an external spin label.

Since superfields are not directly related to induced representations, it is clear that the former do not necessarily give irreducible representations of S_1 (or S_N in the extended case). Indeed our next question will be to analyse the irrep content of a given superfield. We are prepared to work with these combinations of irreps of S_N since they provide a geometrical approach to supergravity, at least for N = 1 (Ogievetsky & Sokatchev 1975, Siegel & Gates 1979, Bedding, Downes-Martin & Taylor 1979, Schwarz 1980).

4.2 The covariant derivative

We mentioned the 'component explosion' difficulty at the beginning, with a scalar superfield having 2^{4N} components. We must therefore analyse superfields into irreps of S_N so that the total number of components may be reduced to manageable proportions. Moreover we may regard these irreps as the 'building blocks' of extended supergravity, and we will see in chapter 6 that their fitting together is a crucial process in building supergravity.

We may analyse irreps by extending S_N to a new algebra $\bar{S}_N$ by the addition of the covariant derivatives

$$D_{\alpha i} = \partial/\partial\bar{\theta}^{\alpha i} - \frac{i}{2}\,(\not\partial\theta)_{\alpha i}. \qquad (12)$$

Direct calculation shows that $[D_{\alpha i},S_{\beta j}]_+ = 0$, (so $D_{\alpha i}$ is covariant under S_N), $[D_{\alpha i},D_{\beta j}]_+ = (\not p\eta)_{\alpha\beta}\,\delta_{ij}$. We may introduce conditions on a superfield in a covariant manner in terms of the $D_{\alpha i}$, such as $D_{\alpha+i}\Phi = 0$. Such a superfield may be (and in fact is) an irrep of S_N.

4.3 Superspin

To proceed with the reduction we introduce the notion of superspin which classifies the irreps in a similar fashion to ordinary spin for the Poincaré group. The superspin operator is (Pickup & Taylor, 1981, Taylor 1980a, Sokatchev 1975)

$$C_\mu = \frac{1}{2}\,\varepsilon_{\mu\nu\lambda\sigma}\,P^\nu J^{\lambda\rho} - \frac{1}{4}\sum_{\ell=1}^{N} \bar{S}_\ell i\gamma_\mu\gamma_5 S_\ell = W_\mu - \Sigma_\mu \tag{13}$$

where the first vector on the r.h.s. of (13) is the Pauli-Lubanski vector and the second is a compensating term chosen so that $[C_\mu,S_{\alpha i}]_- = 0$ (since $[W_\mu,S_{\alpha i}]_- = -\frac{1}{4}\,\varepsilon_{\mu\nu}{}^{\rho\sigma}P^\nu(\sigma_{\rho\sigma}S_i)_\alpha$, $[\Sigma_\mu,S_{\alpha i}]_- = \frac{i}{2}(\not p\gamma_\mu\gamma_5 S_i)_\alpha$ and $\varepsilon^{\mu\nu ab}\,\sigma_{ab} = -2\sigma^{\mu\nu}\gamma_5$). Then with $C_\mu^\perp = C_\mu - p_\mu p^\nu C_\nu (p^2)^{-1}$ we have $[C_\mu^\perp,P_\nu]_- = [C_\mu^\perp,S_{\alpha i}]_- = 0$, so $(C_\mu^\perp)^2$ is a Casimir of S_N (commuting with all the elements of S_N).

In the rest frame (for $p^2 > 0$) we have $C_\mu^\perp = (0,\underline{C})$ with $\underline{C} = m\,\underline{J} - \frac{1}{2}S^+\,\underline{\sigma}\,S$, and $\underline{C}$ satisfies the SU(2) commutation rules. We have therefore by standard arguments that $(C_\mu^\perp)^2 = -2p^2\,Y(Y+1)$ in irreps on a Hilbert space, where Y, the superspin, takes non-negative half-integer values. Since $\frac{1}{2}S^+\,\underline{\sigma}\,S$ corresponds to a spin $-\frac{1}{2}$ operator the values of Poincaré spin in a representation with given Y will therefore by $(Y \pm \frac{1}{2}, Y)$ (for $N = 1$) and similar values for higher N.

4.4 Casimirs for irreps on superfields

We may construct irreps of S_N on any basis state $\Pi\, D_{\alpha+i}|>$, where $|>$ satisfies $D_{\alpha-i}|> = 0$. Since, in the rest frame, $[D_{\alpha i},C_\ell]_- = [S_{\alpha i},J_\ell]$, the superspin values on such basis states will take the same range of values as the spin values in a given irrep of S_N. Thus if $|>$ has spin j the superspins in the set of states $\{\Pi\, D_{\alpha+i}|>\}$ will be $(j + \frac{1}{2}N)$, $(j + \frac{1}{2}N - 1)$, ..., $(j - \frac{1}{2}N)$. We may represent $\{\Pi\, D_{\alpha+i}|>\}$ as a superfield $\Phi_j(X,\theta)$ (since each state corresponds to the component field $\Pi\, D_{\alpha+i}\,\Phi_j(X,\theta)|_{\theta=0}$). By the previous discussion the multiplicities of the various values of Y for a given N are given by the dimensions of the antisymmetric tensor representations of USp(2N), so the superspins of the irreps of S_N and their multiplicities on $\Phi_j(X,\theta)$ for various N are

$$N = 1: \quad (j \pm \tfrac{1}{2}, j^2)$$

$$N = 2: \quad (j \pm 1, (j \pm \tfrac{1}{2})^4, j^{5+1})$$

$$N = 4: \quad (j \pm 2, (j \pm \tfrac{3}{2})^8, (j \pm 1)^{27+1}, (j \pm \tfrac{1}{2})^{48+8}, j^{42+27+1})$$

It is necessary to have further Casimir operators to separate these irreps. As to be expected they arise from the U(N) Casimirs plus one further label associated with separating the USp(2N) multiplets (Ferrara 1981; Rittenberg & Sokatchev 1981, Taylor 1980a; Pickup & Taylor 1981). The former of these are constructed from the U(N) generators T_{ij} with $[T_{ij}, S_{\alpha+\ell}]_- = \delta_{i\ell} S_{\alpha+j}$, $[T_{ij}, S_{\alpha-\ell}]_- = -\delta_{j\ell} S_{\alpha-i}$ and the additional compensating terms $(p^2)^{-1} \bar{S}_{-i}\, i\not{p}\, S_{+j}$, to give $\mathcal{T}_{ij} = T_{ij} + (p^2)^{-1} S_{-i}\, i\not{p}\, S_{+j}$, with $[\mathcal{T}_{ij}, S_{\alpha\ell}]_- = 0$. The appropriate U(N) invariant labelling operators for the irreps of S_N contained in a given superfield are now obtained by constructing the U(N) Casimirs. On a superfield detailed calculation shows that $\mathcal{T}_{ij} = (p^2)^{-1} \bar{D}_{-i} \not{p} D_{+j} + T_{ij}^{(ext)}$, as is expected since the Casimirs should only be constructed from the covariant derivatives and the U(N) operators acting on the U(N)-labels on the superfield being considered. There is also the quadratic Casimir of Sp(2N) whose evaluation and use we will only consider in detail for N = 2.

The next result of the above discussion is a very simple classification of the irreps of S_N contained in a superfield. They are obtained by taking the direct product in spin and U(N) of the vacuum state (with the spin and U(N) labels of the superfield) and those of the product $(\prod_p \bar{D}_{i+} D_{j+}) \prod_{[\ell]} D_{\alpha+\ell}$, where the second factor is in an antisymmetric tensor representation of SU(N). The total irrep of SU(N) created by this product of D's will have a column of P single boxes and p pairs of boxes. The eigenvalues of the remaining U(1) generator (supercharge) are also relevant; they are the number of $D_{\alpha+}$'s in the above product plus the supercharge of the superfield.

We have thus proven that the SU(N) and spin content of the fundamental irrep is, for various N, (with j^m = (spin) $^{\text{dim irrep SU(N)}}$)

$$N = 1: \quad (0^{1+\bar{1}}, \tfrac{1}{2})$$

$$N = 2: \quad (0^{1+3+\bar{1}}, \tfrac{1}{2}^{2+\bar{2}}, 1)$$

$$N = 4: \quad (0^{1+10+20'+\overline{10}+1}, \tfrac{1}{2}^{4+20+\overline{20}+\bar{4}}, 1^{6^2+15+6}, \tfrac{3}{2}^{4+\bar{4}}, 2)$$

and an identical set of values for the U(N) and superspin content of a scalar superfield.

4.5 Projectors onto Irreps

The projectors onto a given irrep in a given superfield may be immediately written down once the eigenvalues of the Casimirs for that irrep are known. These projectors are important in the superfield quantisation programme, as well as for determining the component content of an irrep. The projectors may, however, be very complicated for large N since they could involve a large number of powers of D's. This is because if a Casimir C can have values $\{c_n\}$ the projector onto the subspace with $C = c_{n_o}$ is $\prod_{n\neq n_o} (C - C_n)/(C_{n_o} - C_n)$ and so involves a large number of factors. In the case, for example, of the superspin $C = -(C^{\perp})^2/2p^2$, the projector onto irreps with Y = 0 for N = 4 is, from the entry for N = 4 at the end of the last section, proportional to $(C - \frac{3}{4})(C - 2)(C - \frac{15}{4})(C - 6)$. This product contains up to 16 powers of D's. It thus appears difficult to use this method of constructing projectors directly to analyse, for example, the component content of irreps. A covariant method has been developed most recently (de Wit, & van Holten 1980, Fradkin & Vasiliev 1979) which allows the calculation of projectors without the use of explicit values of Casimir, though it would be too lengthy to describe that here. In any case these projectors still involve many powers of D_α.

An alternative method can be used (Pickup & Taylor 1981) which involves only the SO(N) invariance of S_N. In this approach the eigenvalues of operators defined from the covariant derivatives are calculated. Projectors for these associated objects can then be obtained which are considerably simpler than the fully covariant ones. This is an advantage for some purposes, such as the calculation of field components and their transformation laws. These latter are as important a part of the general analysis of supergravity as is the quantisation aspect, and so we shall spend a little time on the non covariant approach here. We will do this in detail for N = 1 and 2.

4.6 Identities for N = 1

We introduce the tensor C_{ab} in terms of the superspin vector C_b of section (4.3) by $C_{ab} = p_{[a} C_{b]}$, for which $C_{ab}^2 = p^2(C_a^{\perp})^2$. On a scalar superfield the external spin is zero and we therefore expect that the term involving S in C_a combines with the Pauli-Lubanski vector to give $C_{ab} = \frac{1}{2} \bar{D}\, i\, p_{[a} \gamma_{b]} \gamma_5 D$ (and this can be shown by detailed calculation in terms of (6), (7) and (12)). We may then calculate

$$C_{ab} C^{ab} = \frac{1}{8} \{p^2 \bar{D}\, i\, \gamma_a \gamma_5 D\bar{D}\, i\, \gamma^a \gamma_5 D$$

$$- p^a p^b \bar{D}\, i\, \gamma_a \gamma_5 D \bar{D}\, i\, \gamma_b \gamma_5 D\}.$$

We now use the identities (derivable from equ. (12))

$$\bar{D}\, i\, \gamma_a \gamma_5 D \bar{D}\, i\, \gamma_b \gamma_5 D = \eta_{ab}(\bar{D}D)^2 + 2i\, p^c \varepsilon_{abcd} \bar{D}\, i\, \gamma^d \gamma_5 D$$

$$- 4(\eta_{ab} p^2 - p_a p_b)$$

$$(\bar{D}D)^3 = 4p^2 \bar{D}D \tag{14}$$

We may rewrite $C_{ab} C^{ab}$ as $\frac{3}{2} p^4(A - 1)$, where $A = \frac{1}{4p^2} (\bar{D}D)^2$ and $A^2 = A$. Thus we have that A = 0 or 1, and for A = 0, $\frac{3}{4}(1 - A) = \frac{3}{4}$, so $Y = \frac{1}{2}$, and for A = 1, Y = 0. For the multiplets of N = 1 irreps considered in Section (4.4) we know for j = 0 (the scalar superfield) the Y values are $(\frac{1}{2}, 0^2)$, so there are two Y = 0 irreps. These can be separated by the values of the U(1) supercharge, which is generated by the γ_5 transformation $\theta \to e^{a\gamma_5}\theta$. If Γ is the infinitesimal generator of this transformation (so $[S_\alpha,\Gamma]_- = i(\gamma_5 S)_\alpha$) the associated Casimir is $G = \Gamma - \frac{1}{2p^2} \bar{S}\, i\, \not{p}\, \gamma_5 S$. On a scalar superfield this becomes $G = \frac{1}{2p^2} \bar{D}\, i\, \not{p}\, \gamma_5 D$ (as can be shown by explicit calculation using (6) and

(7). Using the previous identities (14) it can easily be shown that $G^2 = A$, so on $Y = 0$ we have $G = \pm 1$. The projector for these irreps are thus:

$$\Pi_{Y=\frac{1}{2}} = (1 - A), \ \Pi_{Y=0,G=\pm 1} = \frac{1}{2} A(1 \pm G) = \frac{1}{2} A \pm \frac{1}{4p^2} \bar{D}\, i \not{p}\, \gamma_5\, D. \tag{15}$$

The latter two irreps can be shown, by means of a Fierz transformation on a product of 3 D_α's, to satisfy $D_{\alpha\pm}\, \Pi_{Y=0,G=\pm 1} = 0$. Thus they are the chiral irreps, so-called because the solution of $D_{\alpha\pm}\, \Phi(X,\theta) = 0$ can be written as a function of $x^\mu - \frac{i}{2}\bar{\theta}\gamma_\mu \gamma_5 \theta$ and $\theta_{\alpha\mp}$ only.

Similar calculations can be performed to obtain projectors for higher spin superfields (Sokatchev 1975).

4.7 Identities for N = 2

On a scalar superfield with N = 2 we have

$C_{ab} = \frac{1}{2} \sum_{\ell=1}^{2} \bar{D}_\ell\, i\, p_{[a}\gamma_{b]}\gamma_5 D_\ell$, so

$$C_{ab}^2 = \frac{3}{2} p^4 (A_1 + A_2) - 3p^4 + J$$

with

$$J = \frac{1}{4} \{p^2 (\bar{D}_1\, i\, \gamma_a \gamma_5 D_1)(\bar{D}_2\, i\, \gamma^a \gamma_5 D_2) - \bar{D}_1\, i \not{p}\, \gamma_5 D_1 \bar{D}_2\, i \not{p}\, \gamma_5 D_2\} .$$

The eigenvalues of J may be obtained from the identities

$$A_1 J = A_2 J = JA_1 = JA_2 = 0, \quad J^2 = 2p^4 J + 3p^8 (1 - A_1)(1 - A_2).$$

We also need the eigenvalues of the supercharge generator $G = \frac{1}{2p^2} \sum_{\ell=1}^{2} \bar{D}_\ell\, i \not{p}\, \gamma_5 D_\ell$, and use the identities $G^2 = A_1 + A_2 + R$, $R = \frac{1}{2p^2} \bar{D}_1\, i \not{p}\, \gamma_5 D_1 \bar{D}_2\, i \not{p}\, \gamma_5 D_2$, $R^2 = 16p^8 A_1 A_2$, so that $[G^2 - 4p^4(A_1 + A_2)]^2 = 64p^8 A_1 A_2$. We thus obtain that $A_1 + A_2 = 0$, $G = 0$; $A_1 + A_2 = 1$, $G = \pm 1$, $A_1 + A_2 = 2$, $G = \pm 2$ or 0. We also need the eigenvalues of the SO(2) rotation Casimir $J = \frac{1}{2} \bar{D}_1 \frac{i \not{p}}{p} D_2$ with $J^2 = 1 - \frac{1}{4p^2} W + \frac{1}{4p^4} (4J - R)$, with $W = \bar{D}_1 D_1 \bar{D}_2 D_2 + \bar{D}_1 \gamma_5 D_1 \bar{D}_2 \gamma_5 D_2$,

$W^2 = 32p^4 A_1A_2 - 8R$. We note that for higher spin superfields the reducible USp(4) $\underline{6} = \underline{5} + \underline{1}$ can only be split by the quadratic USp(4) Casimir, which takes the value $(\frac{1}{4} A + \frac{1}{6p^4} J + \frac{1}{6})$ (Pickup & Taylor 1981).

We may use eigenvalues to write down the projectors

$$\Pi_{Y=1} = \frac{1}{4} (1 - A_1) (1 - A_2) (3 - \frac{J}{p^4}) \tag{16}$$

$$\Pi_{Y=0=G=T} = \frac{1}{16p^4} (4p^4 A_1 A_2 + p^2W - R) \tag{17}$$

The former of these gives the Weyl multiplet (the multiplet containing the graviton) $(\underline{2}, \frac{3}{2}^4, \underline{1}^6, \frac{1}{2}^4, \underline{0})$ whilst the latter gives one of the two multiplets of auxiliary fields $(1, \frac{1}{2}^4, 0^5)$ which are needed for the construction of N = 2 supergravity in its minimal form (deWit & van Holten 1979, 1980). The other multiplet is obtained by taking the real part of the G = ±2, Y = 0 irrep, with projector onto $\frac{1}{4p^2} \bar{D}_1 D_1 \bar{D}_2 D_2 = F = 1$ given by

$$\Pi_{Y=0,F=+1} = A_1A_2 \frac{1}{2}(1 + \frac{R}{4p^4}) \frac{1}{2}(1 + F)$$

$$= \frac{1}{4} A_1A_2 + \frac{1}{16p^4} R + \frac{1}{16p^2} (\bar{D}_1D_1\bar{D}_2D_2 - \bar{D}_1\gamma_5D_1\bar{D}_2\gamma_5D_2) \tag{18}$$

The forms (16), (17) and (18) are convenient in obtaining the component field representations of the corresponding irreps in a scalar superfield. A similar analysis to the above has been performed for N = 4. (Pickup & Taylor 1981).

V COMPONENT FIELDS IN SUPERFIELD REPRESENTATIONS

5.1 Basis functors and components

We have so far avoided the explicit use of expansions of superfields in powers of θ. In order to obtain better physical understanding of the content of a superfield we will analyse the component content of irreps. Let us consider the Y = 0 irrep for N = 1. We can use as projector onto G = +1 the operator $\frac{1}{p^2} \bar{D}_+D_+\bar{D}_-D_-$, and this can be applied to a scalar superfield. We can regard the n-space differential operator $\frac{1}{p^2} \bar{D}_-D_-\bar{D}_+D_+ 1$ (where 1 is acted on only by the θ-derivatives with $\partial/\partial\bar{\theta}.1 = 0$) as a state $|>$ for the Y = 0 chiral irrep, since $D_{\alpha-}|> = 0$. Moreover since $D_\pm 1 = S_\pm 1$, we may write $|> = (\frac{1}{p^2}) \bar{S}_+S_+\bar{S}_-S_- 1$, so that

$S_{\alpha+}|> = 0$ and $|>$ also acts as a vacuum state to construct the complete set of states $|>$, $S_{\alpha-}|>$, $\bar{S}_-S_-|>$ for the Y = 0 irrep. We denote these states or 'basis functions' as e_+, $u_{\alpha+}$, ω_+, these being differential operators in x-space. We may thus expand the Y = 0, G = +1 part of any superfield in basis functions as

$$\Pi_{Y=0,G=+1}\ \Phi_j = e_+\ A_+(x) + \bar{u}^{\alpha+}\ \psi_{\alpha+}(x) + \omega_+\ B_+(x) \qquad (19)$$

where A_+, B_+ are complex scalars and $\psi_{\alpha+}$ is a Weyl spinor. We may extend this to higher Y by attaching the label Y as an external spin index to each of A_+,B_+ and $\psi_{\alpha+}$ and reducing $\psi_{\alpha+,Y}$ to the spins $Y \pm \frac{1}{2}$. This gives the component fields for all possible irreps for N = 1. The detailed values of the operators e_+, ω_+, $\psi_{\alpha+}$ in $D_\pm$ are $e_+ = \frac{1}{p^2}\bar{D}_-D_-\bar{D}_+D_+\ 1$, $\omega_+ = 4\bar{D}_-D_-\ 1$, $u_{\alpha+} = \frac{2}{p^2}\bar{D}_-D_-(\not{p}\ D_-)_{\alpha+}$, and are essentially those introduced earlier (Taylor 1980b).

We may proceed similarly for higher N. The vacuum state $|> = (\frac{1}{p^2})^N \prod_{\ell=1}^{N} \bar{D}_{-\ell}D_{-\ell} \prod_{\ell=1}^{N} \bar{D}_{+\ell}\ D_{+\ell}\ 1$ is again the Y = 0 irrep which is also a vacuum for the $S_{\alpha+\ell}$ operators. We may thus construct the complete set of states $\prod_{r=1}^{p} (\bar{S}_{-\ell_r}\ S_{-m_r}) \prod_{[i]}^{n} S_{\alpha-i}|>$, which may be rewritten as the x-space differential operators(basis functions)

$$(\frac{1}{p^2})^N \prod_{\ell=1}^{N} \bar{D}_{-\ell}\ D_{-\ell} \prod_{\ell=1}^{N} \bar{D}_{+\ell}\ D_{+\ell} \prod_{[i]}^{n} D_{\alpha-i} \prod_{r=1}^{p} (\bar{D}_{-\ell_r}\ D_{-m_r})\ 1$$

$$= e^{(+)}_{n,p} \qquad (20)$$

We note that n/2 is the SL(2,C) spin of the basis function whilst n and p correspond to the SU(N) Young tableau  ^p_n, with a column of single boxes and p pairs of boxes. We may thus expand the Y = 0 irrep of any superfield with G = + n as

$$\Phi_+ = \Pi_{Y=0}\ \Phi = \sum_{n,p} e^{(+)}_{n,p}\ \psi_{n,p} \qquad (21)$$

For N = 1 we have $e_{o,o} = e_+$, $e_{1,o} = u_{\alpha+}$, $e_{o,1} = \omega_+$. For N = 2 the set $\{e_{n,p}\}$ is a covariantisation of basis functions introduced earlier by the author (Taylor 1980b) and similarly for N = 4 (Pickup & Taylor 1981).

The component functions belonging to higher Y and SU(N) irreps of $\mathcal{S}_N$ are given, as for N = 1, by attaching further SL(2,C) and SU(N) labels to $\psi_{n,p}$ and decomposing the resulting products of SL(2,C) and SU(N) irreps to give component functions transforming as irreducible representations of these groups.

For even N a reality condition can be imposed which mixes the fundamental irrep and its complex conjugate. This can be seen by using the operator $F = (\frac{1}{4p^2})^{N/2} \prod_{\ell=1}^{N} \bar{D}_\ell D_\ell$ (with $F^2 = 1$) on a chiral irrep of either chirality which changes the chirality of the vacuum state. In particular for N = 2 (Taylor 1980a) we have $Fe_+(1)\, e_+(2) = \omega_-(1)\, \omega_-(2)$. For general N we can see that

$$Fe^{(+)}_{n,p} = (4p^2)^{p+n-\frac{N}{2}} \prod_{[i]} (2\not{p})\, e^{(-)}_{n,N-n-p} \tag{22}$$

where $\prod_{[i]} (2\not{p})$ indicates that a factor of $(2\not{p})$ acts on each SL(2,C) label associated with the antisymmetised SU(N) labels i in $e^{(-)}_{n,N-n-p}$. Thus if we take the irrep $\Phi_+ + \Phi^*_+$, where Φ_+ is given by (21) with

$\psi^*_{n,N-n-p} = (4p^2)^{p+n-\frac{N}{2}} (\prod_{[i]} 2\not{p})^T \psi_{n,p}$ the irrep has F = +1 and is real and irreducible.

5.2 Supersymmetry transformations

The supersymmetry transformations of the components $\psi_{n,p}$ in (21) can be deduced from those of the basis functions. We can consider this in detail for N = 1 and refer the reader to the literature for more general cases (Bufton & Taylor 1981). As an example we will consider this for the $Y = \frac{1}{2}$ irrep constructed with the basis functions

$$e_o = 2(1 - A)1, \quad \omega_\mu = -\frac{1}{p^2}(1 - A)\,\bar{D}\, i\, \gamma_\mu \gamma_5 D1,$$

$$\bar{u}^\alpha = -\frac{4}{p^2}(1 - A)\,(\eta\not{p}D)^\alpha\, 1 \tag{23}$$

The coefficients have been so chosen that the θ expansion of these basis functions commence with 1, $\frac{1}{4}\bar{\theta}\, i\, \gamma_\mu \gamma_5\, \theta$ and $\bar{\theta}^\alpha$ respectively. Then the $Y = \frac{1}{2}$ irrep $\Phi_{\frac{1}{2}}$ is

$$\Phi_{\frac{1}{2}} = e_o\, D + \bar{u}^{\alpha}\, \psi_{\alpha} + \omega^{\mu}\, A_{\mu}$$

The supersymmetry transformation $\delta_{\varepsilon}\, \Phi_{\frac{1}{2}} = i\, \bar{\varepsilon}^{\alpha}\, S_{\alpha}\, \Phi_{\frac{1}{2}}$ is given by the action of S_{α} on the basis functions. Since $[S_{\alpha}, D_{\beta}]_{+} = 0$ and $S_{\alpha} 1 = - D_{\alpha} 1$ we have

$$S_{\alpha}\, e_o = - 2i(1 - A)\, D_{\alpha} 1 = \frac{i}{2}\, (\not{p} u)_{\alpha} \tag{24}$$

$$S_{\alpha}\, u_{\beta} = \frac{4i}{p^2}\, (1 - A)\, (\not{p} D)_{\beta}\, D_{\alpha} 1 \tag{25}$$

If we use that $D_{\alpha}\, D_{\beta} = \frac{1}{2}\, (\not{p}\eta)_{\alpha\beta} - \frac{1}{4}\, \eta_{\alpha\beta}\, \bar{D}D + \frac{1}{4}\, (\gamma_5 \eta)_{\alpha\beta}\, \bar{D}\, \gamma_5\, D$ $+ \frac{1}{4}\, i(\gamma_{\nu}\, \gamma_5\, \eta)_{\alpha\beta}\, \bar{D}\, i\, \gamma_{\nu}\, \gamma_5\, D$ then (25) becomes $- i\eta_{\alpha\beta}\;\; e_o + (\not{p}\gamma_{\nu}\gamma_5 \eta)_{\beta\alpha}\, \omega^{\nu}$. Finally

$$S_{\alpha}\, \omega_{\mu} = - \frac{i}{2}\, (\gamma'_{\mu}\, \gamma_5\, u)_{\alpha} \qquad \text{(Where } \gamma'_{\mu} = \gamma_{\mu} - p_{\mu}\, \not{p}'\, p^2\text{)} \tag{26}$$

Using (24), (25) and (26)

$$\delta_{\varepsilon}\, \Phi_{\frac{1}{2}} = i\, \bar{\varepsilon}^{\alpha}\, \{\frac{i}{2}\, (\not{p} u)_{\alpha}\, D - i\psi_{\alpha}\;\; e_o$$

$$- (\bar{\psi}\, \not{p}\, \gamma_{\nu}\, \gamma_5\, \eta)_{\alpha}\, \omega^{\nu} - \frac{i}{2}\, (\gamma'_{\mu}\, \gamma_5\, u)_{\alpha}\, A^{\mu}\}$$

$$= e_o\;\; \delta D + \bar{u}^{\alpha}\;\; \delta\psi_{\alpha} + \omega^{\mu}\;\; \delta A_{\mu}$$

so giving

$$\delta D = \bar{\varepsilon}^{\alpha}\, \psi_{\alpha},\;\; \delta\, \psi_{\alpha} = \frac{1}{2}\, (\not{p}\, \varepsilon)_{\alpha}\, D + \frac{1}{2}\, (\gamma_5\, \gamma'_{\mu}\, \varepsilon)_{\alpha}\, A\;,$$

$$\delta A_{\mu} = i(\bar{\varepsilon}\, \gamma_{\mu}\, \gamma_5\, \not{p}\, \psi) \tag{27}$$

The first two of these equations (without the term with A^{μ}) are those we started from, equ. (1), as they should be.

VI LINEARISED LAGRANGIANS AND AUXILIARY FIELDS FOR N = 1 AND 2 SUPERGRAVITIES.

6.1 Linearised N = 1 supergravity

There are three N = 1 multiplets containing the graviton, with

$Y = {}^3/_2$, 2 and ${}^5/_2$. The latter two of these also involve fields of spin greater than 2. Due to the difficulty of constructing interacting theories for such fields (Berhends, deWit, van Holten & van Nieuwenhuizen 1980), and as importantly because the associated high spin particles are not observed experimentally (except as composites), we choose only the $Y = \frac{3}{2}$ irrep, with spin content $(2, \frac{3}{2}, 1)$. The component fields are $(h_{\mu\nu}, \psi_{\mu\alpha}, A_\mu)$ with the usual constraints to give the correct spin values: $\partial_\mu A_\mu = 0 = \partial_\mu \psi_{\mu\alpha} = (\gamma^\mu \psi_\mu)_\alpha = \partial_\mu h_{\mu\nu} = h_{\lambda\lambda} = h_{\mu\nu} - h_{\nu\mu}$. Taking account of the different dimension of the spin 1 and 2 fields (due to two extra S_α's in the spin 2 basis function compared to that for spin 1) this multiplet has the associated Lagrangian

$$- A_\mu^2 + \frac{i}{2} \bar{\psi}_\mu \not{\partial} \psi_\mu - h_{\mu\nu} \Box h_{\mu\nu} \tag{28}$$

This is the linearised Weyl Lagrangian (with an overall reduction by $\Box^{-1}$ in the boson terms).

The Lagrangian (28) is not yet suitable since it involves differentially constrained fields. To describe supergravity with the appropriate gauge invariances we must introduce further degrees of freedom. Thus the vector A_μ becomes unconstrained by addition of a scalar field ϕ. If we write $A_\mu = (B_\mu - \partial_\mu \Box^{-1} \partial_\nu B_\nu)$ for an unconstrained 4-vector B_μ, then $- A_\mu^2 - (\partial_\mu \phi)^2 = - B_\mu^2$, with $\phi = \Box^{-1} \partial_\nu B_\nu$. We notice that the sign of the kinetic term for ϕ in the above is negative, so that the combination $[- A_\mu^2 - (\partial_\mu \phi)^2]$ can be written in a short-hand as $\underline{1}_A - \underline{0}_P$ (where A denotes auxiliary, with dimension L^{-2}, P denotes physical with dimension L^{-1}). Since the field equation for the unconstrained B_μ is $B_\mu = 0$ we can denote this by the 'annihilation' rule $1_A - 0_P \approx 0$. We also need the 'creation' rules $\underline{2}^+_P - \underline{0}^+_P = L_{\text{Einst.}}$, $\frac{3}{2} - \frac{1}{2} = L_{\text{R.S.}}$ (the Rarita-Schwinger Lagrangian for a spin $\frac{3}{2}$ field). These rules are to be expected since the degrees of freedom needed to describe a symmetric tensor $h_{\mu\nu}$ with linearised co-ordinate invariance are 6 (so requiring an extra scalar in addition to pure spin 2) and for the spin $\frac{3}{2}$ field with invariance $\delta\psi_{\mu\alpha} = \partial_\mu \varepsilon_\alpha$ is 12 (so needing an extra spin $\frac{1}{2}$ as well as the pure spin ${}^3/_2$). Thus we need extra spin 0 and $\frac{1}{2}$ fields to combine with the Weyl Lagrangian. In a supersymmetrical invariant theory these fields must occur in irreps of S_1 usually called auxiliary multiplets.

We take for these auxiliary multiplets the chiral multiplets with content $(0_A^+, 0_P^-, \frac{1}{2}^{+i})$ and $(0_A^-, 0_P^+, \frac{1}{2}^{-i})$ (where ± denotes parity and is deduced from the states in section 3 by assuming $S_{\alpha\pm}$ has parity ±i) but with negative energy in their kinetic terms. We can then perform the subtraction sum

$$
\begin{array}{lllll}
 & 2_P^+ & \frac{3}{2}^{\pm i} & 1_A^- & \\
-\,(& 0_P^+ & \frac{1}{2}^{-i} & & 0_A^-) \\
-\,(& & \frac{1}{2}^{+\frac{1}{2}} & 0_P^+ & 0_A^+) \\
\hline
+ & h_{\mu\nu} & +\ \psi_{\mu\alpha} & +\ B_m & -\ (S,P) \\
\hline
\end{array}
$$

where $h_{\mu\nu}$ and $\psi_{\mu\alpha}$ are the graviton and gravitino of supergravity and B_m, S and P auxiliary fields vanishing on-shell (Ferrara & van Nieuwenhuizen 1978, Stelle & West 1978). The Lagrangian is

$$L_Y = \tfrac{3}{2} - (L_{Y=0^+} + L_{Y=0^-}) = L_{Einst}(h_{\mu\nu}) + L_{R.S.}(\psi_\mu) - (B_m^2 + S^2 + P^2) \qquad (29)$$

We note that it is straightforward to non-linearise (29) and the supersymmetry transformations of the multiplets (Ferrara & van Nieuwenhuizen 1978, Stelle & West 1978). The superfield formulation of this theory is discussed elsewhere in this volume and in the references (Ogievetsky & Sokatchev 1975; Siegel & Gates 1979, Bedding, Downes-Martin & Taylor 1979; Schwarz 1980).

6.2 Linearised N = 2 supergravity

We proceed as for N = 1, now with the Y = 1 Weyl multiplet $(2_P^+, 1_A^{+3}, 1_P^+, 1_P^-, 1_A^-, \frac{3}{2}^{\pm i}\ \frac{1}{2}^{\pm i}, 0_P^+)$. The spin $\frac{1}{2}$ fields have to be made auxiliary, and for that we use the new annihilation rule $\frac{1}{2} - \frac{1}{2} \approx 0$ arising from rewriting $i\bar{\psi}\not\partial\psi - i\bar{\phi}\not\partial\phi$ as $\bar{\lambda}_1\lambda_2$, with $\lambda_1 = \psi + \phi$, $\lambda_2 = i\not\partial(\psi - \phi)$. One of the 1_P's must also be removed and for this we can use the rule $(1_P^+ - 1_P^-) \approx 0$ arising from $t_{\mu\nu} = \partial_{[\mu}V_{\nu]} + \varepsilon_{\mu\nu\lambda\sigma}\partial_\lambda A_\sigma$ so $t_{\mu\nu}^2 = (\partial_{[\mu}V_{\nu]})^2 - (\partial_{[\mu}A_{\nu]})^2$. These annihilations require further spin $\frac{1}{2}$ and 1 fields which occur in the Y = 0 multiplet

$(1^-_P, \frac{1}{2}^{\pm i\,2}, 0^\pm_P, 0^+_A, 0^{-2}_A)$. We may set out the recombination in the form of a subtraction sum as for N = 1, taking two Y = 0 multiplets of opposite dimensions, as follows:

$$\begin{array}{lcccccccccc}
(2^+_P & 1^{+\,3}_A & 1^-_P & 1^+_P & 0^+_P & 1^-_A & \frac{3}{2}^{\pm i2} & \frac{1}{2}^{\pm i2}) & & \\
-\,(0^+_P & 0^-_P & & 1^-_P & & & & \frac{1}{2}^{\pm i2} & 0^{-2}_A & 0^+_A) \\
-\,(& 0^{-\,2}_P & & & 1^-_A & 0^+_P & \frac{1}{2}^{\pm i2} & & 0^-_A & 0^+_A) \\
\hline
\end{array}$$

$$h_{\mu\nu} + A^3_m + V_\mu + t_{\mu\nu} - B_m + C_m + \psi_{\mu\alpha} + \lambda_1, \lambda_2 - (P^3, S_1, S_2)$$

with Lagrangian

$$L_{Y=1} - L_{Y=0} - L_{Y=0} = L_{Einst}(h_{\mu\nu}) + L_{R.S.}(\psi_{\mu\alpha}) + L_{Maxwell}(V_\mu)$$
$$+ t^2_{\mu\nu} + B^2_m - C^2_m + \bar{\lambda}_1\lambda_2 - P^2 - S^2_1 - S^2_2$$

where P is an SU(2) triplet. This is the minimal set of auxiliary fields of deWit et al. (deWit, van Holten 1979, 1980, Fradkin & Vasiliev 1979).

VII CENTRAL CHARGES AND SPIN REDUCTION

7.1 Central Charges

We are going to analyse a possible method of spin reduction (Sohnius 1978) which may allow a construction of N = 8 supergravity without the appearance of higher spin fields. It is possible to enlarge the supersymmetry algebra $\mathcal{S}_N$ by the addition of real antisymmetric matrices z_{ij}, z'_{ij} on the r.h.s. of (5) to give

$$[S_{\alpha i}, S_{\beta j}]_+ = (\not{p}\eta)_{\alpha\beta}\,\delta_{ij} + \eta_{\alpha\beta}\,Z_{ij} + (\gamma_5\eta)_{\alpha\beta}\,\bar{Z}_{ij} \qquad (30)$$

If the matrices Z and $\bar{Z}$ commute with $S_{\alpha i}$, P_μ and $J_{\mu\nu}$ then they are termed central charges, so-called because they belong to the centre of $\mathcal{S}_N$. They may arise from global features of the algebra (Witten & Olive 1978) or as extra momenta by dimensional reduction (Cremmer & Julia 1979). This latter possibility is clear from the representation of $S_{\alpha i}$ satisfying (30) on superfields obtained by adding the term $(\frac{1}{2}\,\theta_{\alpha j}\,\partial/\partial z^{ij} + \frac{1}{2}(\gamma_5\theta)_{\alpha j}\,\partial/\partial\bar{z}^{ij})$ to the expression (6), where z^{ij}, $\bar{z}^{ij}$ are a set of

$N(N-1)$ bose variables upon which the superfield is required to depend, and $Z_{ij} = \partial/\partial z^{ij}$, $\bar{Z}_{ij} = \partial/\partial \bar{z}^{ij}$ commute with all the other operators.

For N = 2, both Z_{ij} and $\bar{Z}_{ij}$ must be proportional to ε_{ij}. For N = 4 we may take Z_{ij} and $\bar{Z}_{ij}$ being linear sums of the 6 real anti-symmetric matrix generators $\underline{\alpha}_{ij}$, $\underline{\beta}_{ij}$ which are the generators of $SO(3) \times SO(3) = SO(4)$, with $[\alpha_i,\alpha_j]_- = -2\varepsilon_{ijk}\,\alpha_k$, $[\alpha_i,\alpha_j]_+ = -2\delta_{ij}$. For N = 8 we may similarly take a basis for Z_{ij} and $\bar{Z}_{ij}$ by the real anti-symmetric matrix generators SO(8).

7.2 Spin reduction

The multiplets or irreps of the extended supersymmetry algebra S_{NZ}, with (5) replaced by (30), will in general have the same range of spin as in the case of no central charge. There is one exceptional case when this does not occur. To see this, (Taylor 1980b) we take $\bar{Z}_{ij} = 0$ and write the r.h.s. of (23) as $(\not{p} + Z)\eta$. We then express S as V + W with $V = (2\not{p})^{-1}(\not{p} - z)\,S$, $W = (2\not{p})^{-1}(\not{p} + Z)S$, so that the anticommutation rules for V and W are

$$4p^2[V,V]_+ = (p^2 - Z^2)(\not{p} - z)\eta,\ 4p^4[V,W]_+$$

$$= (p^2 - z^2)\,(\not{p} + Z)\eta\ ,\ 4p^2(W,W) = (p^2 + z^2)(\not{p} + Z)\eta \qquad (31)$$

If $p^2 = Z^2$ then V becomes nilpotent and can be set to zero in non-trivial irreps. We are thus left with W which also satisfies (30). However since W satisfies the Dirac equation $(\not{p} - Z)\,W = 0$ then only half the components of W are independent. The process of constructing irreps with W described in section 3 will have only half the number of W's as expected; spin reduction to half the previous value will have occurred.

The condition $p^2 = Z^2$ will only be independent of internal symmetry labels (so valid for all irreps) if Z^2 is proportional to the identity. This is only possible if $z = \sum_m Z_m \Gamma_m$, where $[\Gamma_m,\Gamma_n]_+ = -2\delta_{mn}$; Γ_m are the elements of a Clifford algebra. For N = 8 this is a 7-dimensional algebra, so there are at most seven central charges; for N = 4 we have a 3-dimensional algebra. In general the condition $p^2 = Z^2$ reduces the corresponding massless wave equation in $(4 + \ell)$ dimensions, where ℓ is the dimension of the algebra of Γ's.

The multiplets which now arise are those of $S_{N/2}$ (N even), but the presence of the Z_m's brings about 'Z-multiplexing' in which a given $S_{N/2}$ multiplet is multiplied by all the independent powers of the Z_m's. For one Z_m there is only one such power, so that only z-doubling occurs. It is possible to arrange this still to be the case even when there are more Z_m's (Taylor 1981b) so we will assume that occurs here. The multiplets of spin-reduced S_N are thus to be written down as pairs of irreps of $S_{N/2}$. Thus the fundamental multiplets for N = 2, 4 and 8 are:

$$(0^2_A, 0^2_P, \tfrac{1}{2}), (0^5_A, 0^5_P, \tfrac{1}{2}^4, 1_A, 1_P)$$

$$\text{and } (0^{42}_A, 0^{42}_P, \tfrac{1}{2}^{48}, 1^{27}_A, 1^{27}_P, \tfrac{3}{2}^8, 2_A, 2_P) \qquad (32)$$

respectively. We have used here the labelling by irreps of the appropriate USp(N), since any unitary matrix U for which $U^T ZU = Z$ is a symmetry of the algebra (30) (with $\bar{Z}_{ij} = 0$). The multiplets (32) may be extended to irreps with higher Y and USp(N) values by multiplication of the components in (32) by the corresponding Y and USp(N) labels and reduction of the resulting direct products of SL(2,C) and USp(N).

VIII N ≥ 3 SUPERGRAVITIES AND CENTRAL CHARGES

8.1 Off-shell no-go theorems

To date there has been no model presented (either in components or in superfields) for N ≥ 3 supergravity which is supersymmetric and for which the supersymmetric transformations close off-shell (without use of the equations of motion). In other words, there has been no success so far in obtaining a suitable set of auxiliary fields for N ≥ 3 supergravity. That this should be so has been shown recently (Rivelles & Taylor 1981b, Taylor 1981b) using the field recombination rules discussed in the last section as annihilation and creation rules. This was achieved by taking an arbitrary linear combination of irreps of S_3 in the linearised Lagrangian of the form $\sum_{Y,n} (a_{Y,n} - b_{Y,n}) \Phi_{Y,n}$ where Y and n denote superspin and isospin and $a_{Y,n}$, $b_{Y,n}$ are non-negative constants. The range of values of Y and n were chosen so as to correspond to irreps occurring in the superfields (V, $V_{\alpha i}$, V_m) in (Rivelles & Taylor 1981b) and to irreps in E_A^M, $\Omega_{AB}{}^C$ in (Taylor 1981b). Use of the annihilation rules $j - j \approx 0$ for half odd-integer j to remove unwanted higher spin and isospin fermion fields caused all fermi fields to occur in pairs, and so annihilate each other. The possibility of the existence of auxiliary fields for $N > 3$ supergravity is then precluded since their existence would lead to auxiliary fields for $N = 3$ supergravity by trivial reduction. Thus no satisfactory off-shell model for $N \geq 3$ supergravity appears to be possible.

8.2 The use of central charges

A crucial ingredient in the above no-go theorems was the absence of central charges. This corresponds to assuming triviality of all fields in a higher dimensional reduction of the theory, for example, from 11 dimensional simple supergravity for $N = 8$ in four dimensions (Cremmer & Julia 1979). When spin reducing central charges are present the above no-go theorems may be avoided since both the SU(N) symmetry and the higher spin components inherent in the central charge-free algebra are removed. This may allow for construction of auxiliary field multiplets using the reduced symmetry. Such multiplets have been shown to exist for all $N \geq 3$ (Taylor 1981c,d), and even to support an internal USp(4) symmetry for $N = 8$ (Taylor 1981f). This may be regarded as the ultimate symmetry, since if supergravity is the ultimate theory of the unification of the forces of nature its largest off-shell symmetry will be that governing all interactions above 10^{19} BeV. The emergence of different symmetries at lower energies is an intriguing problem for further research.

REFERENCES

Bedding, S., Downes-Martin, S. & Taylor, J.G. (1979) Annals of Physics 120, 175.
Berhends, R. deWit, B. van Holten, J & van Nieuwenhuizen, P. (1980) J. Phys. A13, 1643.
Bufton, G. & Taylor, J.G. (1981) "SU(N)-Covariant Basis Functions, Components and Supersymmetry Transformations for Extended Supersymmetry", King's College preprint.
Capper, D. & Leibbrandt, G. (1975) Nucl. Phys. B85, 492.
Cremmer, E. & Julia, B. (1979) Nucl. Phys. B129, 141.
Delbourgo, R. (1975) N. Cim. 25A, 646.
Deser, S. & Zumino, B., Phys. Lett. 62B (1976) 335.
Deser, S., Kay, J. and Stelle, K. Phys. Rev. Lett. 38 (1977) 527.
deWit, B. & van Holten, J.W. (1979), Nucl. Phys. B155, 530; ibid & van Proeyen, A. (1980) Nucl. Phys. B167, 186.
Ferrara, S. in these proceedings.
Ferrara, S. & van Nieuwenhuizen, P. (1978), Phys. Lett. 74B, 333.
Fujikawa, F. & Lang, W. (1975) Nucl. Phys. B88, 61.
Fradkin, E.S. & Vasiliev, M.A. (1979) Lett. N. Cim. 25, 79; (1979) Phys. Lett. 85B, 47.
Freedman, D.Z. van Nieuwenhuizen, P. & Ferrara, S. (1976) Phys. Rev. D13, 3214.

Freund, P.G.O. & Kaplansky, I. (1975) J. Math. Phys. 16,288.
Gates, S.J. (1981) "Towards an Unextended Superfield Formulation of N=2 Supergravity" in "Superspace and Supergravity", ed. S. Hawking & M. Rocek, C.U.P.
Gayduk, A.V., Romanov, V.N. & Schwarz, A.S. (1981) Comm. Math. Phys. 79, 501-528.
Golfand, Yu. A. & Likhtman, E.P.(1971) J.E.T.P. Letters 13, 452.
Grisaru, M.T., van Nieuwenhuizen, P. & Vermaseren, J.A.M. (1976), Phys. Rev. Lett. 37, 1662.
Illiopolous, J. & Zumino, B. (1974), Nucl. Phys. B76, 310.
Kac, V.G. (1977), Commun. Math. Phys. 53, 31-64.
Ogievetsky, V. & Sokatchev, E. (1975) "The gravitational axial superfield and the formalism of differential geometry", Dubna preprint E2-12511.
Pickup, C. & Taylor, J.G. (1981) Nucl. Phys. B188, 577.
Ramond, P. (1971) Phys. Rev. D3, 2415.
Rittenberg, V. (1977) "A Guide to lie Superalgebras", talk at the VI Int. Coll. on Group Theoretical Methods in Physics, Tubingen.
Rittenberg, V. & Sokatchev, E. (1981) "Decomposition of extended superfields into Irreducible Representations of Supersymmetry". Bonn preprint ISSN-0172-8733.
Rivelles, V. & Taylor, J.G. (1981a) "Linearised N = 2 Superfield Supergravity", J. Phys. A. (to appear).
Rivelles, V. & Taylor, J.G. (1981b) Phys. Lett. 104B, 131.
Salam, A. & Strathdee, J. (1974). Nucl Phys. B76, 477; (1975) Phys. Rev. D11, 1521.
Schwarz, A.S. (1980), Nucl. Phys. B171, 154-166.
Siegel, W. & Gates, J. (1979), Nucl. Phys. B147, 77-104.
Sohnius, M. (1978) Nucl. Phys. B138, 109.
Sokatchev, E. (1975), Nucl. Phys. B99, 96-108.
Sokatchev, E. (1981) "Complex Superspace and prepotentials for N = 2 supergravity" in "Superspace and Supergravity",ed. S. Hawking and M. Rocek, C.U.P.
Stelle, K. & West, P.C. (1978), Phys. Lett. 74B 336.
Taylor, J.G.(1980a) Nucl. Phys. B169, 484.
Taylor, J.G., (1981a) "Extended Superfields in linearised supersymmetry and supergravity", in "Superspace and Supergravity", ed. S. Hawking and M. Rocek, Camb. Univ. Press.
Taylor, J.G. (1981b) "A No-Go Theorem for Off-Shell Extended Supergravities", J. Phys. A. (to appear)
Taylor, J.G. (1981c) "Auxiliary Field Candidates for N = 3,4,5 and 6 Supergravities", Phys. Lett. B (to appear)
Taylor, J.G. (1981d) "Auxiliary Fields for Linearised N = 8 Supergravity", Phys. Lett. B (to appear)
Taylor, J.G. (1981e) "Off-Shell central charges and linearised N = 8 supergravity", King's College preprint.
Taylor, J.G. (1981f) "The Internal Symmetry of Off-shell linearised N = 8 Supergravity", King's College preprint.
Taylor, J.G. (1980b) Phys. Lett. 94B, 174.
van Nieuwenhuizen, P. & Vermaseren, J.A.M. (1977) Phys. Rev. D16 298-303.
Volkov, D.V. & Soroka, V.A. (1973), J.E.T.P. Lett. 18, 529.
Wess, J. & Zumino, B. (1974), Nucl. Phys. B70, 39; (1974) Phys. Lett 51B, 239.
Witten, E. & Olive, D. (1978), Phys. Lett. 78B, 97-101.

REPRESENTATIONS OF EXTENDED SUPERSYMMETRY ON ONE-AND TWO-PARTICLE STATES

S. Ferrara and C.A. Savoy
CERN, Geneva, Switzerland

1 INTRODUCTION

The aim of the present lectures is to review the structure of unitary irreducible representations of the extended (Poincaré) supersymmetry algebra (Salam & Strathdee 1974; Gell-Mann & Ne'eman 1974; Nahm 1978; Freedman 1979; Ferrara 1980). Special emphasis will be given to those aspects which are relevant to particle physics. Extended supersymmetry assigns "elementary particles", considered as unitary irreducible representations of the Poincaré group, to supermultiplets whose states are degenerate in mass if supersymmetry is unbroken. However, the correlation between the Wigner spin of these states and their internal symmetry properties is non-trivial owing to the fact that the supersymmetry fermionic charges Q^i_α carry internal quantum numbers as well as half-units of spin. This very fact opens up the possibility of a true unification of internal quantum numbers with space-time symmetries. A remarkable example of a dynamical framework in which these ideas are realized is "extended supergravity" (Ferrara & Van Nieuwenhuizen 1976; Freedman 1979; Ferrara, Scherk & Zumino 1977a, 1977b; Das 1977; Cremmer & Scherk 1977; Cremmer, Scherk & Ferrara 1977, 1978; De Wit & Freedman 1977; De Wit 1979; Cremmer & Julia 1978, 1979) in which a complete unification of all fundamental interactions may eventually be achieved.

These lectures are organized as follows: in Section 2 we will consider massive supermultiplets for N-extended supersymmetry in the absence of central charges. Internal symmetry properties and the spin content of these multiplets will be discussed in detail.

In Section 3, the modification of the massive representations in the presence of central charges will be considered.

In Section 4, the important case of massless representations will be studied. The modification of the internal symmetry assignment and the problem of PCT conjugation for massless one-particle states will be discussed. In addition we will give the relation between massive and massless representations. This relation is especially important in view of the possibility of describing massless bound states in terms of composite operators.

Finally, in Section 5, the representations of supersymmetry on two-particle states will be considered. We will confine our discussion to the case of those two-particle states which are tensor products of one-particle massless states. This analysis is relevant to the quantum number assignment of bound-states of elementary preonic massless fields of Yang-Mills and supergravity theories. The complete structure of these two-particle supermultiplets and their internal symmetry properties will be given for a general N-extended supersymmetry algebra.

2 MASSIVE SUPERMULTIPLETS

In order to be self-contained, we give some preliminaries. We consider in this section the graded extension of the four-dimensional Poincaré algebra P_4 defined as follows

$$\begin{aligned}
\left[M_{\mu\nu},M_{\rho\sigma}\right] &= i(g_{\mu\rho}M_{\nu\sigma} - g_{\nu\rho}M_{\mu\sigma} + g_{\nu\sigma}M_{\mu\rho} - g_{\mu\sigma}M_{\nu\rho}) \\
\left[M_{\mu\nu},P_{\lambda}\right] &= i(g_{\mu\lambda}P_{\nu} - g_{\nu\lambda}P_{\mu}) \qquad (1) \\
\left[P_{\mu},P_{\nu}\right] &= 0
\end{aligned}$$

$$\text{metric:} \quad g_{\mu\nu} = \text{diag}(-1,1,1,1) \text{ i.e. } P^2 = \vec{P}^2 - P_0^2$$

The P_4 algebra is a ten-dimensional Lie algebra. Spinors in four-dimensions are representations of the Clifford algebra

$$\{\gamma^\mu,\gamma^\nu\} = 2g^{\mu\nu} \quad , \quad \gamma_5 = \gamma_0\gamma_1\gamma_2\gamma_3 \ . \qquad (2)$$

The charge conjugation matrix $C = -C^T$ has the property $\gamma_\mu^T = C\gamma_\mu C^T$. Majorana spinors are defined by the Lorentz-invariant condition

$$Q = C\bar{Q} = \bar{Q}C^T = Q^*\gamma^0 C^T = (C\gamma^{0T})Q^* \qquad (3)$$

In four-dimensions there exists a representation (Majorana representation) in which all γ-matrices are real and such that $Q = Q^*$ and $\gamma^0 = C$. We will denote the N-extended super-Poincaré algebra by SP_4^N. It is obtained by adding to P_4 (even sector) N-charges Q_α^i which are Majorana spinors (odd sector) and which define, together with Eq. (1), the following graded Lie algebra

$$\left[Q^i, M^{\mu\nu}\right] = i\sigma^{\mu\nu} Q^i \tag{4}$$

$$\left[Q^i, P^\mu\right] = 0 \tag{5}$$

$$\{Q^i, Q^j\} = (\gamma^\mu C) P_\mu \delta^{ij} . \tag{6}$$

In general, for N > 1 one could add to the left-hand side of Eq. (6) the terms

$$CU^{ij} + (\gamma_5 C) V^{ij} . \tag{7}$$

U^{ij} and V^{ij} are P_4 invariant operators (Haag, Lopuszanski & Sohnius 1975). General requirements from quantum field theory and from the Jacobi identities of the graded Lie algebra SP_4^N then imply that

$$\left[Q^i, U^{k\ell}\right] = \left[Q^i, V^{k\ell}\right] = \left[U^{ij}, U^{k\ell}\right] = \left[U^{ij}, V^{k\ell}\right] = \left[V^{ij}, V^{k\ell}\right] = 0 \tag{8}$$

and $U^{k\ell} = -U^{\ell k}$, $V^{k\ell} = -V^{\ell k}$. In other words U's and V's belong to the centre of SP_4^N and for this reason are called central charges. The introduction of the operators U's and V's on the right-hand side of Eq. (6) has several important consequences for the structure of the representations acting on one-particle states as well as for the invariance properties of SP_4^N. However, in this section we will temporarily consider the case $U^{ij} = V^{ij} = 0$ which is evidently consistent with all the Jacobi identities of SP_4^N. Before considering the structure of the representations it is useful to rewrite the commutator relations given by Eqs. (4) to (8) in the so-called Weyl representation, i.e. a representation in which γ_5 is diagonal. On a Weyl basis we can choose

$$\gamma^\mu = i\begin{pmatrix} 0 & \sigma^\mu \\ -\sigma_\mu & 0 \end{pmatrix}, \quad \gamma_5 = \begin{pmatrix} i & 0 \\ 0 & -i \end{pmatrix}, \quad C = \begin{pmatrix} \sigma_2 & 0 \\ 0 & -\sigma_2 \end{pmatrix} \tag{9}$$

with

$$\sigma_0 = -\sigma^0 = I \begin{pmatrix} 1 & 0 \\ 0 & 1 \end{pmatrix}, \quad \sigma_1 = \begin{pmatrix} 0 & 1 \\ 1 & 0 \end{pmatrix}, \quad \sigma_2 = \begin{pmatrix} 0 & -i \\ i & 0 \end{pmatrix},$$

$$\sigma_3 = \begin{pmatrix} 1 & 0 \\ 0 & -1 \end{pmatrix}.$$

In the Weyl representation a spinor is projected into the $\gamma_5 = \pm i$ components

$$Q^i = \begin{pmatrix} Q^i_L \\ Q^i_R \end{pmatrix} \tag{10}$$

and a Majorana spinor then satisfies the condition

$$Q^i_R = i\sigma_2 Q^{i*}_L \tag{11}$$

so that it is described by a two-component complex Weyl spinor

$$Q^i_\alpha = (Q_L)^i_\alpha, \quad \alpha = 1,2 . \tag{12}$$

It is convenient to use Van der Waerden notations for Weyl spinors:

$$\varepsilon^{\alpha\beta} = -\varepsilon_{\alpha\beta} = \varepsilon^{\dot\alpha\dot\beta} = -\varepsilon_{\dot\alpha\dot\beta} = \begin{pmatrix} 0 & 1 \\ -1 & 0 \end{pmatrix}_{\alpha\beta} = (i\sigma_2)_{\alpha\beta}$$

$$\chi^\alpha = \varepsilon^{\alpha\beta}\chi_\beta, \quad \psi^{\dot\alpha} = \varepsilon^{\dot\alpha\dot\beta}\psi_{\dot\beta} \tag{13}$$

$$\left(Q_R\right)^{\dot\alpha}_i = \varepsilon^{\dot\alpha\dot\beta} Q^{i*}_{\dot\beta} = \left(i\sigma_2 Q^*_{L_i}\right)^{\dot\alpha}$$

With the conventions the basic anticommutators become

$$\{Q^i_\alpha, Q^{*\dot\beta}_j\} = (\sigma_\mu)_\alpha^{\dot\beta} P^\mu \delta^i_j \tag{14}$$

$$\{Q^i_\alpha, Q^j_\beta\} = \{Q^{*\dot\alpha}_i, Q^{*\dot\beta}_j\} = 0 . \tag{15}$$

In the presence of central charges [Eq. (7)] the anticommutator [Eq. (14)] is unchanged while the left-hand side of Eq. (15) acquires a term

$$\varepsilon_{\alpha\beta}(-V^{ij} + iU^{ij}) \ . \tag{16}$$

The SP_4^N algebra, in the absence of central charges, has a U(N) symmetry as a group of invariances. In fact, Eqs. (14) and (15) are invariant under the transformations

$$\begin{aligned} Q_\alpha^i &\to Q_\alpha'^i = U^i_{\cdot j} Q^j_\alpha \\ Q^{*\dot\alpha}_{\cdot i} &\to Q'^{*\dot\alpha}_{\cdot i} = U^{+j}_{\cdot i} Q^{*\dot\alpha}_{\cdot j} \end{aligned} \qquad UU^+ = U^+U = I \tag{17}$$

$Q^i_\alpha, Q^{*\dot\alpha}_{\cdot i}$ transform as the N and $\bar{N}$ representations of U(N), respectively. It is obvious that this U(N) symmetry is relativistic, i.e. $[U(N), P_4] = 0$.

The irreducible representations (irreps.) of SP_4^N are easily obtained by extending the Wigner method of induced representations to the super-Poincaré algebra (Salam & Strathdee 1974) [Eqs. (4) to (6)]. The Poincaré Casimir operator $P^\mu P_\mu$ commutes with SP_4^N so that each irrep corresponds to a given value of $P^2 = M^2$. We also note that the other Casimir operator of P_4, $W^\mu W_\mu$, when $W_\mu = g_{\mu\nu\gamma\rho} M^{\nu\rho} P^\sigma$ is not a SP_4^N Casimir because it does not commute with the Q^i_α operators. We first consider the massive case $M^2 > 0$ and use the fact that we can study the stability subalgebra of SP_4^N which leaves a given time-like momentum $\hat{P}^\mu$ invariant. We can choose $\hat{P}^\mu = (M,0)$ $(M = 1)$ so that Eqs. (14) and (15) become

$$\begin{aligned} \{Q^i_\alpha, Q^{*\beta}_{\cdot j}\} &= \delta^\beta_\alpha \delta^j_{\cdot i} \\ \{Q^i_\alpha, Q^j_\beta\} &= 0 \end{aligned} \tag{18}$$

while the P_4 invariance is reduced to the 0(3) invariance (stability group of $\hat{P}^\mu$) i.e. angular momentum

$$\begin{aligned} \left[J_a, J_b\right] &= i\,\varepsilon_{abc} J_c \ , \quad J_a = i\,\varepsilon_{abc} M^{bc} \\ \left[Q^i_\alpha, J_a\right] &= \left(\sigma_a\right)_{\alpha\beta} Q^i_\beta \ . \end{aligned} \tag{19}$$

The commutator relations given by Eq. (18) define a Clifford algebra containing 2N creation operators $Q^{*\alpha}_{i}$ and 2N destruction operators $Q^{i}_{i\alpha}$. (α = 1,2 , i = 1 ... N .) We have omitted dotted and undotted indices because they are irrelevant in the rest frame.

We can define 4N real operators

$$\Gamma^{i}_{2\alpha-1} = Q^{i}_{\alpha} + Q^{\alpha *}_{i} \qquad \Gamma^{i}_{m} = (\Gamma^{i}_{m})^{+} \qquad (20)$$
$$\Gamma^{i}_{2\alpha} = i(Q^{i}_{\alpha} - Q^{\alpha *}_{i})$$

and the Clifford algebra [Eq. (18)] can be rewritten in real form

$$\{\Gamma^{i}_{m},\Gamma^{j}_{n}\} = 2\delta^{ij}\delta_{mn} , \quad i,j = 1 \ldots N , \quad m,n = 1 \ldots 4 . \qquad (21)$$

This Clifford algebra has one irrep. of dimension 2^{2N}. There is a SO(4N) group of invariance associated with Eq. (21), the Γ^{i}_{m} transforming as the vectorial reps. $\underset{\sim}{4N}$ of SO(4N).

The SO(4N) algebra is realized in terms of bilinear operators

$$O^{ij}_{mn} = \frac{1}{4i}\left[\Gamma^{i}_{m},\Gamma^{j}_{n}\right] \qquad (22)$$

which are the 2N(4N - 1) generators of SO(4N). Then

$$\left[\Gamma^{k}_{\ell},O^{ij}_{mn}\right] = i\delta^{ki}\delta_{\ell m}\Gamma^{j}_{n} - i\delta^{kj}\delta_{\ell n}\Gamma^{i}_{m} . \qquad (23)$$

The irrep. of Eq. (21) with dimension 2^{2N} is reducible into the two spinorial representations of SO(4N) each with dimension 2^{2N-1}. Indeed from Eq. (21) one can construct the SO(4N) invariant operator

$$\Gamma_{4N+1} = \prod_{i=1}^{N}\prod_{m=1}^{4}\Gamma^{i}_{m} , \quad \Gamma^{2}_{4N+1} = 1 , \quad \left[\Gamma_{4N+1},O^{ij}_{mn}\right] = 0 . \qquad (24)$$

The operator Γ_{4N+1} allows us to construct the projections $(1 \pm \Gamma_{4N+1})/2$ which split the irrep. of the Clifford algebra [Eq. (21)] into the two inequivalent 2^{2N-1} irrep. spinorial representations of SO(4N), respectively.

Construction of the spinorial representations: the condition

$$Q^i_\alpha \Omega = 0 \tag{25}$$

defines the "Clifford vacuum" Ω. The orthogonal states

$$\Omega, Q^*_i \Omega, Q^{*\alpha}_i Q^{*\beta}_j \Omega, \; \ldots \; , Q^{*1}_1 Q^{*2}_2 \ldots Q^{*2}_N \Omega \tag{26}$$

define the 2^{2N} states of the reps. of Eq. (21). The two spinorial representations of SO(4N) are given by even and odd products of Q^* applied to Ω, with $\Gamma_{4N+1} = \pm 1$, respectively. They correspond to bosons (fermions) and fermions (bosons) depending on whether the "vacuum" state is assumed to be a boson or a fermion. If the vacuum state is a scalar under rotations and U(N) transformations the irrep. spanned by Eq. (26) has just 2^{2N} states and it will be called the fundamental representation. In order to classify one-particle states with given spin it is useful to exploit those symmetries of Eq. (21) which commute with spin transformations. For this purpose we can group together the Q and Q^* operators and define sympletic spinors (Ferrara 1980)

$$Q^a_\alpha = Q^i_\alpha \delta^a_i \; (a = 1 \ldots N) \; , \qquad \varepsilon_{\alpha\beta} Q^{*\beta}_i \delta^i_{a-N} \; (a = N+1 \ldots 2N)$$
$$\alpha = 1,2 \; , \qquad a = 1 \ldots 2N \; . \tag{27}$$

Then in terms of Eq. (27), Eqs. (18) and (21) become

$$\{Q^a_\alpha, Q^b_\beta\} = \varepsilon_{\alpha\beta} \Omega^{ab} \tag{28}$$

with

$$\Omega^{ab} = \begin{pmatrix} 0 & I \\ -I & 0 \end{pmatrix} \quad \text{and} \quad (Q^a_\alpha)^* = \varepsilon_{\alpha\beta} \Omega^{ab} Q^b_\beta \; .$$

The commutator (28) exhibits an explicit invariance under the subgroup $SU(2)_s \otimes USp(2N)$ of SO(4N), the Q^a_α transforming as the (2,2N) representation. Since $SU(2)_s$ acts as a spin group for the Q's, Eq. (28) shows that states of given spin will be classified by representations of USp(2N). More precisely the 2^{2N}-dimensional (fundamental) massive representation given by Eq. (26) decomposes with respect to $SU(2)_s \otimes USp(2N)$ in completely k-fold antisymmetric irreps. of USp(2N), denoted $[k]$, with spin $(N-k)/2$ $(k = 0 \dots N)$

$$2^{2N} = \left(\frac{N}{2},[0]\right) + \left(\frac{N-1}{2},[1]\right) + \dots \left(\frac{N-k}{2},[k]\right) + \dots \left(0,[N]\right) . \tag{29}$$

The spin range is $0 \leq s \leq N/2$ and Eq. (29) gives the correlation between the spin and the USp(2N) quantum numbers of a given state in the spinorial representations of SO(4N) according to the embedding $(2,2N) \sim 4N$ of $SU(2) \otimes USp(2N)$ into SO(4N). It will be shown below that USp(2N) is the maximal subgroup of SO(4N) which commutes with $SU(2)_s$.

Let us consider the bilinear operators in the $Q^i_\alpha, Q^{*\alpha}_i$ which span the SO(4N) algebra

$$\Lambda^{\alpha j}_{i\beta} = \frac{1}{2}\left[Q^{*\alpha}_i, Q^j_\beta\right] \qquad (4N^2 \text{ operators})$$

$$\left.\begin{aligned} K^{ij}_{\alpha\beta} &= \frac{1}{2}\left[Q^i_\alpha, Q^j_\beta\right] \\ \bar{K}^{\alpha\beta}_{ij} &= \frac{1}{2}\left[Q^{*\alpha}_i, Q^{*\beta}_j\right] = (K^{ji}_{\beta\alpha})^+ \end{aligned}\right\} \; 2N(2N-1) \text{ operators} \tag{30}$$

The spin group $SU(2)_s$ is given by

$$s_k = \frac{1}{4}(\sigma_k)^\beta_\alpha \left[Q^i_\beta, Q^{*\alpha}_i\right] \tag{31}$$

and the maximal U(N) subalgebra of USp(2N) is

$$U(1)_{W_3}: \quad W_3 = \frac{1}{4}\left[Q^{*\alpha}_i, Q^i_\alpha\right] = \frac{1}{2} Q^{*\alpha}_i Q^i_\alpha - \frac{N}{2} \tag{32}$$

$$SU(N): \quad \Lambda^i_j = \frac{1}{2}\left[Q^i_\alpha, Q^{*\alpha}_j\right] + \frac{2}{N}\delta^i_j W_3 . \tag{33}$$

The residual USp(2N) generators are

$$\left.\begin{aligned} \varepsilon^{\alpha\rho} K^{ij}_{\alpha\beta} &= \frac{1}{2}\left[Q^{i\alpha}, Q^j_\alpha\right] = K^{ij} \\ \varepsilon_{\alpha\beta}\bar{K}^{\beta\alpha}_{ji} &= \frac{1}{2}\left[Q^*_{i\alpha}, Q^{*\alpha}_j\right] = \bar{K}_{ji} \end{aligned}\right\} \quad (N(N+1) \text{ operators}) \tag{34}$$

The following properties can be easily checked:

$$\left[Q^i_\alpha, s_k\right] = \frac{1}{2} (\sigma_k)^\beta_\alpha Q^i_\beta$$
$$\left[W_3, Q^{*\alpha}_j\right] = \frac{1}{2} Q^{*\alpha}_j \tag{35}$$

$$W_3 |n\rangle = \frac{n - N}{2} |n\rangle \tag{36}$$

$$\text{for } |n\rangle = Q^{*\alpha_1}_{i_1} \ldots Q^{*\alpha_n}_{i_n} \Omega$$

i.e. $(Q^i_\alpha) \sim (2,N)_{W_3=-1/2}$, $(Q^{*\alpha}_i) \sim (2,\bar{N})_{W_3=1/2}$ under $SU(2)_s \otimes SU(N) \otimes U(1)_{W_3}$.

From Eqs. (32), (33), and (34) we recognize that USp(2N) is indeed the maximal subalgebra of SO(4N) which commutes with $SU(2)_s$. In fact, the generators of USp(2N) are those operators of SO(4N) which are singlets under $SU(2)_s$. There is another interesting subalgebra of USp(2N) given by $SU(2)_W \otimes SO(N)$ where

$$SU(2)_W: \quad W_3 \ , \quad W_+ = \frac{1}{2} \sum_i K^{ii} \ , \quad W_- = \frac{1}{2} \sum_i \bar{K}_{ii} \tag{37}$$

and SO(N) is the orthogonal subalgebra of SU(N) defined in Eq. (33) with generators $(\Lambda^i_j - \Lambda^j_i)/2$.

We are now ready to analyze the spin $[SU(2)_s]$ content of the fundamental massive supermultiplet which transforms as the $2^{2N} = 2^{2N-1} \oplus 2^{2N-1}$ spinor representation of SO(4N). First consider all states with fixed $W_3 = (n - N)/2$, i.e. all states with n Q^*_0's applied to Ω

$$Q^{*\alpha_1}_{i_1} \ldots Q^{*\alpha_n}_{i_n} \Omega \ . \tag{38}$$

If two Q^*'s, e.g. $Q^{*\alpha_1}_{i_1}$ and $Q^{*\alpha_2}_{i_2}$ are antisymmetric in (α_1, α_2) they must be symmetric in (i_1, i_2) and vice versa because of Eq. (18). By using Young-tableaux to represent the symmetries in the indices $i_1 \ldots i_n$, $\alpha_1 \ldots \alpha_n$ of Eq. (37), one then obtains the classification of the completely antisymmetric products in Eq. (38) according to $SU(2)_s \otimes SU(N)$ as follows

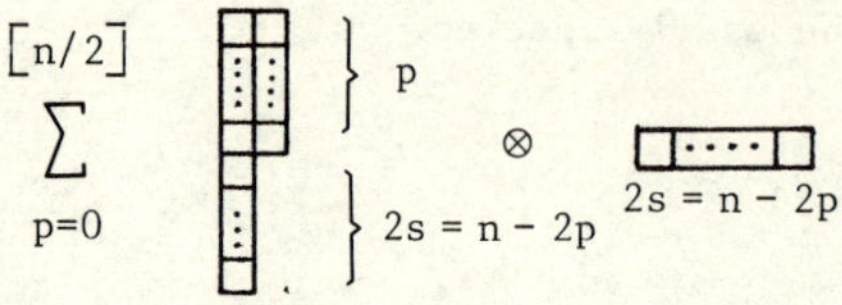

$$s = \frac{N}{2} + W_3 - p \qquad (39)$$

$SU(N) \otimes SU(2)_s$

We note that in SU(2) 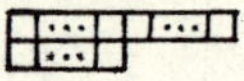 =  since  = 1, while 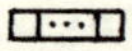 with 2s boxes are the Young-tableaux for the spin-s representation with dimension 2s+1. The diagrammatic expression in Eq. (39) gives the spin content of the $W_3 = (n-N)/2$ states together with their SU(N) representations. The dimension of the SU(N) irrep. of the state with given p,s is

$$D(s,p) = \frac{2s+1}{2s+p+1} \binom{N+1}{p} \binom{N}{2s+p} . \qquad (40)$$

We notice that states corresponding to a given number of Q^*'s, n, have integer (half-integer) spin $s = (n/2) - p$ depending on whether n is even (odd). Also the maximum spin state $s = N/2$ corresponds to $n = N$, i.e. $W_3 = 0$ and $p = 0$ being a singlet under SU(N).

To each irrep. of $SU(N) \times SU(2)_s \times U(1)$

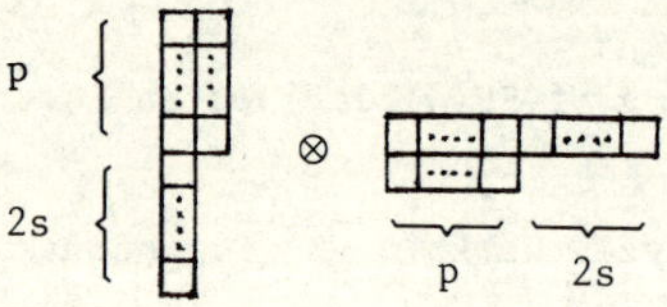

one associates the states

$$\left[Q^{*1}_{i_1} Q^{*2}_{i'_1} - Q^{*2}_{i_1} Q^{*1}_{i'_1}\right] \cdots \left[Q^{*1}_{i_p} Q^{*2}_{i'_p} - Q^{*2}_{i_p} Q^{*1}_{i'_p}\right] Q^{*\alpha_1}_{i_{p+1}} \cdots Q^{*\alpha_{2s}}_{i_{p+2s}} \Omega \qquad (41)$$

where i_k's $[k = 1 \ldots (p+2s)]$ are completely symmetrized, $\{i_k, i'_n\}$ are symmetrized by pairs $(k = 1 \ldots p)$ and the $\alpha_{j,s}$ are completely symmetrized $(i_k, i'_n = 1 \ldots N,\ \alpha_j = 1,2)$. From Eq. (35) one sees that Q^{*1}_i has $s_3 = -\frac{1}{2}$ and Q^{*2}_i has $s_3 = +\frac{1}{2}$ so that the total value of s_3 for the state of Eq. (38) or Eq. (41) is ½ (the number of $Q^{*2}_{i,s}$) − ½ (the number of $Q^{*1}_{i,s}$). Therefore we introduce the notation

$$\left[Q^*\right]_p^{-ss_3}\Omega = \left[Q^{*1}_{i_1}Q^{*2}_{i_1'} - Q^{*2}_{i_1}Q^{*1}_{i_1'}\right]\cdots\left[Q^{*1}_{i_p}Q^{*2}_{i_p'} - Q^{*2}_{i_p}Q^{*1}_{i_p'}\right]$$

$$Q^{*1}_{i_{p+1}}\cdots Q^{*1}_{i_{p+s-s_3}}\,Q^{*2}_{i_{p+s-s_3+1}}\cdots Q^{*2}_{i_{p+2s}}\,\Omega \qquad (42)$$

where Q^{*1}_{1s} (Q^{*2}_{1s}) appears $p+s-s_3$ ($p+s+s_3$) times.

We now put together all the states with $-N/2 \leq W_3 \leq N/2$ which contribute to a given value of spin (s,s_3). From Eq. (39) one gets for the SU(N) representations associated with 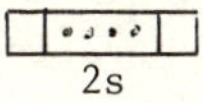 of $SU(2)_s$, the result

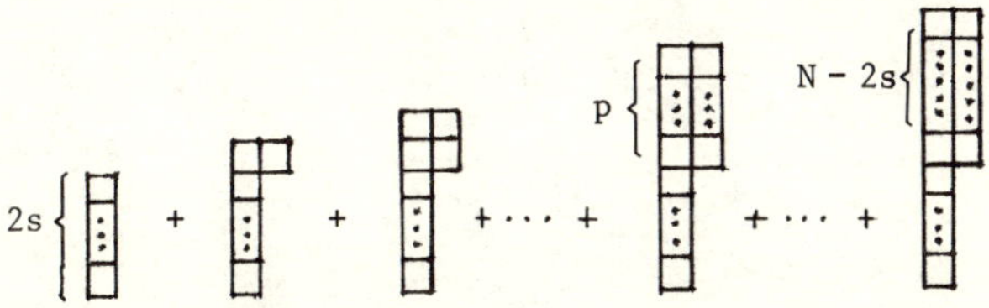

(43)

$$= \sum_{p=0}^{N-2s}$$ 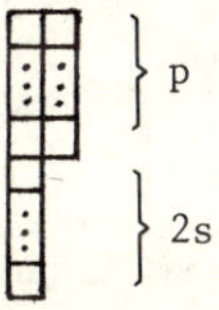

$$W_3 = s + p - \frac{N}{2}$$

This completes the analysis of the $SU(2)_s \otimes SU(N) \otimes U(1)_{W_3}$ content of the states given by Eq. (26). Note that the total number of states with a given value of the spin (s,s_3) is obtained from Eqs. (40) and (43)

$$D(s) = \sum_{p=0}^{N-2s} D(s,p) = \sum_{p=0}^{N-2s} \frac{2s+1}{2s+p+1}\binom{N+1}{p}\binom{N}{2s+p} = \frac{2s+1}{N+1}\binom{2N+2}{N-2s} \qquad (44)$$

which is precisely the dimension of the irrep. $[N-2s]_{2N}$ (traceless antisymmetric) of USp(2N). In order to see that the states given by Eq. (42) are irreducible with respect to USp(2N), one can consider the $SU(2)_W \subset USp(2N)$ as defined by Eq. (37) and observe that the operators $W^{\pm}$ go from a state (p,s) of Eq. (43) to a state ($p \pm 1,s$).

The supermultiplet given by Eq. (43) is characterized by a state of maximum spin $J_{MAX} = N/2$ in a SU(N) singlet. It will be denoted by $[J_{MAX} = N/2, \underset{\sim}{1}]$. It corresponds to a vacuum state Ω with

$$\vec{J}\Omega = 0 \quad \text{and} \quad T^{i}_{j}\Omega = 0 \tag{45}$$

where $\vec{J}$ is the Wigner spin and T^{i}_{j} is the U(N) algebra of automorphisms of Eqs. (14) and (15). Note that Ω is always a singlet with respect to the intrinsic spin $\vec{s}$ and SU(N): Λ^{i}_{j} defined by Eqs. (31) and (32), respectively.

More general representations can be obtained by considering a set of vacua

$$|L,L_3,R\rangle\ , \quad Q^{i}_{\alpha}|L,L_3,R\rangle = 0 \tag{46}$$

such that

$$\begin{aligned} J^2|L,L_3,R\rangle &= L(L+1)|L,L_3,R\rangle \\ J_3|L,L_3,R\rangle &= L_3(L,L_3,R\rangle \end{aligned} \tag{47}$$

and $|L,L_3,R\rangle$ transforms as the components of the representation R of SU(N). In this way one obtains an irrep. of the super-Poincaré algebra SP^N_4 with dimension

$$2^{2N} \times (2L+1) \times \dim R\ . \tag{48}$$

Therefore, these irreducible massive supermultiplets associated with Clifford vacua $|L,L_3,R\rangle$ can be denoted by $[J_{MAX} = N/2 + L, R]$, i.e. by the quantum numbers of the state of highest spin.

In order to give an invariant meaning to (L,R) one can introduce a super spin

$$L_k = J_k - s_k = i\,\varepsilon_{ijk}\,\Gamma_{ij} + \frac{1}{4}\,(\sigma_k)^{\alpha}_{\beta}\left[Q^{*\alpha}_{i}, Q^{i}_{\beta}\right] \tag{49}$$

which commutes with the Q's so that all states constructed from the vacuum $|L,R\rangle$ will be eigenstates of L^2 with the same eigenvalue $L(L+1)$. Correspondingly, one can introduce the relativistic invariant operator

Table 1

Examples of massive representations (without central charges). The multiplicity at each spin is the dimension of the associated antisymmetric irrep. of USp(2N).

J =	5/2	2	3/2	1	1/2	0
N = 1		1	1 2	1 2 1	1 2 1	2 1
N = 2		1	1 4	1 4 5 + 1	4 5 + 1 4	5 4 1
N = 3		1	1 6	6 14 + 1	14 14′ + 6	14′ 14
N = 4		1	8	27	48	42
N = 5	1	10	44	110	165	142

Table 2

SU(4) content of the $J_{MAX} = 2$ multiplet of N = 4 extended supersymmetry

J	SU(4) Young-tableaux	Dimensions in SU(4)	Dimension in USp(8)
2		1	1
3/2	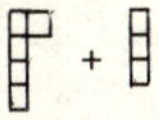	$4 + \bar{4}$	8
1	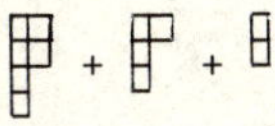	6 + 15 + 6	27
1/2	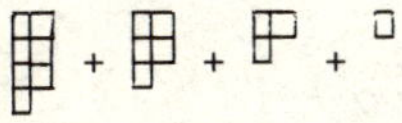	$4 + 20 + \overline{20} + \bar{4}$	48
0	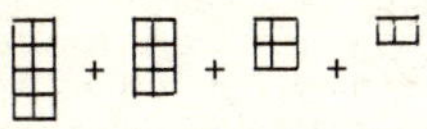 + 1	$1 + 10 + 20' + \overline{10} + 1$	42

$$\tilde{W}_\mu \tilde{W}^\mu = \tilde{W}_\mu{}^2 = \left\{ \varepsilon_{\mu\nu\rho\sigma} P^\nu M^{\rho\sigma} - \frac{1}{4\sqrt{P^2}} \left[\bar{Q}^i \gamma_5 \gamma_\mu Q_i - \frac{1}{P^2} P_\mu \bar{Q}^i \gamma_5 \not{P} Q_i \right] \right\}^2 \tag{50}$$

which is a Casimir operator for SP_4^N. In the same way one can define operators

$$X^i_j = T^i_j - \frac{1}{2} \bar{Q}^i (1 + \gamma_5) \not{P} Q_j / P^2 \tag{51}$$

such that the corresponding U(N) invariants

$$X^{i_1}_{j_2} X^{j_2}_{i_3} \ldots X^{i_p}_{i_1} \tag{52}$$

are invariant under

$$SP_4^N \otimes_s \{T^i_j\} .$$

Table 3

SU(8) representations corresponding to each spin J state of the irreducible massive representation $[J_{MAX} = 4, \underset{\sim}{1}]$ of N = 8 extended supersymmetry, characterized by a maximum spin state with $J_{MAX} = 4$ in an SU(8) singlet. SU(8) representations for a given spin J correspond to an irreducible representation of USp(16).

J	SU(8) representation
4	1
7/2	$8 + \overline{8}$
3	$63 + 28 + \overline{28}$
5/2	56 + 216 + h.c.
2	$70 + 70 + 420 + \overline{420} + 720$
3/2	56 + 504 + 1344 + h.c.
1	(28 + 378 + 1512 + h.c.) + 2352′
1/2	8 + 168 + 1008 + 2352 + h.c.
0	1 + 1 + (36 + 336 + 1176 + h.c.) + 1764

Moreover, PCT invariance requires that $R \simeq \bar{R}$ because under PCT $X^i_j \to -X^j_i$ ($Q^i_\alpha \to Q^{*\alpha}_i$). Finally we observe that the vacuum state can only carry U(N) and not USp(2N) quantum numbers since USp(2N) is an automorphism of Eqs. (14) and (15) only in the rest-frame. Therefore, the spin and SU(N) [U(N)] quantum numbers of the states of an arbitrary irrep. of SP^N_4 are obtained by making the tensor product of the spin and SU(N) quantum numbers of the fundamental representation, as given by Eq. (43), with the spin and SU(N) quantum numbers of the Clifford vacuum $\Omega = |L,R\rangle$.

3 CENTRAL CHARGES

We now consider the modification of the massive representations of N-extended supersymmetry in the presence of central charge operators (Sohnius 1978; Fayet 1979; Taylor 1980; Ferrara, Savoy & Zumino 1981; Lopuszanski & Wolf 1981) as given by Eq. (7) or Eq. (16). In a given irreducible representation the quantity

$$Z^{ij} = -V^{ij} + iU^{ij} \tag{53}$$

is a numerical complex N × N antisymmetric matrix. The supersymmetry algebra given by Eq. (14) and

$$\{Q^i_\alpha, Q^j_\beta\} = \varepsilon_{\alpha\beta} Z^{ij} \tag{54}$$

is not a Clifford algebra. However, we can always perform a U(N) transformation on the $Q^i_{\alpha,s}$ such that the transformed matrix

$$(UZU^T)^{ij} = \tilde{Z}^{ij} \tag{55}$$

has normal form (Zumino 1962)

$$\tilde{Z}^{ij} = i\sigma_2 \otimes \hat{Z}_{N/2} \; ; \quad \hat{Z}_{N/2} = \text{diag}\,(Z_1 \ldots Z_{N/2}) \quad Z_i \geq 0 \; . \tag{56}$$

We consider only even N, the extension to odd N being trivial (Ferrara, Savoy & Zumino 1981). After the unitary transformation which brings Z^{ij} to normal form, the rest frame supersymmetry algebra becomes

$$\{\tilde{Q}^{am}_{\alpha}, \tilde{Q}^{*\beta m}_{b}\} = \varepsilon_{\alpha\beta}\, \varepsilon^{ab}\, \delta^{mn} M \qquad a,b = 1,2 \qquad \alpha,\beta = 1,2$$

$$\{\tilde{Q}^{am}_{\alpha}, \tilde{Q}^{bm}_{\beta}\} = \varepsilon_{\alpha\beta}\, \varepsilon^{ab}\, \delta^{mn} Z_n \qquad m,n = 1 \ldots N/2 \tag{57}$$

where

$$\tilde{Q}^{*\alpha m}_{a} = \varepsilon_{ab}\, \varepsilon^{\alpha\beta} \tilde{Q}^{*bm}_{\beta} , \qquad \varepsilon^{ab} = -\varepsilon_{ab} = (i\sigma_2)_{ab} .$$

If we now define the new spinors

$$S^{am}_{\alpha(1)} = \frac{\tilde{Q}^{am}_{\alpha} + \tilde{Q}^{*\alpha m}_{a}}{\sqrt{2}} , \qquad S^{am}_{\alpha(2)} = \frac{\tilde{Q}^{am}_{\alpha} - \tilde{Q}^{*am}_{\alpha}}{\sqrt{2}} (-)^{a+1} \tag{58}$$

Eqs. (57) become

$$\{S^{am}_{\alpha(1)}, S^{bn}_{\beta(1)}\} = \varepsilon_{ab}\, \varepsilon^{\alpha\beta} \delta^{mn} (M + Z_n)$$

$$\{S^{am}_{\alpha(2)}, S^{bm}_{\beta(2)}\} = \varepsilon_{ab}\, \varepsilon^{\alpha\beta} \delta^{mn} (M - Z_n) \tag{59}$$

$$\{S^{am}_{\alpha(1)}, S^{bm}_{\beta(2)}\} = 0$$

$$S^{am}_{\alpha(i)} = \varepsilon^{ab}\, \varepsilon_{\alpha\beta} S^{*bm}_{\beta(i)} , \qquad i = 1,2 .$$

From Eq. (59) we see that a = b = 2 is redundant so we can consider $S^{m}_{\alpha(i)} = S^{*m}_{\alpha(i)}$, and their complex conjugate

$$\{S^{m}_{\alpha(i)}, S^{*n}_{\beta(j)}\} = \delta_{\alpha\beta} \delta^{mn} \delta_{ij} (M - (-)^{j} Z_n)$$

$$\{S^{m}_{\alpha(i)}, S^{n}_{\beta(j)}\} = \{S^{*m}_{\alpha(i)\phi}, S^{*n}_{\beta(i)}\} = 0 \tag{60}$$

$$\alpha,\beta = 1,2 ; \quad i,j = 1,2 ; \quad m,n = 1 \ldots N/2 .$$

Equations (60) define a Clifford algebra for 2(N - q) creation and destruction operators, q being the number of Z's which fulfil the bound Z = M Positivity requires $Z_n \leq M$. The Clifford vacuum is defined by the condition $S^{m}_{\alpha(i)}\Omega = 0$ and it must be doubled since under PCT $Q^{i}_{\alpha} \to iQ^{*i}_{\alpha}$ which in turn implies $Z^{ij} \to -Z^{*ij}$, i.e. $Z_n \to -Z_n$.

Therefore the one-particle states are those of an $(N-q)$ extended supersymmetry without central charges repeated twice. In the fundamental multiplet the spin will run from $J = 0$ up to $J = (N-q)/2$. If all Z's saturate the bound then $J_{MAX} = N/4$.

In this case the representation will contain the same number of states as a complex massive multiplet of N/2-extended supersymmetry (without central charges) or equivalently the same number of states of a (complex) massless multiplet (see later) of N-extended supersymmetry. From the previous positivity bound it is also clear that if $M = 0$ (massless case) then Z^{ij} must vanish.

We now turn to the construction of general irreps. of N-extended supersymmetry with central charges. The basic irreps. considered above are characterized by $P^2 = M^2$ and by the central charges taking values in the matrix Z^{ij}. In this "basic" irrep. one has implicitly assumed that the vacuum Ω is invariant under the group of automorphisms U(N) defined by Eqs. (17) and (55). More general vacuum states are obtained by considering the algebra $G_Z \subset U(N)$ of the matrix Z^{ij} associated with these vacua. Obviously, G_Z will be determined by the positive members Z_i, the eigenvalues of the $(N/2) \times (N/2)$ matrix $\hat{Z}_{N/2}$ which defines the normal form of Z^{ij} in Eq. (56). One finds that (Ferrara, Savoy & Zumino 1981)

$$G_Z = USp(2q_1) \otimes \ldots \otimes USp(2q_n) \otimes U(2q_0) \qquad (61)$$

Table 4

Examples of massive representations with one central charge ($|Z| = M$). All states are complex, corresponding to $Z = \pm M$.

J =	2	3/2	1	1/2	0
N = 2			1	1 2	2 1
N = 4		1	1 4	4 5 + 1	5 4
N = 6	1	1 6	6 14 + 1	14 14′ + 6	14′ 14
N = 8	1	8	27	48	42

if each eigenvalue Z_i ($i = 1, \ldots, N$) appears with multiplicity q_i, $\sum_{i=0}^{n} q_i = N/2$, in $\hat{Z}_{N/2}$, q_0 being the multiplicity of the eigenvalue 0. In particular, for $n = 1$, one gets $G_Z = USp(2q) \otimes U(N-2q)$, while for $n = N/2$ one has the smallest little group, $G_Z = [SU(2)]^{N/2}$.

Now, general vacuum states defining the irreps. of the supersymmetry algebra with central charges given by the matrix Z^{ij}, can transform as any irrep. R_Z of the little group G_Z. [More precisely, one has to consider the subalgebra G_Z of the U(N) algebra defined in Eq. (51).] These general vacua also have some value L of superspin, defined in Eq. (49).

Summarizing, the general vacuum states of irreducible supermultiplets are defined by $P^2 = M^2$, the superspin L, the matrix Z^{ij}, the irrep. R_Z of the little group G_Z, together with their PCT conjugate states. A general irreducible supermultiplet will then have $2 \times 2^{2(N-q)} \times (2L+1) \times \dim R_Z$ states. These supermultiplets are easily constructed and the internal G_Z quantum numbers and spin content are determined by a straightforward generalization of the formalism in Section 2. This study also applies to N-extended supersymmetry with odd N up to a few modifications (Ferrara, Savoy & Zumino 1981).

4 MASSLESS SUPERMULTIPLETS

In this section we consider the case of massless representations of N-extended supersymmetry (Gell-Mann & Ne'eman 1974; Nahm 1978; Ferrara 1980). In this case we have $P^\mu P_\mu = 0$ and the Wigner method consists of going to a light-like frame, e.g. $P^0 = P^3 = \frac{1}{2}$, $P^1 = P^2 = 0$, $P^\pm = P^0 \pm P^3 = 1(0)$. The supersymmetry algebra [Eq. (6)] then becomes

$$\{Q^i, Q^{jT}\} = (\gamma_+ P^+ C)\, \delta^{ij} \tag{62}$$

In the Weyl representation one gets

$$\begin{aligned} \{Q^i_\alpha, Q^{*\dot\beta}_j\} &= \begin{pmatrix} 1 & 0 \\ 0 & 0 \end{pmatrix}_\alpha^{\dot\beta} \delta^i_j \\ \{Q^i_\alpha, Q^j_\beta\} &= \{Q^{*\dot\alpha}_i, Q^{*\dot\beta}_j\} = 0 \,. \end{aligned} \tag{63}$$

From Eq. (63) we see that

$$\{Q^i_1, Q^{*i}_j\} = \delta^i_j \tag{64}$$

and all other anticommutators vanish. This implies that $Q^i_2 = Q^{*\dot{2}}_{\dot{i}} = 0$ for massless representations (in the chosen frame). As a consequence we use the simplified notation $Q^i_1 = Q^i, Q^{*\dot{i}}_{\dot{j}} = Q^*_{\dot{j}}$ and rewrite Eq. (63) as

$$\{Q^i, Q^*_{\dot{j}}\} = \delta^i_j , \quad \{Q^i, Q^j\} = \{Q^*_{\dot{i}}, Q^*_{\dot{j}}\} = 0 . \tag{65}$$

Now we can proceed as in Section 2 for massive representations and define

$$\Gamma_{2i-1} = Q^i + Q^*_{\dot{i}} = \Gamma^*_{2i-1}$$
$$\Gamma_{2i} = i(Q^i - Q^*_{\dot{i}}) = \Gamma^*_{2i} \tag{66}$$

to obtain Eq. (65) in the real form,

$$\{\Gamma_A, \Gamma_B\} = 2\delta_{AB} , \quad (A,B = 1 \ldots 2N) , \tag{67}$$

with an explicit SO(2N) group of automorphisms. This algebra has one irrep. of dimension 2^N. The SO(2N) generators are

$$O_{AB} = \frac{1}{4i} [\Gamma_A, \Gamma_B] . \tag{68}$$

The irrep. of the supersymmetry algebra is reduced to two irreps. of SO(2N), i.e. the two inequivalent spinorial representations. The two SO(2N) invariant projectors $(1 \pm \Gamma_{2N+1})/2$ with $\Gamma_{2N+1} = i^N \Pi^{2N}_{A=1} \Gamma_A$ project the 2^N states in 2^{N-1} "bosons" and 2^{N-1} fermions depending on the over-all statistics of the Clifford vacuum Ω. These states are constructed, as before, by introducing the "vacuum" state Ω such that

$$Q^i \Omega = 0 \tag{69}$$

and by applying the following $Q^{*i}_{\dot{i},s}$ operators to Ω

$$\Omega, Q^*_{\dot{i}}\Omega, Q^*_{\dot{i}}Q^*_{\dot{j}}\Omega, \ \ldots, \ Q^*_{\dot{1}}Q^*_{\dot{2}}\ldots Q^*_{\dot{N}}\Omega . \tag{70}$$

The generators of SO(2N) can be written as

$$\Lambda^i_{\cdot j} = \frac{1}{2}\left[Q^i, Q^*_j\right] - \frac{1}{2N}\,\delta^i_j\left[Q^k, Q^*_k\right], \quad [SU(N)] \tag{71}$$

$$\Lambda^i_{\cdot i} = \Lambda = \frac{1}{4}\left[Q^i, Q^*_{\cdot i}\right] = -\frac{1}{2}\,Q^*_{\cdot i}Q^i + \frac{N}{4}, \quad U(1) \tag{72}$$

$$K^{ij} = \frac{1}{4}\left[Q^i, Q^j\right] \qquad \bar{K}_{ji} = \frac{1}{4}\left[Q^*_i, Q^*_j\right]. \tag{73}$$

Notice that $\Lambda^i_{\cdot j}\Omega = 0$, $\Lambda\Omega = \frac{N}{4}\,\Omega$.

The SU(N) algebra defined by Eq. (71) can be combined with the SU(N) algebra of invariance of the relativistic supersymmetry algebra, with generators $T^i_{\cdot j}$, in order to define new generators

$$X^i_{\cdot j} = T^i_{\cdot j} - \Lambda^i_{\cdot j} \tag{74}$$

which commute with the Q's.

The U(1) factor defined by Eq. (72) plays, for massless representations, the role of an intrinsic helicity. In fact it is easy to see that the Q^*'s operators transform under Λ in the same way as under M^{12}, the Wigner helicity:

$$\left[M_{12}, Q^*_{\cdot i}\right] = \left[\Lambda, Q^*_{\cdot i}\right] = -\frac{1}{2}\,Q^*_{\cdot i}\,. \tag{75}$$

Therefore, the new operator X (superhelicity)

$$X = M_{12} - \Lambda \tag{76}$$

is a Casimir invariant for a massless representation and together with the SU(N) Casimir operators constructed in terms of the $X^i_{\cdot j}$'s defined in Eq. (74) will completely specify a massless representation. Since $[X, Q^*_{\cdot i}] = 0$ its value is specified on the Clifford vacuum Ω

$$X\Omega = \left(M_{12} - \frac{N}{4}\right)\Omega = \left(\lambda_{MAX} - \frac{N}{4}\right)\Omega\,. \tag{77}$$

An important point comes from the PCT operation on massless supermultiplets. In fact, under PCT $X \rightarrow -X$ so PCT invariance implies $X = 0$ or the doubling of a supermultiplet, namely the addition to a multiplet with superhelicity X of a new multiplet with superhelicity -X. In terms of Wigner helicity (eigenvalue of M_{12}) of the Clifford vacuum this means that a massless supermultiplet is PCT self-conjugate only if X = 0, i.e. $\lambda_{MAX} = N/4$. In this case it has 2^N states. If $X \neq 0$ ($\lambda_{MAX} \neq N/4$) we must add to it a PCT conjugate multiplet with $\lambda'_{MAX} = N/2 - \lambda_{MAX}$ and in this case it will contain 2^{N+1} states. Moreover, if the Clifford vacuum has a non-trivial transormation law under $X^i_j(T^i_j)$, i.e. it belongs to an irrep. R of SU(N) then PCT self-conjugation implies $\lambda_{MAX} = N/4$ and $R = \bar{R}$, while PCT doubling will add a PCT conjugate vacuum Ω' with $\lambda'_{MAX} = N/2 - \lambda_{MAX}$, $R' = \bar{R}$ to the vacuum Ω specified by λ_{MAX} and R. In this more general situation the massless supermultiplet will contain $2^{N+1} \times \dim R$ states (2^N states if $\lambda_{MAX} = N/4$, $R = \bar{R}$). It is clear that if N is odd a representation is never PCT self-conjugate. PCT self-conjugation happens only for N = 4n. In fact, if N = 4n + 2 the multiplet with $\lambda = N/4$ is not PCT self-conjugate unless R = (N/2).

The transformation properties of massless states under the helicity and SU(N) quantum numbers are easily seen by writing them explicitly as given by Eq. (70).

So we see that the generic state

$$Q^*_{i_1} \ldots Q^*_{i_k} \Omega \tag{78}$$

has helicity $\lambda_{MAX} - k/2$ and transforms according to the representation $R \times [k]_N$ of SU(N). Its PCT conjugate state is given by

$$Q^*_{i_1} \ldots Q^*_{i_{N-k}} \Omega' \qquad \left([\overline{N-k}]_N = [k]_N\right) \tag{79}$$

where Ω' is the PCT conjugate Clifford vacuum with quantum numbers $N/2 - \lambda_{MAX}$, $\bar{R}$.

We denote a massless supermultiplet by the helicity λ_{MAX} and the SU(N) representation of the Clifford vacuum Ω

$$\{\lambda_{MAX}, R\} . \tag{80}$$

Because of PCT we have the obvious property

$$\{\lambda_{MAX}, R\} = \{\tfrac{N}{2} - \lambda_{MAX}, \bar{R}\} \tag{81}$$

since PCT doubling is always understood. In other words, the symbol $\{\lambda_{MAX},R\}$ means the following collection of states of given helicity and SU(N) quantum numbers

$$\{\lambda_{MAX},R\} = (\lambda_{MAX},R) \ , \quad (\lambda_{MAX} - \tfrac{1}{2}, R \times [\bar{1}]) \ ,$$
$$(\lambda_{MAX} - 1, R \times [\bar{2}]), \ \ldots \ (\lambda_{MAX} - \tfrac{k}{2}, R \times [\bar{k}]), \ \ldots \ (\lambda_{MAX} - \tfrac{N}{2}, R) \qquad (82)$$

+ PCT conjugate states.

In the last part of this section we consider the decomposition of massive multiplets into massless ones. Group-theoretically this means classifying the states with respect to J_3 rather than J. For example, for a J = 1 state we would have $J_3 = \pm 1$ and $J_3 = 0$ and so on. So for $M \to 0$ a spin 1 particle can describe a massless vector particle and a massless scalar.

We want to answer the following questions: i) How a massive representation decomposes into massless ones; ii) Given a massive state, which in the $M \to 0$ limit decomposes into several helicity states, to which massless representation each helicity state will belong. If we consider the covariant supersymmetry algebra given by Eqs. (14) and (15) we can choose a (collinear) frame such that $P^1 = P^2 = 0$, $P^3 = \sqrt{P_0^2 - M^2}$, $P^0 = \frac{1}{2}$, $P^{\pm} = (1 \pm Z)/2$ with $Z = \sqrt{1 - 4M^2}$.

In this frame one obtains

$$\{Q^i_1, Q^{*i}_j\} = \frac{1+Z}{2}\,\delta^i_j \ , \quad \{Q^i_1, Q^{*\dot{2}}_j\} = 0$$
$$\{Q^i_2, Q^{*\dot{2}}_j\} = \frac{1-Z}{2}\,\delta^i_j \ , \quad \{Q^i_\alpha, Q^j_\beta\} = 0 \qquad (83)$$

when

$$M \to 0 \ , \quad \frac{1+Z}{2} \to 1 \ , \quad \frac{1-Z}{2} \to 0$$

which shows how the massive rest-frame algebra goes to the massless one in the $M/P^0 \to 0$ limit. In fact, for $M/P^0 \to 0$, $Q^i_2 \to 0$ and the algebra reduces to the anticommutator relations given by Eqs. (63) and (64).

Now we consider the 2^{2N} dimensional fundamental massive supermultiplet given by Eqs. (26) and (29). This can be decomposed into "massless" supermultiplets obtained by application of Q^{*1}_i but not Q^{*2}_i. The Q^{*2}_i

Table 5

Massless representations with $\lambda_{MAX} = 1$: helicity and SU(N) content (PCT conjugate states are to be added for $\lambda \neq 0$)

λ =	1	½	0
N = 1	1	1	
N = 2	1	2	1 + 1
N = 3	1	$\bar{3}$ + 1	3 + $\bar{3}$
N = 4	1	$\bar{4}$	6

Table 6

Massless representations with $\lambda_{MAX} = 2$: helicity and SU(N) content (PCT conjugate states are to be added for $\lambda \neq 0$).

λ =	2	3/2	1	½	0
N = 1	1	1			
N = 2	1	2	1		
N = 3	1	$\bar{3}$	3	1	
N = 4	1	$\bar{4}$	6	4	1 + 1
N = 5	1	$\bar{5}$	$\overline{10}$	10 + 1	5 + $\bar{5}$
N = 6	1	$\bar{6}$	$\overline{15}$ + 1	20 + $\bar{6}$	15 + $\overline{15}$
N = 7	1	$\bar{7}$ + 1	$\overline{21}$ + $\bar{7}$	$\overline{35}$ + $\overline{21}$	35 + $\overline{35}$
N = 8	1	$\bar{8}$	$\overline{28}$	$\overline{56}$	70

operators merely classify 2^N Clifford vacua $Q^{*2}_{i_1} \ldots Q^{*2}_{i_k}\Omega$ for the Clifford algebra spanned by the Q^i_1, Q^{*1}_i destruction and creation operators. Note in fact that

$$Q^i_1 \, Q^{*2}_{i_1} \ldots Q^{*2}_{i_k}\Omega = 0 \,, \quad k = 0 \ldots N \,. \tag{84}$$

In this respect the decomposition of the fundamental 2^{2N} massive representation into "massless" representations is (Ferrara 1980):

$$2^{2N} = 2^N \times 2^N = \left\{\frac{N}{2},[0]\right\} + \left\{\frac{N-1}{2},[1]\right\} + \ldots \left\{\frac{N-k}{2},[k]_N\right\} + \ldots \{0,[1]\} \,. \tag{85}$$

[Here {} does not include PCT doubling.]

Note that massless representations {...} in Eq. (85) appear automatically in PCT conjugate pairs because

$$\left\{\frac{N-k}{2},[k]_N\right\} \underset{PCT}{\longrightarrow} \left\{\frac{k}{2},[N-k]_N\right\} . \tag{86}$$

If we collect PCT pair conjugate multiplets together we finally obtain Eq. (85) in the form

$$\left[J_{MAX} = \frac{N}{2}, \underset{\sim}{1}\right] = \left\{\frac{N}{2},[0]\right\} + \left\{\frac{N}{2} - \frac{1}{2},[1]\right\} + \ldots \left\{\frac{N-k}{2},[k]\right\} + \ldots \{last\} \tag{87}$$

where

$$\{last\} = \left\{\frac{N}{4} + \frac{1}{2},\left[\frac{N}{2} - 1\right]\right\} , \quad \text{if } N = 4n + 2$$

$$2 \times \left\{\frac{N}{4},\left[\frac{N}{2}\right]\right\} , \quad \text{if } N = 4n$$

$$\left\{\frac{N+1}{4},\left[\frac{N-1}{2}\right]\right\} , \quad \text{if } N = 2n + 1 \,. \tag{88}$$

For more general massive representations $\{J_{MAX} = L + N/2, R\}$ one must split each massless supermultiplet in Eq. (83) by adding the components of L: $-L, -L+1 \ldots, L-1, L$ to each value of λ_{MAX} and by multiplying the associated $[N-k]$ representation of SU(N) by R:

$$\left[\frac{N-k}{2},[k]\right] \to \left\{\frac{N-k}{2} - L,[k] \otimes R\right\} + \left\{\frac{N-k}{2} - L + 1, [k] \otimes R\right\} + \ldots + \left\{\frac{N-k}{2} + L,[k] \otimes R\right\} \tag{89}$$

for each value of k.

We now want to classify a particular state of the massive supermultiplet into those of the "massless" supermultiplets obtained above (Derendinger, Ferrara & Savoy 1981). In terms of the massive vacuum state $|L,L_3\rangle$ (see previous section), which we assume to be an SU(N) singlet for the moment, and of the product of operators $[Q^*]_p^{ss_3}$ defined by Eq. (42), the component J_3 of the state of spin J with SU(N) quantum numbers defined by (s,p) can be written as follows

$$|J,J_3,sp\rangle = \sum_{s_3} \begin{pmatrix} L & s \\ L_3 & s_3 \end{pmatrix}\begin{vmatrix} J \\ J_3 \end{vmatrix} [Q^*]_p^{ss_3} |L,L_3\rangle \tag{90}$$

Table 7

Decomposition of the N = 4, J_{MAX} = 2 representation into "massless" representations

λ	Multiplicity	$\{2,\underset{\sim}{1}\} = \{0,\underset{\sim}{1}\}$		$\{3/2,\underset{\sim}{4}\} = \{1/2,\underset{\sim}{\bar{4}}\}$		$\{1,\underset{\sim}{6}\}$
2	1	1				
3/2	8	$\bar{4}$		4		
1	27 + 1	6		4 × $\bar{4}$		6
1/2	48 + 8	4		4 × 6	$\bar{4}$	6 × $\bar{4}$
0	42 + 27 + 1	1	1	4 × 4	4 × $\bar{4}$	6 × 6
−1/2	48 + 8		$\bar{4}$	4	$\bar{4}$ × 6	6 × 4
−1	27 + 1		6		$\bar{4}$ × 4	6
−3/2	8		$\bar{4}$		$\bar{4}$	
−2	1		1			

and from the analysis of the previous section Q_i^{*2} appears $(p+s+s_3)$ times in each term of the left-hand side of Eq. (95) so that from the previous discussion the corresponding term will be in a "massless" multiplet with "vacuum" state

$$\left[Q^{*2}_{i_1} \ldots Q^{*2}_{i_n}\right]\Big|L,L_3> , \quad n = s + s_3 + p \tag{91}$$

which is precisely the "massless" supermultiplet

$$\{\lambda_{MAX} = \underbrace{L_3 + \frac{s+s_3+p}{2}}_{J_3 + \frac{s+p-s_3}{2}},\left[N-s-s_3-p\right]\} \tag{92}$$

the projection of $|J,J_3,sp>$ into this "massless" supermultiplet being the Clebsh-Gordan coefficient $\left(\begin{matrix} L & s \\ L_3 & s_3 \end{matrix}\Big|\begin{matrix} J \\ J_3 \end{matrix}\right)$. More generally, if $|L,L_3>$ transforms as a representation R of SU(N), one has to generalize Eq. (90) by including Clebsh-Gordan coefficients for SU(N).

5 TWO MASSLESS PARTICLE SUPERMULTIPLETS

In this section we consider the irreps. of the supersymmetry algebra [Eqs. (14) and (15)] acting on two massless particle states. More precisely we consider the decomposition of the tensor product of the irreducible massless supermultiplets into its irreducible components.

If the two massless supermultiplets are $\{\lambda_1,R_1\}$, $\{\lambda_2,R_2\}$ respectively, the algebra [Eqs. (14) and (15)] is realized on the tensor product of these states (Salam & Strathdee 1974)

$$\{\lambda_1,R_1\}_{p_1^I} \otimes \{\lambda_2,R_2\}_{p_2^{II}} \tag{93}$$

by

$$\begin{aligned} Q_{i\alpha} &= Q^I_{i\alpha} + Q^{II}_{i\alpha} \\ P_\mu &= P^I_\mu + P^{II}_\mu \end{aligned} \tag{94}$$

where $Q^I_{i\alpha}, P^I_\mu$ only act on the supermultiplet $\{\lambda_1,R_1\}$ while $Q^{II}_{i\alpha}, P^{II}_\mu$ only act on $\{\lambda_2,R_2\}$.

Since these multiplets are massless, $p^{I^2} = p^{II^2} = 0$, then $P^2 = 2p^{I} \cdot p^{II}$, corresponding to the obvious fact that the composite state is massive. Our problem here is then to classify the two-particle states in irreps. of the algebra [Eqs. (14) and (15)]. We shall consider the simple case where R_1 and R_2 are the trivial representations of SU(N), $R_1 = R_2 = 1$. The generalization to arbitrary R_1 and R_2 is straightforward. The massless multiplets can also have additional quantum numbers which correspond to one group commuting with the supersymmetry algebra. This is for instance what happens in supersymmetric Yang-Mills theories. The formalism discussed below can be trivially extended to these cases.

We consider the vacuum states $|\lambda_1,\vec{p}_1\rangle$, $|\lambda_2,\vec{p}_2\rangle$ of the massless supermultiplets. In general, $\lambda_i \neq N/4$ so that one has also to consider the other vacuum states $|N/2 - \lambda_i,\vec{p}_i\rangle$. The tensor product $|\lambda_1,\vec{p}_1\rangle|\lambda_2,\vec{p}_2\rangle$ is a Clifford vacuum of the algebra with the $Q_{i\alpha}$'s given by Eq. (94)

$$Q_{i\alpha}|\lambda_1,\vec{p}_1\rangle|\lambda_2,\vec{p}_2\rangle = Q^{I}_{i\alpha}|\lambda_1,\vec{p}_1\rangle|\lambda_2,\vec{p}_2\rangle + (-)^{2\lambda_1}|\lambda_1,\vec{p}_1\rangle Q^{II}_{i\alpha}|\lambda_2,\vec{p}_2\rangle = 0 \ . \tag{95}$$

According to the analysis of Section 2 the vacuum states of the massive irreps. have to be the superposition of tensor products of massless vacuum states, with a given value of the (relative) angular momentum L. A spin-L vacuum in its rest frame (c.m. frame) is given by the familiar expression (Jacob & Wick 1959)

$$|p = |\vec{p}|,L,L_3,\lambda_1,\lambda_2\rangle = \sqrt{\frac{2L+1}{4\pi}} \int d\omega\, \mathfrak{D}^{*L}_{L_3\lambda}(\omega)R(\omega)|p_z\lambda_1\rangle|-p_z\lambda_2\rangle \tag{96}$$

where

$$R(\omega) = R(\theta,\phi) = e^{-i\phi J_3}\, e^{-i\theta J_2}\, e^{i\phi J_3}$$

$$\mathfrak{D}^{*L}_{L_3\lambda}(\omega) = \mathfrak{D}^{*L}_{L_3\lambda}(\phi,\theta,-\phi) \ , \quad (L \geq \lambda)$$

$$\lambda = \lambda_1 - \lambda_2 \ , \quad L_3 = -L \ldots +L \ .$$

Statistics require symmetrization or antisymmetrization of the tensor product of vacuum states in Eq. (96), obtained through the replacement

$$|p_z,\lambda_1\rangle|-p_2,\lambda_2\rangle \to \frac{1}{\sqrt{2}}\Big[|p_z,\lambda_1\rangle|-p_z,\lambda_2\rangle + (-)^{\eta}|-p_z\lambda_2\rangle|p_z,\lambda_1\rangle \tag{97}$$

where $\eta = 1$ if both λ_1,λ_2 are half-integers and $\eta = 0$ otherwise. Notice however that, if the massless supermultiplets have additional quantum numbers, then symmetrization or antisymmetrization with respect to these quantum numbers have also to be taken into account in the usual way. In this case one would find additional supermultiplets besides those discussed here below. This would happen, in particular in supersymmetric Yang-Mills theories where the fundamental multiplet is in the adjoint representation of the gauge group. Instead, in supergravity, the graviton multiplet is an over-all singlet and our list of two-particle states is complete. From the properties of the $\mathfrak{D}$ functions one obtains, by substituting Eq. (97) into Eq. (96),

$$|p,L,L_3,\lambda_1,\lambda_2\rangle = \sqrt{\frac{2L+1}{8\pi}} \int d\omega\ R(\omega) \Big[\mathfrak{D}^{*L}_{L_3\lambda}(\omega)|p_z,\lambda_1\rangle$$
$$|-p_z,\lambda_2\rangle + (-)^{\eta+L+\lambda_1+\lambda_2}\ \mathfrak{D}^{*L}_{\lambda_3,-\lambda}|p_z,\lambda_2\rangle|-p_z,\lambda_1\rangle\Big]. \tag{98}$$

In particular, if $\lambda_1 = \lambda_2$ one has $\lambda = 0$ and $\eta + \lambda_1 + \lambda_2 = 0$ (mod 2) so that one obtains

$$|p,L,L_3,\lambda_1,\lambda_2\rangle = \sqrt{\frac{2L+1}{4\pi}} \int d\omega\ \mathfrak{D}^{*L}_{L_3 0}(\omega)R(\omega)|p_z,\lambda_1\rangle|-p_z,\lambda_1\rangle \tag{99}$$

for even L while it vanishes (i.e. the state is forbidden by statistics) for odd L. Another point to take into account is PCT-conjugation, which, besides the vacuum state [Eq. (96)], requires the PCT conjugate we obtained by replacing

$$|p_z,\lambda_1\rangle|-p_z,\lambda_2\rangle \quad \text{by} \quad |p_z,\tfrac{N}{2} - \lambda_1\rangle|p_z,\tfrac{N}{2} - \lambda_2\rangle$$

in Eqs. (95) to (99). This doubling of states leads to complex supermultiplets.

We now consider the structure of these composite supermultiplets obtained by applying the operators $[Q^*]_q^{ss_3}$, defined in Section 2, to these vacuum states. By combining the angular momenta (s,s_3) of $[Q^*]_q^{ss_3}$ and (L,L_3) of the vacuum one can construct the state of spin (J,J_3) in the massive supermultiplet

$$|J,J_3,L,s,q,\lambda\rangle = \sum_{s_3} \begin{pmatrix} L & s & | & J \\ L_3 & s_3 & | & J_3 \end{pmatrix} [Q^*]_q^{ss_3} |p,L,L_3,\lambda_1,\lambda_2\rangle \tag{100}$$

and by using the property

$$[Q^*]_q^{ss_3} R(\omega) = R(\omega) \sum_{s_3'} \mathfrak{D}^s_{s_3 s_3'}(\omega)\, [Q^*]_q^{ss_3'} \tag{101}$$

and the Clebsh-Gordan series for the $\mathfrak{D}$ functions (Jacob & Wick 1959) one finally obtains

$$|J,J_3,L,s,q,\lambda\rangle = \sqrt{\frac{2L+1}{4\pi}} \sum_{s_3} \int d\omega\, \mathfrak{D}^{*J}_{J_3,\lambda+s_3}(\omega)$$

$$\begin{pmatrix} L & s & | & J \\ \lambda & s_3 & | & \lambda+s_3 \end{pmatrix} R(\omega)\, [Q^*]_q^{ss_3} |p_z,\lambda_1\rangle |-p_z,\lambda_2\rangle . \tag{102}$$

These states transform under the SU(N) group into the irreps. characterized by the numbers (s,q), as shown in Section 2. We can write the explicit expression of $[Q^*]_q^{ss_3}$ in terms of the Q^*_{is} with $J_3 = \pm\frac{1}{2}$ as given by Eqs. (41) to (42). Here we adopt the short notation

$$[Q^*]_q^{ss_3} \sim (Q_1^*)^{s-s_3+q} (Q_2^*)^{s+s_3+q} \tag{103}$$

where $Q_1^{*i}(Q_2^{*i})$ is a spinorial charge with $J_3 = -\frac{1}{2}(+\frac{1}{2})$.

Using the properties of massless representations we have

$$Q_2^{*i} |p_z,\lambda_1\rangle |-p_z,\lambda_2\rangle = (-)^{2\lambda_1} |p_z,\lambda_1\rangle Q_2^{*i^{II}} |-p_z,\lambda_2\rangle$$

$$Q_1^{*i} |p_z,\lambda_1\rangle |-p_z,\lambda_2\rangle = Q_1^{*i^{I}} |p_z,\lambda_1\rangle |-p_z,\lambda_2\rangle . \tag{104}$$

Therefore,

$$[Q^*]_q^{ss_3} |p_z,\lambda_1\rangle |-p_z,\lambda_2\rangle \sim (Q_1^{*I})^{s+q-s_3} |p_z,\lambda_1\rangle (Q_2^*)^{s+q+s_3} |p_z,\lambda_2\rangle$$

$$\sim |p_z,\lambda_1 - \frac{s+q-s_3}{2}\rangle |-p_z,\lambda_2 - \frac{s+q+s_3}{2}\rangle \tag{105}$$

which corresponds to a product of two particular massless states in the antisymmetric representations $[N - s + q \pm s_3]_N$ of SU(N). In this way, we have constructed in Eq. (102) the generic bilinear state with spin J in an SU(N) representation defined by s and q which is in the massive supermultiplet with $J_{MAX} = L + N/2$. Its content in massless states is given in Eqs. (105) and (102).

We can now decide in which irreducible supermultiplets is the particular two-particle state

$$|J,J_3,\lambda_1 - \frac{k_1}{2},\lambda_2 - \frac{k_2}{2}> =$$

$$= \int d\omega \sqrt{\frac{2s+1}{4\pi}}\, \mathfrak{D}^{J*}_{J_3,(k_1-k_2)/2}(\omega)\, |\vec{p},\lambda_1 - \frac{k_1}{2}>\, |-\vec{p},\lambda_2 - \frac{k_2}{2}> \qquad (106)$$

which transforms under SU(N) as the reducible representation $[N - k_1]_N \otimes \otimes [N - k_2]_N$.

By comparing Eq. (106) with Eqs. (105) and (101) we get

$$|J,J_3,L,s,q,\lambda> = C(s,q,k_1,k_2)$$

$$\sum_{s_3} \sqrt{\frac{2L+1}{2J+1}} \begin{pmatrix} L & s & J \\ \lambda & s_3 & \lambda+s_3 \end{pmatrix} |J,J_3,\lambda - \frac{k_1}{2},\lambda_2 - \frac{k_2}{2}> =$$

$$= \sum_{s_3} (-)^{J-L-s_3} \begin{pmatrix} s & J & L \\ s_3 & -\lambda-s_3 & -\lambda \end{pmatrix} C(s,q,k_1,k_2)\, |J,J_3,\lambda_1 - \frac{k_1}{2},\lambda_2 - \frac{k_2}{2}> \qquad (107)$$

where $C(s,q,k_1,k_2)$ are Clebsh-Gordan coefficients of SU(N). By inversion we arrive at

$$C(s,q,k_1,k_2)\, |J,J_3,\lambda_1 - \frac{k_1}{2},\lambda_2 - \frac{k_2}{2}> =$$

$$= \sum_{L} (-)^{J-L-s_3} \begin{pmatrix} s & J & L \\ s_3 & -\lambda-s_3 & -\lambda \end{pmatrix} |J,J_3,L,s,q,\lambda> \qquad (108)$$

which gives the irreducible supermultiplets which contain a definite two-particle state.

We now consider a simple but relevant case in order to illustrate the previous discussion. For N = 4n, the super-Poincaré algebra admits a PCT self-conjugate vacuum with $\lambda_{MAX} = N/4$ which gives rise to the fundamental massless supermultiplet. There are several simplifications which occur for those two-particle states obtained from the supermultiplet $\{N/4,\underset{\sim}{1}\}$. The vacuum states of the massive composite supermultiplets are obtained from the PCT self-conjugate product $|\vec{p},N/4\rangle|-\vec{p},N/4\rangle$. Statistics require the internal angular momentum L to be <u>even</u>, so that the two-particle states are classified in massive multiplets with $J_{MAX} = L + N/2$ which takes only even values $\geq N/2$. It is useful to introduce the "collinear" states:

$$|J,L,ss_3,q\rangle_z = \sqrt{\frac{2L+1}{2J+1}} \begin{pmatrix} L & s & J \\ 0 & s_3 & s_3 \end{pmatrix} *$$

$$* \frac{1}{\sqrt{2}} [Q^*]_q^{ss_3} \left[|p_z,N/4\rangle|-p_z,N/4\rangle + |-p_z,N/4\rangle|p_z,N/4\rangle \right] \qquad (109)$$

so that the generic two-particle state in the case of the PCT self-conjugate massless multiplet $\{N/4,\underset{\sim}{1}\}$ is written as

$$|J,J_3,L,s,q\rangle = \sqrt{\frac{2J+1}{4\pi}} \sum_{s_3} \int d\Omega\, \mathfrak{D}^{*J}_{J_3,s_3}(\Omega)\, R(\Omega)\, |J,L,ss_3,q\rangle_z \, . \qquad (110)$$

In particular, for <u>L = 0</u>, one has the collinear states:

$$|ss_3,q\rangle_z = \sqrt{2s+1}\,|s,L=0,ss_3,q\rangle_z =$$

$$= [Q^*]_q^{ss_3} \left(|p_z,N/4\rangle|-p_z,N/4\rangle + (p_z \to -p_z) \right) . \qquad (111)$$

Then, for a general L = 2,4, ... , the "collinear" states are simply given in terms of the L = 0 fundamental ones by

$$|J,L,ss_3,q\rangle_z = \sqrt{\frac{2L+1}{2J+1}} \begin{pmatrix} L & s & J \\ 0 & s_3 & s_3 \end{pmatrix} |ss_3,q\rangle_z \qquad (112)$$

with $s+L \geq J \geq |s-L|$, $J \geq |s_3|$. Hence, by studying the properties of the L = 0 states $|ss_3,q\rangle_z$ in Eq. (111) one obtains those of the general two-particle state by a trivial addition of angular momenta. From the previous analysis, the massless particle content of $|ss_3,q\rangle_z$ is

$$|ss_3,q>_z \sim |p_z,\frac{N}{4} - \frac{s+q-s_3}{2}>|-p_z,\frac{N}{4} - \frac{s+q+s_3}{2}> \pm (p_z \to -p_z) \tag{113}$$

the massless states being in the SU(N) representations: $[N-s-q\pm s_3]_N$. We now concentrate on "diagonal bilinears", i.e. those containing massless particles of opposite helicity. These are particularly interesting because they are associated with currents with definite spin which conserve the helicity of the massless particles. For these "bilinear" collinears one then has the conditions

$$s + q = N/2$$

$$|ss_3,q>_z \sim |p_z, -\frac{s_3}{2}>|p_z, \frac{s_3}{2}> \tag{114}$$

and since $s_3 \leq s$ only states with $|\text{helicity}| \leq s/2$ may appear in the current with spin s, a well-known selection rule which has been recently discussed by Weinberg & Witten (1980). Note that the states in Eq. (114) are in the SU(N) representations: $[N/2 \pm s_3]_N$. From this one deduces several properties of these "diagonal bilinears" (Derendinger, Ferrara & Savoy 1981). They all transform as real irreducible representations of SU(N) with N boxes arranged in two-column Young-tableaux. Indeed, for each value of s there is only one bilinear and each diagonal bilinear is associated with a different irrep. of SU(N). There is only one "bilinear" which transforms as the adjoint representation of SU(N) for L = 0 and this bilinear is "diagonal" and has spin s = N/2 - 1 (q = 1). By allowing L ≠ 0 one obtains "bilinears" in the adjoint repr. of SU(N) with $|L+1-N/2| \leq J \leq L+N/2-1$. In particular, J = 1 "bilinears" in the adjoint repr. of SU(N) [which are candidates for the SU(N) currents] are obtained for at most two values of L: L = s ± 1 = N/2, N/2 - 2 corresponding to the massive supermultiplet $|J_{MAX} = N,\underset{\sim}{1}>$ and $|J_{MAX} = N-2,\underset{\sim}{1}>$. Massless scalars only appear in "bilinears" with spin J = s (mod 2). Another interesting "diagonal bilinear" is the SU(N) invariant one, which has s = N/2 (q = 0). Hence, J = 2, SU(N) invariant "diagonal bilinears" (candidate for the stress-tensor operator) appear only in the composite supermultiplets with L = N/2 ± 2 (J_{MAX} = N ± 2) and L = N/2 (J_{MAX} = N).

In the N = 4n + 2 case there are two massless supermultiplets with λ = N/4, related to each other by PCT, that can be associated with a charge, say $Q = \pm\frac{1}{2}$. One then has to consider three composite vacuum states:

$$a_1)\quad |\vec{p},N/4\rangle_+ |-\vec{p},N/4\rangle_+ - |-\vec{p},N/4\rangle_+ |\vec{p},N/4\rangle_+$$

$$a_2)\quad |\vec{p},N/4\rangle_- |-\vec{p},N/4\rangle_- - |-\vec{p},N/4\rangle_- |\vec{p},N/4\rangle_-$$

$$b)\quad |\vec{p},N/4\rangle_+ |-\vec{p},N/4\rangle_- - |-\vec{p},N/4\rangle_- |\vec{p},N/4\rangle_+ .$$

The states constructed out of (a_1) and (a_2) are to be considered together as PCT-conjugate charged ($Q = \pm 1$) states. In general, the previous analysis for $N = 4n$ applies to these charged two-particle states. In particular, only even values of L are allowed by statistics. However, it is easy to check that because N/2 is odd and L is even there is no J = 1 composite transforming as the adjoint repres. of SU(N), and no J = 2, SU(N) invariant composites which are bilinear in the scalar massless particles.

The bilinears constructed out of the Q = 0 composite vacuum in (b) are more interesting. The previous analysis applies in general, but L

Table 8

Decomposition of the massive states with $J = J_3 = 1$, in the $\underset{\sim}{15}$ of SU(4), and $J = J_3 = \frac{3}{2}$, in the $\underset{\sim}{4}$ of SU(4), belonging to massive irreps of N = 4 extended supersymmetry with $J_{MAX} = 2 + L$, into states with helicities $J_3 = 1$ and $\frac{3}{2}$, resp., of "massless" supermultiplets. Notice that for L = 0, both states are pure $\{\frac{3}{2},\underset{\sim}{4}\}$ states.

Massless supermultiplets $\{\lambda_{MAX},[k]\}$	$J = 1,\ J_3 = 1$ $R = \underset{\sim}{15}$ (adjoint)		$J = \frac{3}{2},\ J_3 = \frac{3}{2}$ $R = \underset{\sim}{4}$ (fundamental)		
	L = 0 S = 1	L = 2 S = 1	L = 0 $S = \frac{3}{2}$	L = 2 $S = \frac{3}{2}$	L = 2 $S = \frac{1}{2}$
$\{\frac{3}{2},[1]\}$	1	$1/\sqrt{10}$	1	$1/\sqrt{5}$	-
$\{2,[2]\}$	-	$-\sqrt{3/10}$	-	$-\sqrt{2/5}$	-
$\{\frac{5}{2},[3]\}$	-	$\sqrt{3/5}$	-	$\sqrt{2/5}$	-
$\{3,[0]\}$	-	-	-	-	$-1/\sqrt{5}$
$\{\frac{7}{2},[1]\}$	-	-	-	-	$2/\sqrt{5}$

can take both even and odd values. One then obtains J = 1 composites in the adjoint repr. of SU(N) for L = N/2 or L = N/2 - 2 which are bilinear in the scalar massless states (only L = 1 for N = 2).

We now turn to a brief discussion of the N = 1 case in order to illustrate the general situation when the massless supermultiplets are not PCT self-conjugate. We consider the Wess-Zumino multiplet (Wess & Zumino 1974a) which has two Clifford vacua with helicities $\lambda = \frac{1}{2}$ and $\lambda = 0$. In an obvious notation, one has three kinds of vacuum states:

$$\text{a)}\ |0\rangle|0\rangle\ ,\quad \text{b)}\ |\tfrac{1}{2}\rangle|\tfrac{1}{2}\rangle\ ,\quad \text{c)}\ |0\rangle|\tfrac{1}{2}\rangle$$

from which the bilinear multiplets are constructed. The vacua of types (a) and (b) have to be taken together in order to obtain composite supermultiplets which are PCT-conjugate. Since $\lambda_1 = \lambda_2$ only even values of L are

Table 9

Decomposition of the massive states with $J = J_3 = 1$ in the $\underset{\sim}{63}$ of SU(8) and $J = J_3 = \frac{3}{2}$ in the $\underset{\sim}{8}$ of SU(8) belonging to massive irreps of N = 8 extended supersymmetry with $J_{MAX} = 2 + L$, into states with helicities $J_3 = 1$ and $\frac{3}{2}$, resp., of "massless" supermultiplets. Notice that there are no combinations of either $J = J_3 = 1$ or $J = J_3 = \frac{3}{2}$ states in a single massless supermultiplet

Massless supermultiplets $\{\lambda_{MAX}, [k]\}$	$J = 1,\ J_3 = 1$ $R = \underset{\sim}{63}$ (adjoint)		$J = \frac{3}{2},\ J_3 = \frac{3}{2}$ $R = \underset{\sim}{8}$ (fundamental)		
	L = 2 (S = 3)	L = 4 (S = 3)	L = 2 (S = $\frac{7}{2}$)	L = 4 (S = $\frac{7}{2}$)	L = 2 (S = $\frac{1}{2}$)
$\{\frac{3}{2},[1]\}$	$\sqrt{3/7}$	$1/2\sqrt{21}$	$1/\sqrt{2}$	$1/\sqrt{30}$	-
$\{2,[2]\}$	$-\sqrt{2/7}$	$-1/2\sqrt{7}$	$-\sqrt{2/7}$	$-\sqrt{3/35}$	-
$\{\frac{5}{2},[3]\}$	$\sqrt{6/35}$	$1/\sqrt{14}$	$1/\sqrt{7}$	$1/\sqrt{7}$	-
$\{3,[4]\}$	$-\sqrt{3/35}$	$-\sqrt{5/42}$	$-\sqrt{2/35}$	$-2/\sqrt{21}$	-
$\{\frac{7}{2},[5]\}$	$1/\sqrt{35}$	$\sqrt{5}/2\sqrt{7}$	$1/\sqrt{70}$	$\sqrt{3/14}$	-
$\{4,[6]\}$	-	$-1/2$	-	$-1/\sqrt{5}$	-
$\{\frac{9}{2},[7]\}$	-	$1/\sqrt{3}$	-	$\sqrt{2/15}$	-
$\{5,[0]\}$	-	-	-	-	$-1/\sqrt{5}$
$\{\frac{11}{2},[1]\}$	-	-	-	-	$2/\sqrt{5}$

allowed by statistics. There is an irreducible supermultiplet for each value of L which is complex (i.e. charged) because of the PCT-doubling of the vacuum. With the previous formalism one can explicitly construct the bilinears in each complex massive supermultiplet with $J_{MAX} = L + \frac{1}{2}$, which consists of two fermions with spin $L + \frac{1}{2}$ and $L - \frac{1}{2}$ and two complex bosons with spin L.

However, the bilinears obtained from the composite Clifford vacuum $|0\rangle|\frac{1}{2}\rangle$ are rather different from those considered up to now. Since $\lambda = \lambda_1 - \lambda_2 = \pm\frac{1}{2}$, the vacuum state is fermionic with $J = \frac{1}{2}, \frac{3}{2}, \ldots$, giving rise to massive supermultiplets of bilinears with $J_{MAX} = L + \frac{1}{2}$ which includes two real bosons with spin $L + \frac{1}{2}$ and $L - \frac{1}{2}$ and one Dirac fermion with spin L. For instance, for L = 0 and L = 1 one gets two J = 1 operators which coincide exactly with the axial currents discussed by Wess & Zumino (1974b) and Ferrara & Zumino (1975) in the framework of the Wess-Zumino model.

6 CONCLUSIONS

In the present lectures we have reviewed some aspects of unitary irreducible representations of extended supersymmetry acting on one- and two-particle states.

We have emphasized the structure of spin and internal quantum numbers dictated by supersymmetry. The final goal is to use these supermultiplets for particle unification and therefore to identify some of the quantum numbers carried by the supersymmetry generators with the observed symmetry of elementary particle interactions.

We have seen in Section 4 that there is a very limited set of massless representations with a singlet state carrying helicity $\lambda = 2$ to be identified with the graviton. This shows that superunification, in the framework of supersymmetry, puts severe limitations on the possible candidates for the final field theory.

We have considered in some detail the structure of two-particle states, constructed out of two massless representations. Their decomposition into irreducible components has been given. These states give, in particular, the quantum number assignments of bilinear operators (currents) constructed in terms of the elementary fields of the underlying supersymmetric Lagrangian theory. According to some recent speculations (Ellis, Gaillard, Maiani & Zumino 1980; Ellis, Gaillard & Zumino 1980; Curtright & Freund 1979; Zumino 1980; Derendinger, Ferrara & Savoy 1981), the elementary particles of non-gravitational interactions could turn out to be composites of preconstituent fields of an underlying supergravity theory. This was suggested by the discovery of hidden non-compact (Cremmer, Scherk & Ferrara 1978) and local symmetries (Cremmer & Julia 1978, 1979) in $N \geq 4$ extended supergravities.

In this situation the results of Section 5 could be relevant for the quantum number assignment of composite bilinears of the basic constituent fields and would provide a first step toward the "chemistry" of extended supergravity theories.

REFERENCES

Cremmer, E. & Scherk, J. (1977). Nucl. Phys. B127, p. 259.
Cremmer, E., Scherk, J. & Ferrara, S. (1977). Phys. Lett. 68B, p. 234.
Cremmer, E., Scherk, J. & Ferrara, S. (1978). Phys. Lett. 74B, p. 61.
Cremmer, E. & Julia, B. (1978). Phys. Lett. 80B, p. 48.
Cremmer, E. & Julia, B. (1979). Nucl. Phys. B159, p. 141.
Curtright, T.L. & Freund, P.G.O. (1979). In "Supergravity", eds. P. van Nieuwenhuizen & D.Z. Freedman, North Holland, Amsterdam, p. 197.
Das, A. (1977). Phys. Rev. D 15, p. 2805.
Derendinger, J.-P., Ferrara, S. & Savoy, C.A. (1981). Nucl. Phys. B88, p. 77.
De Wit, B. & Freedman, D.Z. (1977). Nucl. Phys. B130, p. 105.
De Wit, B. (1979). Nucl. Phys. B158, p. 189.
Ellis, J., Gaillard, M.K., Maiani, L. & Zumino, B. (1980). In "Unification of the Fundamental Particle Interactions", eds. S. Ferrara, J. Ellis & P. van Nieuwenhuizen, Plenum Press, New York, p. 69.
Ellis, J., Gaillard, M.K. & Zumino, B. (1980). Phys. Lett. 94B, p. 343.
Fayet, P. (1979). Nucl. Phys. B149, p. 137.
Ferrara, S. & Zumino, B. (1975). Nucl. Phys. B87, p. 207.
Ferrara, S. & van Nieuwenhuizen, P. (1976). Phys. Rev. Lett. 37, p. 35.
Ferrara, S., Scherk, J. & Zumino, B. (1977a). Phys. Lett. 66B, p. 35.
Ferrara, S., Scherk, J. & Zumino, B. (1977b). Nucl. Phys. B121, p. 313.
Ferrara, S. (1980). Frascati preprint LNF-80/31, to appear in the Proceedings of the 2nd. Oxford Quantum Gravity Conference, Oxford, April 1980; CERN preprint TH-2957, to appear in the Proceedings of the 9th. Int. Conf. on General Relativity and Gravitation, Jena, July 1980.
Ferrara, S., Savoy, C.A. & Zumino, B. (1981). Phys. Lett. 100B, p. 393.
Freedman, D.Z. (1979). In "Recent Developments in Gravitation", eds. M. Lévy & S. Deser, Plenum Press, New York, p. 549.
Gell-Mann, M. & Ne'eman, Y. (1974). Unpublished.
Haag, R., Lopuszanski, J.T. & Sohnius, M. (1975). Nucl. Phys. B88, p. 257.
Jacob, M. & Wick, G.C. (1959). Ann. of Phys. (NY) 7, p. 404.
Lopuszanski, J.T. & Wolf, M. (1981). Max Planck Institute preprint MPI-PAE/pTH/41/80, Wroclaw preprint No. 533.
Nahm, W. (1978). Nucl. Phys. B135, p. 149.
Salam, A. & Strathdee, J. (1974). Nucl. Phys. B80, p. 499; Nucl. Phys. B84, p. 127.
Sohnius, M. (1978). Nucl. Phys. B138, p. 109.
Taylor, J.G. (1980). Phys. Lett. 94B, p. 174.
Weinberg, S. & Witten, E. (1980). Phys. Lett. 96B, p. 59.
Wess, J. & Zumino, B. (1974a). Nucl. Phys. B70, p. 39.
Wess, J. & Zumino, B. (1974b). Phys. Lett. 49B, p. 310.
Zumino, B. (1962). Journ. Math. Phys. 3, p. 1055.
Zumino, B. (1980). In "Superspace & Supergravity", eds. S.W. Hawking & M. Rocek, Cambridge Univ. Press, Cambridge, p. 423.

FOUR LECTURES ON SUPERGRAPHS

M. T. Grisaru
Department of Physics
Brandeis University, Waltham, MA 02254, USA

Superfield Feynman rules were formulated by Salam and Strathdee (1975) shortly after they introduced superfields (Salam & Strathdee, 1974) to describe the supersymmetry discussed by Wess and Zumino (1974). Superfield perturbation theory was further developed and applied in a number of papers in the middle seventy's. Martin Roček, Warren Siegel and I became interested in them in an attempt to calculate the three loop β-function in O(4) Yang-Mills theory. We soon found that the original rules were somewhat cumbersome and managed to simplify and streamline them so as to turn them into a useful tool for higher loop calculations (Grisaru, Roček & Siegel, 1979). Such calculations have been performed at the three- (Abbott & Grisaru, 1980; Grisaru, Roček & Siegel, 1980; Grisaru, Roček & Siegel, 1981; Caswell & Zanon, 1981) and four-loop (Sen & Sundaresan, 1981) level for globally supersymmetric theories and we are now developing techniques for similar calculations in supergravity (Grisaru & Siegel, 1981).

In these lectures I will attempt to describe some of the techniques we have used so far. As you will see there is still room for improvements and new tricks. Most of what I will talk about is due to a wonderful collaboration with Martin Roček and Warren Siegel, and I would like to thank them for this. I have also used some of their lecture notes from the Nuffield Supergravity Workshop and from CalTech. Finally, I am pleased to present to the parents of supergraphs this somewhat grownup version of their children.

I. NOTATION AND CONVENTIONS

We use two component anticommuting spinors with dotted and undotted indices and the correspondence

$$\tfrac{1}{2}(1+\gamma_5)\Psi \leftrightarrow \psi^{\alpha} \qquad \tfrac{1}{2}(1-\gamma_5)\Psi \leftrightarrow \bar{\psi}_{\dot{\alpha}}$$

We raise and lower indices with

$$\varepsilon_{\alpha\beta} = -\varepsilon_{\beta\alpha} \qquad \varepsilon^{\alpha\beta}\varepsilon_{\gamma\beta} = \delta^{\alpha}_{\gamma}$$

$$\psi_{\alpha} = \psi^{\beta}\varepsilon_{\beta\alpha} \qquad \psi^{\alpha} = \varepsilon^{\alpha\beta}\psi_{\beta}$$

with exactly the same conventions for dotted indices (this differs from other conventions). We define

$$(\psi)^2 = \psi^{\alpha}\psi_{\alpha} \qquad (\bar{\psi})^2 = \bar{\psi}^{\dot{\alpha}}\bar{\psi}_{\dot{\alpha}}$$

$$\psi^{\alpha}\psi_{\beta} = \tfrac{1}{2}\delta^{\alpha}_{\beta}(\psi)^2$$

Tensor indices are related to spinor indices by

$$g_{\alpha\dot{\beta}} = \sigma^{a}_{\alpha\dot{\beta}}\, g_a \ , \quad g_a = \tfrac{1}{2}\sigma_a{}^{\alpha\dot{\beta}} g_{\alpha\dot{\beta}}$$

$$\sigma^a = (1, \vec{\sigma})$$

$$\sigma^{a}_{\alpha\dot{\alpha}}\,\sigma_a{}^{\beta\dot{\beta}} = 2\,\delta_{\alpha}^{\beta}\,\delta_{\dot{\alpha}}^{\dot{\beta}} \ , \quad \sigma^{a}_{\alpha\dot{\beta}}\,\sigma_b{}^{\alpha\dot{\beta}} = 2\,\delta^{a}{}_{b}$$

$$\sigma^{a}_{\alpha\dot{\beta}}\,\sigma^{b\gamma\dot{\beta}}\,\sigma^{c}_{\gamma\dot{\delta}} = [\delta^{ab}\sigma^{c} + \delta^{bc}\sigma^{a} - \delta^{ac}\sigma^{b} - i\,\varepsilon^{abcd}\sigma_d]_{\alpha\dot{\delta}}$$

We work with imaginary (Wick rotated) time components but not really Euclidean space. Our spinors are Minkowski. Note that the metric is $\eta_{ab} = (1,1,1,1)$, but $\varepsilon_{0123} = i$. In practice there is really no need to keep track of this: A scalar product is a scalar product in any convention and all our calculations are covariant.

We shall deal with real superfields $V(x,\theta,\bar{\theta})$, functions of $x^a, \theta^{\alpha}, \bar{\theta}^{\dot{\alpha}}$ and chiral superfields $\phi(x,\theta,\bar{\theta})$ satisfying a differential constraint: We define the covariant derivatives

$$D_{\alpha} = \tfrac{i}{2}(\partial_{\alpha} + i\bar{\theta}^{\dot{\alpha}}\partial_{\alpha\dot{\alpha}})$$

$$\bar{D}_{\dot{\alpha}} = -\tfrac{i}{2}(\partial_{\dot{\alpha}} + i\theta^{\alpha}\partial_{\alpha\dot{\alpha}})$$

$$\{D_{\alpha}, \bar{D}_{\dot{\alpha}}\} = \tfrac{i}{2}\partial_{\alpha\dot{\alpha}}$$

and chiral (ϕ) or antichiral ($\bar{\phi}$) superfields satisfy

$$\bar{D}_{\dot{\alpha}}\phi = 0 \qquad D_{\alpha}\bar{\phi} = 0$$

The spinor derivatives are defined so that

$$\partial_\alpha \theta^\beta = \delta_\alpha{}^\beta \qquad \partial_{\dot\alpha} \bar\theta^{\dot\beta} = \delta_{\dot\alpha}{}^{\dot\beta}$$

Note the factor of i/2 in the definition of the D's. This somewhat unusual convention simplifies some of the manipulations. Our convention for integration is

$$\int d^2\theta\, \theta^2 = 1 \qquad \text{i.e.} \quad \int d^2\theta = -\frac{1}{4}\partial^\alpha \partial_\alpha$$

We note that for complete superspace integrals $i/2\, \partial^\alpha = D^\alpha$ since the difference is a total (space-time) derivative. Therefore

$$\int d^4x\, d^4\theta = \int d^4x\, d^2\theta\, \bar D^2 = \int d^4x\, D^2 \bar D^2$$

The δ-function is defined so that

$$\int d^4\theta\, \delta(\theta-\theta')\, f(\theta) = f(\theta')$$

$$D^2 \bar D^2 \delta^4(\theta-\theta')\big|_{\theta=\theta'} = 1$$

This last relation will be very useful.

Actions will be written as

$$\int d^4x\, d^4\theta\, \mathcal{L}[V, \Phi, \bar\Phi]$$

except that for terms which are purely chiral or antichiral one uses the appropriate measure $d^2\theta$ or $d^2\bar\theta$. However one can always rewrite such terms with the full superspace measure: Noting that if V is a general superfield, D^2V and $\bar D^2 V$ are antichiral and chiral respectively and that if ϕ or $\bar\phi$ are chiral or antichiral

$$\bar D^2 D^2 \Phi = \Box \Phi \qquad D^2 \bar D^2 \bar\Phi = \Box \bar\Phi$$

we have, for example

$$\int d^4x\, d^2\theta\, (\Phi)^n (\bar D^2 V)^m = \int d^4x\, d^2\theta \left(\frac{\bar D^2 D^2}{\Box}\Phi\right)(\Phi)^{n-1}(\bar D^2 V)^m = \int d^4x\, d^4\theta \left(\frac{D^2}{\Box}\Phi\right)(\Phi)^{n-1}(\bar D^2 V)^m$$

Finally we must define functional derivatives: For a general (unconstrained) superfield we have

$$\frac{\delta}{\delta V(x,\theta)} \int d^4x'\, d^4\theta'\, \mathcal{F}(V) = \mathcal{F}'(V(x,\theta))$$

i.e.,

$$\frac{\delta V(x',\theta')}{\delta V(x,\theta)} = \delta^4(x-x')\,\delta^4(\theta-\theta')$$

However, for a chiral superfield if we define

$$\frac{\delta}{\delta\Phi(x,\theta)}\int d^4x' d^2\theta' \mathcal{F}(\Phi) = \frac{\delta}{\delta\Phi}\int d^4x' d^4\theta' \frac{D^2}{\Box}\mathcal{F}(\phi) = \mathcal{F}'(\phi(x,\theta))$$

it is natural to define the chiral δ-function by

$$\frac{\delta\Phi(x',\theta')}{\delta\Phi(x,\theta)} = \bar{D}^2\delta^4(x-x')\delta^4(\theta-\theta')$$

e.g.,

$$\frac{\delta}{\delta\Phi}\int d^4x' d^2\theta' (\Phi)^n = n\int d^4x' d^2\theta' (\Phi)^{n-1}\frac{\delta\Phi(\theta')}{\delta\Phi(\theta)}$$

$$= n\int d^4\theta[\bar{D}^2\delta^4(\theta-\theta')]\frac{D^2}{\Box}\Phi^{n-1} = n\int d^4\theta\,\delta^4(\theta-\theta')\frac{\bar{D}^2D^2}{\Box}\phi^{n-1}$$

$$= n(\Phi)^{n-1}$$

Some final remarks:

a) We will also use superfields with Lorentz indices $\psi_\alpha(x,\theta,\bar\theta)$, $H_a(x,\theta,\bar\theta)$, etc.

b) The following are projectors: $\Pi_+ = \frac{\bar{D}^2D^2}{\Box}$, $\Pi_- = \frac{D^2\bar{D}^2}{\Box}$, $\Pi_{1/2} = -2\frac{D^\alpha\bar{D}^2D_\alpha}{\Box}$

c) We never need be explicit about representations but a few comments might be useful: The general superfield has the representation

$$V = C(x) + \theta^\alpha\chi_\alpha(x) + \bar\theta^{\dot\alpha}\bar\chi_{\dot\alpha}(x) + \theta^2 M + \bar\theta^2\bar{M}$$
$$+ \bar\theta^2\theta^\alpha\psi_\alpha + \theta^2\bar\theta^{\dot\alpha}\bar\psi_{\dot\alpha} + \theta^2\bar\theta^2 D$$

A chiral or antichiral superfield can be written as

$$\Phi(x,\theta,\bar\theta) = e^{U}\varphi(x,\theta) \qquad \bar\Phi(x,\theta,\bar\theta) = e^{-U}\bar\varphi(x,\bar\theta)$$

$$U = i\,\theta^\alpha\sigma^a_{\alpha\dot\beta}\bar\theta^{\dot\beta}\partial_a$$

The integrals we deal with are defined in a (Hermitean) "vector" representation where the fields have the above form but the integrals are invariant under change to a "chiral" or "antichiral" representation, e.g.,

$$\int d^4x d^4\theta\,\mathcal{L}(\Phi,\bar\Phi,V) = \int d^4x d^4\theta\, e^{-U}\mathcal{L}(\Phi,\bar\Phi,V)$$

$$= \int d^4x d^4\theta\,\mathcal{L}(e^{-U}\Phi, e^{-U}\bar\Phi, e^{-U}V) = \int d^4x d^4\theta\,\mathcal{L}(\varphi, e^{-2U}\bar\varphi, e^{-U}V)$$

In this last expression $\phi, \bar\phi$ depend only on $\theta, \bar\theta$ respectively.

II. DERIVATION OF FEYNMAN RULES

We consider first the generating functional

$$Z(J) = \int \mathcal{D}V \exp\int d^4x d^4\theta\left[\tfrac{1}{2}VKV + \mathcal{L}_{int}(V) + JV\right]$$

for a general superfield, where K is some (kinetic) operator. This can be written as

$$Z(J) = \exp\int d^4x\, d^4\theta\, \mathcal{L}_{int}\left(\frac{\delta}{\delta J}\right)\left\{\mathcal{D}V \exp\int d^4x\, d^4\theta\left[\tfrac{1}{2}VKV + JV\right]\right\}$$

$$= \exp\int d^4x\, d^4\theta\, \mathcal{L}_{int}\left(\frac{\delta}{\delta J}\right)\left\{\exp\int d^4x\, d^4\theta\left[-\tfrac{1}{2}JK^{-1}J\right]\right\}$$

where we have done the Gaussian integral in the usual fashion. This leads immediately to the Feynman rules: For example if $K=\Box$ we have

a) Propagator $\langle T\, V(x,\theta)\, V(x',\theta')\rangle \longleftrightarrow -\frac{1}{p^2}\,\delta^4(\theta-\theta')$

b) Vertices: For $\mathcal{L}_{int} \sim V^n$, n lines at each vertex and integration over $d^4\theta$ at each vertex.

c) For each loop an integral $\int d^4q/(2\pi)^4$ and an overall factor $(2\pi)^2\delta(\sum k_{ext})$.

We have generally found it convenient to compute the effective action $\Gamma(V)$ defined by IPI graphs, and

d) For each external line with outgoing momentum k a factor $\int d^4k/(2\pi)^4\; V(-k,\theta)$.

Finally one may have various symmetry factors.

Chiral fields usually have a quadratic action of the form

$$\int d^4x\left[\int d^4\theta\, \bar\Phi\Phi - \tfrac{1}{2}m\int d^2\theta\, \Phi^2 - \tfrac{1}{2}m\int d^2\bar\theta\, \bar\Phi^2\right]$$

$$= \int d^4x\, d^4\theta\left[\bar\Phi\Phi - \tfrac{1}{2}m\Phi\frac{D^2}{\Box}\Phi - \tfrac{1}{2}m\bar\Phi\frac{\bar D^2}{\Box}\bar\Phi\right]$$

with source terms

$$\int d^4x\left[\int d^2\theta\, J\Phi + \int d^2\bar\theta\, \bar J\bar\Phi\right] = \int d^4x\, d^4\theta\left[\Phi\frac{D^2}{\Box}J + \bar\Phi\frac{\bar D^2}{\Box}\bar J\right]$$

The Gaussian integral involves now the expression

$$\int \mathcal{D}\Phi\,\mathcal{D}\bar\Phi\, \exp\int d^4x\, d^4\theta\left[\frac{1}{2}(\Phi\;\bar\Phi)\begin{pmatrix}-m\frac{D^2}{\Box} & 1\\ 1 & -m\frac{\bar D^2}{\Box}\end{pmatrix}\begin{pmatrix}\Phi\\ \bar\Phi\end{pmatrix} + (\Phi\;\bar\Phi)\begin{pmatrix}\frac{D^2}{\Box}J\\ \frac{\bar D^2}{\Box}\bar J\end{pmatrix}\right]$$

The matrix K in this expression has the inverse

$$K^{-1} = \begin{pmatrix}\dfrac{m\bar D^2}{\Box - m^2} & 1 + m\dfrac{\bar D^2 D^2}{\Box(\Box-m^2)}\\ 1 + m\dfrac{D^2\bar D^2}{\Box(\Box-m^2)} & \dfrac{mD^2}{\Box - m^2}\end{pmatrix}$$

giving

$$Z(J) = \exp\int d^4x\, d^4\theta\, \mathcal{L}_{int}\left(\frac{\delta}{\delta J},\frac{\delta}{\delta\bar J}\right)\exp\int d^4\theta\, d^4x\left[-\bar J\frac{1}{\Box - m^2}J + \frac{1}{2}\left(J\frac{mD^2}{\Box(\Box-m^2)}J + h.c.\right)\right]$$

The Feynman rules follow:

a) Propagators $\langle T\,\bar{\Phi}\,\Phi\rangle \;:\; \frac{1}{p^2+m^2}\,\delta^4(\theta-\theta')$

$$\langle T\,\Phi\,\Phi\rangle \;:\; \frac{mD^2}{p^2(p^2+m^2)}\,\delta^4(\theta-\theta')$$

$$\langle T\,\bar{\Phi}\,\bar{\Phi}\rangle \;:\; \frac{m\bar{D}^2}{p^2(p^2+m^2)}\,\delta^4(\theta-\theta')$$

In particular for massless chiral fields there are only $\phi\bar{\phi}$ propagators.

b) Since

$$\frac{\delta J(x,\theta)}{\delta J(x',\theta')} = \bar{D}^2\,\delta^4(x-x')\,\delta^4(\theta-\theta')$$

at vertices one has a factor of $\bar{D}^2(D^2)$ for each chiral (antichiral) field in the interaction, acting on the corresponding propagator. However at a purely chiral vertex, e.g., $\int d^2\theta\phi^n$ one of these factors can be used to convert the $\int d^2\theta$ integral into a $\int d^4\theta$ integral and we always do this. Therefore at such vertices one omits one factor of $\bar{D}^2$. The remaining rules are just like for the general superfield.

Examples:

a) General superfield with no spinor derivative interactions

$$S(V) = \int d^4x\,d^4\theta\left[\tfrac{1}{2}V\Box V + \frac{\lambda}{n!}V^n\right]$$

Consider a loop as shown in Fig. 1

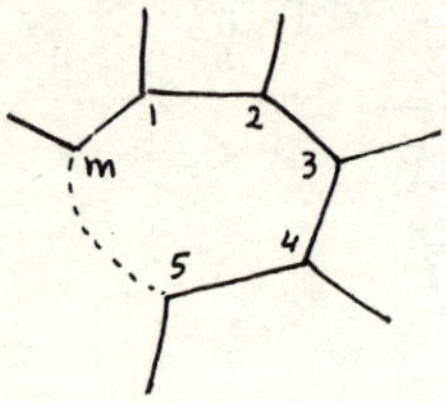

Figure 1

The propagators give factors

$$\delta^4(\theta_1-\theta_2)\,\delta^4(\theta_2-\theta_3)\ldots\;\delta^4(\theta_m-\theta_1)$$

<u>and this product is zero</u> because $\delta^4(\theta)\delta^4(\theta) \simeq (\theta^2\bar{\theta}^2)(\theta^2\bar{\theta}^2) = 0$. Therefore the theory described by the above action receives no quantum corrections. It is of course a sick theory, with ghost component fields which exactly cancel the contributions of physical component fields. However the moral to be drawn from this example is that one needs spinor derivatives to cancel some of the θ-factors resulting from the δ-functions.

b) The Wess-Zumino Model

$$S(\phi,\bar{\phi}) = \int d^4x\left[\int d^4\theta\,\bar{\Phi}\Phi + \frac{\lambda}{3!}\int d^2\theta\,\phi^3 + \frac{\lambda}{3!}\int d^2\bar{\theta}\,\bar{\phi}^3\right]$$

We consider the massless case for simplicity. The one-loop contribution to the self energy is, as shown in Fig. 2

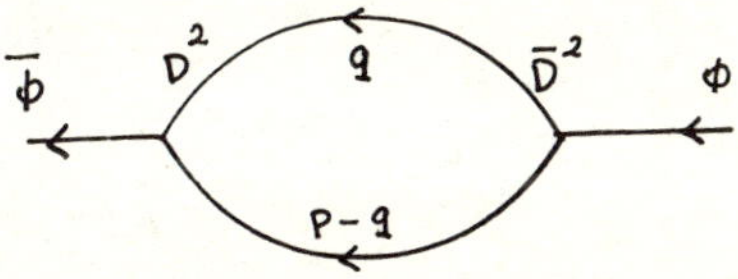

$$\frac{\lambda^2}{2}\int \frac{d^4p\,d^4q}{(2\pi)^8}\, d^4\theta_1\, d^4\theta_2\, \bar\Phi(-p,\theta_1)\Phi(p,\theta_2)\, D_1^2\, \frac{\delta^4(\theta_1-\theta_2)\overleftarrow{\bar D}^2}{q^2}\, \frac{\delta^4(\theta_2-\theta_1)}{(p-q)^2}$$

Figure 2

Since the vertices are chiral only one line is operated on by D^2, $\bar D^2$ factors. Since the usual integration by parts rules apply these factors can be moved onto the other line. Note also that in momentum space

$$\{D_\alpha, \bar D_{\dot\alpha}\} = \frac{1}{2} p_{\alpha\dot\alpha}$$

where p is the momentum of the propagator <u>leaving</u> the end on which the D's are operating. Also $D_\alpha(q,\theta_1)\delta^4(\theta_1-\theta_2) = -D_\alpha(-q,\theta_2)\delta^4(\theta_1-\theta_2)$ which we refer to as a "transfer" operation. Although we do not indicate normally any momentum dependence of the D's or the δ-functions, such dependence is understood in the above sense.

We use now the following rules about the product of two δ-functions in θ-space

$$\delta_{12}\,\delta_{21} = 0$$

$$\delta_{12}\, D\,\delta_{21} = \delta_{12}\, D^2\delta_{21} = \delta_{12}\,\bar D D^2\delta_{21} = \delta_{12}\,\bar D^2\delta_{21} = \delta_{12}\,\bar D\,\delta_{21} = 0$$

$$\delta_{12}\, D^2\bar D^2\delta_{21} = \delta_{12}\,\bar D^2 D^2\delta_{21} = \delta_{12}\, D^\alpha \bar D^2 D_\alpha\,\delta_{21} = \delta_{12}$$

If more factors of D, $\bar D$ are present one can reduce to the cases above using the anticommutation relations and $(D)^3 = (\bar D)^3 = 0$. Therefore in the integral above we immediately obtain for the one-loop, two-particle effective action the expression <u>local in θ</u>

$$\frac{\lambda^2}{2}\int \frac{d^4p}{(2\pi)^4}\, d^4\theta\, \bar\Phi(-p,\theta)\Phi(p,\theta) \int \frac{d^4q}{(2\pi)^4}\, \frac{1}{q^2(q-p)^2}$$

In the massless case the one-loop three point function is zero. For the two loop expression one has the graph in Fig. 3, where we have shown the D-factors and the external fields

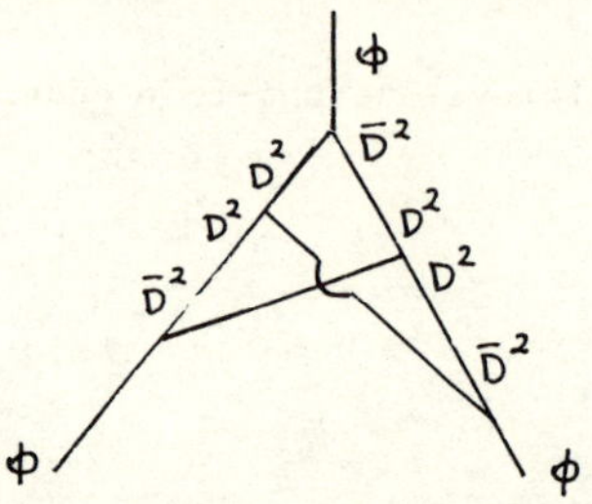

Figure 3

We have taken into account the fact that because of the chiral vertices $\int d^2\theta\phi^3$ a factor of $\bar{D}^2$ is omitted at each vertex; similarly for antichiral vertices. Evaluation of such a diagram is a good exercise that we shall do presently.

The general idea is that one will have loops with D's and $\bar{D}$'s acting on the lines. By integration by parts and transfers one can free one line of D's, after which the $\delta^4(\theta-\theta')$ with no D's acting on it can be used to do a θ-integral. This merely sets $\theta = \theta'$ i.e. shrinks the line to a point in θ space. One repeats now the procedure, working around a loop until one has exactly two lines left, one of which is free of any D's:

$$\delta_{ij}\, D \ldots\ldots \bar{D} \ldots\ldots D\, \delta_{ji}$$

Using the rules given earlier this is zero unless the anticommutation relations can be used to reduce the D factors to one of the forms $D^2\bar{D}^2$, $\bar{D}^2D^2$ or $D^\alpha\bar{D}^2D_\alpha$. If this is the case, $\delta_{ij}D^2\bar{D}^2\delta_{ji} = \delta_{ij}$ and one more θ integral can be done reducing the whole loop to one point in θ-space. The procedure is now repeated loop by loop until any n-loop integral is reduced to a single point in θ-space. Another example in $\lambda\phi^3$ theory, worked out diagrammatically will illustrate the procedure. We use $D^2\bar{D}^2D^2=q^2D^2$ and indicate the factor by a box placed on the line. Lines without any D's on them can be thought of as having been shrunk to a point

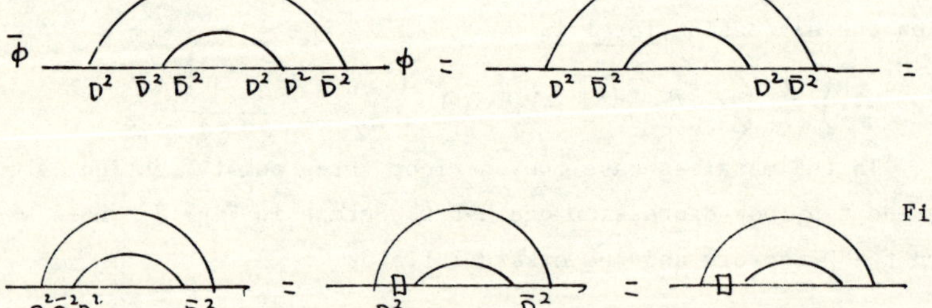

Figure 4

In general one has to work harder and use a certain amount of ingenuity to find shortcuts. For example the three point function of Fig. 3 can be manipulated as follows:

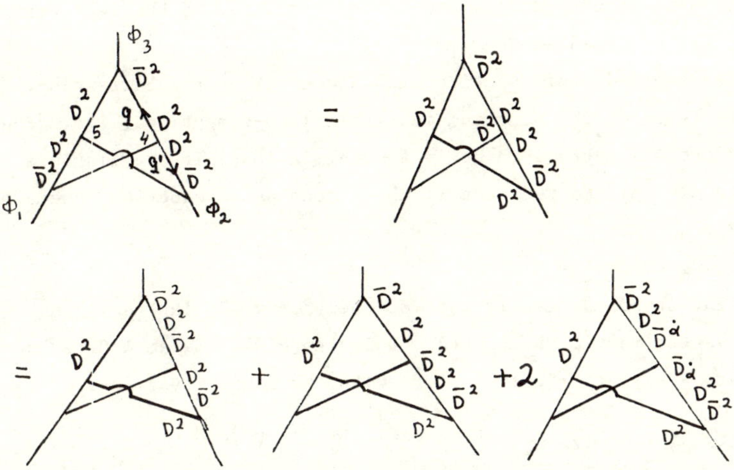

Figure 5

where, after some integration by parts and transfer the D^2 factor on the central line is integrated by parts and can act in three ways giving three terms. The first two are simple: $\bar{D}^2 D^2 \bar{D}^2 = q^2 \bar{D}^2$ or $q'^2 \bar{D}^2$ respectively. In the first diagram, since we must have exactly $D^2 \bar{D}^2$ in each loop the D^2 at the bottom can be integrated by parts and can only act on the external line. Similarly in the second diagram the D^2 on the left can be moved onto the top external line. In the third diagram each D^2 must split to "absorb" (by anticommutation relations) the $\bar{D}_{\dot\alpha}$ factors, e.g. on the bottom right hand side we have

$$D^2 [\phi_2 \bar{D}^2 D^2 \delta_{24} \overleftarrow{\bar{D}}_{\dot\alpha}] = 2 D^\alpha \phi_2 D_\alpha \bar{D}^2 D^2 \delta_{24} \overleftarrow{\bar{D}}_{\dot\alpha}$$

$$= -2 D^\alpha \phi_2 D_\alpha \bar{D}^2 D^2 \bar{D}_{\dot\alpha} \delta_{24} = - D^\alpha \phi_2 q'_{\alpha\dot\alpha} \bar{D}^2 D^2 \delta_{24}$$

Finally the total contribution is (other than propagators, etc.)

$$q^2 \phi_1 D^2 \phi_2 \phi_3 + q'^2 \phi_1 \phi_2 D^2 \phi_3 - 2 \phi_1 D^\alpha \phi_2 D^\beta \phi_3 q'_\alpha{}^{\dot\alpha} q_{\beta\dot\alpha}$$

We draw attention to the minus sign in the last term. It comes from the following rule: We started with $D^2 D^2 \bar{D}^2 = D^\alpha D_\alpha D^\beta D_\beta \bar{D}^{\dot\alpha} \bar{D}_{\dot\alpha}$. However in the expression above the order of some spinor indices has been changed and this accounts for the minus sign. Once this rule is understood one can completely ignore the fact that the D's anticommute and fix up the correct sign at the end of the calculation. On the other hand the direction of the q's has to be paid attention to. Here it is correlated to the fact that the first D^2 factor acted at the 4 vertex.

We conclude this section with a comment on regularization. Our procedure is to do the momentum integrals in n-dimensions, i.e. $\int d^nq/(2\pi)^n$, while the D-algebra is done in four dimensions. If one imagines continuing to n<4 this corresponds to the dimensional reduction scheme of Siegel (1979).

Gauge Theories

The standard case is that of the (non-) Abelian real gauge superfield (vector multiplet) (Ferrara & Zumino, 1974; Salam & Strathdee, 1974 ; Ferrara & Piguet, 1975) $V(x,\theta,\bar\theta)$ with the gauge invariance

$$e^{V} = e^{i\bar\Lambda} e^{V} e^{-i\Lambda} \qquad \bar D\Lambda = D\bar\Lambda = 0$$

This has been discussed in some detail by Ferrara and Piguet (1975). Here $V=V^iG_i$ and G_i are generators of some internal symmetry group. The action is

$$S = -\,\mathrm{tr}\int d^4x\, d^2\theta\, W^\alpha W_\alpha$$

where the chiral field strength is

$$W_\alpha = \bar D^2 [e^{-V}, D_\alpha e^{V}]$$

The action is real up to a total derivative. The quadratic part of the action can be written

$$S_2 = -\,\mathrm{tr}\int d^4x\, d^4\theta\, V D^\alpha \bar D^2 D_\alpha V$$

and needs gauge fixing. Recall that $\Pi_{1/2}=-2D^\alpha\bar D^2 D_\alpha/\Box$ is a projection operator along with $\Pi_0 = (D^2\bar D^2+\bar D^2D^2)/\Box$, $\Pi_0 + \Pi_{1/2} = 1$ so that if one chooses the gauge functions D^2V, $\bar D^2V$ and the gauge fixing term

$$S_{GF} = -\alpha\, \mathrm{tr}\int d^4x\, d^4\theta\, D^2V\, \bar D^2 V$$

one has

$$S_2 + S_{GF} = -\,\mathrm{tr}\int d^4x\, d^4\theta\, \tfrac{1}{2} V \Box(\alpha\Pi_0 + \Pi_{1/2})V$$

In particular for $\alpha=1$ (Fermi-Feynman gauge) one gets an ordinary kinetic term and a propagator

$$\langle T\, V V\rangle = -\frac{1}{p^2}\,\delta^4(\theta-\theta')$$

Other gauges seem to have problems with infrared divergences (propagators $\sim 1/q^4$).

To find the ghosts we note that the gauge transformations can be written as

$$\delta_\Lambda V = -i L_{V/2}\left[\Lambda+\bar\Lambda + \coth(L_{V/2})(\Lambda-\bar\Lambda)\right]$$

$$L_X Y = [X, Y]$$

and this leads to

$$S_{FP} = i\,\mathrm{tr}\int d^4x\, d^2\theta\; c'\,\delta_c(\bar D^2 V) + \mathrm{h.c.}$$

$$= \mathrm{tr}\int d^4x\, d^4\theta\,(c'+\bar c')L_{V/2}\left[(c+\bar c) + \coth L_{V/2}(c-\bar c)\right]$$

with chiral and antichiral ghosts. For $\alpha=1$ the BRS invariance is

$$\delta V = -iL_{V/2}\left[i\varepsilon(c+\bar c) + \coth(L_{V/2})\, i\varepsilon(c-\bar c)\right]$$

$$\delta c = -\varepsilon c^2\,, \quad \delta\bar c = -\varepsilon\bar c^2$$

$$\delta c' = \varepsilon\bar D^2 D^2 V\,, \quad \delta\bar c' = \varepsilon D^2\bar D^2 V$$

To low orders in V, the total action is

$$S = \mathrm{tr}\int d^4x\, d^4\theta\,\Big\{-\tfrac{1}{2}V\Box V + \bar c' c - c'\bar c$$
$$+ V\{D^\alpha V, \bar D^2 D_\alpha V\} + \tfrac{1}{2}(c'+\bar c')[V, c+\bar c)]$$
$$-\tfrac{1}{4}[V, D^\alpha V]\bar D^2[V, D_\alpha V] - \tfrac{1}{3}D^\alpha V\bar D^2[V,[V, D_\alpha V]]$$
$$+\tfrac{1}{12}(c'+\bar c')[V,[V, c-\bar c]] + \cdots\cdots \Big\}$$

For example, the one loop vector contribution to the vector self energy is described by Fig. 6

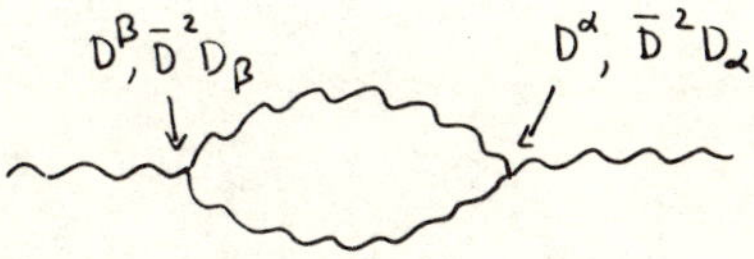

Figure 6

where at the vertices the two operators D^α, $\bar D^2 D_\alpha$ can act on any two of the three lines. Note however that one cannot have both $\bar D^2 D_\alpha$ and $\bar D^2 D_\beta$ act on the external lines: there would be no $\bar D$ left in the loop and the result would be zero.

The gauge superfield interacts with a chiral one through the covariantization of the chiral kinetic term

$$S = \int d^4x\, d^4\theta\, e^{-V}\bar{\Phi} e^{V}\Phi = \int d^4x\, d^4\theta \left[\bar{\Phi}\Phi + \bar{\Phi}V\Phi + \cdots\right]$$

Consequently the one massless chiral loop contribution to the vector self energy is given by Fig. 7 and the resulting expression can be manipulated so as to move the D's off the bottom line. The process is shown below

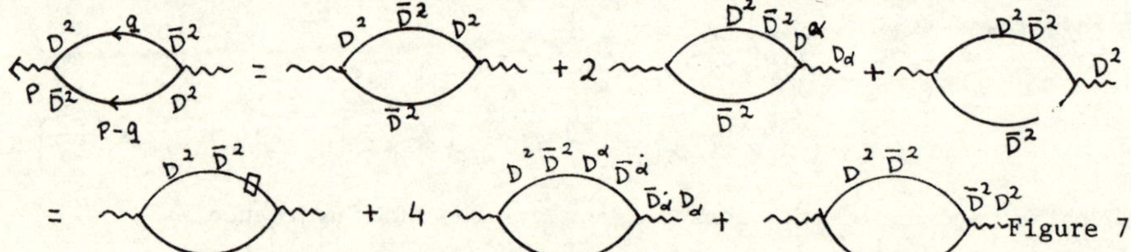

Figure 7

$$\rightarrow V(-p,\theta)\left[q^2 + 2q^{\alpha\dot{\alpha}}\bar{D}_{\dot{\alpha}}D_\alpha + \bar{D}^2D^2\right]V(p,\theta)\ \frac{1}{q^2(p-q)^2}$$

The first term would give zero upon momentum integration (tadpole). The sign of the second one is fixed at the end by paying attention to the order of the spinor indices. For a one loop integral like this a standard trick allows the replacement

$$2q^{\alpha\dot{\alpha}} \rightarrow p^{\alpha\dot{\alpha}} = 2\{D^\alpha, \bar{D}^{\dot{\alpha}}\}V$$

and the whole contribution gives

$$\bar{D}^2D^2 + 2\{D^\alpha, \bar{D}^{\dot{\alpha}}\}\bar{D}_{\dot{\alpha}}D_\alpha = D^\alpha\bar{D}^2D_\alpha$$

after which it can be recast in the gauge invariant form $\int d^2\theta W^\alpha(-p,\theta)$. $W_\alpha(p,\theta)$ times an ordinary (scalar) self energy diagram.

Power Counting

We observe that in renormalizable theories all supersymmetric vertices have four D's (either from the $D^2\bar{D}^2$ of chiral superfields, or the D^α, $\bar{D}^2D_\alpha$ of gauge superfields). In nonrenormalizable theories one may have additional factors but we will not consider this for now. All our vertices carry a $d^4\theta$.

Consider an L-loop integral with P propagators, V vertices and E external lines. We have a total of $(D^2\bar{D}^2)^V \sim q^{2V}$ factors. The propagators are $\sim 1/q^2$ but $\phi\phi$ or $\phi\phi$ propagators give an additional $D^2/q^2 \sim 1/q$. Let there be C of them. Each loop produces a $d^4q \sim q^4$ and uses up a $D^2\bar{D}^2 \sim q^2$ factor from $\delta D^2\bar{D}^2\delta = \delta$. Each external chiral line also accounts for one $\bar{D}^2 \sim q$ missing at the corresponding vertex. Let there be E_C of them. The degree of divergence is then (L-P+V=1)

$$
\begin{aligned}
D &= 4L - 2L - 2P + 2V - C - E_C \\
&= 2 - C - E_C
\end{aligned}
$$

Therefore for graphs with only external V's the degree of divergence is two (but gauge invariance improves this) and zero if there are two external chiral lines. Furthermore if these lines are all chiral (or antichiral) an additional D^2 must come out of the loop: $\int d^4\theta\ \phi^n = 0$ so for a nonzero result we must have at least

$$\int d^4\theta\ \Phi^{n-1} D^2 \Phi = \int d^2\theta\ \Phi^{n-1} \Box \Phi$$

So in this case the convergence is improved and it turns out that only the $\bar{\phi}\phi$ self energy diagram is divergent; mass and coupling terms ϕ^2 and ϕ^3 require no renormalization.

<u>Some General Results</u>

Using these techniques calculations of the β-function to three loops (Abbott & Grisaru, 1980) and four loops (Sen & Sundaresan, 1981) have been done for the Wess-Zumino model and to three loops for the N = 4 Yang-Mills theory (Grisaru, Roček & Siegel, 1980; Grisaru, Roček & Siegel, 1981; Caswell & Zanon, 1981). This latter has the action

$$S = \mathrm{tr}\Big[\int d^4x\, d^4\theta\ e^{-V}\bar{\Phi}_i e^{V}\Phi_i - \int d^4x\, d^2\theta\ W^\alpha W_\alpha + \frac{1}{3!}\int d^4x\, d^2\theta\ i\,\varepsilon_{ijk}\,\Phi_i[\Phi_j, \Phi_k] + \text{h.c.}\Big]$$

In Fermi-Feynman gauge all radiative corrections calculated so far are finite but a rigorous general proof of this fact is still lacking. A nonrigorous argument based on the (unbroken ?) O(4) invariance of this theory uses the fact that the ϕ^3, $\bar{\phi}^3$ vertices are finite to all orders and that the O(4) invariance relates these vertices to the other terms so that the whole action must be finite. However a proof based on the corresponding Ward identities is lacking.

Among other results we mention an old one, namely that if supersymmetry is not broken at the classical level the effective potential vanishes, if evaluated at the classical minimum. This follows easily from the fact that all diagrams give an expression which involves a $\int d^4\theta$ integral. The integrand must have some θ's to give a non zero result but this is possible only if some auxiliary fields (F or D) have non-zero classical values.

While on this subject we might recall that the effective potential in ordinary gauge theories is gauge dependent. Perhaps it has not been appreciated that in supersymmetry there may be an additional kind of gauge dependence, masked by the fact that the potential is usually worked out in Wess-Zumino gauge. Using superfields it is of course possible to obtain results in any gauge.

Some advanced techniques

In our papers (Grisaru, Roček & Siegel, 1979; Grisaru & Siegel, 1981) we have advocated the use of background field techniques to simplify calculations. In particular one loop vector contributions can be simplified by using a convenient background covariant gauge fixing term, and similar simplifications occur in supergravity. Another example is that of a massless chiral loop interacting with a gauge superfield, as shown in Fig. 8.

The D manipulations can get rather cumbersome and the final result quite complicated. A trick due to W. Siegel is to back up to the action $e^{-V}\bar{\phi}e^{V}\phi$ and write $\bar{\phi} = \bar{D}^2\bar{U}$, $\phi = D^2U$, where U is an unconstrained complex superfield. This

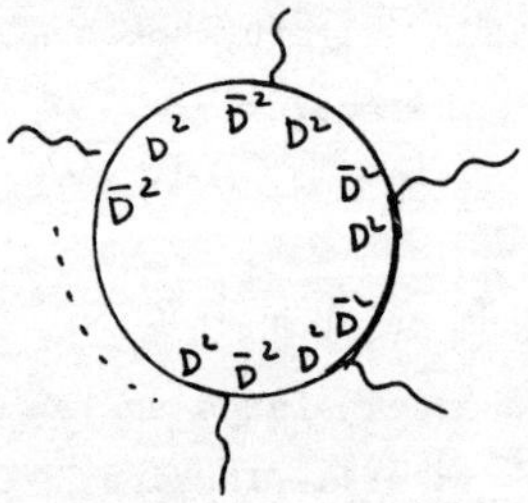

Figure 8

introduces some gauge invariance $\delta U = D^{\alpha}\Lambda_{\alpha}$. One now chooses a suitable gauge fixing term, manipulates the D's some more and obtains new Feynman rules which have one $\bar{D}^2$ in the loop, and at most one D^{α} at each vertex, the external lines being now field strengths or connections. The result, which we quote without explanation is:

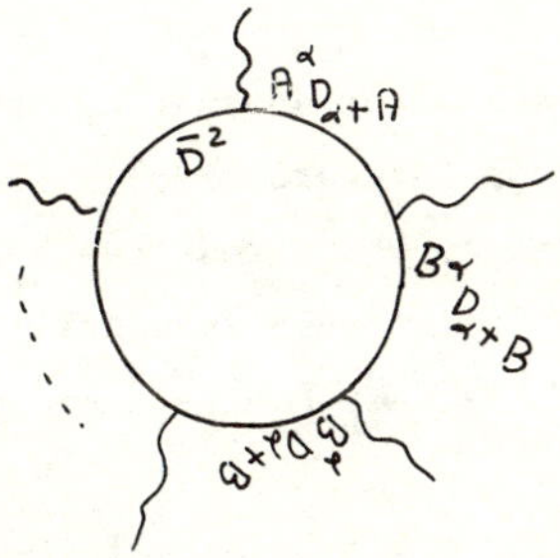

Figure 9

with one A vertex and the rest B vertices,

$$A^\alpha = -2i\Gamma^\alpha \qquad A = -\Gamma^\alpha\Gamma_\alpha + iD^\alpha\Gamma_\alpha$$

$$B^\alpha = -2iW^\alpha \qquad B = -2\Gamma^a\partial_a + (\Gamma^a\Gamma_a - \partial^a\Gamma_a - 2W^\alpha\Gamma_\alpha - iD^\alpha W_\alpha)$$

where

$$e^{-V} D^\alpha e^{V} = D^\alpha - i\Gamma^\alpha$$

$$e^{-V}\partial^a e^{V} = \partial^a - i\Gamma^a$$

While this may appear complicated we emphasize that most of the D-algebra has been done. Other tricks are possible and some of them were described by Grisaru et al. (1980, 1981). In particular we discussed how to treat a special kind of VVV vertex. However, this is one place where further improvement is desirable. This vertex consists of six terms (various permutations of D^α, $\bar{D}^2 D_\alpha$) and some simplifications would be nice.

Further reading and exercises

Redoing the β-function calculation in the Wess-Zumino model is a good exercise. The original rules were given by Grisaru, Roček & Siegel (1979) and are also well summarized in the Caswell & Zanon paper (1981). The paper by Roček and Townsend (1980) has further examples, as do my talks at the Stonybrook Supergravity Workshop and Erice Unification Conference (Grisaru, 1979; Grisaru, 1980). Supergravity rules are being further developed now and have been discussed at the Nuffield Workshop (Grisaru, 1981) and in a recent paper (Grisaru & Siegel, 1981).

III. SUPERGRAPHS IN SUPERGRAVITY

In a certain sense there is nothing to add; and then again... We must write down an action in terms of general and chiral superfields, choose a gauge fixing term appropriately, and derive the Feynman rules. We use the Siegel & Gates (1979) form of supergravity which makes heavy use of the exponential of superfields. My excuse for advocating this form was given at the Nuffield Supergravity Workshop (Grisaru, 1981): For a proper formulation of the background field formalism which requires a "good" splitting between quantum and background fields the exponential appears very suitable. But I have no doubt that the same can be achieved in the formalism of Ogievetsky & Sokatchev (1978).

Feynman rules have already been given by Siegel (1981a), and work has also been carried out by Namazie and Storey (1979). We felt it was important to obtain a formulation which makes full use of the general (super-) covariance of the theory by using the background field method, and of supersymmetry by using supergraphs. The general formulation is discussed in Grisaru and Siegel (1981) and calculations and various results will be published soon. My goal is to prepare the way for a calculation of the three loop divergence in simple supergravity. If someone can prove it is absent this may not be necessary; on the other hand it might make loop calculations more relevant.

The Siegel-Gates formalism

It is not possible to explain here the details of this formulation of supergravity. Besides the original reference, see Roček's Nuffield lectures (Roček, 1981). I will take an engineering approach to the subject: Present a set of definitions and rules and use them to extract what is needed to do perturbation theory.

The basic object is an (axial) vector real superfield which has the expansion (in a particular gauge) and an additional chiral superfield

$$H^a(x,\theta,\bar\theta) = C^a(x) + \dots + \theta\sigma^b\bar\theta h_b^{\ a} + \bar\theta^2\theta^\alpha\psi_\alpha^{\ a} + \dots + \theta^2\bar\theta^2 A^a$$

$$\phi = 1 + \chi = 1 + \dots + \tfrac{1}{3}\theta^2 B$$

which formally makes the theory conformally invariant. Instead one can use the gauge $\phi=1$ but this is not convenient. The supergravity action is

$$S = -\frac{6}{\kappa^2}\int d^4x\, d^4\theta\, (1\cdot e^{-\overleftarrow{H}})^{1/3}\, \hat{E}^{-1/3}(1+\chi)\, e^{-H}(1+\bar\chi)$$

Here $H = H^a\partial_a$, and

$$1\cdot e^{-\overleftarrow{H}} = 1 - i H\cdot\overleftarrow{\partial} - \tfrac{1}{2} H\cdot\overleftarrow{\partial}\, H\cdot\overleftarrow{\partial} + \dots = 1 - i\,\partial\cdot H - \tfrac{1}{2}\partial\cdot(H\,\partial H) + \dots$$

while $\hat{E}$ is a determinant constructed as follows:

Define "semicovariant" derivatives

$$\hat{E}_\alpha = e^{-H} D_\alpha e^{H} \; , \; \hat{E}_{\dot\alpha} = \bar{D}_{\dot\alpha} \; , \; \hat{E}_a = -i\,\sigma_a^{\alpha\dot\alpha}\{\hat{E}_\alpha, \hat{E}_{\dot\alpha}\}$$

and calculate them (to any order in H). Write

$$\hat{E}_A = (\hat{E}_\alpha, \hat{E}_{\dot\alpha}, \hat{E}_a) = \hat{E}_A{}^B D_B$$

where $D_B = (D_\beta, \bar{D}_{\dot\beta}, \partial_b)$. Then $\hat{E} = \det \hat{E}_A{}^B$ can be calculated to any given order in H.

The resulting expression can be used for quantum calculations after suitable gauge fixing. The gauge transformations are

$$e^{H} \to e^{i\bar\Lambda} e^{H} e^{-i\Lambda} \qquad \Phi \to e^{i\Lambda - \frac{1}{3}(\partial_a \Lambda^a - D_\alpha \Lambda^\alpha)} \Phi$$

$$\Lambda = i\,[\Lambda^a \partial_a + \Lambda^\alpha D_\alpha + \bar\Lambda^{\dot\alpha} \bar{D}_{\dot\alpha}]$$

$$\Lambda^a = -i\,\sigma^a_{\alpha\dot\beta} \bar{D}^{\dot\beta} L^\alpha \; , \; \Lambda^\alpha = \bar{D}^2 L^\alpha \; , \; \Lambda^{\dot\alpha} = e^{-H} D^2 \bar{L}^{\dot\alpha}$$

The spinor gauge parameter L_α is the basic object and will lead to corresponding spinor ghosts. It is not difficult to verify that the action written above is invariant under these gauge transformations, and it is certainly a good exercise for familiarizing oneself with the formalism, as is the exercise of writing out the action explicitly. For example after expansion the quadratic part is

$$S_2 = \int d^4x\, d^4\theta \Big[-6\,\bar\chi\chi + 2i(\chi - \bar\chi)\partial . H - \tfrac{1}{2} H^a \Box H_a$$
$$- \tfrac{1}{2}(\partial . H)^2 - \tfrac{1}{6}\left([\bar{D}^{\dot\beta}, D^\alpha] H_{\alpha\dot\beta}\right)^2 + \bar{D}^2 H^a D^2 H_a \Big]$$

where $H^{\alpha\dot\beta} = \sigma_a^{\alpha\dot\beta} H^a$ and one can check the invariance under the infinitesimal transformations

$$\delta H_{\alpha\dot\beta} = -2(\bar{D}_{\dot\beta} L_\alpha - D_\alpha \bar{L}_{\dot\beta})$$

$$\delta\chi = -\bar{D}^2 L^\alpha D_\alpha \chi - \tfrac{1}{3}(\bar{D}^2 D^\alpha L_\alpha)(1 + \chi)$$

Obviously a desirable gauge fixing term would be such as to simplify the kinetic term, e.g., give

$$S_2 + S_{GB} \sim \int d^4x\, d^4\theta \left[\bar\chi\chi - \tfrac{1}{2} H^a \Box H_a \right]$$

which would lead to propagators very similar to the ones we have already discussed. This can be achieved, but I'll discuss it in a more general case when background fields are present.

For completeness I give the formulas for the rest of the quantities of interest. We have already defined $\hat{E}_A$ and $\hat{E}$. In addition we define ψ and $\hat{C}_{AB}{}^C$ through

$$\psi = \phi^{1/2}\bar{\phi}^{-1}(1 \cdot e^{-\overleftarrow{H}})^{-1/3}(\hat{E})^{1/6}$$

$$\{\hat{E}_A, \hat{E}_B\} = \hat{C}_{AB}{}^C \hat{E}_C$$

The covariant supervierbein is

$$E_\alpha = \psi \hat{E}_\alpha \quad , \quad E_{\dot\alpha} = \bar{\psi}\hat{E}_{\dot\alpha}$$

$$E_a = -i\sigma_a^{\alpha\dot\beta}\{\psi\hat{E}_\alpha, \bar{\psi}\hat{E}_{\dot\beta}\} - \frac{1}{2}\psi\bar{\psi}\sigma_{[a}^{\alpha\dot\beta}[\hat{C}_{\alpha c]}{}^c\hat{E}_{\dot\beta} + \hat{C}_{\dot\beta c]}{}^c\hat{E}_\alpha]$$

Then $C_{AB}{}^C$ is defined by

$$[E_A, E_B\} = C_{AB}{}^C E_C$$

and the connection coefficients are

$$\phi_{\alpha bc} = - C_{\alpha[bc]}$$

$$\phi_{abc} = -i\sigma_a^{\alpha\dot\beta}(E_\alpha\phi_{\dot\beta bc} + E_{\dot\beta}\phi_{\alpha bc} + \phi_{\alpha[b}{}^d\phi_{\dot\beta dc]})$$

where [ ] means antisymmetrization. Finally the covariant derivatives are (M = Lorentz transformation)

$$\nabla_A = E_A + \frac{1}{2}\phi_{Ab}{}^c M^b{}_c$$

These expressions look fierce and unfamiliar. However, they provide a step by step procedure for computing all quantities of interest to any order in H and ϕ. Finally the field strengths R, $G_{\alpha\dot\beta}$, $W_{\alpha\beta\gamma}$ can be worked out from their definitions:

$$\{\nabla_\alpha, \nabla_\beta\} = -\bar{R}M_{\alpha\beta}$$

$$\{\nabla_\alpha, \bar{\nabla}_{\dot\beta}\} = \frac{i}{2}\nabla_{\alpha\dot\beta}$$

$$[\nabla_\alpha, \frac{i}{2}\nabla_{\beta\dot\gamma}] = -\frac{1}{2}\varepsilon_{\alpha\beta}[-\bar{R}\bar{\nabla}_{\dot\gamma} + G^\delta{}_{\dot\gamma}\nabla_\delta + (\nabla^\delta G^\varepsilon{}_{\dot\gamma})M_{\delta\varepsilon} + \bar{W}_{\dot\gamma\dot\delta}{}^{\dot\varepsilon}\bar{M}_{\dot\varepsilon}{}^{\dot\delta} - \frac{1}{2}(\bar{\nabla}_{\dot\gamma}\bar{R})M_{\alpha\beta}]$$

A final note on the Lorentz transformations operators M. Their action is ($M_\alpha{}^\beta = \frac{1}{4}\sigma^a_{\alpha\dot\gamma}\sigma_b^{\beta\dot\gamma}M_a{}^b$), for arbitrary antisymmetric $X_b{}^c$

$$[\frac{1}{2}X_b{}^c M_c{}^b, f_a] = X_a{}^b f_b$$

$$[\frac{1}{2}X_b{}^c M_c{}^b, f_\alpha] = X_\alpha{}^\beta f_\beta$$

Background field quantization

Since calculations in supergravity are likely to be very complicated one should take advantage as much as possible of the covariance of the theory. The conventional quantization destroys this covariance because of gauge fixing whereas background field quantization takes full advantage of it. The idea is to split the fields into background and quantum, choose a suitable background covariant gauge fixing term and calculate the background field functional which is fully covariant and from which the effective action can be obtained. This is described for global supersymmetry by Grisaru, Roček & Siegel (1979) and local supersymmetry in my Nuffield lecture (Grisaru, 1981) and by Grisaru & Siegel (1981). Further simplifications can be achieved by choosing a gauge fixing term which cancels some of the background-quantum interactions, and by requiring that the background fields be on shell, i.e., satisfy the classical field equations. Since that is where physics presumably is (with some exceptions), this is a reasonable restriction. Finally one derives an action involving the background and quantum fields in which the background covariance manifests itself as follows: The background fields appear only in gauge invariant forms, i.e., R, $G_{\alpha\dot{\beta}}$, $W_{\alpha\beta\gamma}$ (but the first two are zero on shell) and inside background covariant derivatives $\mathcal{D}_A = \mathcal{E}_A{}^B \mathcal{D}_B + \frac{1}{2}\phi_{Ab}{}^c M^b{}_c$ acting on the quantum fields.

In our approach, before gauge fixing, the quadratic part of the action takes the form (on shell)

$$S = \int d^4x\, d^4\theta\, \mathcal{E}^{-1}\Big[-6\bar{\chi}\chi + 2i(\chi-\bar{\chi})\mathcal{D}.H - \tfrac{1}{2}H^a \Box H_a - \tfrac{1}{2}(\mathcal{D}.H)^2 - \tfrac{1}{6}\big([\bar{\mathcal{D}}^{\dot{\beta}}, \mathcal{D}^{\alpha}] H_{\alpha\dot{\beta}}\big)^2 + \bar{\mathcal{D}}^2 H^a \mathcal{D}^2 H_a - \tfrac{1}{2}H^{\alpha\dot{\beta}}\big(W_{\alpha}{}^{\gamma\delta}\mathcal{D}_{\gamma}H_{\delta\dot{\beta}} + \bar{W}_{\dot{\beta}}{}^{\dot{\gamma}\dot{\delta}}\bar{\mathcal{D}}_{\dot{\gamma}}H_{\alpha\dot{\delta}}\big)\Big]$$

Except for the last term this is just a covariantization of the form given earlier.

The action has the (quantum) gauge invariance which is a direct covariantization of the one given earlier. We fix the gauge in such a way as to reduce the kinetic part of H to the form $H\Box H$ and also to cancel the $H\chi$ cross terms. This can be done in two ways:

a) With the gauge fixing function

$$F_\alpha = \bar{\mathcal{D}}^{\dot{\beta}}\big(H_{\alpha\dot{\beta}} + \tfrac{2i}{5}\mathcal{D}_{\alpha\dot{\beta}}\Box_+^{-1}\bar{\Phi}^3\big)$$

where

$$\Box_+ = \Box + 2\bar{W}^{\dot{\alpha}}{}_{\dot{\beta}}{}^{\dot{\gamma}}\bar{\mathcal{D}}_{\dot{\alpha}}\bar{M}_{\dot{\gamma}}{}^{\dot{\delta}}$$

This is nonlocal and leads to some nonlocalities in the ghost action. However these nonlocalities occur only in the kinetic part of the ghost action and can be eliminated by introducing further ghosts (integrating out such ghosts reproduces the nonlocalities).

b) By first replacing the chiral field ϕ by a real field V,

$$\phi^3 = 1 - \bar{\mathcal{D}}^2 V$$

and using as a gauge fixing function

$$F_\alpha = \bar{\mathcal{D}}^{\dot{\beta}} H_{\alpha\dot{\beta}} + \mathcal{D}_\alpha V$$

The field V has some gauge invariance of its own that requires fixing and new ghosts. The two methods give the same S-matrix but differ in their topological predictions [e.g., anomalies: In the superfield approach only chiral superfields (which may appear as ghosts of general superfields) give anomalies. Changing ϕ to V does have topological consequences.]

The actual details of gauge fixing are rather complicated and will not be discussed here. See Grisaru & Siegel (1981), and Siegel's Nuffield lecture (Siegel, 1981b). The result, for the quadratic quantum action, which is sufficient for one-loop calculations, is

$$\begin{aligned} S = \int d^4x\, d^4\theta\, E^{-1} \Big[& -\tfrac{1}{2} H^a \Box H_a - \tfrac{18}{5} \bar{\chi}\chi \\ & -\tfrac{1}{2} H^{\alpha\dot{\beta}} \left(W_\alpha{}^{\gamma\delta} \mathcal{D}_\gamma H_{\delta\dot{\beta}} + \bar{W}_{\dot{\beta}}{}^{\dot{\gamma}\dot{\delta}} \bar{\mathcal{D}}_{\dot{\gamma}} H_{\alpha\dot{\delta}} \right) \\ & + \sum_{i=1}^{3} \bar{\Psi}_i{}^{\dot{\beta}}\, i\mathcal{D}_{\alpha\dot{\beta}} \Psi_i{}^\alpha + \sum_{i=1}^{2} \phi_i{}^\alpha \mathcal{D}^2 \phi_{i\alpha} + \text{h.c.} \Big] \end{aligned}$$

or

$$\begin{aligned} S = \int d^4x\, d^4\theta\, E^{-1} \Big[& \ldots\ldots\ldots\ldots \\ & \ldots\ldots\ldots\ldots \\ & + \sum_{i=1}^{3} \bar{\Psi}_i{}^{\dot{\beta}}\, i\mathcal{D}_{\alpha\dot{\beta}} \Psi_i{}^\alpha + \sum_{i=1}^{3} V_i \Box V_i + \sum_{i=1}^{7} \bar{\chi}_i \chi_i \Big] \end{aligned}$$

depending on whether one uses the first or second form of gauge fixing.

Some comments are in order:

(a) The $\Psi_{i\alpha}$ and χ_i only have abnormal statistics.

(b) The chiral fields χ, ϕ_i^α are background covariantly chiral. This simply means that, for example,

$$\bar{\chi} = e^{-H(\text{background})}\,\bar{\chi}$$

where the $\bar{\chi}$ on the right is an ordinary antichiral field.

(c) The general spinor ghosts $\Psi_{1\alpha}, \Psi_{2\alpha}$ correspond to the original spinor gauge parameter, L_α, while $\psi_{3\alpha}$ is a Nielsen-Kallosh ghost (Nielsen, 1978 ; Kallosh, 1978). The chiral spinors $\phi_{i\alpha}$ are second generation ghosts associated with the gauge invariance of the first generation ghosts. Various other ghosts mutually cancel or decouple at the one-loop level.

(d) In the second formulation the second generation spinor ghosts are replaced by real ghosts V_i. The χ_i's are third generation ghosts associated with the gauge invariance of the V_i.

IV. A SUPERGRAVITY CALCULATION

I will describe part of a one-loop calculation leading to a determination of the one-loop effective action with up to four external (on-shell) fields. The simplest such calculation is that of the contribution of the general superfields V or ψ_α.

In detail the ψ_α action reads

$$\mathcal{E}^{-1}\bar{\Psi}^{\dot\beta}\, i\mathcal{D}_{\alpha\dot\beta}\,\Psi^{\alpha}$$
$$= \mathcal{E}^{-1} i\sigma^{a}_{\alpha\dot\beta}\,\bar{\Psi}^{\dot\beta}\left[\mathcal{E}_a{}^{c}\partial_c + \mathcal{E}_a{}^{\gamma}D_\gamma + \mathcal{E}_a{}^{\dot\gamma}\bar{D}_{\dot\gamma} + \text{connections}\right]\Psi^{\alpha}$$

with

$$\mathcal{E}^{-1} = 1 + O(H\,\text{background})$$
$$\mathcal{E}_a{}^{c} = \delta_a{}^{c} + O(H\,\text{background}),\quad \mathcal{E}_a{}^{\gamma}, \mathcal{E}_a{}^{\dot\gamma} = O(H\,\text{background})$$

Clearly the propagator comes from $E_a{}^c = \delta_a{}^c$ and is

$$-\frac{\sigma\cdot p}{p^2}\,\delta^4(\theta-\theta')$$

and, since the interaction terms are at most linear in $D, \bar{D}$, the first diagram that can contribute is a box diagram

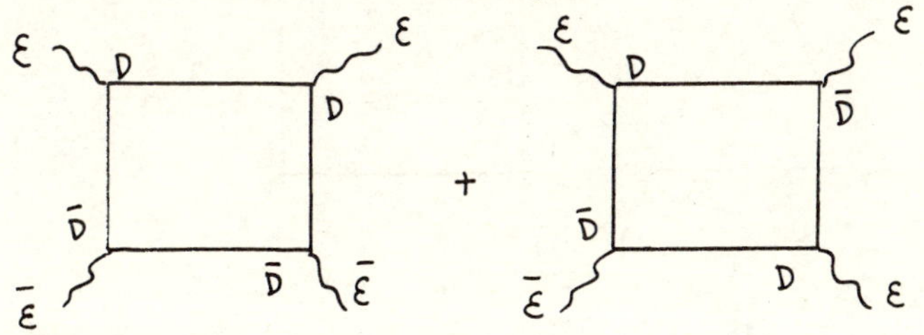

Figure 10

Obviously up to factors the contribution of the first diagram for example is

$$\mathrm{Tr}\ \frac{\sigma.q\ \sigma.\mathcal{E}^{\alpha}\ \sigma.(q+p_1)\ \sigma.\mathcal{E}_{\alpha}\ \sigma.(q+p_1+p_2)\ \sigma.\bar{\mathcal{E}}^{\dot{\alpha}}\ \sigma.(q-p_4)\ \sigma.\bar{\mathcal{E}}_{\dot{\alpha}}}{q^2\,(q+p_1)^2\,(q+p_1+p_2)^2\,(q-p_4)^2}$$

which is just what we would get in QED. One can verify explicitly that the logarithmic divergence cancels and the whole result is expressible in terms of $W^{\alpha\beta\gamma}(x_1\theta)W_{\alpha\beta\gamma}(x_2\theta)\bar{W}^{\dot{\alpha}\dot{\beta}\dot{\gamma}}(x_3\theta)\bar{W}_{\dot{\alpha}\dot{\beta}\dot{\gamma}}(x_4\theta)$ times a nonlocal function of the x's.

The V action reads

$$\mathcal{E}^{-1} V \Box V = \mathcal{E}^{-1} V \mathcal{D}^a \mathcal{D}_a V$$

$$= \mathcal{E}^{-1} V[\mathcal{E}^{a\alpha} D_\alpha + \mathcal{E}^{a\dot{\alpha}} \bar{D}_{\dot{\alpha}} + \mathcal{E}^{ac}\partial_c + \tfrac{1}{2}\phi^{a}{}_{c}{}^{d} M^{c}{}_{d}]\cdot$$

$$\cdot[\mathcal{E}_a{}^{\beta} D_\beta + \mathcal{E}_a{}^{\dot{\beta}} \bar{D}_{\dot{\beta}} + \mathcal{E}_a{}^{d}\partial_d]\, V$$

The part independent of background fields is

$V\partial^a\partial_a V$ and leads to a V propagator

$$\langle T\, V\, V\rangle = \frac{2}{p^2}\,\delta^4(\theta-\theta')$$

Again, one loop contributions can come only from vertices leading to at least two D's and two $\bar{D}$'s in the loop. For diagrams with at most four external lines [$\mathcal{E}_a{}^{\alpha} = \mathcal{O}$ (H background)] the only relevant terms are

$$V[\tfrac{1}{2}\mathcal{E}^{a\alpha}\mathcal{E}_{a\alpha} D^2 + \tfrac{1}{2}\mathcal{E}^{a\dot{\alpha}}\mathcal{E}_{a\dot{\alpha}}\bar{D}^2 + 2(\mathcal{E}^{a\alpha} D_\alpha + \mathcal{E}^{a\dot{\alpha}}\bar{D}_{\dot{\alpha}})\partial_a]\, V$$

and there are no contributions to diagrams with less than four lines. The effective action is given by the diagrams in Fig. 11

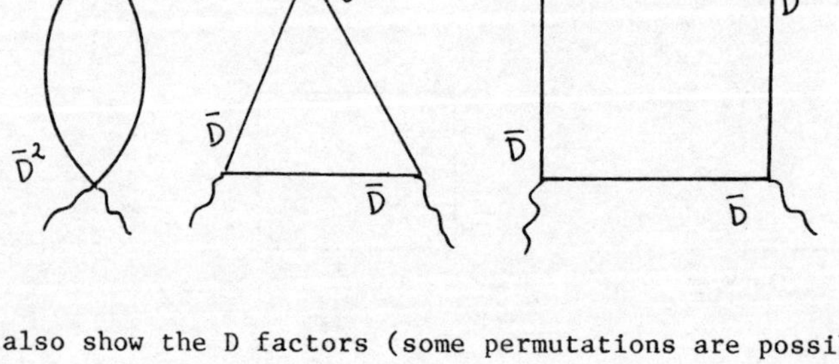

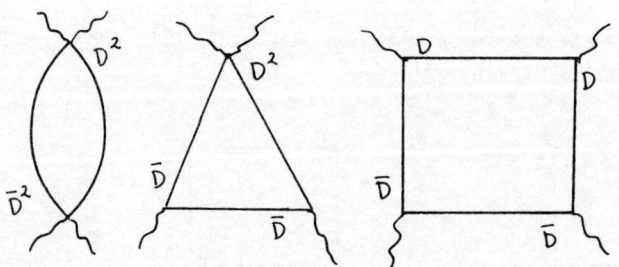

Figure 11

which also show the D factors (some permutations are possible). The

D algebra is trivial and one is left with some ordinary momentum integrals to do. However only the last diagram is relevant as we now discuss:

Because of background covariance and the fact that we are on shell the result can only be a functional of $W_{\alpha\beta\gamma}$, aside from covariantization of nonlocal quantities such as $1/\Box$. We also can argue that this particular contribution to the effective action is finite on shell: We are dealing with a supersymmetric contribution to the action and the divergence which much be proportional to W^2 or $\bar{W}^2$ vanishes on shell (of course we can check this by explicit calculation). Possible terms are then

$$\Gamma \sim \int d^4x_1\, d^4x_2\, d^2\theta\, W^{\alpha\beta\gamma}(x_1,\theta)\, W_{\alpha\beta\gamma}(x_2,\theta)\, G(x_1,x_2) + \text{h.c.}$$

$$+ \int d^4x_1 \ldots d^4x_4\, W(x_1,\theta)\, W(x_2\theta)\, \bar{W}(x_3,\theta)\, \bar{W}(x_4,\theta) \,.$$

$$.\ G(x_1 \ldots x_4)$$

as follows from supersymmetry and a dimensional argument. The G's are nonlocal functions of x_i, covariantized. Now clearly if the first term were present it would receive contributions from a two point function, but we have just pointed out that the V loop gives no such contributions. It turns out then that the role of this diagram, and the second one, is to cancel certain contributions from the third one. The third diagram in momentum space takes the form, after doing the D algebra.

$$\left[2\mathcal{E}^{a\alpha}(p_1)\,\mathcal{E}_\alpha{}^b(p_2)\,\bar{\mathcal{E}}^{c\dot\beta}(p_3)\,\mathcal{E}^d{}_{\dot\beta}(p_4) + \quad p_2 \leftrightarrow p_3,\ b \leftrightarrow c\right]$$

$$\cdot \int \frac{d^4q}{(2\pi)^4} \quad \frac{(q+p_1)_a\,(q+p_1+p_2)_b\,(q-p_4)_c\, q_d}{q^2\,(q+p_1)^2\,(q+p_1+p_2)^2\,(q-p_4)^2}$$

The background gauge invariance here is simply the statement that this expression must depend only on the field strengths

$$W_{\alpha\beta\gamma} \sim \sigma^a{}_\beta{}^{\dot\gamma}\, \sigma^b{}_{\gamma\dot\gamma}\, \mathcal{E}_{\alpha a,b}$$

i.e., on the quantities $\partial_b E_a - \partial_a E_b$. So the calculation is exactly like in QED. Simply extract the finite part by pulling out external momentum factors from the integrand.

The H contributions are very similar. The hardest calculation is that of the contribution from chiral fields, but again using various tricks the D algebra is trivial to do and one is left to evaluate momentum integrals which are no harder than in QED. Details will appear in a future publication.

We feel that the two loop calculations, although complicated, are manageable and that a close (wo)man - computer collaboration could lead to a calculation of the three loop divergence.

REFERENCES

Abbott, L. F. & Grisaru, M. T. (1980), The 3-loop β-function in the Wess-Zumino model, Nucl. Phys. B169, pp. 415-429.

Caswell, W. & Zanon, D. (1981), Zero three-loop beta function in the N=4 supersymmetric Yang-Mills theory, Nucl. Phys. B182, pp. 125-143.

Ferrara, S. & Piguet, O. (1975), Perturbation theory and renormalization of supersymmetric Yang-Mills theories, Nucl. Phys. B93, pp. 261-302.

Ferrara, S. & Zumino, B. (1974), Supergauge invariant Yang-Mills theories, Nucl. Phys. B79, pp. 413-421.

Grisaru, M. T. (1979), Superloops, in Supergravity, eds. D. Z. Freedman & P. van Nieuwenhuizen, North-Holland, pp. 1-10.

Grisaru, M. T. (1980), Superfield perturbation theory, in Unification of the fundamental particle interactions, eds. S. Ferrara, J. Ellis & P. van Nieuwenhuizen, Plenum Press, pp. 545-554.

Grisaru, M. T. (1981), Background superfield method, in Superspace and Supergravity, eds. S. Hawking & M. Rocek, Cambridge University Press, pp. 135-150.

Grisaru, M. T., Roček, M. & Siegel, W. (1979), Improved methods for supergraphs, Nucl. Phys. B159, pp. 429-450.

Grisaru, M. T., Roček, M. & Siegel, W. (1980), Zero value of the 3-loop β-function in N=4 super Yang-Mills theory, Phys. Rev. Lett. 45, pp. 1063-1066.

Grisaru, M. T., Roček, M. & Siegel, W. (1981), Superloops 3, beta 0, Nucl. Phys. B183, pp. 141-156.

Grisaru, M. T. & Siegel, W. (1981), Supergraphity I, Nucl. Phys. B (to be published).

Kallosh, R. E. (1978), Modified Feynman rules in supergravity, Nucl. Phys. B141, pp. 141-152.

Namazie, M. A. & Storey, D. (1979), Supersymmetric quantization of linearized supergravity, Nucl. Phys. B157, pp. 170-188.

Nielsen, N. K. (1978), Ghost counting in supergravity, Nucl. Phys. B140, pp. 499-509.

Ogievetski, V. & Sokatchev, E. (1978), Structure of the supergravity group, Phys. Lett. 79B, pp. 222-224.

Roček, M. (1981), An introduction to superspace and supergravity, in Superspace and Supergravity, eds. S. Hawking & M. Roček, Cambridge University Press, pp. 71-132.

Roček, M. & Townsend, P. (1980), Three-loop finiteness of the N=4 supersymmetric non-linear σ-model, Phys. Lett. 96B, pp. 72-76.

Salam, A. & Strathdee, J. (1974), Super-gauge transformations, Nucl. Phys. B76, pp. 477-482.

Salam A. & Strathdee, J. (1975), Feynman rules for superfields, Nucl. Phys. B86, pp. 142-152.

Sen, A. & Sundaresan, M. K. (1981), The four-loop β-function for the Wess-Zumino model, Phys. Lett. 101B, pp. 61-63.

Siegel, W. (1979), Supersymmetric dimensional regularization via dimensional reduction, Phys. Lett. 93B, pp. 170-172.

Siegel, W. (1981a), Hidden ghosts, Phys. Lett. 93B, pp. 170-172.

Siegel, W. (1981b), The wraiths of graphs, in Superspace and Supergravity, eds. S. Hawking & M. Rocek, Cambridge University Press, pp. 151-164.

Siegel, W. & Gates, S. J. (1979), Superfield supergravity, Nucl. Phys. B147, pp. 77-104.

Wess, J. & Zumino, B. (1974), Supergauge transformations in four dimensions, Nucl. Phys. B70, pp. 39-50.

REPRESENTATIONS OF SUPERSYMMETRY

P.C. West
Mathematics Dept., King's College, Strand, London WC2 England

1 INTRODUCTION

In this contribution I wish to discuss one problem; what are the representations of supersymmetry? I shall not be concerned with any global properties of these representations, but rather be satisfied to find representations of only the infinitesimal elements of the supersymmetry group. This means our final goal is to obtain a set of fields with a set of their infinitesimal transformations under the supersymmetry group. The commutator (anticommutator) of two infinitesimal transformations providing a representation of the commutator (anticommutator) relations of the supersymmetry graded Lie algebra. We will not only give the simplest known examples of representations, but discuss some general techniques that may be used to find new representations.

The wish to find and discuss representations of supersymmetry is of course motivated by the desire to determine whether supersymmetry is realized in nature, in particular in the domain of subnuclear physics. The successfully trodden path for determining if a symmetry is valid in this domain is within the context of Quantum Field Theory. The starting point in this approach is an action, invariant under the symmetry being considered. So it is desirable to find representations of the symmetry in terms of fields out of which the action may be constructed. These fields will not be subject to their equations of motion and preferably will be without any other constraints.

With internal symmetries, a knowledge of the irreducible representations is sufficient to find invariant Lagrangians composed from unconstrained fields. The same however does not apply to space-time groups; the irreducible representations of which often place restrictions on the space-time dependence of the fields on which the representation is carried.

In the case of the Poincaré group all the irreducible representations were found by Wigner in 1939 using his method of induced representations (Wigner 1939). However, these irreducible representations are necessarily eigenstates of P_μ^2 and are carried by fields which are on-shell meaning they are subject to $p^2 = m^2$. From these representations one can construct, in a systematic way, the free-field equations which these fields are required to satisfy (Barut & Raczka 1977). It is a simple matter to write down the Lagrangian from which these field equations follow, the fields in the Lagrangian being unconstrained.

It is this development that one hopes to copy in the case of the supersymmetry space-time group. By a simple application of Wigner's method of induced representations it is easy to find all the irreducible representations of the supersymmetry group. However, these irreducible representations are again eigenstates of P_μ^2, as this operator commutes with all elements of the super Lie algebra. As such they are subject to the constraints $p^2 = m^2$ and so on-shell. Since we know the free-field equations these fields should satisfy, there is little difficulty in realizing the irreducible representations in terms of sets of fields which are subject to their free-field equations. However, it is the next step that presents all the problems. That is the task of extending the irreducible representations to representations carried by unconstrained fields out of which one can construct a Lagrangian.

In the Poincaré group case this step was trivial, for an irreducible representation involves only one spin, which is represented by a given field when subject to its field equation. The Lagrangian is constructed from this field when no longer subject to its field equation. This unconstrained field then carries a reducible representation of the Poincaré group.

For the supersymmetry group, however, the irreducible representations involve more than one spin and must be realized on a set of fields, which are subject to their free field equations; in fact one field for each spin. However, when one wishes to extend these irreducible representations to reducible representations in terms of fields which no longer satisfy their field equations, it is found that one requires more fields than those which appear in the irreducible representations. That is one requires fields which do not correspond to one of the spins of the irreducible representations. These extra fields are called auxiliary fields and are often very difficult to find. The appearance of such fields is a phenomenon unique to supersymmetry, and the fact that we do not know them in several supersymmetric theories represents one of the major obstacles in the development of these theories.

Hence, the problem of finding the representations of supersymmetry boils down to finding the necessary sets of auxiliary fields required to supplement the fields arising in the irreducible representations. Once we have these fields we can construct a representation of supersymmetry carried by fields not subject to their field equations and then use them to construct a Lagrangian.

In this discussion we will consider the representations of the supersymmetry group with only one spinorial generator. We will give the representations necessary to formulate the simplest theories with rigid and local supersymmetry. These are the Wess-Zumino model and N = 1 supergravity. These representations will be presented according to the above scheme. We first find the irreducible representations, realize these in terms of fields subject to their field equations and then find the auxiliary fields.

2 THE SUPERSYMMETRY GRADED LIE ALGEBRA

It is our aim to find representations of the supersymmetry algebra. This algebra occupies a unique place in theoretical physics, for it is the only known algebra that combines the space-time symmetries of the Poincaré group with the symmetries of an internal group in a non-trivial way. In the 1960's it was attempted to achieve this unification of symmetries within the context of a Lie group. However, it was eventually shown that no Lie group could achieve this result (Coleman & Mandula 1967). Therefore, the supersymmetry group must have a structure weaker than that of a Lie group, in particular its algebra must have a more general structure than that of a Lie algebra. The necessary step is to admit in the algebra generators that satisfy anticommutation relations as well as commutator relations. As such, we expect the algebra to be built from the generators of the Poincaré group P_a and J_{cd}, the generators of an internal symmetry T_s, and some generators that satisfy anticommutation relations, $Q_\alpha{}^i$ $(i = 1 \rightarrow N)$. We expect the $Q_\alpha{}^i$ generators to belong to some representation of the Lorentz group. Let us take this to be the spin one half representation. Hence $\alpha = 1 \rightarrow 4$ and the commutation relation between $Q_\alpha{}^i$ and J_{ab} is

$$[Q_\alpha{}^i, J_{ab}] = -i(\sigma_{ab})_\alpha{}^\beta Q_\beta{}^i \tag{2.1}$$

We will for simplicity demand that the spinorial charges $Q_\alpha{}^i$ be Majorana spinors, $\bar{Q}^{\alpha i} = C^{\alpha\beta} Q_\beta{}^i$ where $C^{\alpha\beta} = -C^{\beta\alpha}$ is the charge conjugation matrix. In fact, the algebra also includes some additional generators U^{ij} and V^{ij} called central charges ($U^{ij} = -U^{ji}$, $V^{ij} = -V^{ji}$). We expect the internal symmetry generators, T_s, to act on the generators, $Q_\alpha{}^i$, and leave the Poincaré group generators alone. That is we expect the generators P_a, J_{cd} and T_s when considered alone to be a direct product of the Poincaré group with the internal group. If this were not the case we would violate the no-go theorem of Coleman & Mandula, (1967). Hence we have the commutation relations of the Poincaré group

$$[J_{ab}, J_{cd}] = -i(\eta_{ac}J_{bd} + \eta_{bd}J_{ac} - \eta_{ad}J_{bc} - \eta_{bc}J_{ad})$$

$$[J_{ab}, P_c] = -i(\eta_{ac}P_b - \eta_{cb}P_a), \quad [P_a, P_b] = 0 \tag{2.2}$$

as well as

$$[P_a, T_s] = 0 = [J_{cd}, T_s] \tag{2.3}$$

$$[Q_\alpha{}^i, T_s] = f'_s{}^i{}_j Q_\alpha^j, \quad [T_s, T_r] = f''_{sr}{}^t T_t \tag{2.4}$$

The supersymmetry algebra (Gol'fand & Likhtman 1971 & Wess & Zumino 1974) is, in addition to the relations of equations (2.1 - 2.4), given by

$$\{Q_\alpha{}^i, \bar{Q}^{\beta j}\} = +2\delta^{ij}(\gamma_\mu)_\alpha{}^\beta P_\mu + \delta_\alpha{}^\beta U^{ij} + (\gamma_5)_\alpha{}^\beta V^{ij} \tag{2.5}$$

$$[Q_\alpha{}^i, P_\mu] = 0 \tag{2.6}$$

$$[U^{ij}, \text{anything}] = 0 = [V^{ij}, \text{anything}] \tag{2.7}$$

The possibility of central charges was found in Haag, Lopuszanki, and Sohnius, 1975.

The number N of spinorial charges is arbitrary, the simplest algebra being for N = 1. We can obtain the N = 1 algebra from the general algebra by setting all i,j indices equal to 1, then the central charges and internal symmetries are absent. We then have the Poincaré group as well as $[Q_\alpha, P_\mu] = 0$

$$\{Q_\alpha, \bar{Q}^\beta\} = +\, 2(\gamma^\mu)_\alpha^{\ \beta}\, P_\mu, \quad [Q_\alpha, J_{ab}] = -i(\sigma_{ab})_\alpha^{\ \beta}\, Q_\beta \qquad (2.8)$$

where $Q_\alpha = Q_{\alpha 1}$.

The above algebra is called by mathematicians a Z_2 graded Lie algebra. This means that if one calls the generators $P_a, J_{cd}, T_s, U^{ij}, V^{ij}$ even (bosonic) and the generators $Q_\alpha^{\ i}$ odd (fermionic) then the algebra satisfies the relations

[even, even] = even, [even, odd] = odd {odd, odd} = even

One could envisage algebras with more than one grading, but they will not concern us here. The supersymmetry graded Lie algebras obey generalizations of the Jacobi identities which are satisfied by Lie algebras. For example

$$[\{Q_\alpha^{\ i}, \bar{Q}^{\beta j}\}, T_r] - \{[Q_\alpha^{\ i}, T_r], \bar{Q}^{\beta j}\} - \{[\bar{Q}^{\beta j}, T_r], Q_\alpha^{\ i}\} = 0 \quad \text{etc.} \qquad (2.9)$$

These relations are proved by expanding out the brackets and verifying that the resulting terms cancel against each other. These identities place important restrictions on $f'{}_s{}^i{}_j$ and $f''{}_{rs}{}^t$. In fact, they determine the internal group to be U(N) or some subgroup of U(N). For example, it will be SP(N) if only one central charge is non-trivially realized.

A more pedagogical exposition of the supersymmetry algebra than that given here would be to assume that the algebra contains the generators $P_a, J_{cd}, T_s, U^{ij}, V^{ij}$ and $Q_\alpha^{\ i}$ and has the above Z_2 graded structure. The generalized Jacobi identities would then imply that $Q_\alpha^{\ i}$ belongs to a representation of the Lorentz group, which we would assume to be a spin ½. The rest of the algebra is then fixed entirely by the generalized Jacobi identities.

As the spinorial generators $Q_\alpha^{\ i}$ form a spin ½ representation of the Lorentz group, it follows that the supersymmetry transformations they generate will change the spin of the state on which they act by a half unit of spin. The supersymmetry transformations will therefore rotate a bosonic into a fermionic state and vice-versa. Consequently, any representation of supersymmetry must involve bosons as well as fermions. Another way of arriving at this conclusion is to observe that the fermionic generators, $Q_\alpha^{\ i}$, acting on a commuting (bosonic) state will result in anticommuting (fermionic) state.

An extremely useful rule is that in any representation of supersymmetry for which P_μ is a one to one operator there are the same number of bosonic as fermionic degrees of freedom. This includes field representations where $P_\mu = -i\,\partial_\mu$. This rule follows from the supersymmetry algebra by the following argument. Consider dividing any representation into fermions and bosons and applying the operator $\{Q_\alpha{}^i , \bar{Q}^{\beta j}\}$ to the bosonic set. The first spinorial generator will transform the bosons into the fermions and the second spinorial generator will transform the resulting fermions back into the bosons. However, $\{Q_\alpha{}^i, \bar{Q}^{\beta j}\} = 2\delta^{ij} (\gamma_\mu)_\alpha{}^\beta P_\mu$ and so is a one to one operation. As such one cannot lose degrees of freedom at any step, in particular, the application of the first Q_α^i must be a one to one onto map, implying that the number of bosons and fermions are equal.

In gauge theory, however, it is sometimes desirable to consider the supersymmetry algebra supplemented by gauge transformations. In this case, the above rule will then only apply to the gauge invariant states or fields as only on these will the usual supersymmetry algebra hold.

3 IRREDUCIBLE (ON-SHELL) REPRESENTATIONS

The fact that the four-translations, P_μ, are a vector under the Lorentz group and commute with all other generators of the super algebra, including the spinorial charges $Q_{\alpha i}$ implies that P_μ commutes with all elements of the algebra. Consequently, the states belonging to any irreducible representation of supersymmetry must be on-shell, that is have $p^2 = m^2$. As a result the fermionic and bosonic states of an irreducible representation must have the same mass. An immediate consequence of this fact is that if supersymmetry is realized in nature it must be broken in some way.

The irreducible representations of supersymmetry are easy to find by a straightforward application of the same technique as was used to find the irreducible representations of the Poincaré group. This technique referred to as Wigner's method of induced representations (Wigner 1939) takes a fixed momentum and finds the subgroup H of transformations that leave this chosen momentum inert. Wigner showed that it is sufficient to find a representation of H upon the states of the fixed momentum, in order to induce a representation for the whole group. That is, the behaviour of a particle is determined by its properties at a given momentum (i.e. the rest frame for massive particles).

We now give a simple application of this technique to find the irreducible representations of supersymmetry for massless states (Salam & Strathdee 1974). Let us take, as is usual, the momentum

$q_\mu = (m,0,0,m)$ to be our chosen mementum satisfying $P_\mu^2 = 0$. The generators that lead to symmetry transformations that leave q_μ inert are $J = J_{12}$, the helicity generator $\frac{\underset{\sim}{q} \times \underset{\sim}{J}}{|\underset{\sim}{q}|}$, $J_{20} - J_{23} = L_2$, $J_{10} - J_{13} = L_1$, $Q_{\alpha i}$, P_μ and T_s. That is a supersymmetric extension of the group E_2. The only finite dimensional representations of E_2 are obtained if the action of L_1 and L_2 gives zero on the q_μ states.

To analyse this situation it will prove useful to rewrite the supersymmetry algebra in two component notation. Let us choose a basis in which

$$\gamma^\mu = i\begin{pmatrix} 0 & \sigma^\mu \\ \bar{\sigma}^\mu & 0 \end{pmatrix}, \quad \gamma^5 = \begin{pmatrix} i & 0 \\ 0 & -i \end{pmatrix} \quad C = -i\begin{pmatrix} \varepsilon_{AB} & 0 \\ 0 & \varepsilon^{\dot{A}\dot{B}} \end{pmatrix}$$

In the above $A = 1,2$, $\dot{B} = 1,2$, $(\sigma^\mu)_{A\dot{B}} = (1, \underset{\sim}{\sigma})$ $(\bar{\sigma}^\mu)^{\dot{B}A} = (1,-\sigma)$ where $\underset{\sim}{\alpha}$ are the Pauli matrices and $\varepsilon^{AB} = -\varepsilon_{BA} = -\varepsilon_{\dot{B}\dot{A}} = +\varepsilon^{\dot{A}\dot{B}}$, $\varepsilon_{12} = -1$

$$(3.2)$$

Spinors are written in two components, for example,

$$Q_\alpha{}^i = \begin{pmatrix} Q_A{}^i \\ Q^{\dot{B}i} \end{pmatrix} \tag{3.3}$$

The Majorana condition implies that $Q^{\dot{B}i} = \varepsilon^{\dot{B}\dot{C}} Q_{\dot{C}}{}^i$ where $(Q_A{}^i)^* = Q_{\dot{A}}{}^i$.

(We can raise and lower the internal indices by taking the internal group to be only SO(N).) The supersymmetry algebra now becomes

$$\{Q_A{}^i, Q_{\dot{B}}{}^j\} = 2\,\delta^{ij}(\sigma^\mu)_{A\dot{B}}\,P_\mu, \{Q_A{}^i, Q_B{}^j\} = \{Q_{\dot{A}}{}^i, Q_{\dot{B}}{}^j\} = 0 \tag{3.4}$$

$$[J_{\mu\nu}, Q_A] = \tfrac{1}{2}\,(\sigma_{\mu\nu})_A{}^B\, Q_B$$

$$\text{where } (\sigma_{\mu\nu}) = \frac{i}{2}\,(\sigma_\mu\bar{\sigma}_\nu - \sigma_\nu\bar{\sigma}_\mu) \tag{3.5}$$

On the q_μ states we have

$$\{Q_A{}^i, Q_{\dot{B}}{}^j\} = 2\delta^{ij}(\sigma^\mu)_{A\dot{B}}\,q_\mu = 2m\delta^{ij}((\sigma^0)_{A\dot{B}} + (\sigma^3)_{A\dot{B}}) = 4m\delta^{ij}\begin{pmatrix} 1 & 0 \\ 0 & 0 \end{pmatrix}_{A\dot{B}}$$

$$(3.6)$$

in particular these imply the relations

$$\{Q_1^{i*}\ ,\ Q_2^j\} = 0$$

$$\{Q_2^i,\ Q_2^{j*}\} = 0 \qquad (3.7)$$

which in turn implies that for any state of momentum q_μ say $|q_\mu\rangle$

$$\langle q_\mu|\,(Q_2^i(Q_2^i)^* + (Q_2^i)^*\ Q_2^i)\,|q_\mu\rangle = 0 \qquad (3.8)$$

Demanding that the norm on physical states be positive definite implies that

$$Q_2^{i*}|q_\mu\rangle = Q_2^i|q_\mu\rangle = 0 \qquad (3.9)$$

for all states with momentum q_μ.

Hence, all generators have zero action on q_μ states except J, T_s, P_μ and Q_1^i and (Q_1^{i*}) which obey the algebra

$$[J,\ Q_1^i] = +\frac{1}{2}\,Q_1^i$$

$$[J,\ (Q_1^i)^*] = -\frac{1}{2}\,(Q_1^i)^*$$

and $$\{Q_1^i,\ (Q_1^j)^*\} = 4m\delta^{ij}$$

$$\{Q_1^i,\ Q_1^j\} = \{(Q_1^i)^*,\ (Q_1^j)^*\} = 0 \qquad (3.10)$$

We note that Q_1^i and $(Q_1^j)^*$ form a Clifford algebra and act as raising and lowering operators for the helicity operator.

We find the representations of this algebra in the usual way, we choose a state of given helicity, say λ, and let it be the vacuum state for the lowering operator Q_1^i, i.e.

$$Q_1^i|\lambda\rangle = 0$$

$$J|\lambda\rangle = \lambda|\lambda\rangle \qquad (3.11)$$

The states of this representation are then

$$|\lambda> \;=\; |\lambda>$$

$$|\lambda-\tfrac{1}{2},i> \;=\; (Q_1{}^i)^*|\lambda>$$

$$|\lambda-1,[ij]> \;=\; (Q_1{}^i)^*\,(Q_1^j)^*\,|\lambda>$$

$$\text{etc.} \qquad (3.12)$$

These states have the helicities indicated and belong to the [ijk...] antisymmetric representation of SO(N). The series will terminate after the helicity $\lambda-\frac{N}{2}$, as the next states will be an object antisymmetric in N + 1 indices. The states have helicities from λ to $\lambda - {}^N/_2$, there being $\frac{N!}{m!\,(N-m)!}$ states with helicity $\lambda - \frac{m}{2}$. To obtain a set of states which represent particles of both helicities we must add to the above set, the representation with helicities from $-\lambda$ to $\lambda + \frac{N}{2}$. The exception is the so called CPT self-conjugate sets of states which automatically contain both helicity states.

The representations of the full supersymmetry group are obtained by boosting the above states in accordance with the Wigner method of induced representations.

Hence the massless irreducible representation of N = 1 supersymmetry comprises only the two states

$$|\lambda>$$

$$|\lambda-\tfrac{1}{2}> = (Q_1)^*|\lambda>$$

with helicities λ and $\lambda-\frac{1}{2}$ and

$$(Q_1)^*(Q_1)^* = 0$$

To obtain a C.P.T. invariant theory we must add states of the opposite helicities, i.e. $-\lambda$ and $-\lambda+\frac{1}{2}$. For example, if $\lambda = \frac{1}{2}$ we get on-shell helicity states 0 and $\frac{1}{2}$ and their C.P.T. conjugates with helicities $-\frac{1}{2}$,0; giving a theory with two spin 0's and 1 Majorana spin $\frac{1}{2}$.

Alternatively, if $\lambda = 2$ then we get on-shell helicity states $^3/_2$ and 2 and their C.P.T. self conjugates with helicity $- ^3/_2$ and $- 2$; giving a theory with one spin 2 and one spin $^3/_2$. These on-shell states are those of the Wess-Zumino model and the N = 1 supergravity respectively. Later in this discussion we will give a complete account of these theories.

For N = 4 with $\lambda = 1$ we get the massless states

$$| 1 >, |\tfrac{1}{2}, i> , |0, [ij] >, |-\tfrac{1}{2}, [ijh] >, |-1, [ijhk]>$$

This is a C.P.T. self conjugate theory with 1 spin 1, 4 spin $\frac{1}{2}$ and 6 spin 0.

We refer the reader to the reviews by Freedman 1979, Ferrara 1981, Taylor 1981 for the construction of all irreducible representations of supersymmetry and give here only the results. However, the constructions are so similar to that given above that the reader will be able to prove these results for himself.

Clearly the nature of the on-shell representation depends on the on-shell characteristic being considered, namely, on whether the multiplet is massive or massless, has zero or non-zero central charge, and on the number N of spin and symmetry generators.

If the multiplet is massless and so also has no central charge on-shell then the irreducible representation contains states with helicities from a maximum helicity λ to a minimum helicity $\lambda - \frac{N}{2}$. The demand of a C.P.T. invariant theory requires us to add to this the irreducible representation with helicities from $- \lambda$ to $- (\lambda - ^N/_2)$ if these helicities are not already contained in the first representation.

If the multiplet is massive but has no central charge on-shell, then the irreducible representation with the lowest maximal spin contains a maximal spin of $^N/_2$.

If the multiplet is massive but does possess an on-shell central charge, then the irreducible representation with the lowest maximal spin has a maximal spin of $\frac{N+1}{4}$ if N is odd and $\frac{N}{4}$ if N is even. It is also possible to calculate the number of on-shell states with each helicity.

Although the above gives the spin ranges, the construction of these on-shell states also gives the number of states with each helicity.

The table below gives the results for massless irreducible representations which have maximal helicity 1 or less

N / Spin	1	1	2	2	4
Spin 1	-	1	1	-	1
Spin $\frac{1}{2}$	1	1	2	2	4
Spin 0	2	?	2	4	6

We see that as N increases, the multiplicities of each spin and the number of different types of spin increase. The simplest theories are those for N = 1. The one in the first column is the Wess-Zumino model and the one in the second column is the N = 1 supersymmetric Yang-Mills theory. The latter contains one spin 1 and one spin $\frac{1}{2}$, consistent with the formula for the lowest helicity $\lambda - {}^N/_2$, which in this case gives $1 - \frac{1}{2} = \frac{1}{2}$. The N = 4 multiplet is C.P.T. self conjugate, since in this case we have $\lambda - {}^N/_2 = 1 - {}^4/_2 = -1$. The table has stopped at N equal to 4 since when N is greater than 4 we must have particles of spin greater than 1. (Clearly, N > 4 implies that $\lambda - {}^N/_2 = 1 - {}^N/_2 < -1$.) This leads us to the well-known statement that the N = 4 supersymmetric theory is the maximally extended Yang-Mills theory. It is the supersymmetric Yang-Mills theory which has the greatest number of supersymmetries and which is hoped to be finite.

The content for massless on-shell representations with a maximum helicity 2 is given below

N \ Spin	1	2	3	4	5	6	7	8
Spin 2	1	1	1	1	1	1	1	1
Spin $^3/_2$	1	2	3	4	5	6	8	8
Spin 1		1	3	6	10	16	28	28
Spin $\frac{1}{2}$			1	4	11	26	56	56
Spin 0				2	5	30	70	70

The N = 1 supergravity theory contains only one spin 2 graviton and one spin $^3/_2$ graviton. It is often referred to as simple supergravity theory. For the N = 8 supergravity theory, $\lambda - \frac{N}{2} = 2 - \frac{8}{2} = -2$. Consequently it is C.P.T. self conjugate and contains all particles from spin 2 to spin 0. Clearly, for theories in which N is greater than 8, particles of higher than spin 2 will occur. Thus, the N = 8 theory is the maximally extended supergravity theory.

It is claimed that this theory is in fact the largest possible supergravity theory. This contention rests on the widely held belief that it is impossible to consistently couple particles of spin $\frac{5}{2}$ to other particles.

4 THE WESS-ZUMINO MODEL

In the previous section we discussed the on-shell representations of supersymmetry, the simplest of which contains two spin 0's and one Majorana spin $\frac{1}{2}$. This result obeys our rule concerning the number of fermionic and bosonic degrees of freedom in any representation. The two spin 0's give two helicity states and the one Majorana spin $\frac{1}{2}$ give two helicity states on-shell. We now wish to find a Lagrangian corresponding to this irreducible representation of supersymmetry. This Lagrangian is to be constructed from fields which are no longer on-shell, but still carry a representation of supersymmetry. As explained in the introduction it is this step that is difficult. The two spin 0 on-shell states can be represented by a scalar field A and a pseudo scalar field B and the spin $\frac{1}{2}$ by the field χ_α which is a Majorana spinor and so satisfies

$$\chi_\alpha = C_{\alpha\beta}\,\bar{\chi}^{\beta} \tag{4.1}$$

These fields, as they represent on-shell states, must be subject to their equations of motion

$$\partial^2 A = \partial^2 B = \not{\partial}\chi = 0 \tag{4.2}$$

We can now ask what are the supersymmetry transformations which leave this set of equations of motion invariant. The parameter of supersymmetry is a Majorana spinor of mass dimension $-\frac{1}{2}$ as a consequence of the fact that $\varepsilon^{\alpha}Q_{\alpha}$ is dimensionless. Hence on dimensional grounds we may write down the following set of transformations

$$\delta A = i\,\bar{\varepsilon}\,\chi, \qquad \delta B = i\,\bar{\varepsilon}\,\gamma_5\,\chi$$

$$\delta\chi = \alpha\,\not{\partial}\,(A + \gamma_5 B)\varepsilon \tag{4.3}$$

where α is an undetermined parameter which is the same for A and B due to parity. The variation of δA is straightforward, however the appearance of a derivative in $\delta\chi$ is the only way to match dimensions once the transformations are assumed to be linear. The reader will have no trouble verifying that these transformations do leave the set of field equations of equation (4.2) inert.

We can now test whether the N = 1 supersymmetry algebra of section 1.2 is represented by these transformations. The commutator of two supersymmetries on A gives

$$[\delta_{\varepsilon_1}, \delta_{\varepsilon_2}]A = +\,i\,\bar{\varepsilon}_2\,\alpha\,\not{\partial}\,(A + \gamma_5 B)\varepsilon_1 \quad - (1 \leftrightarrow 2)$$

$$= +\,2i\alpha\,\bar{\varepsilon}_2\,\not{\partial}\,\varepsilon_1 A \tag{4.4}$$

The term involving B drops out due to the properties of Majorana spinors (see the appendix). Provided $\alpha = +1$ this is indeed the four translation ($P_\mu = -i\,\partial_\mu$) required by the algebra. We therefore set $\alpha = +1$. The calculation for B is similar. For the field χ the commutator of two supersymmetries gives the result

$$[\delta_{\varepsilon_1},\delta_{\varepsilon_2}]\,\chi = \not\partial\,(i\,\bar{\varepsilon}_1\chi + \gamma_5\,\bar{\varepsilon}_1\,\gamma_5\,\chi)\,\varepsilon_2 \quad - (1\leftrightarrow 2)$$

$$= -\frac{i}{4}\,\bar{\varepsilon}_1\,\gamma_k\,\varepsilon_2\,\not\partial\,(\gamma_k + \gamma_5\,\gamma_k\,\gamma_5)\,\chi \quad - (1\leftrightarrow 2)$$

$$= +\frac{i}{2}\,\bar{\varepsilon}_2\,\gamma_c\,\varepsilon_1\,2\not\partial\,\gamma_c\,\chi = 2i\,\bar{\varepsilon}_2\,\not\partial\,\varepsilon_1\chi - i\,\bar{\varepsilon}_2\,\gamma^c\,\varepsilon_1\,\gamma_c\,\not\partial\,\chi \tag{4.5}$$

The above calculation uses a Fierz reshuffle (see appendix) as well as the properties of Majorana spinors. However, χ_α is subject to its equation of motion, i.e. $\not\partial\,\chi = 0$, implying the final result

$$[\delta_1,\delta_2]\,\chi = 2i\,\bar{\varepsilon}_2\,\not\partial\,\varepsilon_1\,\chi \tag{4.6}$$

which is the result dictated by the supersymmetry algebra. The reader will have no difficulty verifying that the fields A, B and χ_α and the transformations

$$\delta A = i\,\bar{\varepsilon}\,\chi \quad , \qquad \delta B = i\,\bar{\varepsilon}\,\gamma_5\,\chi$$

$$\delta\chi = \not\partial\,(A + \gamma_5 B)\,\varepsilon \tag{4.7}$$

form a representation of the whole of the supersymmetry algebra provided A,B and χ_α are on-shell (i.e., $\partial^2 A = \partial^2 B = \not\partial\chi = 0$.) Probably this result could be deduced in a more systematic fashion from the supersymmetry algebra by an extension of the technique (see for example (Barut & Raczka 1977)) used to deduce the free field equations from the Poincaré group.

We now wish to consider the fields A, B and χ_α when no longer subject to their field equations. The Lagrangian from which the above field equations follow is

$$L = -\frac{1}{2}\,(\partial_\mu A)^2 - \frac{1}{2}\,(\partial_\mu B)^2 - \frac{i}{2}\,\bar{\chi}\,\not\partial\,\chi \tag{4.8}$$

It is easy to prove that the action $\int d^4x\, L$ is indeed invariant under the transformation of equation (4.7). This invariance is achieved without the use of the field equations. The trouble with this formulation is that the fields A, B and χ_α no longer form a realization of the supersymmetry algebra when they are no longer subject to their field equations; as the last term in equation (4.5) of $[\delta_1, \delta_2]$ demonstrates. It will prove useful to introduce the following terminology. We shall refer to an irreducible representation of supersymmetry carried by fields which are subject to their equations of motion as an on-shell representation. We shall also refer to a Lagrangian as being algebraically on-shell when it is formed from fields which carry an on-shell representation, that is do not carry a representation of supersymmetry off-shell, and the Lagrangian is invariant under these on-shell transformations. The Lagrangian of equation (4.8) is then an algebraically on-shell Lagrangian.

That A, B and χ_α cannot carry a representation of supersymmetry off-shell can be seen without any calculation since these fields do not satisfy the rule of equal numbers of fermions and bosons which was given earlier. Off-shell, A and B have two degrees of freedom, but χ_α has four degrees of freedom. Clearly, the representations of supersymmetry must change radically when enlarged from on-shell to off-shell.

A possible way out of this dilemma would be to add two bosonic fields F and G which would restore the fermi-bose balance. However, these additional fields would have to occur in the Lagrangian so as to give rise to no on-shell states. As such, they must occur in the Lagrangian in the form $+ \frac{1}{2} F^2 + \frac{1}{2} G^2$ and consequently be of mass dimension two. On dimensional grounds their supersymmetry transformations must be of the form

$$\delta F = i \bar{\varepsilon} \partial\!\!\!/ \chi , \qquad \delta G = i \bar{\varepsilon} \gamma_5 \partial\!\!\!/ \chi \tag{4.9}$$

where we have tacitly assumed that F and G are parity even and odd respectively. The fields F and G cannot occur in δA on dimensional grounds, but can occur in $\delta\chi$ in the form

$$\delta\chi = [\beta(F + \gamma_5 G) + \partial\!\!\!/ (A + \gamma_5 B)] \varepsilon \tag{4.10}$$

where β is an undetermined parameter.

We must now test if these new transformations form a realization of the supersymmetry algebra. In fact, straightforward calculation shows they do provided $\beta = +1$. This representation of supersymmetry involving the fields A, B χ_α, F and G was found by Wess and Zumino (Wess and Zumino 1974) and we now summarize this result:

$$\delta A = i\,\bar{\varepsilon}\,\chi, \qquad \delta B = i\,\bar{\varepsilon}\,\gamma_5\,\chi$$

$$\delta\chi = [F + \gamma_5 G + \not{\partial}\,(A + \gamma_5 B)]\,\varepsilon$$

$$\delta F = i\,\bar{\varepsilon}\,\not{\partial}\,\chi, \qquad \delta G = i\,\bar{\varepsilon}\,\gamma_5\,\not{\partial}\,\chi \tag{4.11}$$

The action invariant under these transformations is given by the Lagrangian (Wess and Zumino 1974)

$$L = -\frac{1}{2}(\partial_\mu A)^2 - \frac{1}{2}(\partial_\mu B)^2 - \frac{i}{2}\bar{\chi}\,\not{\partial}\,\chi + \frac{1}{2}F^2 + \frac{1}{2}G^2. \tag{4.12}$$

As expected the F and G fields occur as squares and so lead to no on-shell states. One can recover the irreducible on-shell representation of equation (4.7) by implementing the field equations which set F and G to zero.

The above derivation of the Wess-Zumino model is typical of the general procedure used to find supersymmetric Lagrangians constructed from fields which form a realization of supersymmetry (i.e., from an off-shell representation). The appearance of fields in the Wess-Zumino model that do not lead to on-shell states is, with a few rare exceptions, a universal feature of supersymmetric theories.

This fact is easily seen to be a consequence of our rule for equal number of fermi and bose degrees of freedom in any representation of supersymmetry. It is only spin 0's when represented by scalars that have the same number of field components off-shell as they have on-shell states. For example, a Majorana spin $\frac{1}{2}$ when represented by a spinor χ_α has a jump of 2 degrees of freedom between on and off-shell and a massless spin 1 when represented by a vector A_μ has a jump of 1 degree of freedom. In the latter case it is important to subtract the one gauge degree of freedom from A_μ so leaving 3 field components off-shell. Since the increase in the number of degrees of freedom from an on-shell state to the off-shell field representing it changes by different amounts for fermions and bosons, the fermi-bose balance which holds on-shell will not hold off-shell if we only introduce the fields that describe the on-shell states. The discrepancy must be made up by fields, like F and G, that lead to no on-shell states. These latter type of fields are called auxiliary fields. The whole problem of finding representations of supersymmetry amounts to finding the auxiliary fields.

Unfortunately, it is not at all easy to find the auxiliary fields. Although the fermi-bose counting rule gives a guide to the number of auxiliary fields it does not actually tell you what they are, or how they transform. In fact, the auxiliary fields are only known for almost all N = 1 and 2 supersymmetry theories and for a very few N = 4 theories and not for the higher N theories. In particular, they are not known for the N = 8 supergravity theory.

Theories for which the auxiliary fields are not known can still be described by a Lagrangian in the same way as the Wess-Zumino theory can be described without the use of F and G. Namely, by the so called algebraically on-shell Lagrangian formulation, which for the Wess-Zumino theory was given in equation (4.8). Such 'algebraically on-shell Lagrangians' are not too difficult to find at least at the linearized level. As explained in section (3) we can easily find the relevant on-shell states of the theory. The algebraically on-shell Lagrangian then consists of writing down the known kinetic terms for each spin.

As an example let us consider the case of the supersymmetric N = 4 Abelian theory. For its on-shell states it has one spin 1, four Majorana spin $\frac{1}{2}$'s and 3 parity even spin 0's and 3 parity odd spin 0's. Let us represent these states, according to their SO(4) assignments, by the following fields A_μ, $\lambda_{\alpha i}$, A_{ij} and B_{ij} respectively where $i,j = 1 \to 4$ and A_{ij} and B_{ij} satisfy

$$-A_{ji} = A_{ij} = \varepsilon_{ijk\ell} A^{k\ell}$$

$$B_{ij} = + \varepsilon_{ijk\ell} B^{k\ell} = - B_{ji} \qquad (4.13)$$

Using the same line of reasoning as for the Wess-Zumino theory the reader will find it possible to find the N = 4 on-shell transformations of this theory. That is the set (Brink, Schwarz, Scherk 1977, Gliozzi, Scherk, Olive 1978).

$$\delta A_{ij} = \frac{1}{2}(i\bar{\varepsilon}_i\lambda_j - i\bar{\varepsilon}_j\lambda_i - \varepsilon_{ijk\ell}\, i\bar{\varepsilon}_k\lambda_\ell)$$

$$\delta B_{ij} = \frac{1}{2}(i\bar{\varepsilon}_i\gamma_5\lambda_j - \bar{\varepsilon}_j\gamma_5\lambda_i + \varepsilon_{ijk\ell}\, i\bar{\varepsilon}_k\gamma_5\lambda_\ell)$$

$$\delta A_\mu = i\,\bar{\varepsilon}_i\gamma_\mu\lambda_i$$

$$\delta\lambda_i = -\sigma_{\mu\nu}F^{\mu\nu}\varepsilon_i + 2i\,\partial\!\!\!/\,(A_{ij} + \gamma_5 B_{ij})\,\varepsilon_j \tag{4.14}$$

which form a realization of supersymmetry provided they satisfy their equations of motion. These transformations are an invariance of the algebraically on-shell action (Brink,Schwarz,Scherk 1977, Gliozzi,Scherk, Olive 1978).

$$L = -\frac{1}{4}F^2_{\mu\nu} - \frac{i}{2}\bar{\lambda}^i\,\partial\!\!\!/\,\lambda_i - \frac{1}{2}(\partial_\mu A_{ij})^2 - \frac{1}{2}(\partial_\mu B_{ij})^2 \tag{4.15}$$

A discussion of the symmetries and their consequences can be found in (Sohnius & West 1981).

Of course, we are really interested in the interacting theories. The form of the interactions is however often governed by symmetry principles such as gauge invariance in the above example or general coordinate invariance in the case of gravity theories. When the form of the interactions is dictated by a local symmetry there is a straightforward although maybe very lengthy way of finding the non-linear theory from the linear theory. This method, called Noether coupling, is described in section 7. In one guise or another this technique has been used to construct all non-linear "algebraically on-shell Lagrangians" for all supersymmetric theories.

The reader will now ask himself whether algebraically on-shell Lagrangians may be good enough: "do we really need the auxiliary fields?" This question will be addressed in the next section, but the following example is a warning against over-estimating the importance of a Lagrangian that is invariant under a set of transformations that mix fermi-bose fields, but do not obey any particular algebra.

Consider the Lagrangian

$$L = -\frac{1}{2}(\partial_\mu A)^2 - \frac{i}{2}\bar{\chi}\,\partial\!\!\!/\,\chi \tag{4.16}$$

whose corresponding action is invariant under the transformations

$$\delta A = i\,\bar{\varepsilon}\,\chi$$

$$\delta\chi = \not{\partial} A\,\varepsilon \qquad (4.17)$$

However, this theory is nothing to do with supersymmetry. The algebra of transformations of equation (4.17) does not close on or off- shell without generating transformations which, although invariances of the free theory, can never be generalized to be invariances of an interaction theory. In fact, not even the on-shell states have the correct fermi-bose balance required to form an irreducible representation of supersymmetry. This example illustrates the fact that the "algebraically on-shell Lagrangians" rely for their validity on their on-shell algebra.

As a final remark in this section it is worth pointing out that the problem of finding the representations of any group is a mathematical question not dependent on any dynamical considerations for its resolution. As such which are physical fields and which are auxiliary fields is a model dependent statement. In fact, given an off-shell representation, we can construct several dynamical theories, the auxiliary fields in one theory leading to on-shell states in another theory. As discussed in section (8) this idea enables us to gain information about auxiliary fields from a knowledge of the on-shell states. Of course, we prefer some dynamical theories to others and it is with respect to these preferred theories, such as the Wess-Zumino model, that the auxiliary fields do not propagate.

5 ADVANTAGES AND NECESSITY OF AUXILIARY FIELDS

In the last section we noted that it was easy to find the on-shell states (irreducible representations) of supersymmetry, but difficult to find their extensions to representations in terms of off-shell fields. This results from the difficulty in finding the necessary auxiliary fields. We also discussed how one could find Lagrangians for theories, even where the auxiliary fields were unknown, the so called algebraically on-shell Lagrangians. This leads us to ask if it was necessary to know the auxiliary fields. This question can be tackled in two parts: in practice and in principle. Let us tackle their practical value first.

A knowledge of the auxiliary fields is an enormous advantage in the development of any supersymmetric theory. Only with this knowledge can one find a manifestly supersymmetric formulation of the theory. Clearly such a formulation requires the auxiliary fields since it works with a representation of supersymmetry on fields. The two types of manifestly supersymmetric formalism are superspace and the equivalent component field version, the tensor calculus. The basic advantages can be listed as systematic construction of

(i) supersymmetric invariants

(ii) the quantum Feynman rules of the theory

and (iii) easier calculation of quantum processes using the technique of super-Feynman rules.

The first point is more relevant to locally supersymmetric theories than rigid supersymmetric theories where actions are not too complicated. A similar advantage is obtained by working with four vectors as opposed to individual components when dealing with Maxwell's equations.

The second point is not so much an advantage of a manifestly supersymmetric formalism more that of having an algebra which closes without the use of the field equations. In theories with a local invariance the quantum rules require ghosts which are easiest found by the B.R.S. prescription. However, this prescription requires a closing algebra and hence the use of the auxiliary fields.

The last point refers to the fact that supersymmetric theories can be formulated in superspace and their quantum effects calculated in this enlarged spacetime. The calculations performed in this way are very much quicker than when performed component field by component field. Indeed, in the case of N = 1 supergravity these calculations may be no more difficult than those for ordinary Q.C.D.

The question of whether the auxiliary fields are necessary in principle is an unresolved question. What is clear is that the classical theories with and without auxiliary fields are the same only up to topological effects. Also, in the context of the simplest supersymmetric theory, one can calculate the quantum rules and their effects without the auxiliary fields. However, whether one can do it for all supersymmetric theories is unknown. The reader is referred to (Sohnius, Stelle & West 1980a) and the references therein for a more detailed discussion of this topic.

6 LINEARIZED N = 1 SUPERGRAVITY

Having found the simplest representation of rigid supersymmetry in terms of unconstrained fields - the Wess-Zumino model, we now give the simplest representation of local supersymmetry - N = 1 supergravity. We will follow the same approach as used to obtain the Wess-Zumino model. We will construct the on-shell states, then the algebraically on-shell Lagrangian and finally find the auxiliary fields and the Lagrangian formed from fields that carry a representation of supersymmetry off-shell.

Recalling the result on irreducible massless representations of N = 1 supersymmetry given in section 3, the supersymmetric on-shell representation that includes a spin two graviton must also include an adjacent spin. Supergravity includes a spin $^3/_2$. The spin $^5/_2$ would seem to have considerable problems in coupling to other fields. We first consider the problem at the linearized (free) level. The non-linear theory will be discussed in the next section.

These on-shell states are represented by a symmetric second rank tensor field $h_{\mu\nu}$ ($h_{\mu\nu} = h_{\nu\mu}$) and a Majorana vector spinor, $\psi_{\mu\alpha}$. For these fields to represent a spin 2 and spin $^3/_2$ particle they must possess the gauge transformations

$$\delta h_{\mu\nu} = \partial_\mu \xi_\nu (x) + \partial_\nu \xi_\mu (x)$$

$$\delta\psi_{\mu\alpha} = \partial_\mu \eta_\alpha (x) \tag{6.1}$$

The unique ghost free gauge-invariant, free field equations are

$$E_{\mu\nu} = 0, \qquad R^\mu = 0 \tag{6.2}$$

where $E_{\mu\nu} = R^L_{\mu\nu} - \frac{1}{2}\eta_{\mu\nu} R^L$, $R^{L\ ab}_{\mu\nu}$ being the linearized Riemann tensor given by

$$R^{L\ ab}_{\mu\nu} = -\partial_a\partial_\mu h_{b\nu} + \partial_b\partial_\mu h_{a\nu} + \partial_a\partial_\nu h_{b\mu} - \partial_b\partial_\nu h_{a\mu}$$

and $$R^\mu = \varepsilon^{\mu\nu\rho\kappa}\gamma_5\gamma_\nu\partial_\rho\psi_\kappa,\ R^{Lb}_\mu = R^{L\ ab}_{\mu\nu}\delta^\mu_a,\ R^L = R^{aL}_\mu\,\delta^\mu_a \tag{6.3}$$

For an explanation of this point see van Nieuwenhuizen 1981.

We must now search for the supersymmetry transformations that form an invariance of these field equations and represent the supersymmetry algebra on-shell. On dimensional grounds the most general transformation is

$$\delta h_{\mu\nu} = \frac{1}{2} i(\bar{\varepsilon} \gamma_{\mu} \psi_{\nu} + \bar{\varepsilon} \gamma_{\nu} \psi_{\mu}) + \delta_1 \eta_{\mu\nu} i \bar{\varepsilon} \gamma^{\kappa} \psi_{\kappa}$$

$$\delta\psi_{\mu} = + \delta_2 \sigma^{ab} \partial_a h_{b\mu} \varepsilon + \delta_3 \partial_{\nu} h^{\nu}{}_{\mu} \varepsilon \tag{6.4}$$

The parameters δ_1, δ_2 and δ_3 are parameters whose values will be determined by the demand that the set of transformations which comprise the supersymmetry transformations of equation (6.4) and the gauge transformations of equation (6.1) should form a closing algebra when the field equations of equation (6.2) hold. At the linearized level the supersymmetry transformations are linear rigid transformations, that is they are <u>first order</u> in the fields $h_{\mu\nu}$ and $\psi_{\mu\alpha}$ and parameterized by <u>constant</u> parameters ε^{α}.

Carrying out the commutator of a Rarita-Schwinger gauge transformation, $\eta_{\alpha}(x)$ of equation (6.1) and a supersymmetry transformation, ε of equation (6.4) on $h_{\mu\nu}$

$$[\delta_{\eta}, \delta_{\varepsilon}]\, h_{\mu\nu} = \frac{i}{2} (\bar{\varepsilon} \gamma_{\mu} (\partial_{\nu} \eta) + \bar{\varepsilon} \gamma_{\nu} (\partial_{\mu} \eta)) + \delta_1 \eta_{\mu\nu} i \bar{\varepsilon} \not{\partial} \eta \tag{6.5}$$

This is a gauge transformation, with parameter $\frac{i}{2} \bar{\varepsilon} \gamma_{\mu} \eta$, on $h_{\mu\nu}$ provided $\delta_1 = 0$. Similarly, calculating the commutator of a gauge transformation of $h_{\mu\nu}$ and a supersymmetry transformation on $h_{\mu\nu}$ automatically yields the correct result zero. However, carrying out the commutator of a supersymmetry transformation and an Einstein gauge transformation on $\psi_{\mu\alpha}$ yields

$$[\delta_{\xi_{\mu}}, \delta_{\varepsilon}]\, \psi_{\mu} = + \delta_2 \sigma^{ab} \partial_a (\partial_{\mu} \xi_b) \varepsilon + \delta_3 \partial_{\nu} \partial^{\nu} \xi_{\mu} \varepsilon$$
$$+ \delta_3 \partial_{\nu} \partial_{\mu} \xi^{\nu} \varepsilon \tag{6.6}$$

which is a Rarita-Schwinger gauge transformation on ψ_{μ} provided $\delta_3 = 0$. Hence we take $\delta_1 = \delta_3 = 0$

We must test the commutator of two supersymmetries. On $h_{\mu\nu}$

we find the commutator of two supersymmetries to give

$$[\delta_{\varepsilon_1},\delta_{\varepsilon_2}]\, h_{\mu\nu} = +\frac{i}{2}\bar{\varepsilon}_2\,\gamma_\mu\,\{\delta_2\,\sigma_{ab}\,\partial_a\,h_{b\nu}\,\varepsilon_2 + (\mu\leftrightarrow\nu)\} - (1\leftrightarrow 2)$$

$$= \delta_2\,\{+\frac{i}{2}\bar{\varepsilon}_2\,\gamma^b\,\varepsilon_1\,\partial_\mu\,h_{b\nu} - \frac{i}{2}\bar{\varepsilon}_2\,\gamma^a\,\varepsilon_1\,\partial_a\,h_{\mu\nu} + (\mu\leftrightarrow\nu)\} \qquad (6.7)$$

This is a gauge transformation on $h_{\mu\nu}$ with parameter $(\delta_2\,\frac{i}{2}\,\bar{\varepsilon}_2\,\gamma^b\,\varepsilon_1\,h_{b\nu})$ as well as a space-time translation. The latter coincides with that dictated by the supersymmetry group provided $\delta_2 = -2$ which is the value we now adopt.

It is important to stress that linearized supergravity differs from the Wess-Zumino model in that one must take into account the gauge transformations of equations (6.1) as well as the rigid supersymmetry transformations of equation (6.4) in order to obtain a closed algebra. The resulting algebra is the N = 1 supersymmetry algebra when supplemented by gauge transformations. This algebra reduces to the N = 1 supersymmetry algebra only on gauge invariant states.

For the commutator of two supersymmetries on ψ_μ we find

$$[\delta_{\varepsilon_1},\delta_{\varepsilon_2}]\psi_\mu = -2\,\sigma^{ab}\,\partial_a\,\varepsilon_2\,(\tfrac{i}{2})\,(\bar{\varepsilon}_1\,\gamma_b\,\psi_\mu + \bar{\varepsilon}_1\,\gamma_\mu\,\psi_b) - (1\leftrightarrow 2)$$

$$= +\frac{i}{4}\,\bar{\varepsilon}_1\,\gamma_R\,\varepsilon_2\,\sigma^{ab}\,\partial_a\,\gamma^R\,(\gamma_b\,\psi_\mu + \gamma_\mu\,\psi_b) - (1\leftrightarrow 2)$$

$$= +\frac{i}{4}\,\bar{\varepsilon}_1\,\gamma_R\,\varepsilon_2\,\sigma^{ab}\,\gamma^R\,(\gamma_b\,\psi_{a\mu} + \frac{1}{2}\gamma_\mu\,\psi_{ab}) + \partial_\mu(\frac{i}{4}\,\bar{\varepsilon}_1\,\gamma_R\,\varepsilon_2\,\sigma^{ab}\,\gamma^R\,\gamma_b\,\psi_a) - (1\leftrightarrow 2) \qquad (6.8)$$

where $\psi_{\mu\nu} = \partial_\mu\,\psi_\nu - \partial_\nu\,\psi_\mu$.

Using the different forms of the Rarita-Schwinger equation of motion, given by $R_\mu = 0$

$$\Leftrightarrow \gamma^\mu \psi_{\mu\nu} = 0$$

$$\Leftrightarrow \psi_{\mu\nu} + \frac{1}{2} \gamma_5 \varepsilon_{\mu\nu\rho\kappa} \psi_{\rho\kappa} = 0 \qquad (6.9)$$

we find the final result to be

$$[\delta_{\varepsilon_1}, \delta_{\varepsilon_2}] \psi_\mu = + 2i \bar{\varepsilon}_2 \gamma_c \varepsilon_1 \partial_c \psi_\mu$$

$$+ \partial_\mu \left(- \frac{i}{2} \bar{\varepsilon}_2 \gamma^c \varepsilon_1 \psi_c + \frac{i}{4} \bar{\varepsilon}_1 \gamma_R \varepsilon_2 \sigma^{ab} \gamma^R \gamma_b \psi_a - (1 \leftrightarrow 2) \right)$$

$$(6.10)$$

This is the required result: a translation and a gauge transformation on ψ_μ.

The reader can verify that the transformations of equation (6.4) with the values of $\delta_1 = \delta_3 = 0$, $\delta_2 = -2$ do indeed leave the equations of motion of $h_{\mu\nu}$ and $\psi_{\mu\alpha}$ invariant.

Having obtained an irreducible representation of supersymmetry carried by the fields $h_{\mu\nu}$ and $\psi_{\mu\alpha}$ when subject to their field equations we can now find the algebraically on-shell Lagrangian. The action (Freedman, van Niuwenhuizen & Ferrara 1976, Deser & Zumino 1977) from which the field equations of equation (6.2) follow, is

$$A = \int \left(- \frac{1}{2} h^{\mu\nu} E_{\mu\nu} - \frac{i}{2} \bar{\psi}^\mu R_\mu \right) d^4x \qquad (6.11)$$

It is invariant under the transformations of equation (6.4) provided we adopt the values for the parameters δ_1, δ_2 and δ_3 found above. This invariance holds without use of the field equations, as it did in the Wess-Zumino case.

We now wish to find a linearized formulation which is built from fields which carry a representation of supersymmetry without imposing any restrictions, namely find the auxiliary fields. As a guide to their number we can apply our fermi-bose counting rule which, since the algebra contains gauge transformations, applies only to the gauge invariant states. On-shell $h_{\mu\nu}$ has two helicities and so does $\psi_{\mu\alpha}$, however off-shell $h_{\mu\nu}$ contributes $\frac{4 \times 5}{2} = 10$ degrees of freedom minus 4 gauge degrees of freedom giving 6 bosonic degrees of freedom. On the other hand, off-shell $\psi_{\mu\alpha}$ contributes $4 \times 4 = 16$ degrees of freedom minus 4 gauge degrees of freedom, giving 12 fermionic degrees of freedom. Hence, the auxiliary fields must contribute 6 bosonic degrees of freedom. If there are n auxiliary fermions there must be $4n + 6$ bosonic auxiliary fields.

Let us assume that a minimal formulation exists, that is there are no auxiliary spinors. Let us also assume that the bosonic auxiliary fields occur in the Lagrangian as squares without derivatives (like F and G) and so are of dimension two. Hence we have 6 bosonic auxiliary fields; it only remains to find their Lorentz character and transformations. We will assume that they consist of a scalar M, a pseudo-scalar N and a pseudo-vector b_a, rather than an antisymmetric tensor or 6 spin 0 fields. We will give the motivating arguments for this later.

The transformations of the fields $h_{\mu\nu}$, $\psi_{\mu a}$, M, N and b_μ must reduce on-shell to the on-shell transformations found above. This restriction, dimensional arguments and the fact that if the auxiliary fields are to vanish on-shell they must vary into field equations gives the transformations to be

$$\delta h_{\mu\nu} = \frac{1}{2} (i \bar{\varepsilon} \gamma_\mu \psi_\nu + i \bar{\varepsilon} \gamma_\nu \psi_\mu)$$

$$\delta \psi_{\mu\alpha} = - 2 \sigma^{ab} \partial_a h_{b\mu} \varepsilon - \frac{1}{3} \gamma_\mu (M + \gamma_5 N) \varepsilon + b_\mu \gamma_5 \varepsilon$$

$$+ \delta_6 \gamma_\mu \not{b} \gamma_5 \varepsilon$$

$$\delta M = \delta_4 \, i \bar{\varepsilon} \gamma \cdot R$$

$$\delta N = \delta_5 \, i \bar{\varepsilon} \gamma_5 \gamma \cdot R$$

$$\delta b_\mu = + \delta_7 \, i \bar{\varepsilon} \gamma_5 R_\mu + \delta_8 \, i \bar{\varepsilon} \gamma_5 \gamma_\mu \gamma \cdot R \qquad (6.12)$$

The parameters δ_4, δ_5, δ_6, δ_7 and δ_8 are determined by the restriction that the above transformations of equation (6.12) and the gauge transformations of equation (6.1) should form a closed algebra.

For example, the commutator of two supersymmetries on M gives

$$[\delta_{\varepsilon_1}, \delta_{\varepsilon_2}]\, \check{M}$$

$$= -4_4\, i\, \bar{\varepsilon}_2 \gamma^\mu \varepsilon_1 \partial_\mu M + 8 i\, \bar{\varepsilon}_2 \sigma^{\mu\nu} \gamma_5 \varepsilon_1 \delta_4 (1 + 3\delta_6)\, \partial_\mu b_\nu \tag{6.13}$$

which is the required result provided $\delta_4 = -\frac{1}{2}$ and $\delta_6 = -\frac{1}{3}$. Carrying out the commutator of two supersymmetries on all fields we find a closing algebra provided

$$\delta_4 = -\frac{1}{2},\ \delta_5 = -\frac{1}{2},\ \delta_6 = ,\ \delta_7 = \frac{3}{2} \text{ and } \delta_8 = -\frac{1}{2} \tag{6.14}$$

We henceforth adopt these values for the parameters. An action (Stelle and West 1978, Ferrara & van Nieuwenhuizen 1978) which is constructed from the fields $h_{\mu\nu}$, $\psi_{\mu a}$, M, N and b_μ and is invariant under the transformations (Stelle & West 1978, Ferrara and van Nieuwenhuizen 1978) of equation (6.12) with the above values of the parameters is

$$A = \int d^4x\ \{-\frac{1}{2} h_{\mu\nu}\ E^{\mu\nu} - \frac{i}{2} \bar{\psi}_\mu R^\mu$$
$$- \frac{1}{3}(M^2 + N^2 - b_\mu^2)\} \tag{6.15}$$

This is the action of linearized N = 1 supergravity and upon elimination of the auxiliary field M, N and b_μ it reduces to the algebraically on-shell Lagrangian of equation (6.11).

7 NON-LINEAR N = 1 SUPERGRAVITY - THE NOETHER COUPLING TECHNIQUE

In the previous section we constructed a linearized off-shell representation of supergravity and its corresponding supergravity Lagrangian. We now wish to find the non-linear theory constructed from fields carrying a corresponding representation of local supersymmetry. This result is achieved by the Noether coupling technique, a technique which builds up the result order by order in the gravitation coupling constant, κ. We first demonstrate this procedure in the simpler model of ordinary Yang-Mills theory where the role of gravitational coupling constant is played by the Yang-Mills coupling constant, g.

The linearized (free) SU(2) Yang-Mills theory is given in terms of the fields $\underline{A}_\mu$ (= $(A_{1\mu}\ A_{2\mu}\ A_{3\mu})$) which transform under the constant (rigid) SU(2) transformation with parameter $\underline{T}$ as

$$\delta\underline{A}_\mu = \underline{T} \times \underline{A}_\mu \tag{7.1}$$

and under local abelian transformation

$$\delta\underline{A}_\mu = \partial_\mu \underline{\Lambda}(x) \tag{7.2}$$

The linearized Lagrangian invariant under the above transformations is

$$L_o = -\frac{1}{4}\, \overset{o}{\underline{F}}_{\mu\nu o}\, \overset{o}{\underline{F}}{}^{\mu\nu} \tag{7.3}$$

where $\overset{o}{\underline{F}}_{\mu\nu} = \partial_\mu \underline{A}_\nu - \partial_\nu \underline{A}_\mu$

We now wish to find the non-linear Yang-Mills theory which is invariant under local Yang-Mills transformations. We do this in the following series of steps.

Step 1

Let the rigid SU(2) transformation of equation (7.1) become local, i.e. $\underline{T} = \underline{T}(x)$. Now, L_o is no longer invariant under

$$\delta A_\mu = \underline{T}(x) \times \underline{A}_\mu \tag{7.4}$$

giving the result

$$\delta L_o = + \partial_\mu \underline{T} \cdot \underline{j}^\mu \tag{7.5}$$

where

$$\underline{j}^\mu = - \underline{A}_\nu \times \overset{o}{\underline{F}}{}^{\mu\nu} \tag{7.6}$$

However, consider the Lagrangian L_1 given by

$$L_1 = L_o - \frac{1}{2g}\, \underline{A}_\mu \cdot \underline{j}^\mu \tag{7.7}$$

In fact L_1 is invariant to order g^0 provided we combine the local transformation $\underline{T}(x)$ with the local transformation $\underline{\Lambda}(x)$ with the identification $\underline{\Lambda}(x) = \frac{1}{g} T(x)$. That is the initially separate local and rigid transformations of the linearized theory become knitted together into the one local transformation given by

$$\delta A_\mu = \frac{1}{g} \partial_\mu \underline{T}(x) + \underline{T}(x) \times \underline{A}_\mu(x) \tag{7.8}$$

The first term in the transformation of δA_μ yielding in the last term in L_1 just such a term which cancels the unwanted variation of L^0.

Step Two

We now continue with this process of amending Lagrangian and transformations order by order in g until we have an invariant Lagrangian.

The variation of L_1 under the transformation of equation (7.8) is of order g and is given by

$$\delta L_1 = - g(\underline{A}_\mu \times \underline{A}_\nu) \cdot (\underline{A}^\nu \times \partial^\mu \underline{T}) \tag{7.9}$$

A Lagrangian invariant to order g is

$$L_2 = L_1 + \frac{g}{4}(\underline{A}_\mu \times \underline{A}_\nu) \cdot (\underline{A}^\nu \times \underline{A}^\mu)$$

$$= - \frac{1}{4} F^2_{\mu\nu} \tag{7.10}$$

where

$$F_{\mu\nu} = \quad (\partial_\mu \underline{A}_\nu - \partial_\nu \underline{A}_\mu - g\underline{A}_\mu \times \underline{A}_\nu) \tag{7.11}$$

In fact, the Lagrangian L_2 is invariant under the transformation of equation (7.8) to all orders in g and so represents the final answer, the well known Lagrangian of Yang-Mills theory. The commutator of two transformations on $\underline{A}_\mu$ is

$$[\delta_{T_1}, \delta_{T_2}] \underline{A}_\mu = \underline{T}_2 \times (\frac{1}{g} \partial_\mu \underline{T}_1 + \underline{T}_1 \times \underline{A}_\mu)$$

$$- (1 \leftrightarrow 2)$$

$$= \frac{1}{g} \partial_\mu T_{12} + T_{12} \times \underline{A}_\mu = \delta_{T_{12}} \underline{A}_\mu \tag{7.12}$$

where $T_{12} = \underline{T}_2 \times \underline{T}_1$ and so represents a closing algebra, as it must. Hence we have a set of local transformations of equation (7.8), which have a closing algebra, and are an invariance of the Lagrangian L_2.

We can apply exactly the same technique of Noether coupling to obtain the non-linear supergravity theory. We only outline this calculation which is somewhat lengthy.

The Lagrangian given in equation (6.15), which we now denote L^O, is invariant under the rigid supersymmetry group transformations of equation (6.12) and also the, at this stage unrelated, local gauge transformations of equation (6.1). The reader will notice the similarity with the linearized Yang-Mills case. The first step is to make the rigid supersymmetry transformations local, i.e. set $\varepsilon = \varepsilon(x)$ in equation (6.12). However, L^O is then no longer invariant, but, since it is invariant when ε is a constant, it must vary as follows.

$$\delta L^O = (\partial_\mu \bar{\varepsilon})\, j^\mu$$

where j_μ is an object which is bilinear in the fields $h_{\mu\nu}$, $\psi_{\mu\alpha}$, M, N and b_μ. Consider now a Lagrangian, L_1 given by

$$L_1 = L_o - \frac{\kappa}{4} \bar{\psi}^\mu j_\mu \tag{7.13}$$

where κ is the gravitational constant. The Lagrangian L_1 is invariant to order κ^O provided we combine the now local supersymmetry transformations of equation (6.12) with a local abelian Rarita-Schwinger gauge transformation of equation (6.1) with parameter $\eta(x) = \frac{2}{\kappa}\varepsilon(x)$. That is, we make a transformation

$$\delta\psi_\mu = \frac{2}{\kappa}\partial_\mu \varepsilon - 2\partial_a h_{b\mu}\sigma^{ab}\varepsilon - \frac{1}{3}\gamma_\mu (M + \gamma_5 N)\varepsilon$$
$$+ \gamma_5 (b_\mu - \frac{1}{3}\gamma_\mu \not{b})\varepsilon \tag{7.14}$$

the remaining fields transforming as before except ε is now space-time dependent.

As in the Yang-Mills case the two invariances of the linearized action become knitted together to form one transformation. The addition of the term $-\frac{\kappa}{4}\bar{\psi}^\mu j_\mu$ to L_o does the required job; its variation is

$$- \frac{\kappa}{2} \left(\frac{2}{\kappa}\right) \partial_\mu \bar{\varepsilon} j^\mu + \text{terms of order } \kappa'$$

which we are not at the moment concerned with. (Note $j_{\mu\alpha}$ is linear in $\psi_{\mu\alpha}$.)

In fact, one can carry the Noether procedure out in the context of pure gravity where one finds at the linearized level the rigid translation

$$\delta h_{\mu\nu} = \zeta^\nu \partial_\nu h_{\mu\nu} \tag{7.15}$$

and the local gauge transformation

$$\delta h_{\mu\nu} = \partial_\mu \xi_\nu + \partial_\nu \xi_\mu . \tag{7.16}$$

These become knitted together at the first stage of the Noether procedure to give

$$\delta h_{\mu\nu} = \frac{1}{\kappa} \partial_\mu \xi_\nu + \frac{1}{\kappa} \partial_\nu \xi_\mu + \zeta^\nu \partial_\nu h_{\mu\nu} . \tag{7.17}$$

This variation of $h_{\mu\nu}$ contains the first few terms of an Einstein general coordinate transformation of the veirbein which is given in terms of $h_{\mu\nu}$ by

$$e_\mu{}^a = \eta_\mu{}^a + \kappa h_\mu{}^a \tag{7.18}$$

We proceed in a similar way to the Yang-Mills case. We obtain order by order in κ an invariant Lagrangian by adding terms to the Lagrangian and in this case also adding terms to the transformations of the fields. For example, if we added a term to $\delta\psi_\mu$ say $\delta\bar{\psi}_\mu = \ldots + \bar{\varepsilon} X_\mu \kappa$ then from the linearized action we receive a contribution $- i\kappa \bar{\varepsilon} X_\mu R^\mu$ upon variation of ψ_μ. It is necessary at each step (order of κ) to check that the transformations of the fields form a closed algebra. In fact, any ambiguities that arise in the procedure are resolved by the demand of a closing algebra.

The final set of transformations (Stelle & West 1978, Ferrara & van Nieuwenhuizen 1978) are:

$$\delta e_{\mu a} = i\kappa\, \bar{\varepsilon}\, \gamma_a\, \psi_\mu$$

$$\delta\psi_\mu = 2\kappa^{-1} D_\mu(\omega(e,\psi))\varepsilon + \gamma_5 (b_\mu - \tfrac{1}{3}\gamma_\mu \not{b})\, \varepsilon$$

$$- \tfrac{1}{3}\gamma_\mu\, (M + \gamma_5 N)\, \varepsilon$$

$$\delta M = -\tfrac{i}{2} e^{-1}\, \bar{\varepsilon}\, \gamma_\mu R^\mu - \tfrac{i\kappa}{2}\, \bar{\varepsilon}\, \gamma_5\, \psi_\nu\, b^\nu - i\kappa\, \bar{\varepsilon}\, \gamma^\nu .\, \psi_\nu\, M$$

$$+ \tfrac{i\kappa}{2}\, \bar{\varepsilon}\, (M + \gamma_5 N)\, \gamma^\mu\, \psi_\mu$$

$$\delta N = -\tfrac{i}{2} e^{-1}\, \bar{\varepsilon}\, \gamma_5\, \gamma_\mu R^\mu + \tfrac{i\kappa}{2}\, \bar{\varepsilon}\, \psi_\nu\, b^\nu - i\kappa\, \bar{\varepsilon}\, \gamma^\nu\, \psi_\nu\, N$$

$$- \tfrac{i\kappa}{2}\, \bar{\varepsilon}\, \gamma_5\, (M + \gamma_5 N)\, \gamma^\mu\, \psi_\mu$$

$$\delta b_\mu = \tfrac{3i}{2} e^{-1}\, \bar{\varepsilon}\, \gamma_5\, (g_{\mu\nu} - \tfrac{1}{3}\gamma_\mu\, \gamma_\nu)\, R^\nu + i\kappa\, \bar{\varepsilon}\, \gamma^\nu\, b_\nu\, \psi_\mu$$

$$- \tfrac{i\kappa}{2}\, \bar{\varepsilon}\, \gamma^\nu\, \psi_\nu\, b_\mu - \tfrac{i\kappa}{2}\, \bar{\psi}_\mu\, (M + \gamma_5 N)\, \gamma_5\, \varepsilon$$

$$- \tfrac{i\kappa}{4}\, \varepsilon_\mu{}^{bcd}\, b_b\, \bar{\varepsilon}\, \gamma_5\, \gamma_c\, \psi_d \qquad (7.19)$$

where

$$R^\mu = \varepsilon^{\mu\nu\rho\kappa}\, \gamma_5\, \gamma_\nu\, D_\rho(\omega(e,\psi))\psi_\kappa .$$

$$D_\mu(\omega(e,\psi)) = \partial_\mu + \omega_{\mu\, ab} \frac{\sigma^{ab}}{2}$$

and

$$\omega_{\mu ab} = \tfrac{1}{2} e_a^\nu\, (\partial_\mu\, e_{b\nu} - \partial_\nu\, e_{b\mu}) - \tfrac{1}{2} e_b^\nu\, (\partial_\mu\, e_{a\nu}$$

$$- \partial_\nu\, e_{a\mu}) - \tfrac{1}{2} e_a{}^\rho\, e_b{}^\sigma\, (\partial_\rho e_{\sigma c} - \partial_\sigma e_{\rho c})\, e_\mu^c$$

$$+ \tfrac{i\kappa^2}{4}\, (\bar{\psi}_\mu\, \gamma_a\, \psi_b + \bar{\psi}_a\, \gamma_\mu\, \psi_b - \bar{\psi}_\mu\, \gamma_b\, \psi_a) \qquad (7.20)$$

They form a closing algebra, the commutator of two supersymmetries on any field being

$$[\delta_{\varepsilon_1},\delta_{\varepsilon_2}] = \delta_{\substack{\text{supersy-}\\ \text{mmetry}}}(-\kappa\,\xi^{\nu}\,\psi_{\nu}) + \delta_{\substack{\text{general}\\ \text{coordinate}}}(2\xi_{\mu})$$

$$+ \delta_{\substack{\text{Local}\\ \text{Lorentz}}}\left(-\frac{2\kappa}{3}\,\varepsilon_{ab\lambda\rho}\,b^{\lambda}\,\xi^{\rho} + 4\,\frac{i\kappa}{3}\,\bar{\varepsilon}_1\,\sigma_{ab}\,(M + \gamma_5 N)\,\varepsilon_2\right) \tag{7.21}$$

where

$$\xi_{\mu} = i\,\bar{\varepsilon}_2\,\gamma_{\mu}\,\varepsilon_1.$$

The transformations of equation (7.19) leave invariant the action

$$A = \int d^4x\,\{\frac{e}{2\kappa^2}R - \frac{i}{2}\bar{\psi}_{\mu}R^{\mu} - \frac{1}{3}e(M^2 + N^2 - b_{\mu}b^{\mu})\} \tag{7.22}$$

where

$$R = R_{\mu\nu}{}^{ab}\,e^{\mu}_{a}\,e^{\nu}_{b}$$

and

$$R_{\mu\nu}{}^{ab}\,\frac{\sigma_{ab}}{2} = [D_{\mu},D_{\nu}]$$

A simple proof of this invariance is given using the 1.5 order formalism (see van Nieuwenhuizen 1981). The reader may find it a useful exercise to verify the first few steps in obtaining the above answer.

The auxiliary fields M, N and b_{μ} may be eliminated to obtain the non-linear algebraically on-shell Lagrangian which was the form in which supergravity was originally found (Freedman, van Nieuwenhuizen & Ferrara 1976, Deser & Zumino 1977).

The above account is the usual description of the Noether procedure. However, a more pedagogically sound procedure is to work with the transformations alone. That is to build up a set of closed transformations order by order in κ, using the same knitting of linearized transformations to eliminate $\partial_{\mu}\varepsilon$ terms in the closure of the linearized algebra. Of course, to obtain the local algebra one must know what local transformations are allowed to exist in the algebra. However, only those transformations are allowed which make an appearance, albeit in an abelian guise, at the linearised level. The resulting closed non-linear transformations will then dictate the form of the non linear action.

8 TWO METHODS OF FINDING AUXILIARY FIELDS

In both the Wess-Zumino model and N = 1 supergravity we assumed that the theories had no auxiliary fermions and used the fermi-bose counting rule to give us the number of bosonic auxiliary fields. In the case of N = 1 supergravity we then guessed the Lorentz character of the auxiliary fields. Unfortunately, in more sophisticated supersymmetry theories there are auxiliary fermions and the fermi-bose counting rule provides only a very rough guide to their auxiliary field structure. We now examine two methods of finding auxiliary fields. Although they do not provide ways of finding auxiliary fields for all supersymmetric theories, they are much more useful than the fermi-bose counting rule.

8.1 Currents and auxiliary fields

In Einstein's theory of gravity when coupled to matter the field equation for the graviton $h_{\mu\nu}$ reads

$$R_{\mu\nu} - \frac{1}{2} g_{\mu\nu} R = \kappa\, \theta_{\mu\nu} \qquad (8.1)$$

where $R_{\mu\nu}$ is the Ricci tensor, $R = R_{\mu}{}^{\mu}$ and $\theta_{\mu\nu}$ isthe energy momentum tensor.

It has been shown by Ferrara and Zumino (1975) that in several models of N = 1 supersymmetry the energy-momentum $\theta_{\mu\nu}$ (with suitable improvements) lies in a super-multiplet with the supercurrent, $j_{\mu\alpha}$, the chiral current, $j_{\mu}^{(5)}$ and, if the model is not conformally invariant, two additional fields, A (scalar) and B (pseudo-scalar).

Since in supergravity the right hand side of the equation of motion for $h_{\mu\nu}$ lies in a supermultiplet ($\theta_{\mu\nu}$, $j_{\mu\alpha}$, $j_{\mu}^{(5)}$, A, B), it follows that the equations of motion themselves must lie in this multiplet. As there is one equation of motion for each field in supergravity, the fields of supergravity must lie in a supermultiplet dictated by the multiplet given above. That is we must have fields $h_{\mu\nu}$, $\psi_{\mu\alpha}$, b_{μ}, M and N. These are the fields of supergravity discussed in section (6).

This discussion is closely connected to the Noether coupling procedure of section (7). According to the above argument the coupling of a linearized supersymmetric action A^L to super-matter must contain the following terms at order κ^1

$$A^1 = \int d^4x \, \{\kappa\,(h^{\mu\nu}\, \theta_{\mu\nu} + \bar{\psi}^{\mu\alpha}\, j_{\mu\alpha} + b^{\mu}\, j_{\mu}^{(5)} + AM + BN)\} \qquad (8.2)$$

The second term in equation (8.2) is the term required by the Noether coupling procedure to gain invariance at order κ^{0}, the other terms will follow from the next step, invariance at order κ^{1}. That $j_{\mu\alpha}$ is a conserved current is clear from the above argument, but it also follows from the Noether procedure. The action $A^{L} + A^{1}$ is invariant to order κ^{0}. This applies on-shell when the only surviving term is $\int d^4x\, \bar{\varepsilon}^{\alpha}(x)\, \partial^{\mu} j_{\mu\alpha}$, hence $\partial^{\mu} j_{\mu\alpha} = 0$. The conservation of $\theta_{\mu\nu}$ similarly follows from invariance under general coordinate transformations. The action A^{1} is invariant under rigid supersymmetry and hence knowing the transformations of $\theta_{\mu\nu}$, $j_{\mu\alpha}$, $j_{\mu}^{(5)}$, A and B we can deduce the rigid transformation for $h_{\mu\nu}$, $\psi_{\mu\alpha}$, b_{μ}, M and N.

It was found after these lectures were given that there existed an alternative current multiplet to the one found by Ferrara and Zumino. This lead to an alternative set of auxiliary fields for supergravity (Sohnius & West 1981a).

8.2 Ghost counting

At the end of section (4) we noted that the representations of supersymmetry are determined entirely by the supersymmetry group itself,without regard to any dynamical applications. As such , any claim about which fields are the auxiliary fields in a given representation is a dynamically dependent statement. It is this observation when coupled with the knowledge of all on-shell representations that provides the basis for this method.

Consider a supersymmetric theory in which the physical bosonic fields are denoted generically by h (h may have Lorentz indices which we will not display). The usual dynamical Lagrangian for such a theory is of the generic form

$$L = h\partial\, \partial h + \text{supersymmetric extensions} \qquad (8.3)$$

Let us make a d'Alembertian insertion into the above Lagrangian and add the result to L, obtaining the Lagrangian

$$L^{G} = h\partial\, \partial h + \frac{1}{m^2} h\partial\, \partial^2\, \partial h + \text{supersymmetric extension} \qquad (8.4)$$

The action $A^{G} = \int L^{G}\, d^4x$ is invariant under supersymmetry as the insertion of ∂^2 is a supersymmetric operation (i.e., $[P_\mu^2, Q_\alpha] = 0$). As such the on-shell states of L^G must lie in supermultiplets.

Consider an auxiliary field H that appears in L as H^2 and so will not contribute to the on-shell states of L. In L^G, however, it will appear as $H\partial^2 H \frac{1}{m^2} + H^2$ and so will contribute to the supersymmetric sets of on-shell states of L^G. Since we know all possible supersymmetric on-shell representations we therefore gain some information about the auxiliary fields of L. Most of the power of this technique derives from the fact that L^G contains exactly the same fields as L and so it is the original set of fields that lead to all the on-shell states of L^G. It is important to stress that we are not suggesting some physical significance for L^G, only that it is useful to consider in the above context. Clearly this analysis can only suggest some auxiliary field sets and not prove that they lead to a closing algebra. However it is a technique which has proved useful in the finding of the auxiliary fields for several theories.

Let us apply the above argument to gain information about the auxiliary fields of N = 1 supergravity. We again assume there are no auxiliary fermions , implying that there are six bosonic auxiliary fields. If we add to the linearized supergravity Lagrangian of equation (6.15) the same Lagrangian with the d'Alembertian insertion. The on-shell states of the resulting Lagrangian will divide into a massive and massless representation of supersymmetry. Calculation shows (Ferrara, Grisaru & van Nieuwenhuizen 1978) that the field $h_{\mu\nu}$ leads to one massless spin 2 state, one massive spin 2 ghost state and one massive spin 0 state, whereas the field $\psi_{\mu\alpha}$ leads to one massless spin $^3/_2$ state, two massive spin $^3/_2$ ghost states and two massive spin $^1/_2$ states. These states when supplemented by those coming from the auxiliary fields must fall into supersymmetric multiplets of on-shell states.

Now the massive representation which contains a spin 2 and no higher spin is $(2, {}^3/_2, {}^3/_2, 1)$. Hence it could account for our one massive spin 2 and 2 massive spin $^3/_2$'s, but the spin 1 must come from the auxiliary fields. Hence they cannot be a mixture of 6 scalars or pseudo scalars. Leaving only the two possibilities of 2 scalars (pseudo scalars) and 1 vector or one antisymmetric tensor.

The former possibility can clearly provide the one spin 1 as the vector will generate one spin 1 and 1 spin 0 on-shell state. The two

2 scalars (pseudo scalars) will provide two spin 0's on-shell. These 3 spin 0 states with the spin 0 state arising from $h_{\mu\nu}$ can go together with the two spin $\frac{1}{2}$ states arising from the $\psi_{\mu\alpha}$ to form two massive $(\frac{1}{2},0,0)$ multiplets. A more careful analysis which takes account of parity shows that this set must be one axial vector b_μ, one scalar M and one pseudoscalar N. This analysis only counts states and does not prove that this set is actually realized. To do this one must show they lead to an off-shell closing algebra as was done in section (6).

The possibility of one antisymmetric tensor can be ruled out as it will not lead to states that will fit into multiplets with on-shell states we already have.

It is important to realize in this analysis that two assumptions have been made; firstly that there are no auxiliary fermions and secondly that the auxiliary fields do not possess gauge degrees of freedom. In fact there are known to be two other possible auxiliary fields set for N = 1 supergravity; their existence corresponds to the two assumptions made in the above argument. One set has auxiliary fermions whereas the other set has auxiliary fields with gauge degrees of freedom. A more thorough discussion of the ghost counting technique is given in Sohnius, Stelle & West (1980a).

9 SUMMARY

The above discussion can be summarized as follows. The on-shell representations of supersymmetry can be found by a straightforward application of Wigner's method of induced representations. We can find sets of fields and their transformations which, when subject to their field equations, form a realization of the on-shell states of supersymmetry. From these fields we can find Lagrangians which are invariant under the on-shell transformations without use of field equations. The drawback being that the fields do not carry a representation of the supersymmetry algebra off-shell.

The problem is resolved by the discovery of the auxiliary fields which enable one to find an invariant Lagrangian constructed from fields whose transformations form a realization of supersymmetry without the use of their equations of motion. There is at present no systematic way to find the auxiliary fields but two useful methods were given in section (8).

When one is dealing with theories whose interaction is

determined by some gauge principle it is possible, having found the linearized theory by the above method to obtain the non-linear theory by the technique of Noether coupling described in section (7).

The discussion in this contribution only corresponds to my first lecture and space does not permit a discussion of my other two lectures. The second lecture was addressed to the question of how superspace could help to find auxiliary fields and the third lecture considered the role played by dimensional reduction in particular dimensional reduction by Legendre transformations in these problems. A discussion of this material which is along the lines of that given for the first lecture can be found in the following references as well as the references they contain:- Stelle and West 1980, Sohnius, Stelle & West 1980a, 1980b, Gates, Stelle and West 1980, Sohnius, Stelle and West 1980c and Scherk 1978.

REFERENCES

Barut, A. and Raczka, R. (1977) Theory of Group Representations and Applications pub. P.W.N. Polish Scientific Publishers.

Brink, L., Schwarz, J. and Scherk, J. (1977) Supersymmetric Yang-Mills Theories, Nucl. Phys. B121, p.77.

Coleman, S. and Mandula, J. (1967). All possible symmetries of the S matrix. Phys. Rev. 159, p.1251.

Deser, S. and Zumino, B. (1977). Consistent Supergravity, Phys. Lett. 62B, p.335.

Ferrara, S. and Zumino, B. (1975) Transformation Properties of the Supercurrent, Nucl. Phys. B87, p.207.

Ferrara, S., Grisaru, M. and van Nieuwenhuizen, P. (1978) 'Poincaré and Conformal Supergravity Models with closed algebra', Nucl. Phys. B138, p.430.

Ferrara, S. and van Nieuwenhuizen, P. (1978) The Auxiliary fields of Supergravity, Phys. Lett. B74, p.333.

Ferrara, S. (1981). Contribution to this book.

Freedman, D.Z. (1979). Irreducible Representations of Supersymmetry. In: Recent Developments in Gravitation, Cargesse 1978 eds. M. Levy and S. Deser. Plenum Press.

Freedman, D.Z., van Nieuwenhuizen, P. and Ferrara, S. (1976). Progress towards a theory of Supergravity, Nucl. Phys. B117, p.333.

Gates, J., Stelle, K. and West, P. (1980) Algebraic origins of Superspace constraints in Supergravity , Nucl. Phys. B169, p.347.

Gol'fand, Y.A. and Likhtman, E.P. (1971), J.E.T.P. letters, 13, p.323.

Gliozzi, F., Scherk, J. and Olive, D. (1978). Supersymmetry, Supergravity and the Dual Spinor Model, Nucl. Phys. B122, p.253.

Haag, R, Lopuszanki, J. and Sohnius, M (1975). All Possible Generators of Supersymmetries of the S-Matrix, Nucl. Phys. B88, p.61.

van Nieuwenhuizen, P, Ferrara, S and Grisaru, M.T. (1978). Poincaré and Conformal Supergravity Models with Closed Algebras, Nucl. Phys. B138, p.430.

van Nieuwenhuizen, P. (1981) Supergravity Physics Reports C68 No. 4.

Ogievetsky, V. and Sokatchev, E. (1977). On a Vector Superfield Generated by the Supercurrent, Nucl. Phys. B124, p.309.

Salam, A. and Strathdee, J. (1974), Nucl. Phys. B80, p.499.

Scherk, J. (1978). Extended Supersymmetry and Extended Supergravity, in Recent Developments in Gravitation Cargese Lectures 1978 ed. Levy and Deser. Plenum Press.

Stelle, K. and West, P. (1978), Minimal Auxiliary fields for Supergravity, Phys. Lett. 74B, p.330.

Stelle, K. and West, P. (1980). Realizing the Supersymmetry Algebra, in Second Nuffield Quantum Gravity Conference, eds. Isham, C., Penrose, R. and Sciama, D. O.U.P.

Sohnius, M., Stelle, K. and West, P. (1980a). Supersymmetric Yang-Mills Theories. In: Unification of the Fundamental Interactions eds. S. Ferrara, J. Ellis and P. van Nieuwenhuizen. Plenum.

Sohnius, M., Stelle, K. and West, P. (1980b) Representations of Extended of Supersymmetry. In: Superspace and Supergravity. eds. S.W. Hawking and M. Roček. Cambridge University Press.

Sohnius, M., Stelle, K and West, P. (1980c) Dimensional reduction by Legendre transformation generates off-shell Supersymmetric Yang-Mills theories, Nucl. Phys. B173, p.127.

Sohnius, M. and West, P. (1981a) Alternative Minimal Off-shell version of N = 1 Supergravity, Phys. Lett. 105B, p.353.

Sohnius, M. and West, P. (1981). Conformal Invariance in N=4 Supersymmetric Yang-Mills Theory, Phys. Lett. 100B, p.245.

Taylor, J.G. (1981) Contribution to this book.

Wess, J. and Zumino, B. (1974). Supergauge transformations in four dimensions, Nucl. Phys. B70, p.39.

Wigner, E.P. (1939). On Unitary Representations of the Inhomogeneous Lorentz Group, Ann. of Maths, 40, p.149.

APPENDIX

$$\eta_{\mu\nu} = (-1,+1,+1,+1),\ \sigma_{\mu\nu} = \frac{1}{4}[\gamma_\mu, \gamma_\nu],\ \gamma_5^2 = -1 \qquad \text{(A 1)}$$

Let $C_{\alpha\beta} = -C_{\beta\alpha}$ be the charge conjugation matrix, i.e.

$$C^{-1}\gamma_\mu C = -\tilde{\gamma}_\mu \qquad \text{(A 2)}$$

Then, the matrices C, $\gamma_5 C$, $\gamma_\mu\gamma_5 C$ are anti-symmetric whereas $(\gamma_\mu C)$, $(\sigma_{\mu\nu}C)$ are symmetric.

The Majorana condition is

$$\chi_\alpha = C_{\alpha\beta}\bar{\chi}^\beta \qquad \text{(A 3)}$$

and all spinors anticommute, i.e. $\chi_\alpha \chi_\beta = -\chi_\beta \chi_\alpha$ then

$$\bar{\chi}\,\Gamma\,\psi = \bar{\chi}^\beta(\Gamma C)_{\beta\delta}\,\bar{\psi}^\delta = -\bar{\psi}^\delta(\tilde{\Gamma C})_{\delta\beta}\bar{\chi}^\beta = \pm\,\bar{\psi}\,\Gamma\,\chi \qquad \text{(A 4)}$$

where we get - if $\Gamma = \gamma_\mu$, $\sigma_{\mu\nu}$ and + if $\Gamma = 1, \gamma_5, \gamma_\mu\gamma_5$.

An example of the anticommuting nature of spinors is provided by the commutator of two supersymmetries; using the supersymmetry algebra

APPENDIX

we find

$$[\bar{\varepsilon}_1 Q, \bar{\varepsilon}_2 Q] = -2 \; \bar{\varepsilon}_2 \gamma_\mu \varepsilon_1 P_\mu \tag{A 5}$$

The Fierz identity is

$$\delta_\alpha{}^\gamma \delta_\beta{}^\delta = \frac{1}{4} (\gamma_R)_\alpha{}^\delta (\gamma^R)_\beta{}^\gamma$$

where $\gamma_R = 1, \gamma_5, \gamma_\mu, \gamma_5\gamma_\mu, \sqrt{2}\ \sigma_{\mu\nu}$ and $\gamma^R =$

$$1, -\gamma_5, \gamma^\mu, \gamma_5\gamma^\mu, -\sqrt{2}\ \sigma^{\mu\nu} \tag{A 6}$$

An example of its application is given by

$$\chi_\alpha \bar{\psi} \phi = -\frac{1}{4} (\gamma_R \phi)_\alpha \quad \bar{\psi}\gamma^R \chi \phi \tag{A 7}$$

SIX LECTURES AT THE TRIESTE 1981 SUMMERSCHOOL ON SUPERGRAVITY

P. van Nieuwenhuizen
Institute for Theoretical Physics
State University of New York at Stony Brook
Stony Brook, Long Island, New York, 11794

Contents

I. Elementary simple supergravity

Supergravity is the only known example of a Lagrangian field theory with a local fermionic gauge invariance. The gauge parameter is a spin ½ spinor. Since in $d\neq 4$ dimensions there are no dotted and undotted spinors, we consider in these notes also in d=4 the spinors as four-component, although one also often uses in d=4 the two component (dotted and undotted) formalism of van der Waerden. Thus the gauge parameters are $\varepsilon^a(x)$, a=1, 4 and are anticommuting, as suggested by the spin-statistics connection.

The local symmetry of which $\varepsilon^a(x)$ is the parameter, must transform bosons into fermions, and vice-versa, in order that the variation of a boson B (or fermion F) has the same statistics as the field itself. As explained in detail in [1], this implies

$$\delta B = F\epsilon, \qquad \delta F = (\partial_\mu B)\,\epsilon \tag{1}$$

The presence of the derivative is due to the difference of dimensions of bosons and fermions, which, in turn, is due to the fact that bosons have two, but fermions only one, derivative in their kinetic actions.

Taking the commutator of two <u>global</u> supersymmetry transformations yields always the same result, namely

$$\left[\delta(\epsilon_1), \delta(\epsilon_2)\right] B = \tfrac{1}{2}\,\bar{\epsilon}_2 \gamma^\mu \epsilon_1\, \partial_\mu B \quad + \text{more} \tag{2}$$

For example, in the photon-neutrino system, $\delta A_\mu = -\tfrac{1}{2}\,\bar{\varepsilon}\gamma_\mu \lambda$ and $\delta\lambda = \tfrac{1}{2}\,\sigma^{\mu\nu} F_{\mu\nu}\varepsilon$, and one finds

$$\left[\delta(\epsilon_1), \delta(\epsilon_2)\right] A_\mu = \xi^\nu \partial_\nu A_\mu + \partial_\mu\left(-\xi^\nu A_\nu\right)$$

where $\xi^\nu = \tfrac{1}{2}\,\bar{\epsilon}_2 \gamma^\nu \epsilon_1$

(3)

Clearly, global supersymmetry has produced translations and Maxwell gauge transformations!

Five years ago, Freedman, Ferrara and myself realized that if ε^a becomes local, a term $\xi(x)\partial_\mu B$ would probably still be found on the right hand side of (2). This looks like a local translation, and this, in turn, seemed to us to imply that one would find a general coordinate transformation on the right hand side. That, in turn, would force gravity onto the stage and since we have spinors in supersymmetric theories, one needs that formulation of gravity which can accommodate spinors. The gravitational field is described by (in d=4) the vierbein (vier = 4), or, more general, in arbitrary dimensions d, by a vielbein $\varepsilon_\mu{}^m$(viel=many).*

Since there is a symmetry between bosons and fermions, what is the fermionic companion of the graviton? The spinorial parameter ε^a has spin ½**, hence one expects that the irreducible representations of the symmetry group which is generated by global supersymmetry will contain bosons and fermions with adjacent spins J, J + ½, J + 1, ... Remarkably, one finds (at least for massless theories) that fermion-boson doublets suffice [2]. The question thus arises: is there a global supersymmetry multiplet containing the vierbein, and what is its fermionic partner. The full details were given in [3] , but it is clear that this fermionic partner must have either spin 3/2 or 5/2. The simplest case is 3/2, and it is also the only possible choice**. So the fermionic partner of the graviton $e_\mu{}^m$ is the spin 3/2 gravitino, represented by $\psi_\mu{}^a$. There exist other representations for a spin 3/2 field, for example ψ_{abc} where ψ is symmetric in the three dotted (or undotted) spinor indices, but they give the same results, and are less easy to work with. That the vectorial spinor $\psi_\mu{}^a$ contains spin 3/2 is clear from addition of angular momentum (the same argument is used to show that $e_\mu{}^m$ has spin 2). Why one takes $\psi_\mu{}^a$ (and not ψ_{abc}) is also clear: the gauge field must contain a term $\partial_\mu\varepsilon$ in its transformation law if one wants to use it to make globally supersymmetric matter systems locally supersymmetric. The variation of a globally supersymmetric action vanishes by definition if ε is constant, hence for local $\varepsilon(x)$ one finds in general

$$\delta(\epsilon(x))\int d^4x\,\mathcal{L}(\text{matter}) = \int d^4x\,(j^\mu_N)(\partial_\mu\epsilon(x)) \tag{4}$$

*Another approach, due to Deser and Zumino, used consistency.

**Footnote: One might try to use a spin 3/2 parameter. Theories with such global symmetries have ghosts. Even worse, the corresponding gauge field has spin 5/2 , and cannot be coupled consistently to gravity [18]. Thus only spin ½ seems possible.

where j^{μ}_{N} is called the Noether current. If one has a field ψ_{μ} which transforms as $\delta\psi_{\mu} = \partial_{\mu}\varepsilon$ + more, then one can cancel the $\partial_{\mu}\varepsilon$ terms by adding to the action a term $\mathcal{L}(N) = - j^{\mu}_{N}\psi_{\mu}$. There are of course new terms produced by varying $\mathcal{L}(N)$, but as shown in ref [4] one needs only a finite number of times to add again new terms to the action (except when spin 0 matter fields are present).

Exercise: repeat the above arguments for electromagnetism.

The theory of local supersymmetry thus is at the same time a gauge theory of spin 3/2 fields. In fact, as one should expect from the above, any theory of local supersymmetry is at the same time a theory of gravity. This explains the name supergravity.

Let us turn to the algebra. If one couples for example the photon-neutrino model to the gauge fields e^{m}_{μ} and $\psi_{\mu}{}^{a}$ of supergravity, one must add extra terms both to the action and to the transformation laws of all four fields, in order that the total action (gauge action plus coupling terms, the latter starting with the Maxwell and Dirac actions in curved space, plus Noether couplings, plus more) be invariant. What is new here is that one must add matter fields (λ, A_{μ}) to the laws of the gauge fields (ψ_{μ} in this case). This precludes the possibility to add invariant actions such that the sum is gauge-invariant again. To remedy this, one introduces auxiliary fields.

Auxiliary fields are also needed for another purpose. Consider the commutator in (2), but for local supersymmetry without auxiliary fields. As shown in ref [1] , the local laws differ from the global laws in many ways, and one finds

$$[\delta(\epsilon_1(x)), \delta(\epsilon_2(x))]A_{\mu} = (\partial_{\mu}\xi^{\alpha})A_{\alpha} + \xi^{\alpha}\partial_{\alpha}A_{\mu} + \partial_{\mu}(-\xi^{\alpha}A_{\alpha}) + \tfrac{1}{2}(-\xi^{\alpha}\bar{\psi}_{\alpha})\gamma_{\mu}\lambda \qquad (5)$$

In addition to the expected general coordinate transformation and Maxwell gauge transformation, one finds a local supersymmetry transformation on the right hand side with parameters $-\xi^{\alpha}\psi_{\alpha}$. If one does the same for λ one finds

$$[\delta(\epsilon_1(x)), \delta(\epsilon_2(x))]\lambda = \text{gen. coord. transf.} +$$

local supersymmetry transf. + local Lorentz transf. + terms with $\not{D}\lambda$ (6)

New with respect to A_{μ} are

(i) the local Lorentz transformation, whose parameter is $\xi^{\alpha}\omega_{\alpha}{}^{mn}(e,\psi)$ where $\omega_{\alpha}{}^{mn}$ is the spin connection. It solves its own field equations (as in Palatini formalism of general relativity).

(ii) the terms proportional to the λ field equation. Of course the commutator of two invariances of a given action is again a sum of invariances, hence the $\not\partial\lambda$ terms are new local symmetries of the action. However, if one were to add them to the list of symmetries, one would find in the new commutators new symmetries, etc., and one would presumably end up with infinitely many symmetries. One prefers to work with finite dimensional algebras, even if one is willing to go to superalgebras, hence one would like to eliminate the $\not\partial\lambda$ term. This can usually be done by adding a few fields, which enter in the action usually like pure squares without derivatives and are thus nonphysical. In our example, one finds already for the free-field globally supersymmetric (A_μ, λ) system terms with $\not\partial\lambda$ in the $[\delta(\varepsilon_1), \delta(\varepsilon_2)]$ commutator and in this example it is sufficient to add one auxiliary field D to the theory in order to remove the $\not\partial\lambda$ terms. The complete flat-space free-field globally supersymmetric action reads

$$\mathcal{L} = -\frac{1}{4}(\partial_m A_n - \partial_n A_m)^2 - \frac{1}{2}\bar{\lambda}\not\partial\lambda + \frac{1}{2}D^2 \tag{7}$$

and indeed, the auxiliary field D appears in the form of a square, without derivatives. Note that the dimension of D is one-half unit larger than that of λ. Consequently, possible $\delta\lambda \sim D\varepsilon$ terms will be without derivatives. This a general feature: in $\delta F \sim \partial_\mu B_1 \varepsilon + B_2 \varepsilon$, the boson B_1 will be physical, but B_2 will be an auxiliary field. Since $D = 0$ is a field equation, and field equations rotate into field equations under symmetries, one expects for δD something proportional to $\not\partial\lambda$. The complete transformation rules for this globally supersymmetric system read

$$\delta A_\mu = -\frac{1}{2}\bar{\epsilon}\gamma_\mu\lambda \, , \, \delta\lambda = \sigma^{mn}\partial_m A_n \epsilon + \frac{i}{2}\gamma_5 D\epsilon \, , \, \delta D = \frac{i}{2}\bar{\epsilon}\gamma_5 \not\partial\lambda \tag{8}$$

Let us now turn to the gauge action of supergravity. Since it should contain the vierbein, one is not surprised to see the Hilbert action of ordinary gravity. However, the action is not written in terms of $g_{\mu\nu}$ and the Christoffel symbol because we are dealing with fermions but in terms of $e_\mu{}^m$ and the spin connection $\omega_\mu{}^{mn}$. If the tetrad postulate is used to relate $\{{}^{\lambda}_{\mu\nu}\}$ to $\omega_\mu{}^{mn}$

$$\partial_\mu e_\nu{}^m - \{{}^{\alpha}_{\mu\nu}\} e^m{}_\alpha + \omega_\mu{}^{mn} e_{n\nu} = 0 \tag{9}$$

one finds $\omega_{\mu mn}(e)$, and one can prove the identity

$$R^\alpha{}_{\tau\mu\nu}(\{\})e_{m\alpha}e_n{}^\tau = R_{\mu\nu mn}(\omega) \tag{10}$$

$$R_{\mu\nu ab} = \partial_\mu \omega_{\nu ab} + \omega_{\mu a}{}^c \omega_{\nu cb} - \mu \leftrightarrow \nu \tag{11}$$

In general we do not require that $\omega_{\mu mn}$ be equal to $\omega_{\mu mn}(e)$ but leave it free. In that case the tetrad postulate becomes

$$\partial_\mu e_\nu^m - \Gamma_{\mu\nu}^\alpha e^m{}_\alpha + \omega_\mu{}^{mn} e_{n\nu} = 0 \tag{12}$$

where now Γ could be expressed in terms of ω. We never use Γ, only ω. In that case we still use the above expression for the vierbein action. Thus the bosonic part of the action of simple supergravity reads

$$I = \int d^4x \frac{-e}{2k^2} R_{\mu\nu ab}(\omega) e^{a\nu} e^{b\mu} \equiv \int d^4x \frac{-e}{2k^2} R(e,\omega) \tag{13}$$

We will shortly determine ω.

For the gravitino, the action is unique if one requires that there are no negative energy excitations in the theory. This unique action is a gauge action, and thus one derives gauge invariance from positivity of the energy. (The same is true for all other spins $J \geqslant 1$). This action reads

$$\mathcal{L} = -\frac{1}{2}\bar{\psi}_\mu \gamma^{[\mu}\gamma^\nu\gamma^{5]}\partial_\nu \psi_s \tag{14}$$

where the symbol [ ] means antisymmetrization and γ^μ are Dirac matrices. The bar is Dirac's bar.

Now we are able to guess the action of simple supergravity. One covariantizes the derivative ∂_ν to a gravitationally covariant derivative $D_\nu\psi_\rho = \partial_\nu\psi_\rho + \frac{1}{2}\omega_\nu{}^{mn}\sigma_{mn}\psi_\rho$, where σ_{mn} are the Lorentz generators for spinors. In this way one arrives at

$$I = \int d^4x \left[\frac{-e}{2k^2} R(e,\omega) - \frac{e}{2} \bar{\psi}_\mu \gamma^{[\mu}\gamma^\nu\gamma^{\rho]} D_\nu \psi_\rho \right] \tag{15}$$

In principle we could have added a term $\Gamma_{\rho\sigma}{}^\alpha\psi_\alpha$ to $D_\rho\psi_\sigma$, which would have given a nonvanishing contribution if there were torsion (Torsion = $\Gamma_{\mu\nu}{}^\rho - \Gamma_{\nu\mu}{}^\rho$). Remarkably, the full action is just equal to I, without $\Gamma\psi$ term and without extra four-gravitino couplings, provided one determines ω by solving its (algebraic) field equations. The solution is[1]

$$\omega_{\mu mn}(e,\psi) = \omega_{\mu mn}(e) + \frac{k^2}{4}\left(\bar{\psi}_\mu\gamma_m\psi_n - \bar{\psi}_\mu\gamma_n\psi_m + \bar{\psi}_m\gamma_\mu\psi_n\right) \tag{16}$$

One could expand I by expanding $\omega(e,\psi)$ and would then find certain four-fermion couplings. Their coefficients must have particular values if local supersymmetry is to hold; any change in a coefficient at once destroys supersymmetry. This raises an interesting possibility. Consider, say, a scalar particle ϕ with a Yukawa coupling $\lambda\lambda\phi$ and a ϕ^4 coupling. Just as in the above case, supersymmetry requires the coupling constants to have a particular ratio. Too much ϕ^4 coupling would be too attractive and lead to a free field theory (the ϕ^4 coupling is like a delta-function which does not lead to scattering in 3 dimensions according to exact non-perturbative calculations). Too much Yakawa coupling leads to implosion and no in-and-out states can be defined (cf. black holes).

Thus one might speculate (Feynman, Curtright) that supersymmetry models are the only physical models, and form a "plane" between the two unphysical alternatives.

Having obtained the action, we now must find the transformation laws under which it is invariant. Since bosons vary into fermions, we must find a $\delta e \sim \bar{\varepsilon}\psi$ law. There are only a few possibilities

$$\delta e_\mu{}^m = \bar{\epsilon}\gamma^m\psi_\mu \quad \text{or} \quad \bar{\epsilon}\gamma_\mu\psi^m \quad \text{or} \quad \epsilon_\mu{}^{mkl}\,\bar{\epsilon}\gamma_5\gamma_k\psi_\mu \tag{17}$$

For the gravitino law we have a dilemma. On the one hand we argued that $\delta\psi_\mu = \partial_\mu\varepsilon$ and more. On the other hand, from global supersymmetry we expect $\delta F \sim \partial B\varepsilon$, hence $\delta\psi \sim \partial e\varepsilon$. Both are correct!

To explain this, let us consider first the simpler case of Yang-Mills theory where $\delta W_\mu{}^a = \partial_\mu \Lambda^a + f^a{}_{bc} W_\mu{}^b \Lambda^c$. The homogeneous term is separately an invariance for constant Λ, whereas the $\partial_\mu \Lambda^a$ term is a local invariance of the free field theory (the Maxwell action). In the full theory one must combine both, and this is done by simply covariantizing $\partial_\mu \Lambda^a$. Similarly, one covariantizes in supergravity $\delta\psi_\mu = \partial_\mu \varepsilon$ into

$$\delta\psi_\mu = \frac{1}{k} D_\mu \epsilon = \frac{1}{k}\left(\partial_\mu \epsilon + \frac{1}{2}\omega_\mu{}^{mn}\sigma_{mn}\epsilon\right) \tag{18}$$

One recognizes the two terms: the $\partial_\mu \varepsilon$ term is the local symmetry of the free-field gravitino action, whereas the linearized part of $\omega(e,\psi)$ which is of the form $\omega(e)_{\text{lin}} \sim \partial e$, yields indeed a $\delta F = \partial B \varepsilon$ term.

To settle the remaining vierbein law, we could follow several paths, which, however, all yield the same result.

(i) We could use the supergroup theory, according to which all gauge fields transform as covariant derivatives of parameters.

(ii) We could require invariance of the action. Substitution of $\delta\psi_\mu = D_\mu \varepsilon$ into $D_\rho \psi_\sigma$ yields a commutator of two covariant derivatives, hence a curvature. The general expression has the form $\bar{\varepsilon}\gamma.\psi.R....$ (where dots denote indices), and hence <u>one can only find a contracted Riemann curvature</u>, i.e., a term proportional to the Einstein $G_{\mu\nu}$. Since also the variation of the Hilbert action is proportional to $G_{\mu\nu}$, one can fix $\delta e_\mu{}^m$ such that the total action is invariant.

(iii) We could require that in the commutator of two local supersymmetries on the vierbein, one finds only a sum of local symmetries, but no field equation of the vierbein. (Quite generally, one finds only field equations in the $\{Q,Q\}$ relation if it acts on fermions). In this way one finds

$$\delta e_\mu{}^m = \frac{k}{2}\bar{\epsilon}\gamma^m \psi_\mu \tag{19}$$

One can try to find some auxiliary fields for simple supergravity, such that no gravitino field equations appear in the $(Q,Q)\psi$ commutator. There are actually several sets now known, the simplest consisting of a scalar S, a pseudoscalar P and an axial vector A_m. They enter in the action as

$$\frac{e}{3}\left(S^2 + P^2 - A_m{}^2\right) \tag{20}$$

and their variations are proportional to the gravitino field equations. (But the (Q,Q) commutator is the same as for vierbein and gravitino, and does not contain the gravitino field equations).

SUMMARY: because bosons have one more derivative in their actions than fermions, two local supersymmetry transformations will produce a general coordinate transformation. Since a spin 3/2 field is a gauge field of supersymmetry, one understands why previous endeavours to couple a spin 3/2 field in a consistent way to spins $J \leq 3/2$ failed: because they neglected gravity. (This argument hints at considering the coupling of spin 5/2 not only to spin 2 but also to spin 3). The gauge action is simply the sum of the spin 2 Hilbert action and the spin 3/2 Rarita-Schwinger action, the former written in terms of vielbeins and spin connections, rather than metrics and $\Gamma_{\mu\nu}^{\rho}$ connections because fermions are present. The spin connection is that function of vielbeins and gravitinos which solves the algebraic field equation of the spin connection. In terms of this spin connection $\omega_\mu{}^{mn}(e, \psi)$ the action and transformation rules are minimal and very simple.

In order that the commutators of the three local symmetries (namely: general coordinate invariance, local Lorentz invariance and local supersymmetry) are again a sum of only local symmetries, without field equation, and the same for the fields $(e_\mu{}^m, \psi_\mu{}^a, S, P, A_m)$ on which these commutators act, one introduces the minimal set of auxiliary fields S, P, A_m. Actually, there are now several sets of auxiliary fields which close the gauge algebra of simple supergravity. For the extended supergravities auxiliary fields are known only for N=2 ordinary supergravity and N=4 conformal supergravity in four dimensions. Much efforts are being spent at the moment to find a set of auxiliary fields for the N=8 model, but so far without success. The real problem is not what such a set looks like (a complicated technical problem) but, rather, whether such a set exists at all.

II. Clifford Algebras in d dimensions

In order to describe gravitinos and spin ½ fields in higher dimensions d, one needs the Dirac matrices. These form a Clifford algebra

$$\{\Gamma_m, \Gamma_n\} = 2\delta_{mn} \;;\; m,n = 1,d \tag{1}$$

By taking products of these elements, they generate a finite group with 2^{d+1} elements, namely $\pm I, \pm\Gamma_m, \ldots, \pm\Gamma_1\ldots\Gamma_d$. In the algebra $\pm\Gamma_m$ are not independent, but from the point of group theory, where one only knows multiplication, they are independent. Since every representation of a finite group can be made unitary by a similarity transformation, we may assume that all Γ_m are unitary. Due to $\Gamma_m\Gamma_m = I$, this means that all Γ_m are hermitian. (We could but will not assume that one of these Γ's satisfies $\Gamma_o^2 = -1$. Instead we use the positive Pauli metric, according to which the Dirac operator reads $\gamma_1\partial_x + \gamma_2\partial_y + \gamma_3\partial_z + \gamma_4\,\partial/\partial(ict)$).

How many inequivalent irreducible representations (irrepses) of the group generated by the Clifford algebra exist in d dimensions? The answer can be given by using three simple theorems of finite group theory

(i) the number of irrepses equals the number of classes

(ii) the order of the group equals the sum of the squares of the dimension of all irrepses

$$g = \Sigma\,(d_\alpha)^2$$

(iii) the number of one-dimensional irrepses of a group equals the order of G divided by the order of the commutator subgroup C(G) (the group generated by all elements $aba^{-1}b^{-1}$).

<u>Even dimensions</u>: the number of classes equals $2^d + 1$ (+ I and - I are separate classes). The order of C(G) is equal to 2 (namely C(G) = + I, - I). Hence

$$g = 2^{d+1} = 2^d(1)^2 + (d_\alpha)^2 \qquad \text{hence} \qquad d_\alpha = 2^{\frac{d}{2}} \tag{2}$$

There is thus one irreducible representation which is not one-dimensional.

<u>Odd dimensions</u>: the number of classes is now 2^d+2 because $+\Gamma_1\ldots\Gamma_d$ and $-\Gamma_1\ldots\Gamma_d$ commute with all Γ_m, hence with all group elements, hence each forms a class. Again order C(G) = 2. Thus

$$g = 2^{d+1} = 2^d(1)^2 + (d_{\alpha_1})^2 + (d_{\alpha_2})^2 \tag{3}$$

There are two faithful irrepses, each of dimension $2^{[d]/2}$. It is easy to determine them in terms of the $\Gamma_1, \ldots \Gamma_{d-1}$ of the d-1 = even case, discussed above. Namely: in d dimension the first d-1 matrices Γ_i coincide with the matrices Γ_i of (d-1) dimension, while the last one equals $\Gamma_d = \pm \alpha \Gamma_1 \ldots \Gamma_{d-1}$ where $\alpha = \pm i$ or ± 1. That these two irrepses are inequivalent follows from Schur's lemma: if $S\Gamma_i S^{-1} = \Gamma_i$ for i = 1, d-1 then S commutes with all elements of the group G of d-1 dimensions hence S = I, so that $S\Gamma_d S^{-1}$ cannot be $-\Gamma_d$. From a physicist's point of view changing the sign in front of the matrix which represents Γ_d only means that one replaces one coordinate by minus itself, which is uninteresting in odd dimensions.

All one-dimensional representations of the group violate the algebra. There remain only the $2^{[d/2]}$ dimensional representations which are faithful. (Proof: if they were not faithful, there would be an element $\neq$ identity represented by the unit matrix. Then the representation would be reducible since its character would not be orthogonal to the unit character.)

In which dimensions can one define Weyl spinors, or Majorana spinors, or Majorana-Weyl spinors? A Weyl spinor satisfies.

$$\frac{1}{2}(1+\Gamma_{d+1})\lambda = \lambda \quad \text{or} \quad \frac{1}{2}(1-\Gamma_{d+1})\lambda = \lambda \qquad \text{for d = even} \tag{4}$$

where $\Gamma_{d+1} (:) \Gamma_1 \ldots \Gamma_d$. In the first case one calls λ left-handed, in the second case right-handed. A Weyl spinor at t = 0 remains a Weyl spinor for t > 0 if it is massless, because the Dirac operator $\gamma_4 \Sigma \gamma_k \partial_k$ with $k \neq 4$ commutes with $(1+\Gamma_{d+1})$ if d is even. Hence, Weyl spinors exist in even dimension for massless spinors. The phase of Γ_{d+1} is fixed by $(\Gamma_{d+1})^2 = +1$.

The Majorana conjugate of a spinor λ, denoted by $\bar{\lambda}_M$, is that linear combination of the components λ^a which transforms as the Dirac conjugate. The components of the Dirac conjugate, denoted by $(\bar{\lambda}_D)_a$, are proportional to the components of $(\lambda^\dagger)_a$ such that $\bar{\lambda}\lambda$ is Lorentz invariant. Since $\delta\lambda = \frac{1}{2}\omega^{mn}\sigma_{mn}\lambda$ for small ω, where $\sigma_{mn} = \frac{1}{4}[\gamma_m, \gamma_n]$ and where in our conventions ω^{14} is purely imaginary with ict as fourth coordinate, while as we discussed, all γ_m are hermitian, one has $\bar{\lambda}_D = \lambda^\dagger\gamma_4$ and

$$\delta\bar{\lambda}_D = \bar{\lambda}_D\left(-\frac{1}{2}\omega^{mn}\sigma_{mn}\right), \quad \delta\bar{\lambda}_M = \bar{\lambda}_M\left(-\frac{1}{2}\omega^{mn}\sigma_{mn}\right), \quad \bar{\lambda}_M = \lambda^T C \tag{5}$$

The matrix C is called the charge conjugation matrix.

(Incidentally, in the (½, 0) representation of the Lorentz group, the hermitian conjugate of the elements of the Lorentz group is not equivalent to their inverses, so that no invariant can be constructed. Therefore, one has to add a dual space, which transforms in the adjoint representation $(R)^{\dagger,-1}$, i.e., the representation (0, ½). So in d=4, and only in d=4, one deals with a reducible representation. The Dirac equation mixes both spaces of course).

Hence, in order that $\lambda_{D,a}$ and $\bar{\lambda}_{M,a}$ transform similarly, one has the requirement that

$$(\bar{\lambda}_M)_a = (\lambda^T)^b C_{ba}\,, \quad C\sigma_{mn}C^{-1} = -\sigma_{mn}{}^T \tag{6}$$

A stronger condition follows from requiring that $\bar{\lambda}_M$ satisfies the same Dirac (or Rarita-Schwinger or .....) equation as λ_D. It is easy to show that in this case

<u>Theorem</u>: for <u>massless</u> spinors $C\gamma_\mu C^{-1} = \pm\gamma_\mu{}^T$
for massive spinors $C\gamma_\mu C^{-1} = -\gamma_\mu{}^T$ (6a)

Thus, for massless spinors one has a little bit more freedom, just as in the case of Weyl spinors. This little bit more freedom will allow us to prove that Majorana spinors can exist in d=9, 8. The charge conjugation matrix C is unique up to an overall constant, as follows from Schur's lemma.

Let us define

$$C_+\Gamma_\mu C_+^{-1} = \Gamma_\mu{}^T \quad , \quad C_-\Gamma_\mu C_-^{-1} = -\Gamma_\mu{}^T \tag{7}$$

In even dimensions both C_+ and C_- are possible because $+\Gamma_\mu^T$ and $-\Gamma_\mu^T$ generate the same finite group as Γ_μ (In other words: the group multiplication table is the same). In odd dimensions, however, either C_+ or C_- is possible, but not both, because Γ_{2n+1} is proportional to $\Gamma_1 \ldots \Gamma_{2n}$, so that the C_+ and the C_- of d=2n both give the same result for $C\Gamma_{2n+1}C^{-1}$. One familiar example of this is that in d=3 the Dirac matrices become equal to the Pauli matrices which satisfy $C_- t_i C_-^{-1} = -t_i^T$ with $C_- = t_2$, but no C_+ exists.

Another property of the charge conjugation matrix follows by iterating (7). Using Schur's lemma, one finds that $C_+ = \alpha C_+{}^T$

(idem for C_-) so that (iterating again) $\alpha^2 = 1$. In other words C_+ and C_- are either symmetric or antisymmetric.

Having defined what the Majorana conjugate of a spinor is, we can now proceed and define what a Majorana spinor is.

Definition: A Majorana spinor is a spinor whose Dirac conjugate is proportional to its Majorana conjugate. Hence $\bar{\lambda}_M = \alpha\bar{\lambda}_D$. This means that a Majorana spinor has half as many components and states as a Dirac spinor. Also a Weyl spinor has this property. Iterating $\bar{\lambda}_M = \alpha\bar{\lambda}_D$ one finds a condition on the charge conjugation matrix

$$|\alpha|^2 \gamma_4^* (C^{-1})^* \gamma_4 C^{-1} = 1 \tag{8}$$

Clearly this condition is independent of the phases of α and C. As we shall see, this condition can only be met in certain dimensions, and that is the reason that one cannot define Majorana spinors in every dimension.

Let us now define a Majorana-Weyl spinor. This is a spinor which satisfies simultaneously $\bar{\lambda}_M = \alpha\bar{\lambda}_D$ and $\lambda = \frac{1}{2}(1 + \Gamma_{d+1})\lambda$ (or with a - sign). Recall that the phase of Γ_{d+1} is fixed by $(\Gamma_{d+1})^2 = 1$ (which follows from iterating the Weyl condition). Hence, in even dimensions one can define Majorana-Weyl spinors only if in addition to the Majorana condition one can satisfy

$$C^{-1} \Gamma_{d+1}^T C = \Gamma_4 \Gamma_{d+1}^\dagger \Gamma_4 \tag{9}$$

As we shall see, this relation is representation independent, and only possible in certain dimensions.

Before going on, we want to investigate how the above conditions and the charge conjugative matrix change if one makes a similarity transformation on the Dirac matrices. In other words: is the condition (8) and, in particular, the (anti) symmetry of C dependent on the representation one works in? We will restrict ourselves to unitary basis transformation in order that $\Gamma_\mu \to \Gamma'_\mu = U\Gamma_\mu U^{-1}$ be again hermitian. From $C\Gamma_\mu = +\Gamma_\mu^T C$ for all μ (or $-\Gamma_\mu^T C$) we deduce that $C \to C' = (U^{-1})^T C U^{-1}$. Hence, if C is symmetric (or antisymmetric) in a particular representation of the Dirac matrices, then so it is in any other representation. Moreover, the Majorana condition in (8) is also representation independent. These results are useful, because we can now study the Dirac matrices and C in d dimensions in a particular representation without loss of generality.

Certain representations are of great interest. A Majorana representation is a representation of the Dirac matrices in which (in our conventions, with all Γ_μ hermitian) all Γ matrices are real, except that Γ_4 is purely imaginary. In this case the Dirac operator is real , and hence the real part of a spinor satisfies the Dirac equation separately, as does the imaginary part. Hence, in a Majorana representation, the Majorana condition (8) is automatically satisfied, as follows also from $C = \alpha\gamma_4$ (which, in turn, follows from $\lambda_M = \alpha\lambda_D$). A real representation will be a set of real Γ_μ ; similarly a purely imaginary representation.

We now investigate in which dimensions d between d= 4 and d=11 one can have Majorana and/or Majorana-Weyl spinors. We do this by constructing explicit representations of the Clifford algebras and then checking in each case whether the conditions (8) and (9) hold.

d=4: A Majorana representation exists, for example

$$\Gamma_1 = \begin{pmatrix} 0 & -i\sigma_2 \\ i\sigma_2 & 0 \end{pmatrix}, \quad \Gamma_2 = \begin{pmatrix} 1 & 0 \\ 0 & -1 \end{pmatrix}, \quad \Gamma_3 = \begin{pmatrix} 0 & 1 \\ 1 & 0 \end{pmatrix} \quad \Gamma_4 = \begin{pmatrix} 0 & -i\sigma_1 \\ i\sigma_1 & 0 \end{pmatrix} \tag{10}$$

The Weyl-matrix $\Gamma_5 = \begin{pmatrix} & -i\sigma_3 \\ i\sigma_3 & \end{pmatrix}$ is purely imaginary, hence, as is well known, in d=4 one has Majorana or Weyl spinors but no Majorana-Weyl spinors. In this representation $C_+ = \gamma_4\gamma_5$ and $C_- = \gamma_4$.

d=5: $C^5 = C_+^5 = \gamma_4\gamma_5$. No Majorana spinor exists, except when one adds internal indices a such that

$$(\bar{\lambda}_M)^a = \lambda_b C \Omega^{ab} \quad , \quad \Omega^* \Omega = -1 \tag{11}$$

These matrices Ω are used in the N=8 d=5 model where the internal symmetry group is USp(8). In d=5, $\sigma_+ = -1$.

d=7: A purely imaginary representation is given by

$$\Gamma(7) \equiv \{ \sigma_2\gamma_1, \sigma_2\gamma_2, \sigma_2\gamma_3, \gamma_4, i\sigma_1\gamma_4\gamma_5, i\sigma_3\gamma_4\gamma_5, \gamma_5 \} \tag{12}$$

Clearly. the charge conjugation matrix is the identity, $C = I_8$, and it is symmetric. Since the Dirac matrices in d=7 are antisymmetric, $C = C_-$. Since $\Gamma(7)^* = -\Gamma(7)$ no Majorana spinor exists in d=7.

d=6: Drop γ_5 from (12). Then

$$C_-^{(6)} = I_8 \quad , \quad C_+^{(6)} = \gamma_5 , \quad C_-^{(6)\,T} = C_-^{(6)} \quad , \quad C_+^{(6)\,T} = -C_+^{(6)} \tag{13}$$

No Majorana spinor exists for $C_-^{(6)}$ or $C_+^{(6)}$.

d=9: A purely real representation is given by

$$\Gamma(9) = \Gamma(7) \otimes \sigma_2 \,,\quad I_8 \otimes \sigma_1 \,,\quad I_8 \otimes \sigma_3 \tag{14}$$

Hence

$$C^{(9)} = C_+^{(9)} = I \quad \text{and} \quad C^{(9)T} = + C^{(9)} \tag{15}$$

No matter which Dirac matrix is used as "γ_4" in $\bar{\lambda}_D = \lambda^\dagger \gamma_4$, since they are all real, Majorana spinors exist in d=9.

d=8: Dropping $I_8 \otimes \sigma_3$

$$C_+^{(8)} = I = C_+^{(8)T} \,,\quad C_-^{(8)} = I_8 \otimes \sigma_3 = + C_-^{(8)T} \tag{16}$$

Hence for $C = C_+^{(8)}$ one can define a Majorana spinor. One can also define a Weyl spinor $\lambda = (I_{16} + I_8 \times \sigma_3)\lambda$, but not both: in d=8 one has either a Majorana or a Weyl spinor.

d=11: Define the Dirac matrices by

$$\Gamma(11) \equiv \begin{pmatrix} 0 & \Gamma(9) \\ \Gamma(9) & 0 \end{pmatrix} \,,\quad I_{16} \otimes \sigma_3 \,,\quad I_{16} \otimes \sigma_2 \tag{17}$$

This is actually a Majorana representation, with "γ_4" in $\bar{\lambda}_D$ given by $I_{16} \otimes \sigma_2$. Hence Majorana spinors exist. Also C is antisymmetric.

$$C^{(11)} = C_-^{(11)} = I_{16} \otimes \sigma_2 = - C^{(11)T} \tag{18}$$

d=10. Define "γ_5" for the Weyl condition as $I_{16} \otimes \sigma_3$ (real) and keep "γ_4" $= I_{16} \otimes \sigma_2$. Now

$$C_-^{(10)} = I_{16} \otimes \sigma_2 \quad \text{and} \quad C_+^{(10)} = I_{16} \otimes \sigma_1 \tag{19}$$

and $C_-^{(10)}$ is antisymmetric while $C_+^{(10)}$ is symmetric. Since this is still a Majorana representation, both charge conjugation matrices allow a Majorana spinor or a Weyl spinor or both.

MASSIVE SPINORS AND HIGHER SPINS

If spinors are massive, transposition of the Dirac equation and (6a) show that $C = C_-$ while $C = C_+$ is not allowed. Hence for massive spinors one always has $C\Gamma_\mu C^{-1} = -\Gamma_\mu^T$.

For spin 3/2 the same results hold. For example, transposition of the spin 3/2 field equation leads to (6a).

Comments:

One can interpret Weyl and Majorana spinors also in a purely group-theoretical way. Given an irreducible representation of the group generated by the Clifford algebra, this irreps becomes reducible under the subgroup $\pm I$, $\pm\Gamma_\mu\Gamma_\nu$, $\pm\Gamma_\mu\Gamma_\nu\Gamma_\rho\Gamma_\sigma$ etc. These irrepses are the Weyl spinors.

Given a representation ϕ of the Clifford group, the elements $\phi + C\bar{\phi}_D^T$ and $\phi - C\bar{\phi}_D^T$ transform separately into themselves. This means that the complex C^4 splits into two irreducible real $R^4 + R^4$. (An easy example is the case when one has a Majorana representation, because then $C\bar{\phi}_D^T = \phi^*$). These are the Majorana spinors. However, given ϕ, the projector $P\phi = \phi + C\bar{\phi}_D^T$ must satisfy $P^2 = P$, which is the condition that Majorana spinors exist. This, by the way, shows that when a Majorana spinor exists, there should also exist a Majorana representation.

If Γ_μ is an irreps, then also Γ_μ^* is an irreps. Since the irrepses which are not one-dimensional are unique (see before) up to similarity transformations, $S\Gamma_\mu^* S^{-1} = \Gamma_\mu$. Hence, all irrepses are what is called in group theory pseudo-real.

The results for d=8 and d=9 are new. One could have expected that Majorana spinors exist in d=9, because in d=10 Majorana-Weyl spinors exist.

III. Extended Supergravity

The extended supergravities contain more than one gravitino field. For example, N=2 supergravity contains: one graviton, two gravitini and one photon. It realizes Einstein's dream of unifying gravity with electromagnetism[5]. One can couple the photons minimally to the gravitini, in which case one needs, remarkably enough, also to add to the action a cosmological term. One could again couple (N=2) matter to the gauge action by means of the Noether method.

Irreducible N=2 massless representations consist of: one helicity J, two helicities J - ½ and one helicity J - 1. For example, if J = 2 one gets a (2, 3/2, 3/2, 1) multiplet of helicities, to which one must add the CTP conjugate to find the spin 2, 3/2, 3/2, 1 states. These are exactly the states of the N=2 gauge action. In a similar way one can guess the states of the N=2 vector multiplet, which is that multiplet which starts with a helicity one. It consists of: one spin 1, two spin ½ and a scalar plus a pseudoscalar. The corresponding Lagrange field theory exists, too, the action being given by

$$\mathcal{L} = -\frac{1}{4}(\partial_\mu A_\nu - \partial_\nu A_\mu)^2 - \frac{1}{2}\bar{\lambda}_i \not\partial \lambda^i - \frac{1}{2}(\partial_\mu A)^2 - \frac{1}{2}(\partial_\mu B)^2 \qquad (1)$$

The auxiliary fields for this model are

$$\mathcal{L}(\text{aux}) = \sum_{i=1}^{3} \frac{1}{2}(D^i)^2 \qquad (2)$$

One can couple this N=2 matter model to the N=2 gauge action via the Noether method. In fact one can even couple an arbitrary number of N=2 vector multiplets [6].

One might think that also an N=2 scalar multiplet would exist and consist of: one spin ½ and two spins 0.

However, there is a subtlety which is the cause that the N=2 scalar multiplet is doubled: two helicities ± ½, and two scalars A as well as two pseudoscalars B. If one wants only 2 states, and still have A^i and B^i be SU(2) vectors, one must impose the following SU(2) invariant duality constraint $\phi^i = \pm\ \varepsilon^{ij}\ \phi_i^*$ with $\phi^*_i = (\phi^i)^*$ and $\phi^i = A^i + i\,B^i$. However, after multiplication by another ε^{jk} one finds that this constraint is contradictory, whether one uses the + or the - signs because $\varepsilon^{ij}\varepsilon_{jk} = -\delta^i_k$. In the N=8 model one has scalars ϕ^{ijkl} and there the constraint works since there the product of two 8 -dimensional ε symbols is positive. Thus for scalars with k indices in a 2k - supersymmetry one must double the multiplets when $k = 4\ell+2$. (S. Ferrara, private communication).

In higher dimensions there exist also supergravities, but beyond d=11 there are no supergravities. It is easy to see why. The massless representations of an N-extended supersymmetry in d=4 consist of 1 helicity J, $\binom{N}{1}$ helicities $J - \frac{1}{2}$ ... up to $\binom{N}{N}$ helicities in $J - \frac{N}{2}$. The largest multiplet without J > 2 is for J = 2 and N = 8. This is N = 8 supergravity with 8 gravitini. In higher dimension, spinors have more components. In d=11 spinors have 32 components which is just small enough so that one can compose the 8 gravitini of d=4 into the simple gravitino of d=11, but for d=12 there are not enough fermionic fields available. For example, the gravitino of d=11 splits up under dimensional reduction to d=4 into 8 gravitini and 56 spin ½ fields

$$\Psi_{\Lambda=1,11}^{A=1,32}(x^1,\ldots,x^{11}) = \psi_{\mu=1,4}^{a=1,4;\ i=1,8}(x^1,\ldots,x^4) + \lambda_{\mu=5,11}^{a=1,4;\ i=1,8}(x^1,\ldots,x^4)$$

However, in d=12 there are no Majorana-Weyl spinors, thus the complex spinor (with 128 components) is either chiral (leaving 64 components) or Majorana (again with 64 components), so that one would end up with 16 gravitini in d=4. If there are more than 8 gravitini, the irreducible representations of (the super Poincaré algebra of) supersymmetry contain spin 5/2 fields, and for what is bad about that, see section IX.

In d=10 one can find an N=2 model (by dimensional reduction from d=11: The left-handed and right-handed parts of the d=11 gravitino become the two gravitini in d=10). From this N=2 model one can find an N=1 model by consistent truncation, i.e., by putting the left-handed (or the right-handed) gravitino equal to zero. Consistency means that one then also must put equal to zero the fields in the variation of fields which are put equal to zero. Nahm has conjectured that there exists an equivalent model in d=10 with a <u>four index photon</u> whose Maxwell field strength is selfdual. This model has never been constructed.

For a thorough discussion of extended supergravities we refer to the lectures by E. Cremmer.

IV. Conformal Supergravity

Some years after a theory of ordinary supergravity had been constructed, Kaku, Townsend and the author constructed a theory of what is now called N=1 conformal supergravity [7]. This theory starts with Weyl's $R_{\mu\nu}{}^2 - \frac{1}{3}R^2$ action, plus a higher-derivative gravitino action, plus a Maxwell action, plus coupling terms. Conformal supergravity has since been recognized as being the nucleus of ordinary supergravity models, mostly by the works of B. de Wit and collaborators. Basically the idea is the following: by considering a larger symmetry group, one can use smaller multiplets of fields in order to have a realization of the local algebra and to construct actions. The ordinary supergravity models then emerge (Das, Kaku, Townsend) by taking two or more multiplets of conformal supergravity, namely the gauge multiplet and one or more matter multiplets, and by fixing the conformal symmetries by putting some of the matter and gauge fields equal to zero. After this gauge fixing, one ends up with one large irreducible multiplet, before one has two irreducible multiplets. Not only do the local matter field transformation rules depend on the gauge fields as usual, but in order to stay in the gauges chosen, the parameters of the compensating conformal local symmetries (needed to remain in the gauge) depend on matter as well as gauge fields, and this introduces matter fields into the gauge field transformation rules of the non-conformal theory. Thus one ends up with a multiplet which has not even a reducible subsector (i.e., it is a completely irreducible representation, and not only an irreducible representation).

To see how a larger symmetry group leads to smaller multiplets, we take the following example due to B. de Wit. (For an extensive discussion of conformal supergravity, see his contribution). The action for a massive vector multiplet has no gauge invariance (the smaller group) and an irreducible multiplet of four components (the field V_μ)

$$\mathcal{L} = -\frac{1}{4}\left(\partial_\mu V_\nu - \partial_\nu V_\mu\right)^2 - \frac{1}{2}m^2 V_\mu^2 \qquad (1)$$

If one replaces V_μ by $V_\mu = A_\mu + \frac{1}{m}\partial_\mu\phi$, one obtains a model with a local gauge invariance (the bigger group), in which there are two smaller multiplets (a triplet A_μ (A_μ is a gauge field) and a singlet ϕ).

$$\mathcal{L} = -\frac{1}{4}(\partial_\mu A_\nu - \partial_\nu A_\mu)^2 - \frac{1}{2}(\partial_\mu \phi + m A_\mu)^2$$

$$\delta A_\mu = \partial_\mu \Lambda \,, \quad \delta \phi = -m\Lambda \tag{2}$$

This last action is like conformal supergravity, but fixing the local gauge invariance by $\phi = 0$ leads one back to the original theory (which is the analogue of ordinary supergravity).

Is conformal supergravity interesting in addition to explaining (and providing new results for) ordinary supergravity? We want to make a few very loose speculations. Just as in weak interactions the dimensional Fermi constant was eliminated by adding vector bosons (which in addition made the theory renormalizable), it would be nice to replace the Newton constant by a dimensionless coupling constant. This hints at R^2 theories, and since the more symmetries, the more "miraculous cancellations" of divergences in the quantum S-matrix, one is led to $R_{\mu\nu}^{\ 2} - \frac{1}{3} R^2$ theories. Putting such theories on a lattice (which has at least one dimensioned parameter: the lattice length) might be the bridge between Einstein and Weyl theories. The R^2 theories are renormalizable and not finite, nor unitary, but, as stressed by Hasslacher, Mottola and Kaku, it might be that in a nonperturbative (lattice) approach the unitarity is restored.

In a renormalizable model of (conformal) (super)gravity one could calculate the dependence of the dimensionless coupling constant on Q^2 via the renormalization group, and perhaps, via the lattice link, obtain an energy dependent Newton constant $G(Q^2)$. The Newton constant would be the low energy limit of $G(Q^2)$! Speculating even further, it might be true that all four dimensionless coupling constants become equally strong at the grand unified mass, and not only the usual three (the strong and electroweak coupling constants).

Having a running gravitational coupling constant could lead to the interesting possibility that the Planck mass is lowered to the grand unified mass. It seems that there is little room, at least inside the conventional SU(5) model, to move the grand unified mass up from 10^{15} GeV to the Planck mass at $\hbar^{\frac{1}{2}} G^{-\frac{1}{2}} c^{5/2} = 10^{18}$ GeV. It would be nice for general principles if there were only one mass scale in physics, instead of two, and thus it would seem that the Planck mass would have to come down a few orders of magnitude. (Actually, it would be not nice at all for phenomenologists, who derive comfort in the fact that they do not have to worry about gravity when they work at the grand unified mass).

Of course this is a very speculative program, but it was at the back of our minds some years ago. At that time conformal supergravity was received sceptically, due to its nonunitarity, but developments since have led to a more positive attitude.

Let us give a few details of conformal supergravity. There is simple conformal supergravity, with one gravitino, and N-extended conformal supergravity with N gravitinos. The actions for the gravitinos have 3 rather than one derivative. In fact, their kinetic terms are simple the spin 3/2 projection operators times $\not\partial$

$$\mathcal{L}\,(\text{conformal gravitino}) = \bar{\psi}_\mu \Box P^{3/2}_{\mu\nu} \not\partial \psi_\nu \,, \quad \begin{cases} P^{3/2}_{\mu\nu} = \delta_{\mu\nu} - \omega_\mu \omega_\nu - \frac{1}{3}\hat{\gamma}_\mu \hat{\gamma}_\nu \\ \hat{\gamma}_\mu = \gamma_\mu - \omega_\mu \\ \omega_\mu = \partial_\mu \not\partial \Box^{-1} \end{cases} \tag{3}$$

Before supergravity was invented, it was lore (and thus wrong) that "no consistent field theory exists for spin 3/2". What this meant was that no non-higher derivative field theory exists which only contains spin 3/2. Ordinary supergravity is consistent, has no higher derivatives but contains (in the action, not in the field equations) lower spin ½ as well. Conformal supergravity is also consistent, and contains only spin 3/2, but the price to be paid is higher derivatives. Similarly one can write down consistent field theories containing only spin J for J arbitrarily high, but these theories have more and more derivatives.

Simple conformal supergravity contains the following fields: one vierbein, one gravitino, and one (axial) photon. In the next section we will explain why one expects this photon to be present and to be physical.

There are no auxiliary fields needed for simple conformal supergravity! However, for extended conformal supergravity one needs them to close the gauge algebra.

V. Counting Fields and States

In the N=1 gauge action there are two bosonic states and two fermionic states, hence the number of states is fermi-bose symmetric. How about the fields? If one uses as fields $e_\mu{}^m$, $\psi_\mu{}^a$ and the auxiliary fields S,P,A_m one finds:

$$\begin{aligned} &16\ \psi_\mu{}^m - 6 \text{ local Lorentz} - 4 \text{ gen. coord} &&= 6 \text{ bose} \\ &16\ \psi_\mu{}^a - 4 \text{ local supsym.} &&= 12 \text{ fermi} \\ &\ 6\ \text{ auxiliary fields } S,P,A_m &&= 6 \text{ bose} \end{aligned} \qquad (1)$$

We have used that gauge invariance means that certain fields are absent from the theory. (One can choose a gauge in which as many fields are zero as there are local gauge invariances). Hence, again, we find equal numbers of bose and fermi fields.

The same holds for matter systems. For example, the fields in the photon-neutrino system are (λ, A_μ) and an auxiliary spin 0 field D. Counting:

$$4 \text{ fields } \lambda^a = (4 \text{ fields } A_\mu - 1 \text{ Maxwell inv}) + 1 \text{ aux. field D}$$

In N=1 conformal supergravity, the Hilbert action R is replaced by the Weyl action $R_{\mu\nu}{}^2 - \frac{1}{3} R^2$, and the Rarita-Schwinger action by an action of the higher derivative type , with $\not\partial \Box$ rather than $\not\partial$ as derivatives. One does not need auxiliary fields here since the algebra closes already, a coincidence because for N=2 conformal supergravity one needs auxiliary fields. The fields are: $e_\mu{}^m$, $\psi_\mu{}^a$ and an axial vector field A_μ. Again we have equal numbers of bosonic and fermionic fields

$$\begin{aligned} &16\ \psi_\mu{}^m - 6 \text{ local Lorentz} - 4 \text{ gen. coord.} - 1 \text{ local scale} &&= 5\ e_\mu{}^m \\ &16\ \psi_\mu{}^a - 4 \text{ ordinary supsymm.} - 4 \text{ conf. supsymm.} &&= 8\ \psi_\mu{}^a \\ &\ 4\ A_\mu - 1 \text{ local chiral inv.} &&= 3\ A_\mu \end{aligned} \qquad (2)$$

(The superconformal group contains the usual conformal group SU(2,2) with the following symmetries: translations P_m, scale transf. D, conformal boots K_m, Lorentz transf. M_{ab} (in total fifteen generators of SU(2,2)) In addition one has two supersymmetries, one of which is the square root of P_m, the other of K_m. Finally, there is one more bosonic symmetry: chiral transformations.)

Let us now go to N=2 supergravity. First ordinary supergravity. It contains a graviton, N=2 gravitino fields and one photon . Hence, again, one has equal numbers of states. Among the auxiliary fields there are two spinors and many bosons, but also here one has equal numbers of boson and fermion fields.

A more subtle issue is the counting of states in conformal supergravity. These theories are higher derivative theories, in which the graviton action is Weyl's $R_{\mu\nu}^{2} - \frac{1}{3}R^{2}$, and the gravitino action has three derivatives. For a spin 0 (or spin ½) action, replacing $\Box$ by $\Box^2$ (or $\not\partial$ by $\Box\not\partial$) leads to two (three) states, instead of the original one (two) states: $\Box^2$ acts as twice $\Box$, and $\Box\not\partial$ as three times $\not\partial$. This result is due to Ferrara and Zumino [8]. As stressed by Fradkin, in gauge theories things go differently. For example, if one adds the Einstein action times a parameter β to the Weyl action, one finds a massive spin 2 and a massless spin 2. However, for β tending to zero, one finds one new symmetry (dilatation invariance) which removes one field component (only one because it is an algebraic symmetry). Thus one ends up with six (and not four) states. Similarly for the spin 3/2 states: adding to the conformal higher-derivative action of spin 3/2 an ordinary Rarita-Schwinger action times β (details in the Physics Report, page 260.) one has: two massive spin 3/2 and one massless spin 3/2, in total 10 states. However, for β = 0 one finds one new symmetry: conformal supersymmetry which removes 4 x ½ = 2 states (for fermions one has half as many states as for bosons), leading to eight states (not six).

Now the counting of states in conformal supergravity. For N=1 this is easy:

6 gravitons + 2 photons = 8 gravitinos. (3)

For N = 1 conformal supergravity one does not need auxiliary fields. The only fields are $e_\mu{}^m$, $\psi_\mu{}^a$, A_μ. For N=2 conformal supergravity one has, in addition to the vierbein and two gravitini, four real vectors gauging U(2), two real spinors, and an antisymmetric real non-gauge tensor field, all propagating, plus one real auxiliary field D. Thus the states are

$$6 - 2 \times 8 + 4 \times 2 - 2 \times 2 + 6 = 0 \quad (4)$$

and again one has equal numbers of boson and fermion states.

The counting of states in $R + R^2$ gravitational action was first done by Stelle [9]. The extension to $R + R^2$ supergravity models was made by Ferrara, Grisaru and the author [10]. The counting of states in conformal models (models starting with $R_{\mu\nu}^{2} - \frac{1}{3}R^{2}$) was done by Fradkin [11], using functional methods. For a derivation using canonical formalism, see S.Y. Lee and the author, Stony Brook preprint.

VI. Which Superalgebras Yield Field Theories?

Compare the commutators of two supersymmetry transformations, in the global and in the local case (assuming that one has auxiliary fields, such that both close). For example, in simple supergravity one has

$$[\delta_s(\epsilon_1), \delta_s(\epsilon_2)] = \delta_T(\xi^\mu) \quad , \quad \xi^\mu = \tfrac{1}{2}\bar{\epsilon}_2\gamma^\mu\epsilon_1 \quad \text{and} \quad \delta_T(\xi^\mu)\phi = \xi^\mu\partial_\mu\phi$$

$$[\delta_s(\epsilon_1), \delta_s(\epsilon_2)] = \delta_{gc}(\xi^\mu) + \delta_s(-\xi^\mu\psi_\mu) +$$

$$+ \delta_L(\xi^\mu\omega_{\mu mn}(e,\psi) + \bar{\epsilon}_2\sigma_{mn}(S - i\gamma_5 P) + \epsilon_{mnrs}\xi^r A^s) \tag{1}$$

Clearly, in the local case, the structure constants (the objects between curly brackets on the right hand side without ε_1 and ε_2) depend on fields. One might speak of structure functions. Moreover, and this is often overlooked, the structure functions can be fermionic !

In general one may view a gauge algebra (or superalgebra) as an ordinary global algebra (or superalgebra) with infinitely many constant parameters: n for each point x^μ. Thus a local gauge algebra is a global algebra to the power infinity: infinitely many copies of the same global algebra. The field-dependence of the structure constants means: at each point x^μ one has a different algebra! This is a deep change in gauge theories, and one might ask if one can avoid it. One could: by not choosing in superspace a particular gauge (the so-called Wess-Zumino gauge) but keeping all fields, one would find in ordinary space field-independent structure constants, but, however, terribly many fields. It seems simpler (and more interesting!) to leave the fields in the structure constants.

From a mathematical point of view, one can define a superalgebra as a set of generators (P, Q, M etc.) which one can divide into bosonic (or even) and fermionic (or odd) generators (a so-called Z_2 grading). The bracket relations which define as always the algebra read

$$[X_\alpha, X_\beta] = X_\gamma f^\gamma{}_{\alpha\beta} \tag{2}$$

and preserve "fermion-number" (two odd ones give an even one,etc.) Moreover the brackets are always antisymmetric except that two odd generators have a symmetric bracket. For example $[Q_\alpha, Q_\beta] = [Q_\beta, Q_\alpha]$ where in this case the bracket is the anticommutator, but it need not be in general. Finally, the super- Jacobi identities must be satisfied. They are easily written down. For example

$$[Q_\alpha, [Q_\beta, P_\mu]] - [Q_\beta, [P_\mu, Q_\alpha]] + [P_\mu, [Q_\alpha, Q_\beta]] \equiv 0 \quad (3)$$

One should find identically zero if one were to assume that the multiplication of generators were associative. The signs are then fixed such that all terms cancel pairwise. Substituting (2) into (3) one finds the super- Jacobi identities, and one easily checks that the adjoint representation is indeed a representation of the algebra, while in this representation the bracket is always a commutator or an anticommutator. From a group-theoretic point of view, the new aspect of superalgebras is not their Z_2 grading (after all, the Lorentz group has also a Z_2 grading, namely rotations and boosts), but the symmetry of the bracket of the odd generators. Physicists find it natural to take an anticommutator for two fermionic charges, but the mathematical literature was not accustomed to this idea. When people considered whether the Poincaré group could be part of a bigger group containing also internal symmetries, they considered only commutators, and reached the conclusion that the best one can do is a direct sum of three factors (Coleman and Mandula)

$$\text{Poincaré} \oplus S \oplus A \quad (4)$$

where S is a semisimple and A an abelian ordinary Lie algebra. By allowing anticommutators, one can find an extension of the Poincaré algebra such that one ends up with a semisimple superalgebra which contains the Poincaré algebra. The structure of this superalgebra reads

$$[Q, Q] \sim P + Z \;, \quad [Q, P] = 0 \;, \quad [P, P] = 0$$

$$[Q, M] \sim Q \;, \quad [P, M] \sim P \;, \quad [M, M] \sim M \quad (5)$$

where Z are central charges ($[Z, P] = [\bar{Z}, Q] = [Z, M] = 0$).

There are several interesting other superalgebras. The super Lorentz algebra is $OSp(1C|2C)$ where $Sp(2C) \approx S\ell(2C)$ is the Lorentz group. A simple 3x3 matrix representation is obtained by putting around Sp(2,C) one extra column and row (and a zero in the 3 x 3 entry). The entries in the third row and column define the odd generators, and the bracket relations are commutators and anticommutators.

The super de-Sitter algebra is similarly obtained by a 5 x 5 representation. In the 4 x 4 part one puts the generators of Sp(4) (Sp(4) is isomorphic to SO(5) and SO(5) is the de-Sitter algebra in d=4). In the fifth row one now puts a <u>real</u> spinor Q and in the fifth column one puts puts $\bar{Q}^T$, while the (5, 5) entry is zero again.

There are also super de-Sitter algebras with O(N) (internal) symmetries. Just put Sp(4) and SO(N) along the diagonal, and the fermionic generators off-diagonal. By group-contraction one finds super-Poincaré algebras with O(N) internal charges (see ref.(1) , page 279).

Yet another class of superalgebras are the super unitary algebras: SU(N) x SU(M) x U(1) along the diagonal, complex spinors off-diagonal.

The interesting thing is that two classical algebras of the Cartan family A(n), B(n), C(n), D(n), yield always one superalgebra (namely $OSp(N|2M)$ and $SU(N|M)$ respectively).

In addition to these orthosymplectic and unitary superalgebras, there are other simple superalgebras:

A few more infinite families (whose bosonic subparts are not simple) and even a few exceptional superalgebras. For details we refer to a forth coming book by Bryce S. DeWitt, P. C. West and the author. There also explicit matrix representations of all these algebras will be given.

Do there exist conformal supergravities beyond d=4? The supergroup underlying conformal supergravities must have the structure

$$M = \begin{pmatrix} A & B \\ C & D \end{pmatrix} \tag{6}$$

We assume that the bosonic Lie algebras A and D are simple and that A contains only the spacetime symmetries. (According the Coleman-Mandula theorem, one cannot combine the Poincaré algebra with internal symmetries into a simple Lie algebra). Moreover, since B and C contain fermionic charges Q, and since in superalgebras the generic commutator

$$[\text{bosonic charge}, Q] \sim Q \tag{7}$$

shows that the Q form a representation of A and D, it follows that A must be the spinor representation of SO(d+2) for conformal theories in d dimensions, or of SO(d+1) for de-Sitter theories in d dimensions. (Note that the conformal group in d dimensions is SO(d+2). For d=4 this is SO(6), or, more precisely, SO(4,2) $\sim$ SU(2,2)).

Which superalgebras have an A-part which is spin (n)? (Spin(n) means the matrices given by the Lorentz generators $(\gamma_\mu\gamma_\nu - \gamma_\nu\gamma_\mu)$ in n dimensions). Some cases are well-known. The superalgebras SU(n|N) with n=4 contain the spinor representations of SO(6), namely SU(4), and correspond therefore to N-conformal supergravity in d=4. Another case is Osp(N|4). Now Sp(4) $\sim$ SO(5) is the spacetime group; it is de-Sitter algebra in d=4, and hence these groups correspond to (gauged but) ordinary N-extended supergravity in d=4.

One can also consider SU(2,2) $\tilde{\sim}$ SO(4,2) as the de-Sitter algebra of d=5, and use SU(2,2|1) for simple supergravity in d=5, (as done by the Torino group). We conjecture that N-extended supergravity $(1 \leq N \leq 4)$ in d=5 corresponds to SU(2,2|N). Whether one can gauge the SU(N) symmetries in d=5 we do not know. Note only that by dimensional reduction from d=11 one finds 15 vector fields $A_\mu{}^{ab}$. In addition there are 12 vectors, namely $6A_{\mu\nu a}$ (in d=5 antisymmetric axial tensors are vectors, see the section on 1.5 order formalism) and 6 $e_\mu{}^a$. Perhaps these are central charges.

In d=11 one cannot expect $Osp(1|32)$ as the group for simple supergravity, because Sp(32) contains not only the Poincaré group of d=11 but also internal symmetries. Its $\frac{1}{2}$ 32 x 33 = 528 generators must satisfy $CMC^{-1} = -M^T$ and are $11\Gamma_\mu$, $55\Gamma_{\mu\nu}$, $462\Gamma_{\mu_1 \ldots \mu_5}$.

The Γ_μ could perhaps correspond to the translations, and the $\Gamma_{\mu\nu}$ to the Lorentz rotations, but the five-index generators point to a six-index boson. One would conjecture that only a contraction of $Osp(1|32)$ would correspond to a field theory. This would mean that one could not gauge supergravity in eleven dimensions, which has indeed never been achieved.

Incidentally - one might perhaps use $Osp(N|4)$ for conformal supergravity in d=3 dimensions (Sp(4) = SO(5) = conformal group in d=3).

The above cases seem to be all plausible supergroups one can use. There are only two families of superalgebras, $Osp(N|M)$ and $SU(N|M)$. The only SU(N) vector representations equivalent to a spin(M) are SU(4), and the only vector representations of Sp(N) equivalent to a Spin(M) are Sp(4). Finally, for O(N) the vector and spinor representations are never equivalent (although for N=8 the vector representations and the two spinor representations are all 8 dimensional).

There is one implausible supergroup left.

The algebra F(4) contains spin (7) $\otimes$ $s\ell(2,R)$ in its bosonic sector. Now spin (7) is the 8 x 8 representation of O(7), and since O(7) is de-Sitter for d=6, and spinors have 8 components in d=6, F(4) might yield supergravity in d=6. Or it yields conformal supergravity in d=5, since O(7) (with Σ_{i6} and Σ_{i7} yielding P_i, K_i and $\Sigma_{6,7}$ yielding D) is the conformal in d=5. Since one needs in d=5 complex spinors in order to define a Majorana spinors, spinors have eight components and, indeed, spin (7) is 8 x 8.

VII. The Group Manifold Approach

An interesting approach to understand the relation between groups and gauge theories is what is called the group manifold approach. What it does is to produce a gauge field theory once a group (or a supergroup) has been given. Many groups have been considered and have reproduced existing field theories, but it is not clear whether all gauge field theories can be derived in this way, nor is it clear that every group will always yield a field theory.

The method might in fact be restrictive, and single out certain classes of theories. If this were the case, then this method could serve as a sieve for interesting theories. As we will discuss below, the theories selected have the property that they use coefficients ν_A which satisfy a kind of cohomology equation of the kind first introduced by Chevalley

$$D\nu_A + \frac{\delta}{\delta h^A} \Lambda = 0 \tag{1}$$

where Λ is a cosmological constant in the group manifold. Neglecting the difference between D and d, the forms ν_A are "closed" (if $\Lambda = 0$) but not "exact" ($\nu_A \neq Df_A$) because otherwise the corresponding actions would be total derivatives. Thus the coefficients ν_A are related to spaces with handles, and the group manifold selects only those theories which have something to do with nontrivial topologies.

The method works as follows. One begins by choosing an ordinary Lie algebra A, not necessarily simple or semi-simple; it may even be a superalgebra. If there are n generators, one considers a space with n coordinates: one coordinate per generator. Some of these coordinates will be fermionic if one has a superalgebra. This is the group manifold. The next point is to select a subgroup H and the dimension d of the <u>final</u> bosonic spacetime. In principle, given G, there are a certain number of subgroups H possible, but, as far as we know, the final theory is independent of the subgroup chosen. In this sense, the choice of H is no further input. There is a dependence on d. Example: if one considers $G = SU(2, 2|1)$ = the superunitary group with $SU(2, 2) \otimes U(1)$ as bosonic sector, then the choice d=5 leads to simple ordinary supergravity in five dimensions, but the same group leads in d=4 to simple conformal supergravity. ("Simple" means: the simplest theory. In d=5 this is actually an N=2 theory because one needs in d=5 two gravitini to define a Majorana spinor, while in d=4 simple implies N=1).

Usually one makes in gauge theories a clear distinction between the gauge algebra, and particular dynamical models which are or are not invariant under a given algebra. Thus, a certain given set of fields, say $e_\mu{}^m$, $\psi_\mu{}^a$, S, P, A_m satisfy a certain algebra, but may be used for quite different actions (N=1 ordinary supergravity, or N=1 conformal supergravity, or supersymmetric R^2 theories, or N=1 gauged supergravity, etc.) In the group manifold the opposite is true: here one starts from an action, and derives the transformation rules of the final theory (the theory in d dimensions) from the transformation rules and the action in the group manifold. Hence, actions are more important in the group manifold than elsewhere.

The action in the group manifold is a d-form, constructed from all curvature two-forms and the G - H one-forms. The curvatures are G - curvatures: curls of connections plus terms quadratic in connections multiplied by the structure constants of G. The G - H one-forms are the one-forms of those connections which correspond to the generators in G but not in H. So a typical action looks like

$$I = \int_M R^a \wedge R^b \wedge \ldots \wedge h^c \wedge h^d \; \nu_{ab\ldots cd} \tag{2}$$

The manifold M is an arbitrary curved d-dimensional hypersurface in the n-dimensional group manifold. It is parametrized by d (bosonic) coordinates in the surface, and hence, although it is a hypersurface in a bosonic - fermionic space in general, the surface M itself is bosonic. The $\nu_{ab\ldots cd}$ are purely numerical objects like Kronecker deltas, Dirac matrices etc. One can have terms in the action without a curvature; these are thus d forms constructed entirely from the one-forms $h^A \equiv Dx^\Lambda h_\Lambda{}^A$ where $A \varepsilon G - H$, and are the cosmological constants Λ mentioned above. Or one has terms linear in curvatures; these actions read thus

$$I = \int_M R^A \wedge \nu_A \tag{3}$$

where ν_A are d-2 forms consisting of h^A with A not in H. One now requires that I be H invariant. This means that the Λ, ν_A, $\nu_{ab\ldots cd}$ are H-invariant tensors. One does not require that I be G - gauge invariant (this would yield a total derivative as action). Thus at this point one splits G into H and G - H, and this will lead, via the coset space G/H, to the final space R^d. Thus one constructs actions using only exterior derivatives and H-invariant tensors. Glaringly absent are the dual forms: they cannot be used on a d-dimensional hypersurface in n dimensions.

The very use of differential forms ensures that the theory is automatically invariant under diffeomorphisms (= general coordinate transformations) in the group manifold. Thus the actions are invariant under two kinds of local symmetries: diffeomorphisms and H-gauge transformation. The former read

$$\delta h^A = (D\epsilon)^A + \epsilon^B h_B \wedge R_{\Lambda\pi}{}^A \, dx^\pi \tag{4}$$

and are a sum of a G-gauge transformation plus a curvature term, where ε^B is related to the diffeomorphism parameters ξ^Λ by $\varepsilon^B \equiv \xi^\Lambda h_\Lambda{}^B$. The local H gauge transformations are G-gauge transformations

$$\delta h^A = (D\epsilon)^A = d\epsilon^A + h^C \epsilon^B f_{BC}{}^A \tag{5}$$

but where ε^B is nonzero only if B lies in H.

<u>Exercise:</u> derive (4).

Before going on, let us give an example. Ordinary Einstein gravity in d dimensions can be handled in this way. Choose G = Poincaré group in d dimensions. The dimension of the group manifold is then n = d + ½ d(d-1) = ½ d(d+1). The surface M is d dimensional, and the action is unique (see below) and given by the Riemann curvature two-form R^{ab}, wedged to d-2 vielbein one-forms $V^c \wedge \ldots \wedge V^n$, everything contracted with the Lorentz -invariant ε tensor of d-dimensions. This example shows that there is no roof on the dimenensionality of d.

The field equations are now obtained by a new kind of Euler-Lagrange variational principle: one varies both the connection one-forms

h^A and the surface M. This means in practice that one varies independently with respect to all h^A, i.e., with respect to all n^2 connections $h_\Lambda{}^A$, without requiring constraints which would be due to the fact that only those $h^A = dx^\Lambda h_\Lambda{}^A$ are actually present, whose dx^Λ lie in M.

The n^2 field equations now pose a Cauchy problem, which can be solved in principle, just as one can propagate gravity off a 3-dimensional spacelike hypersurface into the time directions. The reason that this is possible is the following. Due to the field equations, derivatives of the connections w.r.t. coordinates not in M (namely: the curvatures $R_{BC}{}^A$ with B and/or C in H) are (algebraic) functions of derivatives of the connections w. r.t. coordinates in M(namely: the curvatures $R_{BC}{}^A$ with both B and C in G - H). The situation is not unlike a Dirac equation which is linear in ∂/∂_t and for which the knowledge of ψ on a spacelike hypersurface defines a Cauchy problem.

For further details we refer to refs (12, 13).

At this point one makes two extra postulates which seem so weak that one may doubt that they could be of any use, but which in fact are so strong that most starting points (G, H, d) do not satisfy them.

(i) flat space is a solution

(ii) there are more solutions than flat space.

Flat space is defined by the vanishing of all curvatures: $R^A = 0$ for all $A\epsilon G$. Thus if the h^A are pure gauge, the field equations must be satisfied. Obviously actions with more than one curvature automatically satisfy this requirement, but the integrands $R^A{}_\wedge \nu_A$ lead to a term $D\nu_A$ in the field equation because $\delta R^A = D\delta h^A$. In this way one arrives at (1). Some interesting mathematics concerning the cohomology solutions of (1) can be found in the appendix of ref (12b). Often (1) fixes most of the free constants in $\Lambda + R^A{}_\wedge \nu_A$. Requirement (ii) usually fixes all the rest. Thus, the requirement that flat space be a solution of the field equations leads to manifolds with a nontrivial topology, in the sense explained above.

Theories with only flat-space as a solution are not unknown even outside the group manifold. For example, pure gravity in three dimensions is an example. However, in the group manifold this is the usual case, rather than an exception. (In three dimensions, the Riemann tensor is proportional to the Ricci tensor.)

VIII. The 1.5 Order Formalism

Sometimes one can solve a field χ (or a set of fields) algebraically from their own field equation (s). Let the solution be denoted by $\chi(\phi)$ where ϕ are the other fields. The action then reads $I(\chi(\phi),\phi)$, and if one treats χ as if it were an independent field, and writes down the Euler-Lagrange equations for χ, but replaces in the result χ again by its solution $\chi(\phi)$, then these field equations vanish identically. Thus

$$\frac{\delta}{\delta\chi} I(\chi,\phi)\Big|_{\chi=\chi(\phi)} \equiv 0 \tag{1}$$

Hence, in varying $I(\chi(\phi),\phi)$ with respect to ϕ, one only needs to vary the explicit ϕ, but for $\delta\chi(\phi)$ one may choose anything (and not only what one would get from the chain rule) since $\delta\chi(\phi)$ is always multiplied by zero, namely by (1).

This observation is of invaluable practical use. For example, in N=1 supergravity, the action depends on vierbeins, gravitinos and the spin connection. The latter is expressed in terms of $e_\mu{}^m$ and $\psi_\mu{}^a$ by solving its own field equation. Consequently, when one varies the action one may choose $\delta\omega(e,\psi) = 0$. Thus only the fields indicated by arrows need be varied

$$\mathcal{L} = -\frac{e}{2}\, \overset{\downarrow}{e}{}^{a\mu} \overset{\downarrow}{e}{}^{b\nu} R_{\mu\nu ab}(\omega) - \frac{1}{2}\varepsilon^{\mu\nu\rho\sigma}\, \overset{\downarrow}{\bar\psi}_\mu \gamma_5 \overset{\downarrow}{\gamma}_\nu D_\rho(\omega) \overset{\downarrow}{\psi}_\sigma \tag{2}$$

but not the e,ψ fields inside $\omega_{\mu mn}(e,\psi)$. The first three variations are proportional to the Einstein tensor, while the $\delta\psi_\mu = D_\mu\varepsilon$ variation yields a commutator of two D_μ derivatives, hence a curvature again. Since there are not enough derivatives one only finds a contracted curvature tensor, hence again the Einstein tensor. Both terms with an Einstein tensor cancel, provided one chooses $\delta e_\mu{}^m$ appropriately (Note that the Einstein tensors contain $\omega(e,\psi)$ and not $\omega(e)$ as spin connection). There remains the variation of the vierbein in $\gamma^\nu = \gamma_m e^{m\nu}$. This variation cancels against a $D_\mu\gamma^\nu$ term due to partially integrating the

$\delta\bar{\psi}_\mu = D_\mu\bar{\varepsilon}$ term. In this way one very easily proves the invariance of the action under local supersymmetry. Incidentally, since one needs to partially integrate, the Lagrangian itself varies into a total derivative

$$\delta\mathcal{L} = -\tfrac{1}{2}\partial_\mu\left[\epsilon^{\mu\nu\rho\sigma}\;\bar{\epsilon}\gamma_5\gamma_\nu D_\rho\psi_\sigma\right] \tag{3}$$

and this shows that (local) supersymmetry is not an internal symmetry but rather a spacetime symmetry. Compare with general coordinate transformations where

$$\delta\mathcal{L} = \partial_\mu(\xi^\mu\mathcal{L}) \tag{4}$$

This 1.5 order formalism was discovered by Townsend and the author [14] some years ago when they tried to understand a result by Chamseddine and West [14] who found from group theory that $\delta\omega_{\mu mn} = 0$ which, it seemed, was in contradiction with the rule for $\delta\omega$ as found either in first or in second order formalism. As we have seen, the results $\delta\omega = 0$ and $\delta\omega - 0$ are both correct <u>in second order formalism</u>. The 1.5 order formalism is just second order formalism, plus the observation that it pays not to expand $\omega(e,\psi)$ in the action.

Let us now give a second example of the use of 1.5 order formalism due to Nicolai and Townsend [15]. Suppose $\mathcal{L}(B,\phi)$ is invariant under δB, $\delta\phi$ while B appears only as $\partial_\mu B$. We can then replace $\partial_\mu B$ by L_μ provided we add a Lagrange multiplier $M_{\mu\nu}$ which ensures that L_μ is really a gradient

$$\mathcal{L}(\text{extra}) = \epsilon^{\mu\nu\rho\sigma}\;M_{\mu\nu}\,\partial_\rho L_\sigma \tag{5}$$

We define $\delta L_\mu = \partial_\mu\delta B$. Then the variation of the action reads

$$\delta I = \int\tfrac{1}{2}(\partial_\mu L_\nu - \partial_\nu L_\mu)\,S^{\mu\nu} \tag{6}$$

because the action is invariant if L_ν is replaced by $\partial_\nu B$. By defining

$$\delta M_{\mu\nu} = - \epsilon_{\mu\nu\rho\sigma} \; S^{\rho\sigma} \tag{7}$$

the action $I(M_{\mu\nu}, L_\mu, \phi)$ (where ϕ are all other fields) is invariant. Note that the variation of L_σ in $\mathcal{L}$(extra) cancels.

Since L_μ only appears algebraically, we can solve it from its field equation. The solution for L_μ reads

$$L_\mu = \epsilon_{\mu\nu\rho\sigma} \partial_\nu M_{\rho\sigma} + \phi\text{-terms} \tag{8}$$

if the original action contained only terms quadratic and linear in $\partial_\mu B$. If one eliminates L_μ in this way from the action, one must forget that $\delta L_\mu = \partial_\mu \delta B$, but instead use for δL_μ what one gets if one varies the fields on the right hand side of the result above. This is allowed by 1.5 order formalism (see the beginning of this section).

Thus, using 1.5 order formalism, one has swapped a scalar for an antisymmetric tensor field. Similarly one can swap axial vectors and vectors in d=4 dimensions, and antisymmetric tensors $A_{\mu\nu}$ with vectors B_μ in d=5. A final swap is between $A_{\mu\nu\rho}$ and "nothing" : one can introduce $\mathcal{L}$(extra) = $\phi\varepsilon^{\mu\nu\rho\sigma}\partial_\mu A_{\nu\rho\sigma}$ and find then that ϕ is an auxiliary field.

One should be careful about claiming that fields related by such duality transformations are equivalent. They give the same S-matrix [16], but different anomalies [17]. This is due to the mathematical theorem that any k-form in differential geometry is a sum of an exact, a coexact and an harmonic form. It is this last part which contributes only to the anomalies.

IX. Problems With Spins Larger Than 2

For spin 0, ½, 1, 3/2 and 2, one can write down consistent Lagrangian field theories with interactions. One begins by choosing a field representation for a field with spin J, for example, the fields ϕ, λ^a (a=1, 4), A_μ, $\psi_\mu{}^a$ and $h_{\mu\nu} = h_{\nu\mu}$. Requiring that the free field theories have positive energy then reveals two things:

(i) the action for the given field representation is unique

(ii) for spins $J \geqslant 1$ the actions have a local gauge invariance. Thus gauge invariance folows from positive energy.

In order to couple certain free fields to each other, one may use the familiar order-by-order Noether method, and arrive in this way at, for example, QED, Yang-Mills theory, supergravity and gravity. This was first done, for the case of gravity, by Gupta and Thirring. In supergravity the Noether method has yielded the coupling of matter to gauge actions and the nonlinearities in gauge actions, and these results finally gave rise to a tensor calculus.

For spins exceeding two there seem to be problems. Let us analyze the case of spin 5/2, and take as field representation a symmetric tensor-spinor $\psi_{\mu\nu}{}^a = \psi_{\nu\mu}{}^a$. General arguments concerning what field representations for spin J are possible can be found in ref [18]. One writes down the most general action of the form $\bar\psi\partial\psi$, i.e, bilinear in the spinors, and with one derivative. The bar is Dirac's bar. As for spin 3/2 we consider massless and real spinors. Real in the sense of Majorana, meaning that if $\gamma_1, \gamma_2, \gamma_3, \gamma_0$ are real, then also $\psi_{\mu\nu}$ is real. In a general representation of the Dirac matrices, $\psi_{\mu\nu}$ is not real, but satisfies a reality conditon. This has as a consequence that one has the usual symmetries, such as

$$\bar\psi_{\mu\nu}\gamma_\nu\partial_\rho\psi_\mu = -\partial_\rho\bar\psi_\mu\gamma_\nu\psi_{\mu\nu} \tag{1}$$

One next requires positive energy. What one technically does is equivalent but easier: one inverts the field equations, using spin projection operators. One then sandwiches this propagator with external sources which satisfy only those constraints which follow from the field equations (cf in QED: from $\partial_\mu F_{\mu\nu} = j_\nu$ one finds the constraints $\partial_\nu j_\nu = 0$). Although the original free-field equations were singular, and the propagator therefore ambiguous, the sandwiched propagators are unambiguous. Requiring that there be only first-order poles with positive residues completely fixes the undetermined parameters in the orinigal action. The result is

$$\mathcal{L} = -\tfrac{1}{2}\bar{\psi}_{\mu\nu}\not{\partial}\psi_{\mu\nu} + \bar{\psi}\cdot\gamma_\mu\left(2\partial\cdot\psi_\mu - \not{\partial}\gamma\cdot\psi_\mu - \partial_\mu\psi\right) + \tfrac{1}{4}\bar{\psi}\not{\partial}\psi \tag{2}$$

where

$$\psi = \psi_{\mu\mu}\,,\quad \partial\cdot\psi_\mu = \partial_\nu\psi_{\mu\nu} \quad \text{and} \quad \gamma\cdot\psi_\mu = \gamma_\nu\psi_{\mu\nu} \tag{3}$$

This action is invariant under local fermionic gauge transformation with a spin 3/2 Majorana vector-spinor $\varepsilon_\mu^{\ a}$ as gauge parameter

$$\delta\psi_{\mu\nu} = \partial_\mu\epsilon_\nu + \partial_\nu\epsilon_\mu \qquad \text{with} \quad \gamma^\mu\epsilon_\mu = 0 \tag{4}$$

One may check that

$$\begin{aligned}\delta\mathcal{L} = {} & -2\bar{\psi}_{\mu\nu}\not{\partial}\partial_\mu\epsilon_\nu + \bar{\psi}\cdot\gamma_\mu\left(2\Box\epsilon_\mu + 2\partial_\nu\partial_\mu\epsilon_\nu - \not{\partial}\not{\partial}\epsilon_\mu\right. \\ & \left.-\not{\partial}\partial_\mu\gamma\cdot\epsilon - 2\partial_\mu\partial\cdot\epsilon\right) + \left(2\bar{\psi}\cdot\partial_\mu - \bar{\psi}\cdot\gamma_\mu\not{\partial} - \bar{\psi}\partial_\mu\right)\left(\not{\partial}\epsilon_\mu + \partial_\mu\gamma\cdot\epsilon\right) \\ & + \bar{\psi}\not{\partial}\partial\cdot\epsilon = 0 \qquad \text{if} \quad \gamma\cdot\epsilon = 0\end{aligned} \tag{5}$$

However, in curved space, one expects

$$\delta\psi_{\mu\nu} = D_\mu\epsilon_\nu + D_\nu\epsilon_\mu \qquad \text{with} \quad \gamma^\mu\epsilon_\mu = 0 \tag{6}$$

and in this case one finds

$$\delta\mathcal{L} = R(\ldots.) + R_{\mu\nu}(\ldots.) + R_{\mu\nu\rho\sigma}(\ldots.) \tag{7}$$

The terms with dots are linear in ε and $\psi_{\mu\nu}$ and without the derivatives. The terms with R and $R_{\mu\nu}$ can be cancelled by adding the Einstein action (whose variation is proportional to $R_{\mu\nu} - \frac{1}{2}g_{\mu\nu}R$) and choosing $\delta e_\mu^{\ m}$ appropriately, but the terms with $R_{\mu\nu\rho\sigma}$ cannot be cancelled.

Exercise: Check that substituting (6) into (5) the coefficient of $R_{\mu\nu\rho\sigma}$ in (7) is nonvanishing.

In N=2 extended supergravity, the first discovered extended supergravity and the realization of Einstein's dream of unifying gravity with electromagnetism, the action reads

$$\mathcal{L} = \mathcal{L}^{E} + \mathcal{L}^{RS}(\psi^{i}_{,i=1,2}) + \mathcal{L}(\text{photon}) + \bar{\psi}\psi F\,(\text{Pauli terms}) + \psi^{4}\text{ terms} \tag{8}$$

So one needs gravity and two gravitini to make the (1, 3/2) coupling consistent . Does this idea work for spin 3? In other words, could one make the coupling of spin 5/2 consistent by constructing something like N=2 hypergravity? Probably not, because in N=2 supergravity the gravitini have no electric charge (no minimal EM coupling), whereas it seems not possible to turn off the gravitational charge (the energy) for the spin 5/2 fields. However, one can add a minimal EM coupling of the photon to the (2 real = 1 complex) ψ's, so that there is a (slim) possibility that the spin (3, 5/2, 2 and more?) coupling exists. From the N=2 model in supergravity one would expect that a nonminimal gravitational coupling like $\bar{\psi}\psi R$.... would be needed between the spin 5/2 and spin 2 fields , because these terms would be the analogue of the Pauli terms in (8) . There are many good arguments why one cannot couple massless spin 3 fields in a consistent way to other fields, but there were also many good arguments why one could not couple spin 3/2 in a consistent way. One gets the impression that here is an interesting problem waiting to be solved.

X. Quantization of Supergravity

One can quantize supergravity in three ways:

(i) by using BRST invariance

(ii) by using path-integrals with p,q integration variables (the Hamiltonian approach)

(iii) by using directly Feynman diagrams, and finding the extra vertices needed for unitarity and gauge-invariance of the S-matrix. Quantum supergravity reveals a number of interesting properties not present in simpler gauge theories like Yang-Mills theory or ordinary gravity such as

(i) if the gauge algebra is open (i.e., without auxiliary fields) one needs 4-ghost couplings in order that the theory be unitary and gauge-invariant

(ii) in supergravity antisymmetric tensor fields occur as gauge fields. The ghosts, introduced by fixing the gauges of these gauge fields, are themselves gauge fields, so that one needs ghosts for ghosts for ....

(iii) in certain gauges, such as $\mathcal{L}(\text{fix}) = \frac{1}{4}\bar{\psi}\cdot\gamma \not{\partial}\gamma\cdot\psi$, one finds new ghosts, so-called Nielsen-Kallosh ghosts. In general these ghosts appear in whenever in the gauge fixing term

$$\mathcal{L}(\text{fix}) = \frac{1}{2} F_\alpha \gamma^{\alpha\beta} F_\beta \tag{1}$$

the matrix $\gamma^{\alpha\beta}$ is field-dependent.

Let us begin with BRST invariance (Becchi-Rouet-Stora-Tyutin). The quantum action reads for field-independent $\gamma^{\alpha\beta}$ [1]

$$\mathcal{L}(\text{qu}) = \mathcal{L}(\text{class}) + F_\alpha \gamma^{\alpha\beta} d_\beta - \frac{1}{2} d_\alpha \gamma^{\alpha\beta} d_\beta + C^*_\alpha \gamma^{\alpha\beta} F_{\beta,\delta} C^\delta \tag{2}$$

The symbol $F_{\alpha,\beta}$ means: vary F_α under gauge transformations with parameters ξ^β, and omit ξ^β. The C^δ and $C_\alpha{}^*$ are the Faddeev-Popov ghosts and antighosts.

The d_α are the Nielsen-Kallosh ghosts, and if one eliminates them from the action by using their field equations $d_\alpha = F_\alpha$, one finds the gauge fixing term in (1) back. The action is invariant under BRST transformations, which are the remnant of the classical gauge invariance in the quantum action. They have a constant anticommuting parameter Λ , and are thus global instead of local symmetries. These rules read

$$\delta\phi^i = R^i{}_\alpha C^\alpha \Lambda \quad , \quad \delta C^*_\alpha = \Lambda d_\alpha$$

$$\delta C^\alpha = -\tfrac{1}{2} f^\alpha{}_{\beta\gamma} C^\gamma \Lambda C^\beta \quad , \quad \delta d_\alpha = 0 \tag{3}$$

The ϕ^i are the classical gauge fields and the classical gauge invariances are denoted by $\delta\phi^i = R^i{}_\alpha \xi^\alpha$. The $R^i{}_\alpha$ depend in general on ϕ^i and derivatives. Thus one finds the BRST rules for ϕ^i by replacing ξ^α by $C^\alpha\Lambda$, and in order to keep the same statistics, Λ must be anticommuting.

The BRST transformations are nilpotent. In fact, it was known that d_α has to be added to the theory in order that $\delta\delta C^*_\alpha$ be vanishing, before it was known that this d_α is the same as the Nielsen-Kallosh ghost, due to exponentiating a determinant (section XI).

One may check that $\mathcal{L}$(qu) is invariant under BRST transformations, provided the gauge algebra is closed. Closure means that the commutator of two local (clasical) gauge transformations on the gauge fields ϕ^i is (uniformly in ϕ^i) equal to a sum of local gauge transformations

$$\begin{aligned}[\delta(\eta), \delta(\xi)]\phi^i &= R^i{}_{\alpha,j} R^j{}_\beta \eta^\beta \xi^\alpha - \eta \leftrightarrow \xi \\ &= R^i{}_\gamma f^\gamma{}_{\alpha\beta} \eta^\beta \xi^\alpha \end{aligned} \tag{4}$$

The structure functions $f^\gamma{}_{\alpha\beta}$ depend in general on ϕ^i. Thus, in order to prove nilpotency of BRST transformations on C^α , one needs an extension of the Jacobi identities for the case that $f^\alpha{}_{\beta\gamma,i}$ (which means: right-differentiation w.r.t.ϕ^i) is nonvanishing. Evaluating the identity

$$\delta(\zeta)[\delta(\eta), \delta(\xi)]\phi^i - [\delta(\eta), \delta(\xi)]\delta(\zeta)\phi^i + \text{cyclic in } \zeta\eta\xi = 0 \tag{5}$$

and using the closure relations twice, one finds

$$R^i{}_\lambda A^\lambda{}_{\alpha\beta\gamma} \zeta^\gamma \eta^\beta \xi^\alpha + \text{cyclic in } \zeta\eta\xi = 0$$

$$A^\lambda{}_{\alpha\beta\gamma} = f^\lambda{}_{\alpha\delta} f^\delta{}_{\beta\gamma} + f^\lambda{}_{\alpha\beta,k} R^k{}_\gamma = 0 \tag{6}$$

If the local gauge invariances are linearly independent, one may omit the $R^i{}_\lambda$. This relation holds for bosonic as well as fermionic local symmetries.

It is believed (and true in all cases considered) that BRST invariance implies unitarity. Thus BRST invariance is a short-cut for obtaining the correct quantum action, and hence the correct Feynman rules. One can, however, derive the above result for $\mathcal{L}$(qu) from a path-integral approach. Strictly speaking, this should be done by using Dirac's formalism for constrained Hamiltonian systems and integrating out the p's. For simplicity one often follows a path-integral approach in configuration space, although this gives incorrect results in some instances.

For practical calculations one diagonalizes the term with F_α and d_α as follows

$$F_\alpha \gamma^{\alpha\beta} d_\beta - \frac{1}{2} d_\alpha \gamma^{\alpha\beta} d_\beta = -\frac{1}{2} d_\alpha' \gamma^{\alpha\beta} d_\beta' + \frac{1}{2} F_\alpha \gamma^{\alpha\beta} F_\beta \tag{7}$$

The best choices for the gauge fixing terms are

$$\mathcal{L}(\text{fix}) = \frac{1}{4}\left[\partial_\mu(\sqrt{g}\, g^{\mu\nu})\right]^2 + \frac{1}{4} \bar{\psi}\cdot\gamma \not\partial \gamma\cdot\psi + \alpha(e_{a\mu} - e_{\mu a})^2 \tag{8}$$

in which case the propagators become very simple. From here on one can do Feynman graph calculations, and find that in supergravity the <u>S-matrix</u> (not the Green's function) are one-loop finite. This has been explained theoretically, and two-loop finiteness has been proven theoretically. Three-loop finiteness of the N=8 model seems not impossible, in view of the vanishing of the β-function in the brother of the N=8 model in global supersymmetric Yang-Mills model [19]. At the conference we also presented work done with E. Sezgin on spontaneously broken N=8 supergravity. Also here the S-matrix is one-loop finite [20].

When $\gamma^{\alpha\beta}$ is field-dependent, one can find the extension of the action by adding all possible terms with $\gamma^{\alpha\beta}{}_{,i}$ which have the correct dimension and statistics. Since the algebra should be independent of dynamical models, one does not want $\gamma^{\alpha\beta}$ and F_α terms in the BRST rules, so that one keeps the same BRST rules. Requiring invariance of the action then fixes the coefficients of the extra terms with $\gamma^{\alpha\beta}{}_{,i}$ in the action. For details see Nielsen's article in Physics Letters [21] , and F. Ore and the author, to be published. The result is that the transformation rules in (3) are unchanged, while the quantum additions are a total BRST - derivative

$$\mathcal{L} = \mathcal{L}\text{ (class)} + \frac{\delta}{\delta\Lambda}\left(-\tfrac{1}{2} C^*_\alpha \gamma^{\alpha\beta} d_\beta + C^*_\alpha \gamma^{\alpha\beta} F_\beta\right) \tag{9}$$

where the variation w.r.t. Λ is a right-derivative. In fact, one could add to the term within brackets any other terms, and still keep invariance under BRST transformations, because the BRST laws are nilpotent. The extra terms in the action due to the field-dependence of $\gamma^{\alpha\beta}$ are

$$\mathcal{L}\text{ (extra)} = C^*_\alpha \gamma^{\alpha\beta}{}_{,i} R^i{}_\gamma C^\gamma \left(F_\beta - \tfrac{1}{2} d_\beta\right)_{(-)}{}^{\gamma} \tag{10}$$

For field - independent $\gamma^{\alpha\beta}$, the action (9) reduces to the action in (2).

XI. Ghosts in path-integrals

In the process of using path-integrals for the quantization of gauge theories, one encounters superdeterminants which one wants to write as an exponent in order to be able to do perturbation theory. Let us recall that a supermatrix is given by

$$M = \begin{pmatrix} A & B \\ C & D \end{pmatrix} \qquad \begin{matrix} A \text{ and } D \text{ bosonic} \\ B \text{ and } C \text{ fermionic} \end{matrix} \tag{1}$$

The superdeterminant can be written in two ways: sdet $M = \det A \det (D - CA^{-1} B)^{-1} = \det(A - BD^{-1} C) (\det D)^{-1}$.

One can derive this result from sdet $M = \exp(\mathrm{str}\,\ell n M)$ if one defines the supertrace of a supermatrix. In order that the supertrace satisfies str MN = str NM, the supertrace must be defined by str $N = \Sigma (-)^a N_a{}^a$, i.e., the trace over fermi-fermi elements should acquire an extra minus sign. One may prove that $\delta(\text{sdet } M) = (\text{sdet } M)\ \text{str}\,(M^{-1}\,\delta M)$

Superintegration over fermonic coordinates is defined by $\int d\theta\, \theta = 1$, $\int d\theta = 0$. These rules follow by requiring translational invariance

$$\int d\theta f(\theta) = \int d\theta f(\theta + a) \tag{2}$$

where the most general function $f(\theta)$ has the form $f(\theta) = A + \theta B$. (Since the θ anticommute, $\theta^2 = 0$). One can now derive two theorems concerning superintegration

Theorem I: $$\text{sdet } M = \int \prod_{k,\ell} dC_2{}^k dC_1{}^\ell \exp\left(C_1{}^i M_{ij} C_2{}^j\right)$$

Theorem II: $$(\text{sdet } M)^{1/2} = \int \prod_k dN^k \exp\left(N^i M_{ij} N^j\right) \tag{3}$$

where $C_1{}^k$, $C_2{}^k$ and N^k are commuting or anticommuting depending on whether the indices of M_{ij} with which they are contracted are fermionic or bosonic. ("Ghosts have opposite statistics").

To show a little bit why these results are true, consider first the case that only A in M is nonvanishing. In that case all $C_2{}^k$

and C_1^{ℓ} are anticommuting, and only one term in the expansion of the integrand of Theorem I contributes, namely the term with each ghost appearing once. (Terms without a given ghosts vanish due to $\int d\, C = 0$, while terms with a given ghost appearing more than once vanish due to $C\,C = 0$). Hence, only certain terms in $(C_1^{\ i} M_{ij}\, C_2^{\ j})^n$ contribute, and one easily convinces oneself that due to the antisymmetry of the $C_1^{\ i}$ and $C_2^{\ j}$ we end up with det M = det A. When only D in M is nonzero, the ghosts commute and one finds of course $(\det D)^{-1}$. In both these special cases one therefore ends up with sdet M.

Let us now comment on theorem II. Of course one cannot obtain a square root of a determinant by putting a factor ½ in the exponent of the Gaussian integral (think of ordinary Gaussian integrals). Let us again consider the case that only A in M is nonzero, so that the N^k are anticommuting. The matrix M is then antisymmetric. It is a well-known theorem for antisymmetric matrices that their determinant is a pure square. For example, in QED one has $\det F_{\mu\nu} = (\vec{E}\cdot\vec{B})^2$. If there are $n = 2\ell$ ghosts N^k in Theorem II, only the term with $(N^i M_{ij} N^j)^{n/2}$ contributes, and one obtains indeed $\sqrt{\det A}$. As already mentioned, $\sqrt{\det A}$ is an algebraic expression in the matrix elements A_{ij}.

<u>Exercise</u>: prove that

$$\det M = (a\,f - eb + cd)^2 \qquad M = \begin{pmatrix} o & a & b & c \\ & o & d & e \\ & & o & f \\ & & & o \end{pmatrix}$$

where

$M = -M^T$ and

I thank B. Zumino for this example.

We have discussed in section I that auxiliary fields are necessary in the classical theory in order to have a closed gauge algebra off-shell. It might even be true that certain globally supersymmetric matter systems can only be coupled to supergravity if auxiliary fields are present (in this case one presumably could not eliminate them through their field equation from the quantum action). One can quantize the theory without auxiliary fields. However, the naive ideas then do not hold. For example, in order to prove that the path-integral is independent of the gauge fixing term F_α, one usually tries to show that $\int d\phi$ times the Faddeev-Popov determinant is invariant under certain (nonlinear) gauge transformation. However, this is not true when the gauge algebra is open. (One can prove gauge independence in other ways, for example using Hamiltonian path-integrals or BRST techniques).

When one quantizes using path integrals, one adds "unity" $\sim$ $\int \delta(F_\alpha - b_\alpha)\Delta_{FP}$. Following 't Hooft, one exponentiates the delta-function $\delta(F_\alpha - b_\alpha)$ by multiplying again by "unity"

$$\text{"unity"} = (\text{sdet}\,\gamma)^{1/2} \int db_\alpha \; e^{\frac{1}{2} b_\alpha \gamma^{\alpha\beta} b_\beta} \tag{4}$$

(recall "Theorem II"). Combining with the Faddeev-Popov determinant $\Delta_{FP} = \text{sdet}\, F_{\alpha,\beta}$ one finds

$$(\text{sdet}\,\gamma)^{1/2}\; (\text{sdet}\, F_{\alpha,\beta}) = (\text{sdet}\,\gamma)^{-1/2}\; \text{sdet}\,(\gamma^{\alpha\beta} F_{\beta,\gamma}) \tag{5}$$

The sdet $\gamma^{-\frac{1}{2}}$ is exponentiated, using "Theorem II" with real Nielsen-Kallosh ghosts, and one arrives at the $\mathcal{L}$(qu) of Section X, eq. (2).

Considering a closed fermion triangle loop in a background gravitational field one clearly must consider three contributions

(i) the real anticommuting gravitino itself

(ii) the complex commuting supersymmetry Faddeev-Popov ghosts

(iii) the real anticommuting (d_α has the same statistics as F_α) Nielsen-Kallosh ghosts. These loops yield the axial anomaly.

We must now determine their axial weights before we can obtain the axial anomaly. The theory has a global chiral symmetry

$$\delta\psi_\mu = i\gamma_5\psi_\mu \,, \quad \delta C^*_\alpha = i\gamma_5 C^*_\alpha \,, \quad \delta C^\alpha = (i\gamma_5 C)^\alpha \tag{6}$$

Choosing $\delta\psi_\mu = + i\gamma_5\psi_\mu$, one finds the weights of the Faddeev-Popov ghosts by looking at $C^*_\alpha\, \psi\, C_\nu$ vertices (where C_ν is the coordinate ghosts). Finally, the weight of d_α follows from the extra term with $\gamma^{\alpha\beta}{}_{,j}$ in $\mathcal{L}$(qu): $\delta d_\alpha = - i\gamma_5 d_\alpha$. Thus one calculates and finds for the axial anomaly - 21A, where 2A is the anomaly for the electron [22].

In some topological approaches one only considers the two modes in the gravitino, but no ghosts. This amounts to not exponentiating $\delta(F_\alpha - b_\alpha)$, so that no NK ghosts are needed, while one works in the unweighted gauge $\gamma\cdot\psi = 0$. The same answer is obtained [23], in fact it was obtained before the Feynman graph calculation in [22] which used the Adler-Rosenberg method.

REFERENCES

[1]P. van Nieuwenhuizen, Physics Reports 68, 189-398(1981).

[2]A. Salam and J. Strathdee, Nucl. Phys. B80, 499(1974).

[3]M. T. Grisaru, H. Pendleton and P. van Nieuwenhuizen, Phys. Rev. D15, 996 (1977).

[4]S. Ferrara, F. Gliozzi, J. Scherk and P. van Nieuwenhuizen, Nucl. Phys. B117, 333(1976).

[5]S. Ferrara and P. van Nieuwenhuizen, Phys. Rev. Lett. 37, 1669(1976).

[6]J. F. Luciani, Nucl. Phys. B135, 111 (1978)

[7]M. Kaku, P.K. Townsend and P. van Nieuwenhuizen, Phys. Rev. D17, 3179(1978).

[8]S. Ferrara and B. Zumino, Nucl. Phys. B134, 301(1978).

[9]K. S. Stelle, Phys. Rev. D 16, 953(1977).

[10]S. Ferrara, M. T. Grisaru and P. van Nieuwenhuizen, Nucl.Phys. B138, 403(1978)

[11]E. Fradkin, Lebedev preprint.

[12]A. D'Adda, R. D'Auria, P. Fre' and T. Regge. Riv. del Nuovo Cim. 6(1980).
L. Castellani, P. Fre' and P. van Nieuwenhuizen, Ann. of Physics.

[13]The earliest paper is: Y. Ne'eman and T. Regge, Riv. del Nuovo Cim. 1, 5, 1(1980).

[14]P. K. Townsend and P. van Nieuwenhuizen, Phys. Lett. 67B, 439 (1977).
A. H. Chamseddine and P. C. West, Nucl. Phys. B129, 39(1977).

[15]H. Nicolai and P. K. Townsend , Phys. Lett. 98B, 257(1981).

[16]E. Sezgin and P. van Nieuwenhuizen, Phys. Rev. D22, 301(1980).

[17]M. J. Duff and P. van Nieuwenhuizen, Phys. Lett. 94B, 179(1980).

[18]F. A. Berends, J. W. van Holten, B. deWit and P. van Nieuwenhuizen, J. Phys. A13, 1643(1980), Nucl. Phys. B154, 261(1979).

[19]L. V. Avdeev, O. V. Terasov, A. A. Vladimirov, Phys. Lett 69B, 94(1980)
M. T. Grisaru, M. Rocek and W. Siegel, Nucl. Phys B183, 141(1981)
W. E. Caswell and D. Zanon, Nucl. Phys. B181, 125(1981)

[20]E. Sezgin and P. van Nieuwenhuizen, Nucl. Phys. B(1981)

[21]N. K. Nielsen, Phys. Lett. 103B, 197(1981).

[22]M. T. Grisaru, N. K. Nielsen, H. Römer and P. van Nieuwenhuizen, Nucl. Phys. B140, 477(1978).

[23]S. Christensen and M. J. Duff, Phys. Lett 76B, 571(1978).

M.J. Duff†
CERN, Geneva, Switzerland

Abstract. Recent calculations have confirmed the belief that the ultraviolet behaviour of N-extended supergravities improves with increasing N. We give a comprehensive review.

CONTENTS

---

† Permanent address: Blackett Laboratory, Imperial College, London.

1 INTRODUCTION

1.1 Quantum gravity and grand unification

These lectures are about the ultraviolet problem in gravity, but I would like to begin with some thoughts on the grand unification of strong, weak, and electromagnetic interactions. Here, the question is "How many of the observed, and yet to be observed, particles are truly elementary and what are their quantum numbers?" The answer might lie in SU(5) or SO(10) or maybe some more primitive theory of "preons", but it remains a mystery why Nature should single out one symmetry from among the many mathematical possibilities. To begin with, how does one count the number of particle species? Well, there is a sense in which gravity does just that. Consider, for example, closed-loop corrections to the graviton self-energy. By the Equivalence Principle, gravity couples to everything (including itself) with equal strength. Thus no matter whether the particle going round the loop is a quark, a W boson, a photon, or whatever; each contributes to the self-energy with the same order of magnitude. For consistency, therefore, one must include all the elementary particles irrespective of their masses or internal quantum numbers. In this way gravity cares, in a way which no other force does, just how many particles there are. Thus it is not inconceivable that gravity may have something to say about the spectrum of the elementary particles.

Now let us turn to the other side of the coin, to the problem of constructing a consistent quantum theory of gravitation. Here, the question is "How does one make sense of a theory which, by power-counting at least, is non-renormalizable?" As we shall recall in Section 2, the superficial degree of divergence of a Feynman graph is given by $D = (d - 2)L + 2$, where d is the dimension of space-time and L the number of closed loops. For $d = 4$, this increases with increasing loop order and leads to a non-renormalizable theory. One way out might be to couple gravity to matter fields and to look for a mutual cancellation of ultraviolet divergences. Note that this represents a shift in philosophy away from conventional renormalizability, and away also from the old ideas of an ultraviolet cut-off at the Planck length. Rather one hopes that on-shell S-matrix elements will be finite order by order in perturbation theory. It should be clear, however, that if this idea is to work at all it could work only for some very special assignment of masses, spins, and internal quantum number of the matter fields. Thus it is not inconceivable that the spectrum of the elementary particles may, in its turn, have something to say about gravity.

With these premises, the most economical conclusion is that the problem of constructing a grand unified strong electro-weak theory and the problem of constructing a quantum theory of gravity are really one and the same problem: only with the right elementary particles will the theory be finite; finiteness determines the right elementary particles. And before dismissing the whole idea as being too fantastic, one should recall, as we shall in Section 2, some of the alternative proposals for solving the renormalizability problem in gravity in comparison to which, one could argue, the present idea is rather conservative. [After all, we are already used to the anomaly-free criterion in grand unified theories (GUTs), whereby the absence of certain divergences is invoked to restrict the allowed numbers of quarks and leptons.]

This idea is not new and was pursued before the discovery of supergravity, albeit without much success. First it was realized from general positivity arguments that contributions to the graviton self-energy from particles of spin 0, $\frac{1}{2}$, and 1 entered not only with the same order of magnitude but also with the same sign (Capper et al. 1974; Capper & Duff 1974 a; Deser & van Nieuwenhuizen 1974 b.) So hopes of infinity cancellations in off-shell Green's functions seemed hopeless. More promising was the idea of finite on-shell S-matrix elements, but although pure gravity was found to be all right at one-loop ('t Hooft & Veltman 1974), gravity coupled to various combinations of spins 0, $\frac{1}{2}$ and 1 gave infinite results. (A list of references is given in Section 2.) And the prospect for higher loop order, with or without matter, seemed even bleaker. Even at the time, one was aware of two shortcomings in this approach. First, in no sense was there a unification of gravity and matter; one simply picked one's favourite matter theory and only afterwards grafted on the gravity. Secondly, it was completely hit-and-miss in its attempts to find the magic combinations of matter fields; there was simply no guiding principle. This state of affairs was reviewed by Duff (1975) and Deser (1975) in the Proceedings of the 1974 Oxford Quantum Gravity Conference. With characteristic foresight, Salam (1975) pointed out in the same volume a third possible shortcoming: the neglect of spin $\frac{3}{2}$.

1.2 Supergravity

By consistently coupling spin-2 gravitons to spin-$\frac{3}{2}$ "gravitinos" simple (N = 1) supergravity (Freedman et al. 1976; Deser & Zumino 1976) provided the first example of a gravity-matter system yielding finite on-shell S-matrix elements at one loop (Grisaru et al. 1976). But it did much more. Here, for the first time, was the dreamed-of unification: gravity and matter as merely different components of the same symmetry multiplet, and Bose-Fermi symmetry as an obvious candidate for that missing guiding principle. These features become even more striking when one considers the extended ($1 < N \leq 8$) supergravities and especially the N = 8 theory which combines one spin-2, eight spin-$^3/_2$, twenty-eight spin-1, fifty-six spin-$^1/_2$, and seventy spin-0 particles in one supermultiplet. Although the internal symmetry assignments prevent these particles from being identified with those observed at present energies, they might possibly be the preons from which bound-state quarks and so on are formed (Ellis et al. 1980; Derendinger et al. 1981). Now conventional model-building via preons is prejudiced by the economical requirement that the number of pre-particles be small; a criterion obviously violated by N = 8 supergravity. However, if one thinks in terms of pre-fields rather than pre-particles then it is the ultimate in economy since there is but one single superfield.

Naturally, two questions now arise (a) Is there a finite theory of extended supergravity? (b) Does it describe the right particles? These lectures will summarize what we know in response to (a). The reader is also referred to the lectures by Dr. Kallosh at this school. Other recent reviews on ultraviolet divergences may be found in Weinberg (1979) and van Nieuwenhuizen (1981 a). As described in detail by Dr Kallosh, and summarized here in Section 2, it has now been established that supergravity theories are on-shell finite both at one and two loops, but that superinvariants exist as possible counterterms at three loops and beyond. [We are assuming here (a) trivial space-time topology, (b) no cosmological constant, and (c) no breaking of supersymmetry. We shall deal with cases (a) and (b) later on. Case (c) is discussed by van Nieuwenhuizen (1981 b).] As Kallosh explains, moreover, going to higher N (e.g. N = 8) does not avoid the problem of invariants as potentially dangerous counterterms. The only hope remaining, it seems, is that the coefficient of such invariants must vanish in the counterterm. (And here we remind the reader that, despite great efforts in quantum gravity, no explicit calculations yet exist beyond one loop.) This might seem like clutching at straws were it not for the fact that we already have concrete examples in supersymmetry and supergravity where this non-renormalization phenomenon actually occurs!

The first example, discussed in Section 3, concerns extended supergravity with local SO(N) invariance. When the internal SO(N) symmetry is gauged, the usual arguments for one-loop finiteness cease to apply because of the appearance of a cosmological constant related to the gauge coupling constant e. Indeed, for $N \leq 4$ one finds that infinite renormalizations are required. Remarkably, the particle content of theories with $N > 4$ results in a cancellation of these infinities implying, in particular, a vanishing one-loop $\beta(e)$ function (Christensen et al. 1980). This is reminiscent of the vanishing β-function in supersymmetric Yang-Mills theories for $N > 2$, which is now known to hold to at least three-loop order (Avdeev et al. 1980; Grisaru et al. 1980; Caswell & Zanon 1980). The point I wish to emphasize is that in both cases candidate counterterms do exist but nevertheless appear with zero coefficient.

The purpose of Section 3 will be to show how, to one loop at least, these otherwise "miraculous" cancellations have a common explanation in terms of certain "spin-moment sum rules" (Curtwright 1981). These sum rules provide at last concrete evidence that higher N means better ultraviolet behaviour, as had long been hoped for. In showing how these spin sum rules are relevant to ultraviolet divergences and anomalies, we shall revive some earlier work of Christensen & Duff (1978 a, 1979) on counterterms, axial anomalies, and conformal anomalies for fields of arbitrary spin.

Another point of technical interest concerns the cosmological constant Λ related to the Yang-Mills coupling constant e of the gauged extended supergravities by $\kappa^2\Lambda = -6e^2$, where $\kappa^2 = 16\pi G$ and G is Newton's constant. To calculate the $\beta(e)$ function, therefore, one may either (a) compute the usual charge renormalization effects, i.e. the coefficient of the Yang-Mills Tr $\sqrt{g}\, F^{\mu\nu}F_{\mu\nu}$ counterterm, which receives contributions from spins 0, ½, 1, and 3/2 but not 2, since the graviton is a singlet, or (b) compute the cosmological renormalization i.e. the coefficient of the $\sqrt{g}$ counterterm, which receives contributions from all spins including gravity. By supersymmetry, one coefficient determines the other. (Incidentally, it is amusing to note that, whether or not one believes in supergravity, this enables one to deduce the magnitude of graviton loop effects in pure gravity from knowledge of flat-space Yang-Mills theories.) In Section 3 we shall calculate $\beta(e)$ both ways and demonstrate their equivalence. In order to carry out method (b), however, one must first understand how to handle quantum effects of gravity with a non-vanishing cosmological constant (Christensen & Duff 1980), which we shall also briefly describe.

As we have discussed, it is our purpose in these lectures to concentrate on those ultraviolet properties which are peculiar to particular values of N, rather than dwell on those common to all N. In Section 4 we return to ordinary (i.e. ungauged) supergravity and examine another aspect of N dependence, namely anomalies and topological counterterms. In gravity theories, the anomalous contribution to the trace of the effective energy momentum tensor receives a one-loop contribution proportional (on-shell) to $\varepsilon^{\mu\nu\rho\sigma}R_{\rho\sigma\alpha\beta}\varepsilon^{\alpha\beta\gamma\delta}R_{\mu\nu\gamma\delta}$ which, when integrated over all space, yields a topological invariant: the Euler number χ (Duff 1977). Now anomalies arise because of divergences, and there is a corresponding counterterm proportional to χ which is non-zero in spaces with non-trivial topology. The numerical coefficient of this anomaly, call it A, has been calculated for extended supergravity theories and found to be non-vanishing, and non-integer, for N = 1 and 2 but equal to an integer, A = 3-N, for $N \geq 3$ (Christensen & Duff 1978 a; Christensen et al. 1980). These calculations assume the usual field representation assignments for each spin. Recently, however, it was shown that A depends not only on spin but also on choice of representation (Duff & van Nieuwenhuizen 1980). Thus, contrary to naïve expectations, the gauge theory of a rank-two antisymmetric tensor field $\phi_{\mu\nu}$ is not equivalent to a scalar ϕ, even though both describe spin-0. Similarly, the gauge theory of a rank-three antisymmetric field $\phi_{\mu\nu\rho}$ is not equivalent to nothing.

These results might be of only academic interest were it not for the appearance of such unusual representations in the auxiliary fields of simple supergravity and in the versions of extended supergravity obtained via dimensional reduction. For example, N = 1 supergravity in d = 11 dimensions contains a rank-three field $\phi_{\mu\nu\rho}$. After dimensional reduction to d = 4 dimensions, one obtains an N = 8 theory, not with seventy scalars, but with sixty-three ϕ, seven $\phi_{\mu\nu}$, and one $\phi_{\mu\nu\rho}$. Remarkably, the A coefficient for N = 8 with these representations now vanishes! A similar phenomenon happens for the d = 4, N = 4 theory obtained from d = 10, N = 1, where the two spin-0 fields appear as one ϕ and one $\phi_{\mu\nu}$. So now we have A = 0 not only for N = 3, but also N = 4 and N = 7, 8. By extrapolating the representation assignments to N = 5 and 6, one can arrange for A = 0 for all $N \geq 3$ (Duff 1981 b,c; Nicolai & Townsend 1981). The derivation of these results has been discussed at length elsewhere (Duff 1981 c) and so in Section 4, I shall instead concentrate on their interpretation within the framework of superfield quantization and superfield Feynman graphs (Grisaru & Siegel 1981 a). In particular, we shall note that of the two kinds of superfield, chiral and non-chiral, only closed loops of chiral superfields contribute to the A coefficient. By analysing the extended theories in terms of N = 1 superfields, therefore, one can explain the absence of anomalies and the finiteness by noting that the net number of chiral superfields (i.e. physical minus ghost) is zero for $N \geq 3$.

We also discuss in Section 4 how anomaly coefficients can change not only with a change of physical and/or auxiliary field assignments but also with a change of boundary conditions.

Since these lectures were delivered, there have been several interesting new developments in the subject. These are summarized in Section 5.

1.3 Kaluza-Klein?

Finally, I should mention that there has recently been a renewed interest in higher dimensional theories of the Kaluza-Klein type. Here one begins with a gravity theory on $M \times B$, where M is four-dimensional space-time and B is some compact space, and ends up with a gravity-Yang-Mills theory on M with a gauge group determined by the symmetries of B. There are two reasons why N = 1 supergravity in d = 11 is particularly interesting in this connection. First, as pointed out by Witten (1981), eleven dimensions is both the minimum number to accommodate $SU(3) \times SU(2) \times U(1)$ as a gauge group and the maximum number allowed by supersymmetry. (Witten's paper contains many other interesting results.) Secondly, as pointed out by Freund and Rubin (1980), preferential compactification to four (or seven) dimensions is found to occur dynamically. This is because a rank-three gauge potential $\phi_{\mu\nu\rho}$ can give rise to a cosmological constant (Duff & van Nieuwenhuizen 1980; Aurelia et al. 1980) and its appearance in d = 11 supergravity is just what is required to make $M \times B$ a solution of the field equations with B compact with d = 7 and M non-compact with d = 4. Note that in this Kaluza-Klein picture, the extra dimensions must be taken seriously. This is to be contrasted with the dimensional reduction discussed earlier, which is merely a cunning device for determining the N = 8 Lagrangian in d = 4, and corresponds to discarding all but the massless modes.

In this Kaluza-Klein picture, therefore, the question is not whether N = 8 supergravity is finite in four dimensions, but whether N = 1 supergravity is finite in eleven dimensions!

In d = 11 the degree of divergence is 9L + 2, which is of course worse than in d = 4. Remember, however, that we are not looking for power-counting renormalizability but finiteness due to a cancellation of ultraviolet divergences. *A priori*, this seems to me just as likely in d = 11 as in d = 4. Moreover, in odd dimensions, gravity theories are automatically finite at odd loop order since there are no invariants formed from the metric involving an odd number of derivatives. In this respect, we are already half-way there!

In any event, the lectures presented here are based on the prejudice that we live in four dimensions. Ultraviolet divergences in Kaluza-Klein theories will be treated elsewhere (Duff & Toms 1981).

2 REVIEW OF RENORMALIZABILITY PROBLEM

2.1 Pure gravity

Consider the Lagrangian for pure gravity with zero cosmological constant

$$\mathcal{L} = - \frac{1}{16\pi G} \sqrt{g}\, R \ . \tag{2.1}$$

Since the curvature scalar R involves two derivatives of the metric, the corresponding momentum-space vertex functions will behave like p^2, and the propagator like $1/p^2$. In d dimensions each loop integral will contribute p^d, so with L loops, V vertices, and P internal lines, the superficial degree of divergence of a Feynman diagram is given by

$$D = dL + 2V - 2P \ . \tag{2.2}$$

Combined with the topological relation

$$L = 1 - V + P \tag{2.3}$$

this yields

$$D = (d - 2)\, L + 2 \ . \tag{2.4}$$

Note that D does not depend on the number of external lines. The crucial point is that D increases with increasing loop order for $d > 2$ and leads to a non-renormalizable theory. For $d = 2$, Eq. (2.1) ceases to have any dynamical content since $\mathcal{L}$ is a total derivative. Let us see what this means in practice for $d = 4$ within the framework of some specific regularization scheme. We shall use dimensional regularization which means working in $4 + \varepsilon$ dimensions and then letting $\varepsilon \to 0$ at the end. At one loop, $D = 4$, which means we expect the one-loop counterterms $\mathcal{L}_{(1)}$ to depend on four derivatives or less. On dimensional grounds, the only generally covariant scalars available are $R_{\mu\nu\rho\sigma}R^{\mu\nu\rho\sigma}$, $R_{\mu\nu}R^{\mu\nu}$, and R^2. Hence

$$\mathcal{L}_{(1)} = \frac{1}{\varepsilon} \sqrt{g} \left[\alpha R_{\mu\nu\rho\sigma}R^{\mu\nu\rho\sigma} + \beta R_{\mu\nu}R^{\mu\nu} + \gamma R^2 \right] . \tag{2.5}$$

We are assuming here that the background field method has been employed, (see Section 3) so that the counterterm depends only on the background metric and not on the gravitons or ghosts going round the loop. The constant α is independent of one's choice of gauge for the quantum graviton field but the constants β and γ are not. This corresponds to the fact that off-mass-shell Green's functions may be gauge-dependent. It is the on-shell S-matrix elements which correspond to the gauge-invariant physics. Within the framework of the background field method, putting the external lines on mass shell corresponds to using the classical equations of motion for the background field, i.e. $R_{\mu\nu} = 0$. Before doing so, however, we first note that the combination $R_{\mu\nu\rho\sigma}R^{\mu\nu\rho\sigma} - 4R_{\mu\nu}R^{\mu\nu} + R^2$ is a total divergence and its integral over all space may be neglected provided space-time has trivial topology, which we assume for the moment (otherwise it yields a topological invariant, the Euler number, which takes on integer values in spaces with non-trivial topology: see Section 4). Consequently $\int d^4x\, \mathcal{L}_{(1)}$ vanishes on-shell, and hence at one loop order on-shell S-matrix elements are actually finite.

The real problem arises when we go beyond one loop. At two loops, for example, D = 6 and in addition to terms vanishing with the field equations we anticipate terms like $\sqrt{g}\, R_{\mu\nu}{}^{\alpha\beta} R_{\alpha\beta}{}^{\gamma\delta} R_{\gamma\delta}{}^{\mu\nu}$ which do not. Consequently divergences would survive even for on-shell S-matrix elements which can be removed only by counterterms of a kind not present in the original Lagrangian. In general, we anticipate an infinite number of distinct counterterms and correspondingly an infinite number of undetermined parameters: the disaster of non-renormalizability.

In the literature one may find several possible responses to this disaster:

a) Quantum gravity makes no sense. In other words, one should quantize all matter fields but keep the gravitational field classical. See, for example, Kibble (1981).

b) Quantum gravity is all right, but the problem is perturbation theory, i.e. sum all graphs, or some appropriate subset of all graphs, in the hope that the result will be finite and unambiguous. This was the idea behind the old non-polynomial Lagrangian approach (Isham et al. 1971). Again, the problem was not so much in obtaining a finite answer, but in obtaining a finite answer which was unambiguous.

c) Perhaps renormalizability is, after all, the wrong criterion. A recent proposal which falls into this category is that of Weinberg's "Asymptotic Safety" programme (Weinberg 1979).

d) Modify Einstein's Lagrangian to include terms quadratic in the curvature and hence depending on four derivatives of the metric (DeWitt 1965; Stelle 1977). Now the dominant behaviour of the vertices is p^4 and that of the propagators $1/p^4$. Hence, in four dimensions $D = 4L + 4V - 4P = 4$. Thus no counterterms are required beyond those of the kind already present in the original Lagrangian. However, renormalizability has been bought at the expense of apparent lack of unitarity, since four derivative theories contain unphysical ghost-poles in the propagators. Various arguments have been put forward to circumvent this unitarity problem but none with complete success. Summing bubble graphs, for example, seems to lead to a lack of causality (Tomboulis 1980); whereas propagating torsion theories can be either unitary or renormalizable but not both simultaneously (Sezgin & van Nieuwenhuizen 1980 a). It remains unclear whether this apparent lack of unitarity is an artefact of one's approximation scheme or whether it would disappear in the exact theory. The reader is referred to the literature (Julve & Tonin 1978; Salam & Strathdee 1978 a; Weinberg 1979, Fradkin & Tseytlin 1981 a,b,c; Christensen 1981). Under the category of fourth-order theories we should also include the "induced gravity" approach, initiated by Sakharov (1968), whereby the Einstein Lagrangian, plus fourth-order terms, is induced by quantum matter effects. A review of induced gravity theories has recently been given by Adler (1981).

e) The problem is not with Einstein's theory *per se*, nor with perturbation theory, but with the failure to include precisely the correct set of matter fields. In other words there exists a, possibly unique, choice of matter fields for which a mutual cancellation of infinities occurs leading to finite on-shell S-matrix elements order by order in perturbation theory.

It is not my intention here to attack or defend proposals (a) to (d) except to repeat the comment made in the Introduction that, according to current ideas in quantum field theory, proposal (e), though bold, seems to me conservative by comparison. My own objections to semi-classical approaches may be found elsewhere (Duff 1981 a).

2.2 Gravity plus matter

Whether or not we adopt viewpoint (e) above, it is natural to ask what happens when coupling to matter is allowed. Again one can write down the most general one-loop counterterm consistent with dimensional analysis, general covariance, and whatever other symmetries are present in the theory. Thus, in addition to $\sqrt{g}\, R_{\mu\nu}R^{\mu\nu}$ and $\sqrt{g}\, R^2$ terms, one might expect $\kappa^2\sqrt{g}\, R_{\mu\nu}T^{\mu\nu}$, $\kappa^2\sqrt{g}\, RT^{\mu}{}_{\mu}$ terms or $\kappa^4\sqrt{g}\, T_{\mu\nu}T^{\mu\nu}$, $\kappa^4\sqrt{g}\, T^{\mu}{}_{\mu}T^{\nu}{}_{\nu}$ terms, where $T_{\mu\nu}$ is the energy-momentum tensor of the matter fields. Other, more complicated terms involving matter fields will also be present in general. Unlike pure gravity, however, explicit calculations are now required to fix the numerical coefficients. These calculations have now been carried out for various gravity-matter systems ('t Hooft & Veltman 1974; Deser & van Nieuwenhuizen, 1974 a,b; Deser et al. 1974; Nouri-Moghadam & Taylor 1975; Sezgin & van Nieuwenhuizen 1980 b; Duff & van Nieuwenhuizen 1980; van Proeyen 1980; Barvinsky & Vilkovisky 1981). In general, all terms which are allowed by the symmetry appear with non-vanishing coefficients. If the matter fields are massive ($m \neq 0$), moreover, then one also acquires new divergences like $m^4\sqrt{g}$ and $m^2\sqrt{g}\, R$. (Such terms disappear in the massless limit if, as we are assuming, one employs a regularization scheme with a dimensionless regularizing parameter. With a dimensionful cut-off Λ, terms like $\Lambda^4\sqrt{g}$ and $\Lambda^2\sqrt{g}\, R$ survive even for massless theories.) Occasionally, however, there are some surprises when *a priori* allowed counterterms do not appear, for example the vanishing of $\sqrt{g}\, R_{\mu\nu\rho\sigma}F^{\mu\nu}F^{\rho\sigma}$ in Einstein-Maxwell theory (Deser & van Nieuwenhuizen 1974 b) or the finiteness of the gravitational modification to the anomalous magnetic moment of the electron in gravity-modified QED (Berends & Gastmans 1975). These can be explained by invoking some non-obvious symmetry (like duality invariance) or else by embedding in supergravity (Deser 1981; van Proeyen 1980).

The crucial question, of course, is whether such one-loop counterterms vanish on-shell. By "on-shell" we now mean on using the non-vacuum Einstein equations $R_{\mu\nu} - \frac{1}{2} g_{\mu\nu}R = \kappa^2 T_{\mu\nu}$ plus the equations of motion for the matter fields. For all combinations and representations of fields with spin 0, $\frac{1}{2}$, and 1 which have been tried to date, the answer is no! Thus otherwise "respectable" theories like QED, QCD, Weinberg-Salam, or GUTS (all of which were obtained by requiring renormalizability) cease to make sense when gravity is present.

It is perhaps hardly surprising that the coupling of a renormalizable theory like QED to a non-renormalizable theory like gravity, leads to divergent results. Should one, in order to obtain a finite theory, couple gravity to another non-renormalizable theory? An example would be gravity plus non-linear σ-model (Duff & Goldthorpe 1981). This theory is not finite, either. [The motivation for the calculation of one-loop counterterms was rather to demonstrate the inconsistency of quantum field theory in curved space-time (Duff 1981 a).] However, it is interesting to note that the one-loop counterterms for the σ-model with coupling constant κ' are already in flat space of the form $\kappa'^4 T_{\mu\nu}T^{\mu\nu}$ and $\kappa'^4 T^{\mu}{}_{\mu} T^{\nu}{}_{\nu}$, thus increasing the probability of cancellations with the previously mentioned curved-space counterterms if κ' is chosen to equal κ. So this may be a step in the right direction, and such couplings do in fact occur in extended supergravity. It is to supergravity that we now turn.

2.3 Simple supergravity

The Lagrangian for simple supergravity (Freedman et al. 1976; Deser & Zumino 1976)

$$\mathcal{L} = -\frac{1}{2\kappa^2} eR + \frac{1}{2} \varepsilon^{\mu\nu\rho\sigma}\bar{\psi}_\mu \gamma_5 \gamma_\nu D_\rho \psi_\sigma + \frac{e}{3}\left(S^2 + P^2 - b^\mu b_\mu\right) \qquad (2.6)$$

describes the coupling of gravity $e_\mu{}^a$ to a single spin-3/2 Majorana spinor ψ_μ and does not fall into the category of gravity-matter systems discussed previously. We have also included in Eq. (2.6) the minimal set of auxiliary fields S, P, and b_μ which vanish on-shell but are necessary for the closure of the supersymmetry algebra off-shell (Stelle & West 1978; Ferrara & van Nieuwenhuizen 1978 a). We shall return to these in a moment. The first quantum-loop calculations in supergravity were in fact carried out before these auxiliary fields were known. As is by now well-known, supergravity provided the first example of a gravity-matter system which was one-loop finite on-shell (Grisaru et al. 1976). In fact, one can show that the one-loop counterterm vanishes when both the Einstein and Rarita-Schwinger field equations are satisfied because it takes the symbolic form

$$\mathcal{L}_{(1)} = \frac{1}{\varepsilon} \text{(field equations)}^2 \ . \qquad (2.7)$$

(Once again, we have ignored topological effects. See Section 4.) The reason why supergravity succeeded where all other gravity-matter couplings had failed was not an accident, of course, but due to the extra symmetry of Eq. (2.6), i.e. the supersymmetry: there are no supersymmetric quantities of the right dimension which can contribute to $\mathcal{L}_{(1)}$ without also vanishing on-shell. In this sense, it matches pure gravity.

Remarkably, however, supergravity goes one better than pure gravity in being finite on-shell at two loops! (Grisaru 1977). There are no supersymmetric quantities of the right dimension which can contribute to $\mathcal{L}_{(2)}$ without also vanishing on-shell. In other words: a) there is no fermionic partner of the (Riemann)^3 invariant discussed previously; b) no new fermionic invariants, without bosonic partners, appear.

Unfortunately, this pattern breaks down at three loop order. As was shown by Deser et al. (1977) an on-shell superinvariant exists, at least at the linearized level, which might act as a three-loop counterterm. It corresponds, in fact, to the supersymmetric completion of the square of the Bel-Robinson tensor. Thus invariance arguments alone are not sufficient to rule out counterterms in simple supergravity and the case for finiteness, although not lost, remains unproved.

It remains, of course, to confirm that the three-loop invariant survives at the non-linear level and to investigate four loops and beyond. This requires a more systematic approach, using either the tensor calculus (Ferrara & van Nieuwenhuizen 1978 b) or else superspace. For an introduction to superspace, see Salam & Strathdee (1974) and Strathdee (this volume). To analyse higher loop counterterms, the Wess-Zumino superfields (Wess & Zumino 1977) prove most useful, because all supertorsions and supercurvatures can be expressed in terms of just three superfields

$$R\ , \quad G_{\alpha\dot{\alpha}}\ , \quad W_{\alpha\beta\gamma}\ ,$$

where we have used two-component spinor notation. All the component field equations (Einstein, Rarita-Schwinger, and auxiliary field) are contained in

$$R = 0\ , \quad G_{\alpha\dot{\alpha}} = 0\ . \tag{2.8}$$

$W_{\alpha\beta\gamma}$, on the other hand, survives on-shell and is given by

$$W_{\alpha\beta\gamma} \sim F_{\alpha\beta\gamma} + C_{\alpha\beta\gamma\delta}\theta^{\delta}\ , \tag{2.9}$$

where $F_{\alpha\beta\gamma}$ is the spin-$\frac{3}{2}$ field strength, $C_{\alpha\beta\gamma\delta}$ is the Weyl tensor, and θ^{δ} are the fermionic superspace coordinates. Invariants can be built only from these fields and their covariant derivatives. The structure of the one-loop counterterm Eq. (2.7) and both one- and two-loop finiteness now follows almost immediately since only W survives on-shell and no invariants exist to this order.

It is amusing to note a point that has so far gone unnoticed in the literature. In ordinary gravity there are 14 algebraically independent scalars that can be formed from the Riemann tensor, of which 14 - 10 = 4 survive on-shell [e.g. Weinberg (1979)]. In supergravity there are just half as many! Seven may be formed from R, G, and W, of which 7 - 5 = 2 survive on-shell (Duff & Stelle, unpublished). They are

$$W_{\alpha\beta\gamma}W^{\alpha\beta\gamma} \quad \text{and} \quad \bar{W}_{\dot{\alpha}\dot{\beta}\dot{\gamma}}\bar{W}^{\dot{\alpha}\dot{\beta}\dot{\gamma}} ,$$

where $\bar{W}^{\dot{\alpha}\dot{\beta}\dot{\gamma}}$ is the complex conjugate of $W_{\alpha\beta\gamma}$. When integrated over superspace, these yield the two topological invariants $\chi \pm iP/2$ where χ is the Euler number, and P the Pontryagin number (see Section 4).

Algebraic independence, of course, is not sufficient for enumerating possible counterterms. New counterterms may be formed by taking products. For example (Ferrara & Zumino 1978 a)

$$W_{\alpha\beta\gamma}W^{\alpha\beta\gamma}\bar{W}_{\dot{\alpha}\dot{\beta}\dot{\gamma}}\bar{W}^{\dot{\alpha}\dot{\beta}\dot{\gamma}}$$

is just the three-loop (square of Bel-Robinson tensor) invariant discussed previously. Note, however, that since $W_{\alpha\beta\gamma}$ is anticommuting with three symmetric indices, $W^n = 0$ for $n > 4$ (Christensen et al. 1979). The real proliferation of possible counterterms occurs because new invariants can be formed from the supercovariant derivatives D_α and $\bar{D}_\alpha$. Their number can be considerably reduced, however, by appealing to another symmetry of our Lagrangian (2.6), namely γ_5 invariance. The remainder may then be classified, and one discovers in the process the interesting property that all vanish when the fields are self-dual or anti-self-dual, i.e. when either W or $\bar{W}$ vanishes (Christensen & Duff 1979; Christensen et al. 1979; Kallosh 1979 a,b). This is intimately connected with the phenomenon of helicity conservation in supergravity (Grisaru & Pendleton 1977; Christensen et al. 1979; Duff & Isham 1979, 1980; Duff 1979). Despite this reduction in number, non-vanishing invariants survive at every loop order ≥ 3.

What about further matter couplings? Just as coupling random matter to pure gravity only made things worse, so coupling random numbers of N = 1 supermultiplets to N = 1 supergravity spoils even one loop finiteness (Fischler 1979; van Nieuwenhuizen & Vermaseren 1977). In ordinary gravity we learned that things improved only when the matter coupling was fixed by supersymmetry. Might the coupling of N = 1 supergravity to N = 1 supermatter lead to finiteness if this coupling is constrained by more supersymmetry? This leads us naturally to extended supergravity.

2.4 Extended supergravity

A detailed discussion of higher loop invariants which might act as counterterms in extended supergravity is given by Dr. Kallosh in this volume and here we confine ourselves to a few remarks. First we note that in spite of the extra supersymmetry, and in spite of the generalization of γ_5 invariance to combined γ_5 and duality invariance, higher loop invariants still exist which might act as counterterms (Deser & Kay 1978; Howe & Lindström 1981; Kallosh 1981; Howe et al. 1981; Stelle 1981 a). This is true, moreover, even for N > 4 (for which no matter multiplets exist) and including N = 8 which naturally stands out as the favourite candidate for finiteness. This seems a good time to pause, and question some of the assumptions which have underlied this programme so far.

Before looking for ways of reducing, or completely eliminating, counterterms, let us first look on the negative side and ask whether we have not been too slick in simply listing on-shell invariants. (We recall first of all that we have ignored topological effects, ignored cosmological additions in our Lagrangian, ignored also the effects of boundary terms, and ignored spontaneously broken versions of supergravity; more of all this later.) To begin with, we have assumed that our regularization scheme preserves supersymmetry. Is this justified? Although conventional dimensional regularization is known not to, "dimensional reduction" seems better in this respect. See, for example, Siegel et al. (1981). However, it has recently been argued that dimensional reduction can be made consistent only at the expense of losing manifest supersymmetry at high loop order, for example at eight loops in N = 4 Yang-Mills (Avdeev et al. 1981). It may be that supersymmetry is in fact preserved but this issue has not yet been completely resolved.

Given that we have a good regularization scheme (see also Slavnov 1981) are we now justified in looking only at on-shell invariants as possible counterterms? This presupposes a background field method which manifestly preserved the supersymmetry, which in turn requires supersymmetric gauge-fixing and ghosts. Such a scheme must, in principle, exist even if we do not resort to it in actual loop calculations. Some of the problems involved have been discussed by De Wit and Grisaru (1979). Fortunately such a scheme does exist, at least for N = 1 supergravity where the superfield formalism (which incorporates the auxiliary fields) is completely known. The most comprehensive reference is Grisaru & Siegel (1981 a), and the techniques described there will probably turn out to be most efficient in practice as well as in principle. However, this brings us back to the question we have postponed so far: that of the auxiliary fields. Already for simple supergravity there is an *embarras de richesse* of different auxiliary field formulations (Breitenlohner 1977; Stelle & West 1978; Ferrara & van Nieuwenhuizen 1978 a; Sohnius & West 1981 b). Do they yield different quantum theories? They certainly yield different anomalies, as explained in Section 4, though there is a general belief that, anomalies aside, different auxiliary fields yield the same on-shell S-matrix elements. But what about extended supergravity with $N \geq 3$, and Yang-Mills with N = 4 where the auxiliary field structure is not only unknown but for which there is not even an existence proof? (See, for example, Taylor 1981 a,b,c.) Is it fair to list on-shell invariants as possible counterterms when their off-shell extension might not exist? The need for a thorough understanding of the auxiliary field structure in extended theories as a prerequisite for understanding the ultraviolet problem has been stressed by several authors. This said, however, it seems unlikely that the beautiful cancellations discussed in Section 3 would suddenly disappear when the auxiliary fields are included! Few would dispute, of course, that a complete auxiliary-field/superspace approach would make the task of investigating finiteness a whole lot easier.

In summary, it seems that listing on-shell invariants as possible counterterms is probably correct as far as it goes. An entirely different question, of course, is whether it is sufficient. Turning now to a more positive approach we ask "Might the numerical coefficients of such counterterms turn out to be zero?".

3 SPIN SUM RULES AND VANISHING β-FUNCTIONS

3.1 Spin sum rules

In this section we concentrate on infinity cancellations known to occur in supersymmetric theories and supergravity which are not explicable by the "absence-of-invariant-as-counterterm arguments" discussed so far, i.e. the so-called "miraculous" cancellations whereby an invariant counterterm might exist but nevertheless appears with zero coefficient.

The most startling example of this phenomenon in supergravity is the vanishing of the one loop β function in gauged N > 4 extended theories (Christensen et al. 1980). At the time this result was published, the reason for this "miraculous" cancellation was unknown, but one was immediately reminded of the equally "miraculous" vanishing of the β-function in N > 2 supersymmetric Yang-Mills theories which is now known to hold to at least three loops (Avdeev et al. 1980; Grisaru et al. 1980; Caswell & Zanon 1981). Indeed, at the time, the complete Lagrangians for the gauged N > 4 theories had not yet been constructed. There was even doubt in some quarters whether gauged N > 4 supergravities actually existed; doubt which has now been dispelled by De Wit & Nicolai (1981 a,b).

The question now arises "What is so special about higher N values?" Since we do not yet have a complete superfield description of extended theories, let us examine the spin content in components. See Tables 1 and 2. As is well known, all supermultiplets share the property of having an equal number of Bose and Fermi degrees of freedom, i.e.

$$\sum_{\lambda} (-1)^{2\lambda} \, d(\lambda) = 0 \qquad (3.1)$$

where λ is the helicity, ranging over 1, ½, 0, -½, -1 in Yang-Mills and 2, 3/2, 1, ½, 0, -½, -1, -3/2, -2 in supergravity, and $d(\lambda)$ is the number of states with helicity λ in a supermultiplet. More remarkable, however, are the following generalizations, first published by Curtwright (1981):

$$\sum_{\lambda} (-1)^{2\lambda} \, d(\lambda) \, \lambda^k = 0 \, , \quad N > k \, . \qquad (3.2)$$

These rules were first made known to me in 1980 by A. D'Adda, who conjectured that they were related to the vanishing β function. The

Tables 1 and 2: Number of states d(λ) with helicity λ in supersymmetric Yang-Mills (1) and extended supergravity (2). The CPT conjugate multiplets must also be counted except for the self-conjugate N = 4 Yang-Mills and N = 8 supergravity. The numbers in brackets denote the quadratic Casimir invariant C(λ).

Table 1

λ \ N	1	2	3	4
1	1(c)	1(c)	1(c)	1(c)
½	1(c)	2(c)	3(c)	4(c)
0		1(c)	3(c)	6(c)
-½			1(c)	4(c)
-1				1(c)

Table 2

λ \ N	1	2	3	4	5	6	7	8
2	1(0)	1(0)	1(0)	1(0)	1(0)	1(0)	1(0)	1(0)
3/2	1(0)	2(1)	3(1)	4(1)	5(1)	6(1)	7(1)	8(1)
1		1(0)	3(1)	6(2)	10(3)	15(4)	21(5)	28(6)
½			1(0)	4(1)	10(3)	20(6)	35(10)	56(15)
0				1(0)	5(1)	15(4)	35(10)	70(20)
-½					1(0)	6(1)	21(5)	56(15)
-1						1(0)	7(1)	28(6)
-3/2							1(0)	8(1)
-2								1(0)

proof of this relation was carried out by Duff and Gibbons (unpublished) both for Yang-Mills and supergravity, independently of Curtwright. Since Curtwright's explanation of the supergravity result differs somewhat from ours, we shall here present both explanations and demonstrate their equivalence. There is one further difference, of a purely technical nature. Curtwright bases his β function calculations on a Feynman graph analysis due to Hughes (1980). We shall instead adopt the arbitrary-spin background-field formalism of Christensen and Duff (1978 a, 1979, 1980).

3.2 Background field method

Consider a general field theory with fields denoted by the generic symbol $\hat{\phi}^i(x)$ and action denoted by $S[\hat{\phi}]$. If we make the background field split

$$\hat{\phi}(x) = \phi(x) + h(x) \tag{3.3}$$

and Taylor expand the action about the background field $\phi(x)$, we obtain

$$S[\phi + h] = S[\phi] + \int d^4x \left.\frac{\delta S}{\delta\hat{\phi}^i(x)}\right|_{\hat{\phi}=\phi} h^i(x) + \int d^4x\, d^4y\, h^i(x) \left.\frac{\delta^2 S}{\delta\hat{\phi}^i(x)\delta\hat{\phi}^j(y)}\right|_{\hat{\phi}=\phi} h^j(y) + O(h^3) \,. \tag{3.4}$$

Note that terms linear in the quantum field $h(x)$ are absent when ϕ is chosen to be a solution of the classical background field equations

$$\left.\frac{\delta S}{\delta\hat{\phi}}\right|_{\hat{\phi}=\phi} = 0 \,.$$

This will be important when we consider quantum gravity with a cosmological constant, as we must do in order to analyse gauged supergravity. To one loop order we need retain in Eq. (3.4) only these terms quadratic in h.

The quantum effective action $\Gamma[\phi]$ is now given by

$$e^{i\Gamma[\phi]} = e^{iS[\phi]}\, e^{iW[\phi]} , \tag{3.5}$$

where W is given by the functional integral

$$e^{iW[\phi]} = \int dh \exp\left(i \int d^4x\ \tfrac{1}{2} h^i\, \Delta_{ij} h^j\right) . \tag{3.6}$$

If h is a boson,

$$e^{iW[\phi]} = (\det \Delta)^{-1/2} . \tag{3.7}$$

Δ_{ij} is a second-order differential operator determined by the second functional derivative of S in Eq. (3.4) and is a functional of the background field ϕ. Graphically, the effective action $W[\phi]$ describes all one-loop graphs with closed h loops and external ϕ lines. Setting ϕ to be a solution of the classical field equations corresponds to going on-shell. It turns out that for the theories in which we are interested one can always arrange (e.g. by gauge fixing) that Δ_{ij} takes the simple form

$$\Delta^{ij} = -\delta^{ij}\, \nabla^\mu \nabla_\mu + X^{ij} , \tag{3.8}$$

where

$$\nabla_\mu h^i = \partial_\mu h^i + N_\mu{}^{ij}\, h_j \tag{3.9}$$

and where the matrices X^{ij} and N^{ij} are functionals of ϕ such that

$$N_\mu^{ij} = -N_\mu^{ji} , \quad X^{ij} = X^{ji} . \tag{3.10}$$

If, on the other hand, h is a fermion, we have

$$e^{iW[\phi]} = \det \not{D} = (\det \Delta)^{1/2} , \tag{3.11}$$

where

$$\Delta = \not{D}^2 = -\nabla^\mu \nabla_\mu + X \, , \tag{3.12}$$

and

$$\nabla_\mu = \partial_\mu + N_\mu \tag{3.13}$$

where $\not{D}$ is the Dirac operator in the ϕ background. For most purposes ϕ will be bosonic. Thus the fermion calculation can be converted into the same as the boson. In both cases we study operators of the form (3.8), but there is a change in sign in going from Eq. (3.7) to Eq. (3.11). [The exponent ½ in Eq. (3.11) is replaced by ¼ if the spinor is real, i.e. Majorana.]

We are now in a position to write down, without proof, the one-loop counterterm $\mathcal{L}_{(1)}$ which must be added to render W finite. Defining the matrix

$$Y_{\mu\nu}{}^{ij} = -Y_{\mu\nu}{}^{ji} \tag{3.14}$$

by

$$\left[\nabla_\mu, \nabla_\nu\right] h^i = Y_{\mu\nu}{}^{ij} \, h_j \, , \tag{3.15}$$

i.e.

$$Y_{\mu\nu} = \partial_\mu N_\nu - \partial_\nu N_\mu + \left[N_\mu, N_\nu\right] \, , \tag{3.16}$$

then

$$\mathcal{L}_{(1)} = \pm \frac{1}{\varepsilon} \frac{1}{180(4\pi)^2} \sqrt{g} \; \mathrm{Tr} \left[\mathbb{1} \left(R_{\mu\nu\rho\sigma} R^{\mu\nu\rho\sigma} - R_{\mu\nu} R^{\mu\nu} + \frac{5}{2} R^2 \right) - 30RX + 90X^2 + 15Y_{\mu\nu} Y^{\mu\nu} \right] , \tag{3.17}$$

where ± refers to boson or fermion, respectively. In Eq. (3.17) we have allowed for the possibility of a non-vanishing gravitational background field in addition to any other background fields. This result can be obtained either by writing down the most general counterterm allowed and then fixing the numerical coefficients using momentum-space Feynman graphs or better (especially for the cosmological, topological and boundary terms) by the coordinate space "heat-kernel" expansion. A list of references may be found in Christensen and Duff (1979).

From Eq. (3.17), we see that the problem of computing one-loop counterterms is reduced to the problem of determining the matrices X and $Y_{\mu\nu}$ for the system in question. In the case where the gauge group is the internal Yang-Mills group the quantum field h^i transforms like

$$h^i \to \Omega^{ij}\, h_j \,, \tag{3.18}$$

where

$$\Omega^{ij} = \exp \varepsilon^a(x)\, T_a{}^{ij} \tag{3.19}$$

and where the generators T_a obey

$$\left[T_a, T_b\right] = f_{ab}{}^c T_c \,. \tag{3.20}$$

The covariant derivative is

$$\nabla_\mu h^i = \partial_\mu h^i + A_\mu{}^a T_a{}^{ij}\, h_j \tag{3.21}$$

and

$$Y_{\mu\nu} = \partial_\mu A_\nu - \partial_\nu A_\mu + \left[A_\mu, A_\nu\right] \tag{3.22}$$

$$= F_{\mu\nu}{}^a T_a \,, \tag{3.23}$$

with $F_{\mu\nu}{}^a$ the Yang-Mills field strength. In the case where the gauge group is the external Lorentz group, we have

$$h^\alpha \to \Omega^{\alpha\beta} h_\beta \,, \tag{3.24}$$

where α is an arbitrary spin index, and where

$$\Omega^{\alpha\beta} = \exp \omega^{ab}(x)\, \Sigma_{ab}{}^{\alpha\beta} . \tag{3.25}$$

The Lorentz generators Σ_{ab} obey

$$\left[\Sigma_{ab}, \Sigma_{cd}\right] = \frac{1}{2}\left(\eta_{cb}\Sigma_{ad} - \eta_{ca}\Sigma_{bd} + \eta_{bd}\Sigma_{ca} - \eta_{da}\Sigma_{cb}\right) \tag{3.26}$$

with

$$\Sigma_{ab} = - \Sigma_{ba} . \tag{3.27}$$

The covariant derivative is now

$$\nabla_\mu h^\alpha = \partial_\mu h^\alpha + \omega_\mu{}^{ab}\Sigma_{ab}{}^{\alpha\beta} h_\beta \tag{3.28}$$

and

$$Y_{\mu\nu} = \partial_\mu\omega_\nu - \partial_\nu\omega_\mu + \left[\omega_\mu, \omega_\nu\right] \tag{3.29}$$

$$= R_{\mu\nu}{}^{ab}\Sigma_{ab} \tag{3.30}$$

with $R_{\mu\nu}{}^{ab}$ the Riemann tensor.

In Yang-Mills theories $Y_{\mu\nu}$ depends on the internal representation of whatever quantum field if going around the loop as in Eq. (3.23) but is independent of the spin. The spin dependence enters via

$$X = \Sigma^{\mu\nu} F_{\mu\nu}{}^a T_a . \tag{3.31}$$

Remarkably, this formula is valid whatever the spin, i.e. h^i could be spin-0, spin-½, or else the spin-1 gauge field itself (or its spin-0 ghosts). Similarly in gravity $Y_{\mu\nu}$ is already linear in $\Sigma_{\mu\nu}$ as in Eq. (3.30), whereas X is quadratic, typically

$$X = \Sigma^{\mu\nu} R_{\mu\nu}{}^{ab}\Sigma_{ab} . \tag{3.32}$$

It remains to evaluate the Tr operation in Eq. (3.17). These arbitrary spin formulae and the evaluation of the necessary traces may be found in Christensen and Duff (1979) and we shall now summarize the results.

3.3 Arbitrary spin formalism

Irreducible representations of the Lorentz group are labelled (A,B) where the non-negative numbers A and B take on integer or half-integer values. The dimensionality of the representation, or number of degrees of freedom, is

$$
\begin{aligned}
D(A,B) &= (2A+1)(2B+1) \\
&= (s+1)^2 - t^2 ,
\end{aligned}
$$

where the spin s is given by

$$s \equiv A + B \tag{3.33}$$

and where

$$t \equiv A - B , \quad -s \leq t \leq s . \tag{3.34}$$

However, the particles which appear in Tables 1 and 2 are not in general described by a single irreducible representation. They each correspond to two degrees of freedom (counting scalars as complex) and for $s \geq 1$ it is necessary to include Fadeev-Popov ghost subtractions in order to arrive at just two helicity states. The general rule (Duff 1979) is first to compute the contributions to Eq. (3.17) from a representation (A,B), add to it the contribution from (A-1, B-1) and then subtract twice the contribution from $(A-\frac{1}{2}, B-\frac{1}{2})$. For example, the correct degree-of-freedom count is given not by

$$\mathrm{Tr}\, \mathbb{1} = D(A,B) = (2A+1)(2B+1) \tag{3.35}$$

but rather by

$$
\begin{aligned}
D'(A,B) &= D(A,B) + D(A-1, B-1) - 2D(A-\tfrac{1}{2}, B-\tfrac{1}{2}) \\
&= 2 .
\end{aligned} \tag{3.36}
$$

In tracing products of Σ's, one also encounters the functions

$$E_{\pm}(A,B) = D\left[A(A+1) \pm B(B+1)\right] \tag{3.37}$$

from two Σ's, and

$$F_{\pm}(A,B) = D\left[A(A+1)(2A-1)(2A+3) \pm B(B+1)(2B-1)(2B+3)\right] \tag{3.38}$$

$$G(A,B) = D\left[A(A+1) + B(B+1)\right]^2 \tag{3.39}$$

from four Σ's (which is as many as we ever need, at least to this one-loop order). If we translate these into functions of s and t and then calculate the corresponding primed quantities as in Eq. (3.36), one obtains

$$D' = 2$$

$$E'_+ = 6s^2$$

$$F'_+ = -15s^2 + 15s^4 + t^2\ (5 - 5t^2 + 30s^2)$$

$$E'_- = 6st$$

$$F'_- = -10st + 40s^3t$$

$$G' = \frac{3}{2}\, s^2 + \frac{15}{2}\, s^4 + t^2 \left(\frac{1}{2} - \frac{t^2}{2} + 3s^2\right) . \tag{3.40}$$

As we shall see, both in this section and in Section 4, these functions are all we need to write down the β-functions, the one-loop counterterms, the conformal anomalies, and the axial anomalies for fields of arbitrary spin, in both Yang-Mills and gravity.

3.4 Supersymmetric Yang-Mills: $\beta = 0$ for $N > 2$

In flat space, the one-loop counterterm Eq. (3.17) reduces to

$$\mathcal{L}_{(1)} = \pm \frac{1}{\varepsilon} \frac{1}{(4\pi)^2}\, \mathrm{Tr} \left[\frac{1}{2} X^2 + \frac{1}{12} Y_{\mu\nu} Y^{\mu\nu}\right] . \tag{3.41}$$

For supersymmetric Yang-Mills, the particles going around the loop will be the spin-1 quantum gauge fields, their spin-0 ghosts, spin-½ fermions, and physical spin-0 scalars. In each case X and $Y_{\mu\nu}$ are given by Eq. (3.23) and Eq. (3.31). Hence

$$\mathrm{Tr}\, Y_{\mu\nu}Y^{\mu\nu} = -D\, F_{\mu\nu}{}^{a}F^{\mu\nu a}C \tag{3.42}$$

$$\mathrm{Tr}\, X^2 = \frac{2}{3}\left(E_+F_{\mu\nu}{}^{a}F^{\mu\nu a} + E_-{}^*F_{\mu\nu}{}^{a}F^{\mu\nu a}\right)C\ , \tag{3.43}$$

where $F_{\mu\nu}{}^{a}$ is the background field strength and ${}^*F_{\mu\nu}{}^{a}$ is its dual

$${}^*F_{\mu\nu}{}^{a} \equiv \frac{1}{2}\,\varepsilon_{\mu\nu\rho\sigma}F^{\rho\sigma a}\ , \tag{3.44}$$

where D and $E_\pm$ are given by Eqs. (3.35) and (3.37), and where C is the second-order Casimir

$$-\delta^{ab}C = \mathrm{Tr}\, T^aT^b\ . \tag{3.45}$$

For physical fields we replace D and $E_\pm$ by D' and $E'_\pm$ of Eq. (3.40) to obtain the counterterm

$$\mathcal{L}_{(1)} = \frac{1}{\varepsilon}\frac{(-1)^{2s}}{(4\pi)^2}\,e^2\left[\left(-\frac{1}{6}+2s^2\right)F_{\mu\nu}{}^{a}F^{\mu\nu a} + 2st\,{}^*F_{\mu\nu}{}^{a}F^{\mu\nu a}\right]C\ , \tag{3.46}$$

where we have rescaled $A_\mu \rightarrow eA_\mu$ to introduce the gauge coupling constant e. The topological counterterm

$$P = \frac{e^2}{16\pi^2}\int d^4x\ {}^*F_{\mu\nu}{}^{a}F^{\mu\nu a} \tag{3.47}$$

need not concern us yet and we turn instead to the one-loop $\beta(e)$ function. The contribution to $\beta(e)$ from a particle of spin s is given by the coefficient of $-\varepsilon^{-1}e^{-1}F_{\mu\nu}{}^{a}F^{\mu\nu a}$ in Eq. (3.46), i.e.

$$\beta(s) = \frac{e^3}{96\pi^2}(-1)^{2s}(1-12s^2)C\ . \tag{3.48}$$

This is the same result as that of Hughes (1980) and Curtwright (1981), who interpret the first and second terms in Eq. (3.48) as the "convective charge" and "magnetic moment" contributions, respectively. The asymptotic freedom of pure Yang-Mills (i.e. $\beta < 0$ for $s = 1$) then follows immediately as a consequence of the negative magnetic moment term dominating the positive convective charge term; an interpretation which had, in fact, already been anticipated by Salam and Strathdee (1975). If there are several fields in the theory, the complete one-loop β-function is obtained by multiplying Eq. (3.48) by d(s), the number of fields of spin s, and summing over spins

$$\beta = \sum_s d(s)\beta(s) \ . \tag{3.49}$$

In the case of supersymmetric Yang-Mills, the internal symmetry factor C is the same for all spins since each belongs to the same (adjoint) representation of the gauge group. Now, of course, we may invoke the spin sum rules [Eq. (3.2)], in particular

$$\sum_s (-1)^{2s} d(s) = 0 \ , \quad \forall N \tag{3.50}$$

$$\sum_s (-1)^{2s} d(s) s^2 = 0 \ , \quad N > 2 \ . \tag{3.51}$$

(Note that for sum rules with even powers of λ we may replace the helicity λ by the spin s provided we include the CPT conjugates and sum over s = 0, ½, and 1 with the understanding that d(0) counts the number of complex scalars.)

The crucial observation is that the vanishing of the one-loop β-function for N = 4 is no longer miraculous but an obvious consequence of the sum rules [Eqs. (3.50) and (3.51)] applied to the arbitrary spin results [Eqs. (3.48) and (3.49)]. Note that the first term in Eq. (3.48) cancels for all N, whereas the second term cancels only for N > 2. (For N = 1 and N = 2 this second term demonstrates asymptotic freedom.)

What about higher loops? Sum rules apart, we know that the N = 4 theory is finite to at least three loops and arguments have been put forward to suggest that this is true to all orders (Ferrara & Zumino 1978 b; Sohnius & West 1981 a). Can we therefore give a sum-rule explanation? First we note that beyond one loop, it no longer makes sense to attribute a contribution to β from each individual spin as in Eq. (3.48), since the contributions from different spins will mix. So we would expect to have to sum over spins as in Eq. (3.49). Thus one might make an all-orders guess of

$$\beta = \sum_{\lambda} (-1)^{2\lambda} d(\lambda)\left[a + b\lambda^2\right] , \tag{3.52}$$

with a and b universal functions of the coupling constant, since this reduces to the correct result at one loop with

$$a = \frac{e^3}{96\pi^2} C , \quad b = -\frac{e^3}{6\pi^2} C . \tag{3.53}$$

We have included no powers of λ greater than two under the summation; otherwise the sum rules [Eqs. (3.50) and (3.51)] could not explain the vanishing β at 2 and 3 loops in N = 4. Secondly, we would not expect such a formula to hold for non-supersymmetric theories beyond one loop, because the gauge, Yukawa, and quartic scalar coupling constants (which coincide for supersymmetric Yang-Mills theories) are in general different. In this supersymmetric case, moreover, the a term now vanishes for all N by virtue of Eq. (3.50). Having said all this, however, the resulting formula

$$\beta = b \sum_{\lambda} (-1)^{2\lambda} \lambda^2 ,$$

though consistent with the vanishing β for N = 4, fails to account for another curious result, namely the vanishing of the two-loop contribution to β for N = 2 but not N = 1 (Jones 1981). To explain this, we would have to give up the universality of b and instead introduce another factor $\Sigma_{\lambda}(-1)^{2\lambda}\lambda$ into its two-loop contribution which from Eq. (3.2) would vanish for N = 2 but not N = 1. Thus it seems that we must abandon guess-work about higher loop order for the time being since the simplest guess Eq. (3.52) does not seem to work. Let us instead remain at one loop and examine supergravity.

3.5 Gauged extended supergravity: $\beta = 0$ for $N > 4$

Extended supergravities with N gravitini exhibit a global SO(N) symmetry. Moreover, the N(N-1)/2 spin-1 fields lie in the adjoint representation, see Table 2. (For N = 6, there is one extra vector.) This suggests a possible gauging of this SO(N) symmetry whereby ordinary derivatives are replaced by SO(N) Yang-Mills covariant derivatives with a corresponding covariantization of the spin-1 kinetic term. In this way we acquire a new dimensionless coupling constant e in addition to the dimensionful coupling constant κ already present. It remains to show, however, that one can make other e-dependent additions to the Lagrangian in such a way as to maintain (with e-dependent corrections to the transformation laws) the N-fold supersymmetry. That this can indeed be carried out in a consistent fashion was demonstrated by Freedman & Das (1977) and Fradkin & Vasiliev (1976) for N = 2 and N = 3; by Das et al. (1977) and Freedman & Schwarz (1978) for N = 4; and most recently by De Wit & Nicolai (1981 a,b) for N = 5, 6, 7, and 8. The graded Poincaré algebra gets replaced by the graded de Sitter algebra, and the Lagrangian acquires a cosmological constant Λ given by

$$\Lambda = -\frac{6e^2}{\kappa^2} \tag{3.54}$$

and gravitino "mass" term given by

$$m^2 = \frac{2e^2}{\kappa^2}\ . \tag{3.55}$$

Equation (3.54) has an interesting consequence. By linking the gauge coupling constant to the cosmological constant, the renormalization of $\kappa^2\Lambda$ determines the $\beta(e)$-function. Thus to calculate β one may proceed in one of two ways. Either (a) compute the usual charge renormalization effects, i.e. the coefficient of the Yang-Mills Tr $\sqrt{g}\, F_{\mu\nu}F^{\mu\nu}$ counterterm which receives contributions from spins 0, ½, 1, and 3/2, but not 2, since the graviton is a singlet, or (b) compute the cosmological renormalization, i.e. the coefficient of the $\sqrt{g}$ counterterm, which receives contributions from all spins including gravity. By supersymmetry, one determines the other.

Since the one-loop cosmological renormalization had already been carried out by Christensen and Duff (1980) for spins 0, ½, 1, and 2, it was natural to attempt the latter of these approaches first by extending these calculations to spin $3/2$ and hence to gauged extended supergravity (Christensen et al. 1980). Remarkably, we found that while $\beta > 0$ for $N \leq 4$, $\beta = 0$ for $N > 4$! Subsequently Curtwright (1981) arrived at the same result using the first of these approaches.

We shall now describe both calculations, show how they lead to equivalent results, and in each case give a spin sum rule explanation for the vanishing β-function when $N > 4$. First, however, it is necessary to say a few words about quantizing gravity with a cosmological constant.

3.6 The cosmological constant

Although the renormalizability properties of quantum gravity and supergravity have received considerable attention over the last few years, almost all these investigations have confined their attention to theories with vanishing cosmological constant, Λ. Contrary to a popular school of thought, however, the calculation of the quantum effective action when $\Lambda \neq 0$, i.e. when the gravitational action is

$$S = -\frac{1}{2\kappa^2}\int d^4x\,\sqrt{g}\,(R - 2\Lambda)\ , \qquad (3.56)$$

is no more difficult than when $\Lambda = 0$, provided one consistently expands about a background field satisfying the Einstein equation with a Λ term

$$R_{\mu\nu} = \Lambda g_{\mu\nu}\ . \qquad (3.57)$$

See Christensen and Duff (1980). [In particular one must avoid an expansion about flat space. This is not the correct ground-state when $\Lambda \neq 0$. Attempting the expansion $g_{\mu\nu} = \eta_{\mu\nu} + \kappa h_{\mu\nu}$ leads to problems both in the term linear in $h_{\mu\nu}$ and the term quadratic in $h_{\mu\nu}$ which appear in the expansion of a $\sqrt{g}$ Lagrangian. The former gives rise to awkward ill-defined tadpole graphs and the latter to massive ghosts.] As explained in Section 3.2, terms linear in the quantum field h are absent when the background field is a solution of the classical field equations [see Eq. (3.4)]. Terms quadratic in h, which govern the one-loop calculations, are then determined in a suitable gauge by the operators

$$\Delta h_{\mu\nu} = -\nabla^{\rho}\nabla_{\rho}h_{\mu\nu} - 2R_{\mu\rho\nu\sigma}h^{\rho\sigma} \tag{3.58}$$

in the case of the graviton and

$$\Delta h_{\mu} = -\nabla^{\ell}\nabla_{\rho}h_{\mu} - \Lambda h_{\mu} \tag{3.59}$$

in the case of the spin-1 Fadeev-Popov ghosts. The one-loop counterterms then follow from Eq. (3.17). They will be of the form $R_{\mu\nu\rho\sigma}R^{\mu\nu\rho\sigma}$, $R_{\mu\nu}R^{\mu\nu}$, R^2, ΛR, and Λ^2 with gauge-dependent coefficients. Gauge independence is achieved after going "on-shell" by use of Eq. (3.57), resulting in the one-loop counterterm

$$S_{(1)} = \frac{1}{\varepsilon}\left[A\chi - \frac{B\kappa^2\Lambda}{12\pi^2}S\right], \tag{3.60}$$

where A and B are numerical coefficients,

$$\chi \equiv \frac{1}{32\pi^2}\int d^4x\,\sqrt{g}\left(R_{\mu\nu\rho\sigma}R^{\mu\nu\rho\sigma} - 4R_{\mu\nu}R^{\mu\nu} + R^2\right) \tag{3.61}$$

is the Euler number (see Section 4), and S is the classical action on-shell. Explicit calculation (Christensen & Duff 1980) yields

$$A = \frac{106}{45}, \quad B = -\frac{87}{10}. \tag{3.62}$$

Thus, in contrast to the case $\Lambda = 0$ discussed in Section 2, pure gravity with a Λ term is no longer one-loop finite (in the non-topological sense) because $B \neq 0$.

One may now repeat the exercise for simple (N = 1) supergravity with a gravitino mass term

$$S = \int d^4x \det e_{\mu}{}^{a}\left[-\frac{1}{2\kappa^2}R + \frac{1}{2}\varepsilon^{\mu\nu\rho\sigma}\bar{\psi}_{\mu}\gamma_5\gamma_{\nu}D_{\rho}\psi_{\sigma}\right.$$
$$\left. + m\,\bar{\psi}_{\mu}\sigma^{\mu\nu}\psi_{\nu} + \frac{1}{3}(S^2 + P^2 - b^{\mu}b_{\mu}) + \frac{2m}{\kappa}S\right]. \tag{3.63}$$

Elimination of the auxiliary field S yields a cosmological constant $\Lambda = -3m^2$. See, for example, van Nieuwenhuizen (1979). Previous discussions of supergravity with a cosmological constant may be found in MacDowell & Mansouri (1977), Deser & Zumino (1977), and Townsend (1977). The one-loop counterterm will again be of the form (3.60), where S is now given by Eq. (3.63) on-shell. The coefficients A and B will now receive contributions from both the graviton and the gravitino (with its appropriate mass parameter). Explicit calculation (Christensen et al. 1980) yields

$$A = \frac{41}{24}, \quad B = -\frac{77}{12}, \tag{3.64}$$

and once again in contrast to the case $\Lambda = 0 = m$ discussed in Section 2, simple supergravity is no longer one-loop finite.

We may now combine these results with those for spins 1, ½, and 0 (Christensen & Duff 1980) and apply them to the extended SO(N) theories with gauged internal symmetry. The cosmological coefficient B now takes on a new significance: by supersymmetry it also determines the renormalization of the gauge coupling constant e. Combining Eqs. (3.54) and (3.60) we have

$$S_{(1)} = \frac{1}{\varepsilon}\left[A\chi + B\frac{e^2}{2\pi^2}S\right], \tag{3.65}$$

where the classical action S will now contain a spin-1 gauge field contribution $\mathrm{Tr}\, F_{\mu\nu}F^{\mu\nu}$. The signal for asymptotic freedom is $B > 0$.

Only the kinetic terms in the classical Lagrangian are needed to fix the contributions to A from fields of different spin. See Christensen and Duff (1978) and references therein. To calculate B we also require knowledge of the mass terms. All particles must be massless for all N if, as we are assuming, supersymmetry is not spontaneously broken. There is, however, an "apparent mass" parameter for the gravitinos given by Eq. (3.55). No such terms are present for the spin-1 or spin-½ fields but the spin-0 fields, which make their appearance for $N \geq 4$, require greater care. The spin-2, spin-0 coupling is known to be of the form

$$\mathcal{L} = -\frac{1}{2\kappa^2}\sqrt{g}\,(R-2\Lambda) + \frac{1}{2}\sqrt{g}\,\phi^i\left(-\Box + \frac{2\Lambda}{3}\right)\phi^i + O(\phi^3)\ , \qquad (3.66)$$

i.e. minimal coupling with a mass term. (The range of the index i is given by Table 2.) However, one could equally well use

$$\mathcal{L} = -\frac{1}{2\kappa^2}\sqrt{g}\,(R-2\Lambda) + \frac{1}{2}\sqrt{g}\,\phi^i\left(-\Box + \frac{R}{6}\right)\phi^i + O(\phi^3)\ , \qquad (3.67)$$

i.e. conformal coupling with no mass term. The equivalence is seen by making a Weyl rescaling in the Lagrangian (3.67) of the form

$$g_{\mu\nu} \to \Omega^2 g_{\mu\nu}\ , \quad \phi^i \to \Omega^{-1}\phi^i\ ; \qquad \Omega^2 \equiv 1 + \frac{\kappa^2}{6}\phi^i\phi^i\ , \qquad (3.68)$$

which yields the Lagrangian (3.66). Both Lagrangians yield the same B coefficient since the field equations imply $R = 4\Lambda + \ldots$. We note in passing that the mass-term properties discussed above were, at the time the B coefficients were first calculated (Christensen et al. 1980), assumed for $N > 4$ on the basis of an extrapolation from the known $N = 4$ mass terms (Das et al. 1977; Freedman & Schwarz 1978). The existence of these gauged $N > 4$ theories together with their mass-term properties has subsequently been confirmed by De Wit & Nicolai (1981 a,b). [In none of the theories discussed here does the scalar potential contain terms linear in ϕ. In this respect they differ from the alternative $N = 4$ theory of Freedman & Schwarz (1978), which has a chiral $SU(2) \times SU(2)$ symmetry with two independent coupling constants, and which we shall not discuss. However, all the gauged theories with scalars apparently suffer from a potential $V(\phi)$ which is not bounded below. But since $V(\phi)$ is intimately connected with Λ, it may be that the criterion for stability is different when $\Lambda \neq 0$. After all, we have already argued that an x-independent vacuum expectation value for $g_{\mu\nu}$ cannot provide the correct ground state when $\Lambda \neq 0$ and the same may well be true of the scalars $\phi^i(x)$. This is deserving of further study.]

The results of calculating the B coefficient for fields of spin s are summarized in Table 3. The combined results for extended supergravity then follow from the particle content of Table 2 and are listed in Table 4. The vanishing of B for $N > 4$ seems, at first sight, miraculous.

Table 3

Contributions to the cosmological B coefficient and charge renormalization β-function for fields of spin s. (For convenience, the spin-0 result is quoted for a two-component, i.e. complex, scalar)

S	60B	$96\pi^2\beta e^{-3}$
0	-2	C(0)
$\frac{1}{2}$	-3	2 C($\frac{1}{2}$)
1	-12	-11 C(1)
$\frac{3}{2}$	137	26 C($\frac{3}{2}$)
2	-522	0

Table 4

The values of B and β in extended supergravity, demonstrating the equivalence of the two calculational methods

N	B	$-16\pi^2\beta e^{-3}$
1	$-\frac{77}{12}$	-
2	$-\frac{13}{3}$	$-\frac{13}{3}$
3	$-\frac{5}{2}$	$-\frac{5}{2}$
4	-1	-1
5	0	0
6	0	0
7	0	0
8	0	0

However, we note that the B's of Table 3 are described by the quartic polynominal

$$60B(s) = (-1)^{2s}\left[-2 + 30s^2 - 40\ s^4\right] \tag{3.69}$$

$$= \frac{1}{3}\ (-3D' + 25E'_+ - 20G' + 2F'_+)\ , \tag{3.70}$$

where D', E'_+, G', and F'_+ are given by Eq. (3.40), as may be verified by an arbitrary spin background field calculation along the lines of that already described for the Yang-Mills field. The spin sum rules

$$\sum_s (-1)^{2s} d(s) s^k = 0\ , \quad N > k\ , \tag{3.71}$$

may now be invoked to explain the vanishing of B for N > 4. We note incidentally that we encounter a polynomial in spin of degree 4 in gravity in contrast to one of degree 2 in Yang-Mills. This is due to the extra Lorentz generators in the matrices X and Y of Eqs. (3.30) and (3.32) compared with those of Eqs. (3.23) and (3.31). A naïve argument for higher loops might be to note that, since Yang-Mills is renormalizable, one could never encounter more than $X^2 \sim s^2$ in the higher loop counterterm and that the sum rules could then keep N > 2 finite to all orders. The same naïve power counting argument applied to gravity, on the other hand, would yield $X^3 \sim s^6$ at two loops, $X^4 \sim s^8$ at three loops, etc. And even the N = 8 sum rule fails at s^8. However, we have already seen the danger of extrapolation to higher loops based on guess-work, and a deeper understanding of the infinity cancellations is still required.

3.7 Charge renormalization

Having derived the supergravity β function via the cosmological method, we now discuss the alternative method due to Curtwright (1981) who simply extrapolated the β(s) of Eq. (3.48) to the case of spin $^3/_2$ (an extrapolation which can be justified by the arbitrary spin background field calculations). This yields the β coefficients of Table 3. As is well-known, the spin-0 and spin-$^1/_2$ contributions enter with the opposite sign to the spin-1 gauge field. However, we learn that the spin-$^3/_2$ field also enters with the opposite sign and works against asymptotic freedom.

When applied to extended supergravity multiplets we obtain the results of Table 4 in complete agreement with the cosmological method. What is not so obvious, however, is the spin sum rule explanation of vanishing β for $N > 4$. To give such an explanation we need the new rules (Curtwright 1981)

$$\sum_{\lambda} (-1)^{2\lambda} d(\lambda)\lambda^{k} C(\lambda) = 0 , \quad N > k + 2 , \tag{3.72}$$

which, it can be shown, follow as a consequence of the old rules (3.2). Thus although the $\beta(s)$ of Eq. (3.48) is only quadratic in spin, the Casimir invariant is now spin dependent and we again find $\beta = 0$ for $N > 4$.

To understand a little better how two such apparently different approaches lead satisfactorily to the same result for all N, we note that in the cosmological method we are calculating the $\sqrt{g}$ counterterm, whereas in the charge renormalization method we are calculating the $\sqrt{g}\ \mathrm{Tr}\ F_{\mu\nu}F^{\mu\nu}$ counterterm but, by supersymmetry, they enter in a fixed ratio. In fact, apart from χ, the only counterterm surviving on-shell is the classical action itself [see Eq. (3.65)]. In the cosmological method it appears with coefficient $B\ e^2(2\pi^2\varepsilon)^{-1}$ and in the charge renormalization method with coefficient $-8\beta(e\varepsilon)^{-1}$. Hence

$$B = -\frac{16\pi^2}{e^3}\beta , \tag{3.73}$$

a result confirmed by the explicit calculations summarized in Table 4. Remember, moreover, that graviton loops were used to compute the left-hand side of this equation but not the right-hand side!

There remains one possible question mark concerning the validity of the charge renormalization method and the extrapolation of the Yang-Mills $\beta(s)$ of Eq. (3.48) to the case of curved space theories with a cosmological constant. The β-function, i.e. that which is given by the coefficient of the $e^2\ \mathrm{Tr}\ F_{\mu\nu}F^{\mu\nu}$ counterterm, in principle receives contributions from two different sources. In addition to the usual charge renormalization effects, there might also be a one-loop counterterm of the form $\kappa^2 R\ \mathrm{Tr}\ F_{\mu\nu}F^{\mu\nu}$. By using the field equations $R = 4\Lambda + \ldots$, together with $\kappa^2\Lambda = -6e^2$ this is converted into an extra $e^2\ \mathrm{Tr}\ F_{\mu\nu}F^{\mu\nu}$ term. The fact that agreement with the cosmological method has been achieved without taking this into account means that, in one-loop supergravity at least, such counterterms must be absent. In fact, all Einstein-Yang-Mills theories avoid this problem at one loop order as can be seen by taking the $e \to 0$ limit and exploiting the Einstein-Maxwell duality arguments (Deser 1981). However, the $e \to 0$ argument cannot be invoked at higher loops where counterterms involving higher powers of the curvature with e dependent coefficients might yield $F^{\mu\nu}F_{\mu\nu}$ on-shell. These might lead to a bizarre situation where flat space theories change their β function when coupled to gravity with a cosmological constant. One suspects that such terms are always absent in supergravity.

It is sometimes said that gravity theories, if they are to make sense, cannot be renormalizable but can only be finite owing to the dimensionful coupling constant κ. This ceases to be true, of course, when one allows for another dimensionless coupling constant e. Indeed the $N \leq 4$ theories discussed here are one-loop renormalizable in the sense that the non-vanishing counterterm is proportional to the classical action. This will continue to hold at two loops since the ungauged ($e = 0$) theories are known to be two-loop finite. Of course, the $N > 4$ theories stand a chance of being completely finite even after gauging.

Finally, although we have applied the spin sum rules (3.2) and (3.72) to the known Lagrangian field theories of supersymmetric Yang-Mills and gauged SO(N) supergravity, we should point out that from the purely group theoretical point of view, they have a much wider validity. As discussed by Curtwright (1981), they apply to all supermultiplets with either SO(N) or SU(N) internal symmetry, for example to the massive $N = 8$ supercurrent multiplet (Ellis et al. 1980; Deredinger et al. 1981), which features in the bound-state interpretation of presently observed quarks, leptons, and bosons.

3.8 Mass sum rules and symmetry breaking

The spin sum rules discussed above bear a remarkable resemblance to the mass sum rules which were already known in supergravity spontaneously broken via dimensional reduction. Consider the ungauged extended supergravities spontaneously broken through reduction from five dimensions (Cremmer et al. 1979; Ferrara & Zumino 1979). One has

$$\sum_{s} (-1)^{2s}(2s+1)M_s^{2k} = 0 \;, \quad N > 2k \;, \tag{3.74}$$

where M_s is the mass of the particle with spin s.

In contrast to the (unbroken) gauged supergravities, these theories have zero Λ at the tree level but will acquire one through radiative corrections. Owing to the mass sum rules, however, the induced cosmological constant is finite for $N \geq 6$. For further applications to spontaneously broken N = 8, see the lectures by van Nieuwenhuizen (1981 b).

4 ANOMALIES AND CHIRAL SUPERFIELDS

4.1 Axial and conformal anomalies for arbitrary spin

The arbitrary spin background field formalism discussed in Section 3 may also be used to determine the one-loop anomalies in ordinary (i.e. ungauged) supergravity. Consider, for example, the conformal anomaly in the trace of the stress tensor (Capper & Duff 1974 b; Deser et al. 1976; Duff 1977)

$$T^{\mu}{}_{\mu} = A \frac{1}{32\pi^2} {}^{*}R^{*}{}_{\mu\nu\rho\sigma} R^{\mu\nu\rho\sigma} + a \frac{e^2}{16\pi^2} F^{*}{}_{\mu\nu}{}^{a} F^{\mu\nu}{}_{a} C \; , \tag{4.1}$$

where A is the same coefficient that appears in the topological counter-terms (3.60) and (3.65) and a determines the Yang-Mills counterterm (3.46). One can show (Chistensen & Duff 1979) that

$$\begin{aligned} 12a &= (-1)^{2s}[-D' + 4E'_{+}] \\ 360A &= (-1)^{2s}[4D' - 10E'_{+} + 6F'_{+}] \; , \end{aligned} \tag{4.2}$$

where D', E'_{+}, and F'_{+} are given by Eq. (3.40). These yield

$$\begin{aligned} 6a &= (-1)^{2s}[-1 + 12s^2] \\ 360A &= (-1)^{2s}[8 - 150s^2 + 90s^4 + 30t^2(1 - t^2 + 6s^2)] \; . \end{aligned} \tag{4.3}$$

This A coefficient corresponds to the entries $\ell_{\mu a}$, ψ_{μ}, ϕ_{μ}, χ, and ϕ in the first column of Table 5. [Note, incidentally, that if we drop the F'_{+} term in Eq. (4.2), we obtain the simpler expression

$$360A' = (-1)^{2s}(8 - 60s^2) \; , \tag{4.4}$$

which from the sum rules of Section 4 would yield a vanishing result for supermultiplets with $N > 2$. The corresponding numbers are given in the second column of Table 5. Note that $\Delta A = A - A' =$ integer. We shall return to the significance of this shortly.]

Table 5

Contributions to the A and A′ coefficients from the component fields of supergravity. The numbers correspond to those of physical fields (i.e. after gauge-fixing and after ghost subtractions).

	360A	360A′	ΔA
$e_{\mu a}$	848	-232	3
ψ_μ	-233	127	-1
ϕ_μ	- 52	- 52	0
χ	7	7	0
ϕ	4	4	0
$\phi_{\mu\nu}$	364	4	1
$\phi_{\mu\nu\rho}$	-720	0	-2

Similarly, the axial anomalies are given by the same combinations of E'_- in the case of Yang-Mills and E'_- and F'_- in the case of gravity

$$\nabla^\mu J^5_\mu = \left(\frac{-E'_-}{6} + \frac{F'_-}{10}\right) \frac{1}{48\pi^2} R^*_{\mu\nu\rho\sigma} R^{\mu\nu\rho\sigma} + \frac{2E'_-}{3} \frac{e^2}{8\pi^2} \mathrm{Tr}\, F^*_{\mu\nu} F^{\mu\nu} \quad . \quad (4.5)$$

In particular, spin-3/2 fermions yield an axial anomaly 3 times that of spin-1/2 fermions in the case of Yang-Mills, and 3 - 24 = -21 times in the case of gravity. As an aside, we note that the relative size of the Yang-Mills and gravitational contributions, which at first sight seem unrelated, can be checked by topological arguments. The integrated version of Eq. (4.5) yields the index theorem

$$n_+ - n_- = \left[-\frac{s}{2} + s^3\right] \frac{1}{48\pi^2} \int d^4x \sqrt{g}\, R^*_{\mu\nu\rho\sigma} R^{\mu\nu\rho\sigma}$$

$$+ s \frac{e^2}{8\pi^2} \int d^4x \sqrt{g}\, F^*_{\mu\nu} F^{\mu\nu} \, , \quad (4.6)$$

where n_+ and n_- are the number of right- and left-handed zero modes of the arbitrary spin Dirac operator (minus ghosts for $s > \frac{1}{2}$) and for simplicity we consider an Abelian gauge field. Each term on the right of Eq. (4.6) is proportional to a topological invariant and normally each term is separately an integer. On spaces which do not admit of an ordinary spin structure, however, only their sum will be an integer (Hawking & Pope 1978 b; Pope 1980). For example, on CP^2

$$n_+ - n_- = \left[-\frac{s}{2} + s^3\right] + s\, e^2 \, , \quad (4.7)$$

where e = odd half-integer, in which case one finds, consistently, that the right-hand side is indeed an integer for all odd half-integer spin. Indeed, the argument could be turned around to deduce the gravitational anomaly from knowledge of the gauge-field anomaly, but only up to an integer (e.g. in the case of spin-3/2, one could not distinguish the factor -21 from +3, at least not without further detailed knowledge of CP^2). A discussion of arbitrary spin axial anomalies may also be found in the papers of Römer (1979, 1981).

Note that the trace anomaly and β-function calculations always involved even powers of spin, whereas the axial results involve odd powers of spin. [Strangely enough, we have been unable as yet to find any application of the spin sum rules (Eq. (3.2)) when k is odd.]

If one now applies these results to supersymmetric Yang-Mills theories one finds: a) The conformal anomalies and the axial anomalies form a multiplet along with the anomaly in $\gamma^\mu S_\mu$, where S_μ is the supercurrent, as one expects they should (Ferrara & Zumino 1975). For a review, see Grisaru (1978). b) The anomalies vanish for $N > 2$. This merely reflects the vanishing β-function for $N > 2$ discussed in Section 3.

A naïve application of these results to extended supergravity theories (with the usual field representations) leads to a more curious state of affairs. One finds: a) The anomalies do not appear to form a supermultiplet (Christensen & Duff 1978 a); b) The anomalies do not vanish for $N > 2$. Rather one finds that for $N > 2$, the A coefficient is an integer $A = 3 - N$ (Christensen et al. 1980). See Table 7.

Before enlarging on these remarks, we first discuss antisymmetric tensor fields.

4.2 Antisymmetric tensors: quantum inequivalence

The results just quoted for the A coefficient in supergravity are valid provided one makes the (ostensibly innocent) assumptions that: a) the auxiliary fields are irrelevant, b) the spin-0 particles are described by scalar fields $\phi(x)$. However, one can show that the A coefficient depends not only on the spin but also on the choice of field representation (Duff & van Nieuwenhuizen 1980). Thus the gauge theory of a rank-2 antisymmetric tensor $\phi_{\mu\nu}$ (with one degree of freedom) differs from that of a real scalar ϕ; and that of an antisymmetric rank-3 tensor $\phi_{\mu\nu\rho}$ (with 0 degrees of freedom) differs from nothing. One finds

$$A[\phi_{\mu\nu}] = A[\phi] + 1$$

$$A[\phi_{\mu\nu\rho}] = -2 , \qquad (4.8)$$

hence the values quoted in Table 5. The reasons for this quantum inequivalence have been discussed at length elsewhere (Duff 1981 b,c). The important point is that such representations can, in fact, occur either as auxiliary fields in N = 1 supergravity or as physical fields in the versions of extended supergravity obtained by dimensional reduction. [Incidentally, the study of higher-rank antisymmetric tensor fields in particle physics has a distinguished history. I am very grateful to Professor Kemmer for drawing my attention to his 1938 paper (Kemmer 1938). Another reference is Ogievetsky & Pulubarinov (1967).] For example, the N = 8 theory obtained by dimensional reduction from N = 1 in d = 11, taking the extra dimensions to be a 7-torus yields (Cremmer & Julia 1979) $63\phi + 7\phi_{\mu\nu} + \phi_{\mu\nu\rho}$. Only after making (topologically non-trivial) duality transformations does one obtain the theory with 70 ϕ fields. The importance of this, as noted by Siegel (private communication) is that the A coefficient of -5 in Table 7 becomes, on using Eq. (4.8)

$$A' = -5 + 7 - 2 = 0 \,. \tag{4.9}$$

Indeed, one can arrange that the vanishing of the A coefficient for all N > 2 (Duff 1981 b,c) with the choice of representations such that

$$3 - N[\psi_\mu] + N[\phi_{\mu\nu}] - 2N[\phi_{\mu\nu\rho}] = 0 \,, \tag{4.10}$$

where $N[\psi_\mu]$, $N[\phi_{\mu\nu}]$, and $N[\phi_{\mu\nu\rho}]$ are the number of gravitini, rank-two and rank-three antisymmetric tensors, respectively. Nicolai & Townsend (1981) have shows that $N \geq 3$ supergravity theories obeying Eq. (4.10) may be constructed from three basic N = 3 multiplets, one of which contains the antisymmetric tensor gauge field.

The idea that anomalies vanish in supergravity for N > 2 just as in Yang-Mills is certainly appealing, but at this stage several important questions remain unanswered:

1) Why does the antisymmetric tensor version of N = 8 with only SO(7) symmetry seem to have better ultraviolet behaviour than the scalar version with $E_7 \times SU(8)$ symmetry?

2) In order to fulfil Eq. (4.10), does one consider only physical fields or should one also consider auxiliary fields? Even if we confine our attention to the traditional minimal formulation of N = 1 auxiliary fields (Stelle & West 1978; Ferrara & van Nieuwenhuizen 1978 a) there is still some ambiguity in how one does the algebra. One or both of the scalars S and P could be replaced by $\partial^\mu S_\mu$ or $\partial^\mu P_\mu$ (Stelle & West

1978; Ogievetsky & Sokatchev 1980). Setting $\phi_{\mu\nu\rho} = \varepsilon_{\mu\nu\rho\sigma} S^{\sigma}$, etc., then yields the gauge theory of the rank-three field discussed above, and hence different anomalies. This problem becomes even more severe for N > 2 because no-one yet knows what the auxiliary fields are, or even if they exist!

3) Having allowed for antisymmetric tensors, do the anomalies now form a multiplet? We have seen that the inclusion of such representations can change the gravitational conformal anomaly. Interestingly enough, they can also contribute to the gravitational axial anomaly (Dowker 1978; Nielsen et al. 1978) if they are described by a first-order Lagrangian but not, apparently, if they are described by a second-order Lagrangian, since one cannot construct triangle graphs from the available vertices. [It may seem odd that one should ever describe bosons with a first-order Lagrangian, but such kinetic terms do in fact appear in the Lagrangian when one uses a manifestly supersymmetric gauge-fixing procedure (Grisaru & Siegel 1981 a).] One simple illustration of this puzzle is provided by the Wess-Zumino scalar multiplet, which describes a spinor, a scalar, and a pseudoscalar. When coupled to supergravity, the induced anomalies form a multiplet. If one now swaps the pseudoscalar for a gauge antisymmetric tensor one obtains a new multiplet (Siegel 1979). The conformal anomaly is now different but the axial anomaly seems not to have changed!

4) The quantum inequivalence discussed above was obtained using a background field gauge (Duff & van Nieuwenhuizen 1980). What is one to make of Siegel's claim (Siegel 1980) that there exists another gauge choice, e.g. a flat-space gauge, in which the inequivalences never occur? This would lead to the A′ coefficients for $\phi_{\mu\nu}$ and $\phi_{\mu\nu\rho}$ quoted in the second column of Table 5.

5) Is it significant that one can arrange for vanishing anomalies for N > 2, even with the usual representations, by taking the A′ coefficient of Eq. (4.4) and invoking the spin sum rules?

6) Why do these A′ coefficients differ from the A coefficients by integers? See Table 5.

7) Why does the barrier for vanishing anomalies occur for N > 2 both in Yang-Mills and supergravity? What does this imply for the N > 2 auxiliary field quest?

A partial set of answers to these questions can be obtained by re-examining the problem in terms of N = 1 superfields, and using the

Grisaru-Siegel superfield Feynman rules (Grisaru & Siegel 1981 a), to which we now turn.

4.3 Chiral superfields: vanishing anomalies for $N \geq 2$

In this subsection, which was written with the help of Marc Grisaru, I shall summarize some results to be described in more detail in a future publication (Duff et al. 1981). The starting point is the form of the on-shell background field N = 1 supergravity action to second order in the quantum fields, after fixing the gauge and including the Fadeev-Popov ghosts (Grisaru & Siegel 1981 a):

$$S = \int d^4x\, d^4\theta\, E^{-1} \Big[-\frac{1}{2} H \,\Box\, H - \frac{1}{2} H^{\alpha\dot{\beta}} \Big(W_\alpha{}^{\gamma\delta} D_\gamma H_{\delta\dot{\beta}} + \bar{W}_{\dot{\beta}}{}^{\dot{\gamma}\dot{\delta}} \bar{D}_{\dot{\gamma}} H_{\alpha\dot{\delta}} \Big) - \frac{18}{5} X\bar{X} + \sum_{i=1}^{3} \bar{\Psi}_i{}^{\dot{\beta}} i D_{\alpha\dot{\beta}} \Psi_i{}^{\alpha} + \sum_{i=1}^{2} \Big(\Phi_i{}^{\alpha} D^2 \Phi_{i\alpha} + \text{h.c.} \Big) \Big], \tag{4.11}$$

where the physical axial-vector superfields $H_{\alpha\dot{\alpha}}$ is real, the general spinors $\Psi_{\alpha i}$ have abnormal (ghost) statistics, the chiral spinors $\Phi_{\alpha i}$ have normal (ghost-for-ghost) statistics, and the chiral scalar X is a physical compensating field. The on-shell background supergravity fields appear in the determinant of the superveilbein E and in $W_{\alpha\beta\gamma}$ of Eq. (2.9). This action corresponds in components to the usual (S and P) auxiliary field choice. For N-extended supergravity one has to add contributions from superfields representing the {3/2, 1}, {1, 1/2}, and {1/2, 0} matter multiplets, taking into account gauge-fixing and ghosts where necessary. The number of superfields of each type is given in Table 7 (V is a real scalar superfield) where the minus signs denote abnormal statistics. Again, the representation content is the usual one, e.g. N = 8 has 70 scalars.

To calculate the contribution to the one-loop anomalies from each superfield one may either compute a one-loop supergraph using the Feynman rules for superfields, or else infer it from the known component results. In the latter case, it is first necessary to replace the component contributions of Table 5, which correspond to physical fields after ghost subtraction, by those of Table 6 which are valid before ghost subtractions have been made. The remarkable result, as shown in Table 6, is that only chiral superfields contribute to anomalies. [This result may

Table 6

The A and A′ coefficients for both component fields and superfields after gauge fixing but before ghost subtraction. Note that (as in Yang-Mills) only chiral superfields yield anomalies.

COMPONENTS:

	360A	360A′	ΔA
$e_{\mu a}$	760	-320	3
ψ_μ	-212	148	-1
$\phi_\mu = \phi_{\mu\nu\rho}$	- 44	- 44	0
χ	7	7	0
ϕ	4	4	0
$\phi_{\mu\nu}$	264	- 96	1

N = 1 SUPERFIELDS:

	A	A′	ΔA	CHIRAL?
$H_a = e_{\mu a} + 4\psi_\mu + 4\phi_\mu + \phi_{\mu\nu}$	0	0	0	NO
$\Psi_\alpha = \psi_\mu + 2\phi_\mu + 4\chi + 2\phi + \phi_{\mu\nu}$	0	0	0	NO
$V = \phi_\mu + 4\chi + 4\phi$	0	0	0	NO
$X = \chi + 2\phi$	$\frac{1}{24}$	$\frac{1}{24}$	0	YES
$\Phi_\alpha = 4\chi + 2\phi + \phi_{\mu\nu}$	$\frac{5}{6}$	$-\frac{1}{6}$	1	YES

Table 7

Extended supergravities in terms of N = 1 superfields corresponding to the usual assignment of physical and auxiliary fields.

N	H	ψ	V	X	φ	A	A′	ΔA
1	1	-3	0	1	2	$^{41}/_{24}$	$-^{7}/_{24}$	2
2	1	-2	-1	2	1	$^{11}/_{12}$	$-^{1}/_{12}$	1
3	1	-1	-1	0	0	0	0	1
4	1	0	0	-4	-1	-1	0	-1
5	1	1	2	-8	-2	-2	0	-2
6	1	2	6	-12	-3	-3	0	-3
7	1	4	14	-20	-5	-5	0	-5
8	1	4	14	-20	-5	-5	0	-5

Table 8

A choice of N = 1 superfields yielding A = A′. The absence of chiral superfields for N > 2 yields vanishing anomalies.

N	H	ψ	V	X	φ	A = A′
1	1	-3	4	-7	0	$-^{7}/_{24}$
2	1	-2	1	-2	0	$-^{1}/_{12}$
3	1	-1	-1	0	0	0
4	1	0	-2	0	0	0
5	1	1	-2	0	0	0
6	1	2	0	0	0	0
7	1	4	4	0	0	0
8	1	4	4	0	0	0

be verified directly from the superspace Feynman graphs by counting powers of θ in the vertices and propagators (see the lectures by Grisaru in this volume).] With these representations, therefore, one simply reproduces the results already quoted (Christensen & Duff 1978). In particular, A is an integer for $N > 2$.

However, an alternative action to Eq. (4.11), with a different choice of representations, can be obtained by employing a real scalar compensating field V, in which case (Grisaru & Siegel 1981 a):

$$S' = \int d^4x\, d^4\theta\, E^{-1}\Bigg[-\frac{1}{2}H\,\Box\,H - \frac{1}{2}H^{\alpha\dot\beta}\left(W_\alpha{}^{\gamma\delta}D_\gamma H_{\delta\dot\beta} + \bar W_{\dot\beta}{}^{\dot\gamma\dot\delta}\bar D_{\dot\gamma}H_{\alpha\dot\delta}\right) + \sum_{i=1}^{4} V_i\,\Box\,V_i + \sum_{i=1}^{3}\bar\Psi_i{}^{\dot\beta} iD_{\alpha\dot\beta}\Psi_i{}^{\alpha} + \sum_{i=1}^{7} X_i\bar X_i\Bigg], \qquad (4.12)$$

where Ψ and X have abnormal statistics. Note that the chiral spinor Φ_α never appears. This action corresponds in components to exchanging one of the scalar auxiliary fields for a rank-three gauge antisymmetric tensor. With this starting point, one is led naturally to the superfield assignments of Table 8. One now finds that the components satisfy Eq. (4.10) and hence yield vanishing anomalies for $N > 2$. [Curiously, this seems to imply a connection between the $\phi_{\mu\nu\rho}$ field as an auxiliary field for $N = 1$ and the $\phi_{\mu\nu\rho}$ field obtained in $N = 8$ by dimensional reduction from $N = 1$ in $d = 11$.]

However, in terms of superfields, Table 8 yields a much simpler explanation for vanishing anomalies: The anomalies vanish for $N > 2$ because the net number of chiral superfields (i.e. physical minus ghost) is zero for $N > 2$. This provides another explanation, incidentally, for the vanishing one-loop β function in $N > 2$ Yang-Mills. In terms of $N = 1$ superfields: $N = 1$ Yang-Mills requires one physical non-chiral field V and three ghost chiral fields X; $N = 2$ Yang-Mills requires one physical V, one physical X and three X ghosts; $N = 4$ Yang-Mills requires one physical V, three physical X, and three X ghosts. Hence the β functions for $N = 1, 2, 4$ are in the ratio 3:2:0; the zero result for $N = 4$ being due to vanishing of the net number of chiral superfields.

In the absence of a complete formalism for N-extended superfields, we cannot give a fundamental explanation of why the net number of N = 1 chiral superfields should be zero for N > 2 in these models. The answer to this question may, however, lie in the results of Dr. Kallosh (this volume) in which she examines the consistency conditions for extended (as opposed to N = 1) superfields, and finds a discrete difference between $N \leq 2$ and N > 2 (due in the case of supergravity to the presence of spin-½ components for N = 3 onwards). Her results imply: i) no extended chiral matter superfields can exist in an extended superfield supergravity background for N > 2 (which might suggest that there should be no net number of N = 1 chiral superfields in any correct N = 1 analysis of N > 2 theories using the background field method and hence no anomalies for N > 2). This would certainly be consistent with her second claim: ii) no super extension of the Gauss-Bonnet invariant exists for N > 2.

4.4 Zero modes, gauge fixing, and boundary conditions

We have still not given an explanation for the different between the A and A′ coefficients. Let us first consider the fields $\phi_{\mu\nu}$ and $\phi_{\mu\nu\rho}$. By using a background covariant gauge Duff and van Nieuwenhuizen (1980) found "quantum inequivalence of different field representations", see Eq. (4.8). In a subsequent paper "Quantum equivalence of different field representations" Siegel (1980) found, using a different gauge

$$A'[\phi_{\mu\nu}] = A'[\phi]$$

$$A'[\phi_{\mu\nu\rho}] = 0 \ . \qquad (4.13)$$

On the basis of this, Siegel concluded that the trace anomalies are gauge dependent and hence not physically observable. Here we give a different interpretation. A fuller explanation will be given elsewhere and here we summarize the salient points.

As explained in Duff and van Nieuwenhuizen (1980) and in Duff (1981 c), the reason for the quantum inequivalence was a topological one. Consider a p-form A, its exterior derivative dA and an action functional

$$S = \frac{1}{2}\,(dA,dA) \ . \qquad (4.14)$$

S is invariant under the symmetry

$$A \rightarrow A + V \ , \tag{4.15}$$

where

$$dV = 0 \ . \tag{4.16}$$

In other words,

$$A \rightarrow A + V_H + d\alpha \ , \tag{4.17}$$

where V_H is harmonic

$$\delta V = dV = 0 \ ; \qquad \delta = {}^*d^* \ . \tag{4.18}$$

Now consider the gauge-fixing addition S′

$$S' = \frac{1}{2} (\delta A, \delta A) \ . \tag{4.19}$$

The new action, given by

$$S + S' = \frac{1}{2} (A, \Delta A) \ ; \qquad \Delta = d\delta + \delta d \ , \tag{4.20}$$

although no longer invariant under the usual "small" gauge transformations $A \rightarrow A + d\alpha$, is still invariant under the "large" gauge transformations $A \rightarrow A + V_H$. The number of harmonic p-forms V_H will be the number of zero eigenvalue modes of the corresponding Laplacian Δ_p which in a compact manifold is given by the Betti numbers, n_p. These are related to the Euler number Eq. (3.61) by

$$\chi = \sum_{p=0}^{4} (-1)^p \, n_p \ . \tag{4.21}$$

It is these zero modes in the functional integral which are responsible for the quantum inequivalence; the ghosts and ghosts-for-ghosts, etc., enter with alternating signs in the trace anomaly as in Eq. (4.21), yielding a coefficient of χ which differs by an integer from that expected on the grounds of naïve equivalence.

Suppose, on the other hand, one had chosen a gauge-fixing addition which broke both small and large gauge invariances. These zero-modes would now be absent and one would recover "quantum equivalence". This is our explanation for Siegel's result. Question: Which is correct? Answer: It depends on the boundary conditions!

The following simple example nicely illustrates the problem (and I am grateful to G.W. Gibbons, C.J. Isham, M. Peskin & E. Witten for discussions in it). Consider free Maxwell theory at finite temperature, i.e. on a manifold $M_3 \times S^1$ where M_3 is three-dimensional space. In a co-variant gauge like Eq. (4.19) the partition function is given by

$$Z = \frac{\det \Delta_0}{(\det \Delta_1)^{1/2}} , \tag{4.22}$$

where Δ_1 is the Laplacian on 1-forms (the physical potential A_μ) and Δ_0 is the Laplacian on 0-forms (the two scalar ghosts). However the operator Δ_1 has a zero-mode (S^1 gives rise to a non-vanishing first Betti number, i.e. the space is not simply connected) corresponding to the field configuration

$$\begin{aligned} A_0 &= c = \text{constant} \\ &= \Omega \, i\partial_0 \Omega^{-1} , \end{aligned} \tag{4.23}$$

where

$$\Omega = \exp ict . \tag{4.24}$$

Now Eq. (4.23) is simply pure gauge and in Minkowski space R^4 would be trivial. At finite temperature, however,

$$\frac{1}{2\pi} \int_{S^1} A_\mu \, dx^\mu = n \tag{4.25}$$

[n = integer if $\Omega \subset U(1)$] is non zero. Suppose, on the other hand, one had chosen the gauge

$$A_0 = 0 \tag{4.26}$$

one would automatically have ruled out all the non-trivial $n \neq 0$ configurations. We thus (apparently) would reach different partition functions in different gauges. Including charged matter via an $A_\mu J^\mu$ coupling would lead to a delta-function of total charge Q

$$Q = \int d^3x\, J^0 \tag{4.27}$$

in the gauge where one integrates over the zero mode but not in the gauge where the zero mode is omitted. Again one asks: which is correct? The answer depends on M_3. If, as is usual, M_3 is chosen to be just flat Euclidean space R^3 then Eq. (4.23) is not square integrable and this mode is omitted from the functional integral. If, however, M_3 were chosen to be compact, say S^3, then Eq. (4.23) would be square integrable and should be included. The lesson is that the answer is not gauge-dependent *per se*. Rather it is theory dependent: one obtains different answers for different physical boundary conditions. [It is simply that for a given boundary condition, there are good and bad gauges. For example Eq. (4.26) is a forbidden gauge in the compact case.] So it is with the antisymmetric tensors: one can have either "quantum inequivalence" or "quantum equivalence" depending on the physics. We emphasize, however, that for a given theory (i.e. Lagrangian plus boundary conditions) the trace anomaly is unique and gauge independent. For example, if the space-time is compact [as in Hawking's space-time foam approach (Hawking 1978)] one must obtain the A coefficient of Table 5, whereas the A′ coefficient would be appropriate for asymptotically flat space-times. It seems that similar remarks apply to spin-$3/2$ and spin-2: one can have pure gauge configurations

$$h_{\mu\nu} = \nabla_\mu \xi_\nu + \nabla_\nu \xi_\mu$$

$$\psi_\mu = D_\mu \varepsilon \tag{4.28}$$

which are still non-trivial. The omission of zero-modes when they cease to be square integrable also explains why ΔA is positive for commuting fields and negative for anticommuting. [Pure gauge modes of this kind have been discussed before in a somewhat different context by Hawking & Pope (1978 a). See also a recent paper by Gibbons (1981).] This newly found realization of boundary-condition dependence in anomaly calculations may also help to clear up some murky areas in the literature. For example, a Maxwell gauge field A_μ has zero degrees of freedom in two dimensions. By extrapolating the dimensional regularization (and flat-space gauge-fixing) calculations of Capper et al. (1974) to d = 2 one finds a vanishing trace for its one-loop quantum stress tensor (Duff 1977). The heat kernel method using a background field gauge, however, would yield

$$T^\mu{}_\mu = -\frac{1}{4\pi} R \tag{4.29}$$

or

$$\int d^4x \sqrt{g}\, T^\mu{}_\mu = -\chi , \tag{4.30}$$

where χ is the two dimensional Euler number. Indeed this example is precisely the d = 2 analogue of $\phi_{\mu\nu\rho}$ in d = 4. The whole debate about quantum inequivalence could have taken place years ago. In fact, it did. The coefficient that governs the R term in $T^\mu{}_\mu$ for d = 2 is the one that governs the $\Box R$ term in $T^\mu{}_\mu$ for d = 4. And there is a long-standing (and until now unresolved) debate about the coefficient of $\Box R$ in the trace anomaly for photons. See Duff (1977) for a list of references. [There is a somewhat different sense, of course, in which boundary conditions are important for anomalies. In space-times with boundary, the anomalies and counterterms will receive boundary term corrections. As explained in Duff (1981 c), moreover, these also cancel in supersymmetric theories obeying Eq. (4.10). See also Fradkin and Tseytlin (1981 a).]

In summary, therefore, one can obtain different anomalies in supergravity according to the choice of (a) physical field representations (b) auxiliary field representations, and (c) boundary conditions. For example, Table 7 shows that with suitable boundary conditions, the anomalies vanish for N > 2 even with the <u>usual</u> representations (thus restoring the $E_7 \times SU(8)$ symmetric version to its former glory). But what about that choice of fields and/or boundary conditions for which the anomalies do not vanish for N > 2 and for which, moreover, they do not appear to form a multiplet? It seems to me that, in this case, the zero-modes are actually breaking the supersymmetry (i.e. symmetry breaking via gravitational instantons). The argument which says anomalies <u>must</u> form a multiplet if one calculates with superfields now needs modification: the supersymmetric gauge-fixing inherent in the superfield formalism may simply be incompatible with the boundary or periodicity conditions. Similar phenomena are, in fact, already familiar from somewhat different contexts. For example, finite temperature for which the bosons are periodic in imaginary time but fermions antiperiodic is not compatible with rigid supersymmetry. Similarly, the vanishing of the vacuum energy density in supersymmetry (Zumino 1975) does not imply zero Casimir effect since perfectly conducting plates would place different boundary conditions on different components of a supermultiplet.

Finally, for those who have not found this section too confusing already, a few more words on auxiliary fields. Although we have discussed different representations in this section on anomalies, they each correspond to the old minimal set (Stelle & West 1978; Ferrara & van Nieuwenhuizen 1978 a). In other words, they correspond to $n = -1/3$ in the language of Gates et al. (1981). However, these authors have recently examined the "new-minimal" auxiliary fields of Sohnius and West (1981 b) corresponding to $n = 0$, as well as the old non-minimal set of Breitenlohner corresponding to $n \neq 0, -1/3$, in the context of anomalies [in this case the coupling to matter is via the "new deteriorated energy momentum tensor" (Duff & Townsend 1981)] and find the presence of trace anomalies will in general lead to <u>anomalies in the Ward identities for local supersymmetry</u>. They conclude that for $N = 1$ supergravity, at least, only the $n = -1/3$ set lead to a consistent theory.

We close this section on anomalies with another unanswered question. At one loop we have learned that no chiral superfields means no trace anomalies and hence, by supersymmetry, no chiral anomalies either. This is a very satisfactory result. Chiral anomalies rely on an imbalance between left- and right-handed fields and so such imbalance is possible in the absence of chiral superfields. Is this merely a one-loop phenomenon, or does it have ramifications for the exact theory?

5 RECENT DEVELOPMENTS

Since the bulk of these lectures was written, there have been several interesting developments in our understanding of the ultraviolet divergences in extended supergravity. The most important are summarized below.

Sum rules

Although the spin sum rules amd mass sum rules discussed in Section 3 may be verified by inspection of the mass and spin assignments of extended supersymmetry multiplets, it would obviously be advantageous to derive these relations from first principles. Such a derivation has now been given by Ferrara et al. (1981), starting from the fundamental algebra of extended supersymmetry. Indeed this work confirms one's suspicions that the spin and mass rules have the same origin. Moreover, they find that the rules used to demonstrate one-loop ultraviolet divergence cancellations are merely special cases of more general rules which are suggestive of new, as yet undiscovered, cancellation phenomena. See also Ferrara (1981).

N = 8 supergravity with local SO(8) × SU(8)

As discussed in Section 3, the vanishing of the one-loop β-function in gauged N > 4 supergravity was discovered before the Lagrangians for these models had been completely constructed and hence before their actual existence was established. Now, De Wit & Nicolai (1981 a,b) have succeeded not only in constructing gauged N = 8 (and hence by truncation N < 8) but in exhibiting a local, non-linearly realized, SU(8) invariance in addition to the local linear SO(8). Their assignments are

	$e_\mu{}^a$	$\psi_\mu{}^i$	$A_\mu{}^{IJ}$	χ^{ijk}	$U_{ij}{}^{IJ}$	V^{ijIJ}
SO(8)	1	1	28	1	28	28
SU(8)	1	8	1	56	$\overline{28}$	28

[One may recover the anticipated 1, 8, 28, 56, and 70 representations of SO(8) by imposing a special SU(8) gauge.] As one switches off the gauge coupling one recovers the version of ungauged N = 8 with E_7(rigid) × × SU(8) (local) invariance (Cremmer & Julia 1979).

This gauged model of N = 8 offers different possibilities for superunification from the ungauged theory. One possibility, suggested by De Wit & Nicolai (1981 a,b) is that only the SO(8) singlets of spin 2, $^{3}/_{2}$, and $^{1}/_{2}$ are unconfined, though it remains unclear whether this is compatible with the vanishing β-function. An analysis of possible counterterms in this model has been given by Howe & Nicolai (1981).

Conformal supergravity: finite for N = 4

Although fourth-order theories were mentioned in the Introduction, we have not yet discussed those encountered in supergravity, for example, the conformal supergravities with N = 1, 2, 3, and 4, which are the supersymmetric extensions of the (Weyl tensor)^2 conformal gravity. [For a review, see De Wit (1981).] One would not expect such theories to be consistent at the quantum level because the existence of conformal anomalies at one loop would force non-conformal modifications to the Lagrangian at two loops and beyond (Capper & Duff 1975), and the conformal invariance is necessary to maintain the number of propagating degrees of freedom. This objection could be avoided, of course, if the theory were ultraviolet finite. Recently, Fradkin & Tseytlin (1981 a,b,c) have calculated the one-loop counterterms for N = 1, 2, 3, and 4 and found that N = 4 is one-loop finite! This is yet another example of a "miraculous" cancellation not expected on the grounds of invariants-as-counterterm arguments.

Topologically massive gravity

Whilst on the subject of higher derivatives, we mention the recent work of Deser et al. (1981) who consider a three-dimensional gravity theory with a third derivative "topological" mass term. This term is enough to yield power counting renormalizability; yet surprisingly, in contrast to most higher derivative theories, yields no ghosts. A supersymmetric version no doubt exists also, but the implications for four-dimensional supergravity, if any, remain unknown.

Non-renormalization theorems:
proofs of finiteness to all orders

We have emphasized in Section 3 the inadequacy of the absence-of-invariants-as-counterterms arguments in explaining divergence cancellations in theories with extended supersymmetry. Useful though the knowledge of higher derivative invariants may be, this line of thought may even be seen ultimately to have delayed progress in supergravity by several years!

It was already known in the early days of global supersymmetry that the ultraviolet behaviour of supersymmetric theories was better than might be expected from naïve power counting, even for N = 1. For example, the absence of quadratic mass renormalization for scalar partners of chiral fermions in the Wess-Zumino model was known some time ago (Wess & Zumino 1974 a; Illiopoulos & Zumino 1974). Similar non-renormalization theorems occur in Abelian (Wess & Zumino 1974 b) and non-Abelian (Ferrara & Piguet 1975) gauge theories. This latter phenomenon has stimulated much of the recent interest in supersymmetric GUTS. See, for example, Ferrara (1981).

These non-renormalization phenomena are but special cases of an even stronger result proved by Grisaru et al. (1979) using N = 1 superspace Feynman rules: all quantum corrections to the effective action can be written as a single $\int d^4\theta$ integral over the full N = 1 superspace. This means that terms in the tree Lagrangian which cannot be so written (e.g. the mass and self interaction terms for chiral scalar superfields, $\int d^2\theta\, \phi^n$, in the Wess-Zumino model) will never be renormalized!

However, the far-reaching implications of this result have only recently become fully appreciated. In a recent paper Stelle (1981b) exploits the presence of such an $\int d^2\theta$ chiral interaction in the N = 4 Yang-Mills theory written in terms of N = 1 superfields and then invokes the fourfold supersymmetry to prove that, since this term is non-renormalized, all terms are non-renormalized. Thus the β-function remains zero to all orders of perturbation theory! It also appears that this same N = 1 superfield non-renormalization theorem can be invoked to prove the vanishing of the β-function to all orders in gauged supergravity for N > 4 (Stelle & Townsend 1981).

String theories

The historical foundations of supersymmetry are intimately tied up with the dual string. Recently there have been two interesting, though separate, developments in this connection. The first, due to Polyakov (1981 a,b) relates the critical dimensions of d = 26 for the bosonic string and d = 10 for the fermionic string to anomalies in the trace of the stress tensor (Section 4) in two-dimensional gravity and supergravity theories, respectively. It seems that the topological and boundary corrections to the anomalies discussed in Section 4 will also be important, when one considers possible holes and boundaries for the two-dimensionsal sheet swept out by the string. After these trace anomalies are taken into account, the resulting theory is described by an interacting Liouville Lagrangian, the mass parameter being related to the cosmological constant which Polyakov introduces on the grounds of renormalizability. In the author's opinion, incidentally, no such interaction is required for the fermionic string because no cosmological term is induced by quantum corrections. Indeed, this is a trivial example of the non-renormalization theorem discussed above. The absence of an induced cosmological term has been confirmed in explicit calculations by Fradkin & Tseytlin (1981 d).

A second interesting development, due to Green et al. (1981) concerns the emergence of N = 1 Yang-Mills in d = 10 and N = 2 supergravity in d = 10 as the zero-slope limits of the fermionic string models. By considering the extra dimensions to be compactified as circles and then letting the radii of the circles tend to zero, they are able to analyse the ultraviolet and infrared divergences of the theories for different values of d. Interestingly, they find that at one loop both theories are ultraviolet (UV) finite for $d < 8$ and infrared (IR) finite for $d > 4$, the IR divergence being milder in the gravitational case. They speculate that if certain kinematical features persist at higher loops the N = 4 Yang-Mills in d = 4 will be UV finite to all orders but N = 8 supergravity would diverge at three loops and beyond. Interestingly, their results for two-, three-, and four-particle Green's function and thus speculations on the UV behaviour of these theories for various values of N exactly parallel those in Section 3 made on the basis of the sum rules and the rough guide of at most one power of spin at each vertex for Yang-Mills and at most two powers of spin for gravity.

These authors also stress, of course, that the original string theories might yet be UV finite even if the zero-slope limits are not.

More supergraphs

Grisaru & Siegel (1981 b) have continued to exploit their Feynman rules for N = 1 superfields. A recent application is the calculation of the one-loop four-particle S matrix in extended supergravity. They find the calculations hardest for N = 1 supergravity, becoming progressively easier with higher N culminating in an almost trivial result for N = 8, in agreement with the above-mentioned string calculations. The absence of chiral superfields for $N \geq 3$ discussed in Section 4 considerably simplifies the results, and the vanishing of various terms in the amplitudes for increasing N again strongly suggests that the spin sum rules of Section 3 are at work.

[We have not discussed in these lectures the theory of "gauge supersymmetry" due to Arnowitt and Nath, since the particle content is difficult to extract and in any event includes unphysical ghost states. However, these authors were among the first to exploit the power of superspace power counting. Indeed, their theory, albeit unphysical, can be shown to be finite to all orders. See, for example Arnowitt & Nath (1979). Further discussions of UV divergences and superfields may be found in Taylor (1979).]

In a more recent paper on superspace power counting rules Grisaru & Siegel (1982) are able (subject to the existence of an extended superfield formalism): a) to confirm the finiteness of N = 4 Yang-Mills to all orders, b) to extend the vanishing two-loop contribution to β for N = 2 (Section 3) to all loops greater than one, c) to rule out counterterms in N = 8 supergravity in the first six loop orders (this means, in particular, that the suggested three one-loop counterterm is not in fact present).

At this stage, one might regard the present inability to rule out all counterterms for N = 8 as a disappointment, but time and again we have learned never to underestimate the power of supersymmetry!

Kaluza-Klein

One possibility remaining is the one suggested in the Introduction: perhaps N = 1 supergravity is finite in d = 11 (Duff & Toms 1981). Incidentally, the Brink-Green-Schwarz results do not include this possibility. They consider a theory which corresponds to N = 2 in d = 10 and this is obtained from N = 1 in d = 11 only after discarding an infinite tower of massive states. [An analysis of the kinds of massive states which arise in Kaluza-Klein theories has recently been given by Salam & Strathdee (1981). Interestingly, they find that they belong to infinite dimensional representations of non-compact groups.]

Although Cremmer & Julia (1979) were aware of the possibility of choosing the extra seven dimensions to be something other than the 7-torus ($S^1 \times S^1 \times \ldots S^1$), they restricted their explicit calculations to this case. In this way, they obtained an N = 8 supergravity in d = 4 with a rigid SO(7) invariance. (This is the antisymmetric tensor version discussed in Section 4.) Only after performing duality transformations (which convert antisymmetric tensors to scalars) did they obtain the version with E_7(rigid) × SU(8)(local). The geometric origin of these symmetries is therefore obscured.

However, in a forthcoming paper (Duff & Pope 1982) it will be shown that a gauged version of N = 8 may be obtained from N = 1 in d = 11, but by taking the extra dimensions to be a 7-sphere (S^7), which guarantees that the four-dimensional Lagrangian for the massless states will have a local SO(8) invariance and a (negative) cosmological constant. The resulting theory is therefore almost certainly that of De Wit & Nicolai (1981) discussed above. Further confirmation comes from a counting of the massless states: we find 1 graviton, 8 gravitini, 28 vectors and 56 spin-½ fermions. A direct count of the number of massless scalars is rather more difficult and is presently under way, but indirectly one can argue that the number of physical scalars must be 70 owing to the 8-fold supersymmetry, which in turn is guaranteed by the presence of eight massless spin-³⁄₂ particles. What can be said with certainty is that they are genuine scalars and not antisymmetric tensors. (This is due to the difference in Betti numbers between a 7-sphere and a 7-torus.) Consequently no duality transformations are required to obtain the symmetries of the De Wit-Nicolai Lagrangian. It seems likely therefore that the hidden SU(8) (local) in the gauged version and the hidden E_7(rigid) one obtains on switching off the gauge coupling (shrinking the sphere to zero radius) may have a more transparent geometrical explanation in the picture with the 7-sphere than in the picture with the 7-torus.

This Kaluza-Klein framework puts into perspective the difference between the two N = 8 theories discussed in Section 4. Indeed, there could be an infinite number of d = 4 theories corresponding to the infinite number of ways of compactifying the extra seven dimensions, though N = 8 supersymmetry (eight massless gravitini) would be the exception rather than the rule. The question of ground-state stability still remains open but this large number of different theories might even be reinterpreted as many different phases of the same theory. Some of them would display the SU(3) × SU(2) × U(1) symmetry of the Witten (1981) scheme. The crucial question is still the number of fermions. Rules for counting the numbers of massless states in Kaluza-Klein supergravity will be given elsewhere (Duff & Pope 1982).

Can the vanishing β-functions of Section 3 and the vanishing anomalies of Section 4 be understood from higher dimensions? What is the connection between the non-renormalization theorems discussed above, the spin and mass sum rules of Section 3, and the absence of chiral superfields of Section 4? Can we find off-shell formulations of extended supergravity in $d = 4$, or $N = 1$ in $d = 11$ and, if so, how will they affect the renormalization?

Conclusion

Our understanding of the ultraviolet divergences in extended supergravity is only just beginning.

REFERENCES

Adler, S.L. (1981). Einstein gravity as a symmetry breaking effect in quantum field theory. Princeton preprint.

Alvarez-Gaumé, L. & Freedman, D.Z. (1980). Geometrical structure and ultraviolet finiteness in the supersymmetric σ-model. MIT preprint.

Arnowitt, R. & Nath, P. (1979). Supergraphs and ultraviolet finiteness in gauge supersymmetry. In Supergravity, eds. D.Z. Freedman and P. van Nieuwenhuizen. Amsterdam: North Holland.

Aurelia, A., Nicolai, H., & Townsend, P.K. (1980). Hidden constants: The θ parameter of QCD and the cosmological constant of N = 8 supergravity. Nucl. Phys. B176, 509.

Avdeev, L.V., Chochia, G.A. & Vladimirov, A.A. (1981). On the scope of supersymmetric dimensional regularization. Dubna preprint E2-81-370.

Avdeev, L.V., Tarasov, O.V. & Vladimirov, A.A. (1980). Vanishing of the three-loop charge renormalization function in a supersymmetric gauge theory. Phys. Lett. 96B, 94.

Barvinsky, A.O. & Vilkovisky, G.A. (1981). Divergences and anomalies for coupled gravitational and Majorana spin-½ fields. Nucl. Phys. B191, 237.

Berends, F.A. & Gastmans, R. (1975). Quantum gravity and the electron and muon magnetic moments. Phys. Lett. 55B, 311.

Breitenlohner, P. (1977). A geometric interpretation of local supersymmetry. Nucl. Phys. B182, 125.

Capper, D.M. & Duff, M.J. (1974 a). One loop neutrino contribution to the graviton propagator. Nucl. Phys. B82, 147.

Capper, D.M. & Duff, M.J. (1974 b). Trace anomalies in dimensional regularization. Nuovo Cimento 23A, 173.

Capper, D.M. & Duff, M.J. (1975). Conformal anomalies and the renormalizability problem in quantum gravity. Phys. Lett. 53A, 361.

Capper, D.M., Duff, M.J. & Halpern, L. (1974). Photon corrections to the graviton propagator. Phys. Rev. D 10, 461.

Caswell, W.E. & Zanon, D. (1981). Zero three-loop beta function in the N = 4 supersymmetric Yang-Mills theory. Nucl. Phys. B182, 125.

Christensen, S.M. (1981). Quantizing fourth order gravity theories. In Quantum Structure of Space and Time, eds. M.J. Duff and C.J. Isham. Cambridge: Cambridge University Press.

Christensen, S.M., Deser, S., Duff, M.J. & Grisaru, M.T. (1979). Chirality, self-duality, and supergravity counterterms. Phys. Lett. 84B, 411.

Christensen, S.M. & Duff, M.J. (1978 a). Axial and conformal anomalies for arbitrary spin in quantum gravity and supergravity. Phys. Lett. 76B, 571.

Christensen, S.M. & Duff, M.J. (1978 b). Quantum gravity in 2 + ε dimensions. Phys. Lett. 79B, 213.

Christensen, S.M. & Duff, M.J. (1979). New gravitational index theorems and supertheorems. Nucl. Phys. B154, 301.

Christensen, S.M. & Duff, M.J. (1980). Quantizing gravity with a cosmological constant. Nucl. Phys. B170, 480.

Christensen, S.M., Duff, M.J., Gibbons, G.W. & Roček, M. (1980). Vanishing one-loop β function in gauged N > 4 supergravity. Phys. Rev. Lett. 45, 161.

Cremmer, E. & Julia, B. (1979). The SO(8) supergravity. Nucl. Phys. B159, 141.

Cremmer, E., Scherk, J. & Schwarz, J. (1979). Spontaneously broken N = 8 supergravity. Phys. Lett. 84B, 83.

Curtwright, T. (1981). Charge renormalization and high spin fields. Phys. Lett. 102B, 17.

Das, A., Fischler, M. & Roček, M. (1977). Super-Higgs effect in a new class of scalar models and a model of super QED. Phys. Rev. D 16, 3427.

Derendinger, J.P., Ferrara, S. & Savoy, C.A. (1981). Flavour and super-unification. Nucl. Phys. B188, 77.

Deser, S. (1975). Trees, loops and renormalizability. In Quantum Gravity: An Oxford Symposium, eds. C.J. Isham, R. Penrose and D.W. Sciama. Oxford: Oxford University Press.

Deser, S. (1981). Divergence cancellations in gravity-matter systems from supergravity embedding. Phys. Lett. 101B, 311.

Deser, S., Duff, M.J. & Isham, C.J. (1976). Non-local conformal anomalies. Nucl. Phys. B114, 29.

Deser, S., Jackiw , R. & Templeton, S. (1981). Topologically massive gauge theories. MIT preprint CTP 964.

Deser, S. & Kay, J.H. (1978). Three-loop counterterms for extended supergravity. Phys. Lett. 76B, 400.

Deser, S., Kay, J.H. & Stelle, K.S. (1977). Renormalizability properties of supergravity. Phys. Rev. Lett. 38, 527.

Deser, S., Tsao, H-S & Van Nieuwenhuizen, P. (1974). One-loop divergences of the Einstein-Yang-Mills system. Phys. Rev. D 10, 3337.

Deser, S. & Van Nieuwenhuizen, P. (1974 a). Non-renormalizability of the quantized Einstein-Maxwell system. Phys. Rev. Lett. 32, 245.

Deser, S. & van Nieuwenhuizen, P. (1974 b). One-loop divergences of quantized Einstein-Maxwell fields. Phys. Rev. D 10, 401, 411.

Deser, S. & Zumino, B. (1976). Consistent supergravity. Phys. Lett. 62B, 335.

Deser, S. & Zumino, B. (1977). Broken supersymmetry and supergravity. Phys. Rev. Lett. 38, 1433.

De Wit, B. (1981). Conformal invariance in extended supergravity. This volume.

De Wit, B. & Grisaru, M.T. (1979). On-shell counterterms and non-linear invariances. Phys. Rev. D 8, 2082.

De Wit, B. & Nicolai, H. (1981 a). Extended supergravity with local SO(5) invariance. Nucl. Phys. B188, 98.

De Wit, B. & Nicolai, H. (1981 b). N = 8 supergravity with local SO(8) × SU(8) invariance. CERN preprint TH-3183.

De Witt, B.S. (1965). Dynamical theory of groups and fields. New York: Gordon and Breach.

Dowker, J. (1978). Another discussion of the axial vector anomaly and the index theorem. J. Phys. A11, 347.

Duff, M.J. (1975). Covariant quantization of gravity. In Quantum Gravity: An Oxford Symposium, eds. C.J. Isham, R. Penrose and D.W. Sciama. Oxford: Oxford University Press.

Duff, M.J. (1977). Observations on conformal anomalies. Nucl. Phys. B125, 334.

Duff, M.J. (1979). Self-duality, helicity, and supergravity. In Supergravity, eds. P. van Nieuwenhuizen and D.Z. Freedman. Amsterdam: North Holland.

Duff, M.J. (1980 a). Quantization of supergravity with a cosmological constant. In Unification of the Fundamental Interactions, eds. J. Ellis, S. Ferrara and P. van Nieuwenhuizen. New York and London: Plenum.

Duff, M.J. (1981 a). Inconsistency of quantum field theory in curved spacetime. In Quantum Gravity II: A Second Oxford Symposium, eds. C.J. Isham, R. Penrose, and D.W. Sciama. Oxford: Oxford University Press.

Duff, M.J. (1981 b). The cosmological constant in quantum gravity and supergravity, in Quantum Gravity II: A Second Oxford Symposium, eds. C.J. Isham, R. Penrose, and D.W. Sciama. Oxford: Oxford University Press.

Duff, M.J. (1981 c). Antisymmetric tensors and supergravity. In Superspace and Supergravity, eds. S.W. Hawking and M. Roček. Cambridge: Cambridge University Press.

Duff, M.J. & Goldthorpe, A. (1981). Inconsistency of semiclassical gravity: the 1/N expansion. To be published.

Duff, M.J., Grisaru, M.T. & Siegel, W. (1981). Superanomalies. To be published.

Duff, M.J. & Isham, C.J. (1979). Self-duality, helicity and supersymmetry: the scattering of light by light. Phys. Lett. 86B, 157.

Duff, M.J. & Isham, C.J. (1980). Self-duality, helicity and coherent states in non-Abelian gauge theories. Nucl. Phys. B162, 271.

Duff, M.J. & Pope, C.N. (1982). Counting massless states in Kaluza-Klein supergravity. To be published.

Duff, M.J. & Toms, D. (1981). Ultraviolet divergences in Kaluza-Klein theories. To be published.

Duff, M.J. & Townsend, P.K. (1981). A new deteriorated energy momentum tensor. In Quantum Structure of Space & Time, eds. M.J. Duff and C.J. Isham. Cambridge: Cambridge University Press.

Duff, M.J. & Van Nieuwenhuizen, P. (1980). Quantum inequivalence of different field representations. Phys. Lett. 94B, 179.

Ellis, J., Gaillard, M.K., Maiani, L. & Zumino, B. (1980). Attempts at superunification. In Unification of the Fundamental Interactions, eds. J. Ellis, S. Ferrara and P. van Nieuwenhuizen. New York and London: Plenum.

Ferrara, S. (1981). Ultraviolet properties of supersymmetric gauge theories. In Quantum Structure of Space and Time, eds. M.J. Duff and C.J. Isham. Cambridge: Cambridge University Press.

Ferrara, S. & Piguet, O. (1975). Perturbation theory and renormalization of supersymmetric Yang-Mills theories. Nucl. Phys. B93, 261.

Ferrara, S., Savoy, C.A. & Girardello, L. (1981). Spin sum rules in extended supergravity. Phys. Lett. 105B, 363.

Ferrara, S. & Van Nieuwenhuizen, P. (1978 a). The auxiliary fields for supergravity. Phys. Lett. 74B, 333.

Ferrara, S. & Van Nieuwenhuizen, P. (1978 b). Structure of supergravity. Phys. Lett. 78B, 573.

Ferrara, S. & Zumino, B. (1975). Transformation properties of the supercurrent. Nucl. Phys. B87, 207.

Ferrara, S. & Zumino, B. (1978 a). Structure of linearized supergravity and conformal supergravity. Nucl. Phys. B134, 301.

Ferrara, S. & Zumino, B. (1978 b). Unpublished.

Ferrara, S. & Zumino, B. (1979). The mass matrix of N = 8 supergravities. Phys. Lett. 86B, 279.

Fischler, M. (1979). Finiteness calculations for O(4) through O(8) extended supergravity and O(4) supergravity coupled to self-dual O(4) matter. Phys. Rev. D 20, 396.

Fradkin, E.S. & Tseytlin, A.A. (1981 a). Renormalizable asymptotically free quantum theory of gravity. Phys. Lett. 104B, 377.

Fradkin, E.S. & Tseytlin, A.A. (1981 b). Higher derivative quantum gravity: One-loop counterterms and asymptotic freedom. Lebedev preprint N70.

Fradkin, E.S. & Tseytlin, A.A. (1981 c). Asymptotic freedom in extended conformal supergravities. Lebedev preprint.

Fradkin, E.S. & Tseytlin, A.A. (1981 d). Quantization of two-dimensional supergravity and critical dimensions for string models. Phys. Lett. 106B, 63.

Fradkin, E.S. & Vasiliev, M.A. (1976). SO(2) supergravity with minimal electromagnetic interaction. Lebedev preprint N197.

Freedman, D.Z. & Das, A. (1977). Gauge internal symmetry in extended supergravity. Phys. Lett. 74B, 333.

Freedman, D.Z. & Schwarz, J. (1978). N = 4 supergravity with local SU(2) × SU(2) invariance. Nucl. Phys. B137, 333.

Freedman, D.Z., Van Nieuwenhuizen, P. & Ferrara, S. (1976). Progress toward a theory of supergravity. Phys. Rev. D 13, 3214.

Freund, P.G.O. & Rubin, M.A. (1980). Dynamics of dimensional reduction. Phys. Lett. 97B, 233.

Gates, G.J, Jr., Grisaru, M.T. & Siegel, W. (1981). Auxiliary field anomalies. CalTech preprint CALT-68-79.

Gibbons, G.W. (1981). The multiplet structure of solitons in the O(2) supergravity theory. In Quantum Structure of Space & Time, eds. M.J. Duff and C.J. Isham. Cambridge: Cambridge University Press.

Green, M.B., Schwarz, J.H. & Brink, L. (1981). N = 4 Yang-Mills and N = 8 supergravity as limits of string theories. CalTech preprint CALT-68-880.

Grisaru, M. (1977). Two-loop renormalizability of supergravity. Phys. Lett. 66B, 75.

Grisaru, M.T. (1978). Anomalies in supersymmetric theories. In Recent Developments in Gravitation, eds. S. Deser and M. Levy. New York and London: Plenum.

Grisaru, M. & Pendleton, H. (1977). Some properties of scattering amplitudes in supersymmetric theories. Nucl. Phys. B124, 81.

Grisaru, M.T., Roček, M. & Siegel, W. (1979). Improved methods for supergraphs. Nucl. Phys. B159, 429.

Grisaru, M.T., Roček, M. & Siegel, W. (1980). Zero 3-loop β-function in N = 4 super Yang-Mills theory. Phys. Rev. Lett. 45, 1063.

Grisaru, M.T. & Siegel, W. (1981 a). Supergraphity. Nucl. Phys. B187, 149.

Grisaru, M.T. & Siegel, W. (1981 b). The one-loop, four particle S-matrix in extended supergravity. CalTech preprint CALT-68-882.

Grisaru, M.T. & Siegel, W. (1982). Supergraphity (II). Manifestly covariant rules and higher-loop finiteness. CalTech preprint.

Grisaru, M.T., Van Nieuwenhuizen, P. & Vermaseren, J.A.M. (1976). One-loop renormalizability of pure supergravity and of Maxwell-Einstein theory in extended supergravity. Phys. Rev. Lett. 37, 1662.

Hawking, S.W. (1978). Spacetime foam. Nucl. Phys. B144, 349.

Hawking, S.W. & Pope, C.N. (1978 a). Symmetry breaking by instantons in supergravity. Nucl. Phys. B146, 381.

Hawking, S.W. & Pope, C.N. (1978 b). Generalized spin structures in quantum gravity. Phys. Lett. 73B, 42.

Howe, P. & Lindström, U. (1981). Counterterms for extended supergravity. In Superspace and Supergravity, eds. S.W. Hawking and M. Roček. Cambridge: Cambridge University Press.

Howe, P. & Nicolai, H. (1981). Gauging N = 8 supergravity in superspace. CERN preprint TH-3211.

Howe, P., Townsend, P. & Stelle, K.S. (1981). Superactions. Nucl. Phys. B191, 445.

Hughes, R.J. (1980). Some comments on asymptotic freedom. Phys. Lett. 97B, 246.

Illiopoulos, J. & Zumino, B. (1974). Broken supergauge symmetry and renormalization. Nucl. Phys. B76, 310.

Isham, C.J., Salam, A. & Strathdee, J. (1971). Infinity suppression in gravity-modified electrodynamics. Phys. Rev. D 3, 1805.

Jones, D.R.T. (1981). Private communication.

Julve, J. & Tonin, M. (1978). Quantum gravity with higher derivative terms. Nuovo Cimento 46B, 137.

Kallosh, R. (1979 a). Super self-duality. JETP Lett. 29, 172.

Kallosh, R. (1979 b). The structure of divergences in supergravitation. JETP Lett. 29, 449.

Kallosh, R. (1980). Self duality in superspace. Nucl. Phys. B165, 119.

Kallosh, R.E. (1981). Counterterms in extended supergravities. Phys. Lett. 99B, 122.

Kemmer, N. (1938). Quantum theory of Einstein-Bose particles and nuclear interaction. Proc. Roy. Soc. CLXVI, No. A924, 127.

Kibble, T.W.B. (1981). Is semi-classical gravity theory viable? In Quantum Gravity II: A Second Oxford Symposium, eds. C.J. Isham, R. Penrose, and D.W. Sciama. Oxford: Oxford University Press.

MacDowell, S. & Mansouri, F. (1977). Unified geometric theory of gravity and supergravity. Phys. Rev. Lett. 38, 739, 1376 (E).

Nicolai, H. & Townsend, P.K. (1981). N = 3 supersymmetry multiplets with vanishing trace anomaly: building blocks of the N > 3 supergravities. Phys. Lett. 98B, 257.

Nielsen, N.K., Romer, H. & Schroer, B. (1978). Anomalous currents in curved space. Nucl. Phys. B136, 475.

Nouri-Moghadam & Taylor, J.G. (1975). One loop divergences for the Einstein-charged meson system. Proc. Roy. Soc. A344, 87.

Ogievetsky, V. & Polubarinov, I. (1967). The notoph and its possible interactions. Sov. J. Nucl. Phys. 4, 156.

Ogievetsky, V. & Sokatchev, E. (1980). Equation of motion for the axial gravitational superfield. Dubna preprint E2-80-139.

Polyakov, A.M. (1981 a). Quantum geometry of bosonic strings. Phys. Lett. 103B, 207.

Polyakov, A.M. (1981 b). Quantum geometry of fermionic strings. Phys. Lett. 103B, 211.

Pope, C.N. (1980). Eigenfunctions and spin structures in CP^2. Phys. Lett. 97B, 417.

Römer, H. (1979). Axial anomaly and boundary terms for general spinor fields. Phys. Lett. B83, 172.

Römer, H. (1981). A universality property of axial anomalies. Phys. Lett. 101B, 55.

Sakharov, A. (1968). Vacuum quantum fluctuations in curved space and the theory of gravitation. Sov. Phys. Doklady 12, 1040.

Salam, Abdus (1975). The impact of quantum gravity theory on particle physics. In Quantum Gravity: An Oxford Symposium, eds. C.J. Isham, R. Penrose, and D.W. Sciama. Oxford: Oxford University Press.

Salam, A. & Strathdee, J. (1974). Supergauge transformations. Nucl. Phys. B79, 477.

Salam, A. & Strathdee, J. (1975). Transition electromagnetic fields in particle physics. Nucl. Phys. B90, 203.

Salam, A. & Strathdee, J. (1978 a). Remarks on high-energy stability and renormalizability of gravity theory. Phys. Rev. D 18, 4480.

Salam, A. & Strathdee, J. (1978 b). Supersymmetry and superfields. Fortschr. Phys. 26, 57.

Salam, A. & Strathdee, J. (1981). On Kaluza-Klein theories. Trieste preprint.

Sezgin, E. & Van Nieuwenhuizen, P. (1980 a). New ghost-free gravity Lagrangians with propagating torsion. Phys. Rev. D 21, 3269.

Sezgin, E. & Van Nieuwenhuizen, P. (1980 b). Renormalizability properties of antisymmetric tensor fields coupled to gravity. Phys. Rev. D 22, 301.

Siegel, W. (1979). Gauge spinor superfield as scalar multiplet. Phys. Lett. 85B, 333.

Siegel, W. (1980). Quantum equivalence of different field representations. Phys. Lett. B138, 107.

Siegel, W., Townsend, P.K. & Van Nieuwenhuizen, P. (1981). Supersymmetric dimensional regularization. In Superspace and Supergravity, eds. S.W. Hawking and M. Roček. Cambridge: Cambridge University Press.

Slavnov, A.A. (1981). Symmetry preserving regularization for gauge and supergauge theories. In Superspace and Supergravity, eds. S.W. Hawking and M. Roček. Cambridge: Cambridge University Press.

Sohnius, M.F. & West, P.C. (1981 a). Conformal invariance in N = 4 supersymmetric Yang-Mills theory. Phys. Lett. 100B, 245.

Sohnius, M.F. & West, P.C. (1981 b). An alternative minimal off-shell version of N = 1 supergravity. Phys. Lett. 105B, 353.

Stelle, K.S. (1977). Renormalization of higher derivative quantum gravity. Phys. Rev. D 16, 953.

Stelle, K.S. (1981 a). Manifest internal symmetry and supersymmetry. CERN preprint TH-3127 and in Proc. 18th Winter School of Theoretical Physics, Karpacz, Poland.

Stelle, K.S. (1981 b). Extended supercurrents and the ultraviolet finiteness of N = 4 supersymmetric Yang-Mills theory. In Quantum Structure of Space and Time, eds. M.J. Duff and C.J. Isham. Cambridge: Cambridge University Press.

Stelle, K.S. & Townsend, P.K. (1981). To be published.

Stelle, K.S. & West, P.C. (1978). Minimal set of auxiliary fields for supergravity. Phys. Lett. 74B, 330.

Taylor, J.G. (1979). The ultraviolet divergences of superfield supergravity. In Proc. EPS Int. Conf. on High-Energy Physics. Geneva, Vol. 2, p. 969. Geneva: CERN.

Taylor, J.G. (1981 a). A no-go theorem for off-shell extended supergravities. King's College preprint.

Taylor, J.G. (1981 b). Auxiliary field candidates for N = 3, 4, 5, and 6 supergravities. King's College preprint.

Taylor, J.G. (1981 c). Building linearised extended supergravities, to appear in Quantum Structure of Space & Time, eds. M.J. Duff and C.J. Isham. Cambridge: Cambridge University Press.

't Hooft, G. & Veltman, M. (1974). One loop divergences in the theory of gravitation. Ann. Inst. Henri Poincaré 20, 69.

Tomboulis, E. (1980). Renormalizability and asymptotic freedom in quantum gravity. Princeton preprint.

Townsend, P.K. (1977). Cosmological constant in supergravity. Phys. Rev. D 15, 2802.

Townsend, P.K. (1981 a). Finite field theory? CERN preprint 3066 and in Proc. 18th Winter School of Theoretical Physics, Karpacz, Poland.

Townsend, P.K. (1981 b). Classical properties of antisymmetric tensor gauge fields. CERN preprint TH-3067 and in Proc. 18th Winter School of Theoretical Physics, Karpacz, Poland.

Van Nieuwenhuizen, P. (1979). Proceedings of the Summer Institute on Recent Developments in Gravitation, Cargèse France, 1978, eds. M. Levy & S. Deser. New York and London: Plenum.

Van Nieuwenhuizen, P. (1981 a). Supergravity. Phys. Rep. 68, 189.

Van Nieuwenhuizen, P. (1981 b). Renormalizability properties of broken N = 8 supergravity. Lectures at this school.

Van Nieuwenhuizen, P. & Vermaseren, J.A.M. (1976). One-loop divergences in quantum theory of supergravity. Phys. Lett. 65B, 263.

Van Proeyen, A. (1980). Gravitational divergences of the electromagnetic interactions of massive vector particles. Nucl. Phys. B174, 189.

Weinberg, S. (1979). Ultraviolet divergences in quantum gravity. In General Relativity: an Einstein Centenary Survey, eds. S.W. Hawking and W. Israel. Cambridge: Cambridge University Press.

Wess, J. & Zumino, B. (1974 a). A Lagrangian model invariant under supergauge transformations. Phys. Lett. 49B, 52.

Wess, J. & Zumino, B. (1974 b). Supergauge invariant extension of quantum electrodynamics. Nucl. Phys. B78, 1.

Wess, J. & Zumino, B. (1977). Superspace formalism of supergravity. Phys. Lett. 66B, 361.

Witten, E. (1981). Search for a realistic Kaluza-Klein theory. Nucl. Phys. B. 186, 412.

Zumino, B. (1975). Supersymmetry and the vacuum. Nucl. Phys. B89, 535.

CONFORMAL INVARIANCE IN EXTENDED SUPERGRAVITY

B. de Wit,
NIKHEF-H, Amsterdam

1 INTRODUCTION

The group of conformal transformations leaves the lightcone invariant. The presence of masslike parameters spoils the invariance under these transformations, which seems to limit their possible physical significance. In fact, the same problem arises for theories that are conformally invariant at the classical level, since quantum effects introduce a mass scale which invalidates results based on conformal invariance. Although conformal transformations thus do not seem to have any direct physical consequences, they are a useful tool to analyse the structure of gravity and its coupling to matter. By the same token this approach can be extended to supergravity, where it can be used to systematically study the off-shell structure of these theories. In these lectures we will introduce these techniques and discuss their application to extended supergravity.

Notwithstanding their intrinsic beauty, supergravity theories are rather complex. It is therefore important to find techniques that can clarify their structure as a classical field theory. One outstanding problem is that the off-shell structure is not known for the extended theories with N>2. Beyond N=2, supergravity is formulated mainly in terms of the fields that correspond directly to physical degrees of freedom. However, these fields do not yet define a representation of the supersymmetry gauge algebra; the (anti)commutators close only modulo the equation of motions. Thus such a field representation is only relevant within the context of a given action. These formulations are therefore called "on-shell", since they define a representation of the algebra for the on-shell degrees of freedom. This is in contrast to off-shell representations which make no reference to any action, and which realize the gauge algebra without the need for imposing generalized wave equations on the fields. For such representations the group-theoretic structure is obviously much more transparent, which greatly facilitates the quantization of the theory or the discussion of invariant counterterms.

The fields in an off-shell formulation that can be eliminated by imposing the field equations are called auxiliary fields. These fields are of limited significance for the dynamics of the theory in question. Nevertheless, it is possible to have lagrangians based on different off-shell representations which lead to the same physics at the classical level, but differ in their quantum-mechanical aspects.

However, even when an off-shell version of the theory is available, there are still substantial complications caused by the nonlinearities in the transformation rules and the large number of fields that is involved. In particular for the higher-extended theories these difficulties are extremely cumbersome. In addition one has to cope with the fact that different and inequivalent off-shell formulations of these theories may exist, which would require separate studies.

All these considerations have motivated the study of conformally invariant formulations. The standard conformal symmetries are fused with supersymmetry transformations, and one considers off-shell representations of the resulting superconformal algebra. The higher degree of symmetry restricts the complexity of the transformation rules, and allows the decomposition of the supersymmetry gauge fields into smaller irreducible subsets. Therefore the superconformal transformations and the invariants have a simpler form. The multiplet that underlies conformal supergravity is the backbone of any supergravity theory, and by systematically introducing other multiplets one can describe any possible off-shell formulation of Poincaré or de Sitter supergravity. In this way one framework encompasses all supergravity theories, so that there is no need for separate studies of all possible formulations. Hence, the superconformal framework still leaves the option of discussing Poincaré or de Sitter theories; their symmetries are a subset of the superconformal ones, and their underlying multiplets can be found by combining superconformal multiplets. The resulting multiplet is then simply gauge equivalent to Poincaré or de Sitter supergravity.

In these lectures we will explain these ideas using the "component" language. This is only a matter of strategy; in fact parallel techniques are being used in the superspace framework. It may be that these concepts are more easily appreciated in terms of ordinary fields, but the interested reader is encouraged to consult the literature. The lectures are organized as follows. Section 2 discusses a formulation of simple supergravity within the context of the super-Poincaré algebra in the presence of conventional constraints. The idea of using gauge-equivalent formulations is presented in a simple example in section 3. The gauge theory of the superconformal algebra is the subject of section 4; it leads to an incomplete field representation of conformal supergravity. Conformally invariant field theories in d dimensions are derived in section 5; furthermore the use of compensating fields in gravity is explained. A more general approach is described in section 6, where the matter currents are discussed that may couple to the superconformal gauge fields. As an example the supermultiplet of currents for the N=4 supersymmetric Maxwell theory is constructed. Section 7 contains a discussion of N=4 conformal supergravity, and section 8 describes various applications of the conformal methods to N=2 Poincaré supergravity. Some useful references are collected in section 9.

2 GAUGE THEORY OF THE SUPER-POINCARÉ ALGEBRA

It is straightforward to construct a gauge theory for any finite-dimensional Lie algebra. One starts by assigning a gauge field (connection field) to each of the generators of the algebra; for a superalgebra we make the distinction between bosonic and fermionic generators, which have corresponding commuting and anticommuting gauge fields. These generators close under (anti)commutation: the (anti)commutator of two generators should be proportional to a linear combination of generators, whose proportionality is determined by the structure constants $f_{AB}{}^{C}$

$$[t_A, t_B\} = f_{AB}{}^{C}\, t_C \,, \tag{2.1}$$

where the t_A denote the group generators. The notation $[\ ,\ \}$ indicates that one has an anticommutator when both t_A and t_B are fermionic; otherwise one has a commutator. The gauge fields $W_\mu{}^A$ assigned to t_A transform then according to

$$\delta W_\mu{}^A = \partial_\mu \xi^A - f_{BC}{}^{A}\, W_\mu{}^B\, \xi^C \,, \tag{2.2}$$

with ξ^A the space-time dependent parameters of the infinitesimal gauge transformations. The structure constants also enter in the definition of the curvature tensors (field strengths)

$$R_{\mu\nu}{}^A = \partial_\mu W_\nu{}^A - \partial_\nu W_\mu{}^A - f_{BC}{}^{A}\, W_\mu{}^B\, W_\nu{}^C \,, \tag{2.3}$$

which transform covariantly under the group:

$$\delta R_{\mu\nu}{}^A = -\, f_{BC}{}^{A}\, R_{\mu\nu}{}^B\, \xi^C \,. \tag{2.4}$$

The gauge symmetries of supergravity are general coordinate transformations (local translations), Lorentz, and supersymmetry transformations. Corresponding transformations are contained in the super-Poincaré algebra, on the basis of which one may thus attempt to formulate supergravity by following to the procedure outlined above. We denote the generators of the super-Poincaré algebra by M_{ab}, P_a, and Q_α, and give their nonvanishing (anti)commutators schematically:

$$\begin{aligned} [M_{ab},M_{cd}] &\propto \eta_{ac} M_{bd} + \eta_{bd} M_{ac} - \eta_{bc} M_{ad} - \eta_{ad} M_{bc} , \\ [M_{ab},P_c] &\propto \eta_{ac} P_b - \eta_{bc} P_a , \\ [M_{ab},Q_\alpha] &\propto (\sigma_{ab} Q)_\alpha , \\ \{Q_\alpha,\bar{Q}_\beta\} &\propto \gamma^a{}_{\alpha\beta} P_a . \end{aligned} \tag{2.5}$$

We now assign gauge fields $\omega_\mu{}^{ab}$, $e_\mu{}^a$ and ψ_μ to the generators M_{ab}, P_a, and Q_α, respectively. Their transformation rules are in correspondence to (2.5)

$$\begin{aligned} \delta\omega_\mu{}^{ab} &= \mathcal{D}_\mu \varepsilon^{ab} , \\ \delta e_\mu{}^a &= \varepsilon^{ab} e_\mu{}^b + \mathcal{D}_\mu \xi_P{}^a + \bar{\varepsilon}\gamma^a \psi_\mu , \\ \delta\psi_\mu &= \tfrac{1}{2} \varepsilon^{ab} \sigma_{ab} \psi_\mu + 2\mathcal{D}_\mu \varepsilon , \end{aligned} \tag{2.6}$$

with ε^{ab}, $\xi_P{}^a$ and ε_α the infinitesimal parameters of the M, P, and Q transformations, and $\mathcal{D}_\mu$ the Lorentz-covariant derivative. The curvature tensors take the form

$$\begin{aligned} R_{\mu\nu}{}^{ab}(M) &= \partial_\mu \omega_\nu{}^{ab} - \partial_\nu \omega_\mu{}^{ab} - \omega_\mu{}^a{}_c \omega_\nu{}^{cb} - \omega_\mu{}^b{}_c \omega_\nu{}^{ac} , \\ R_{\mu\nu}{}^a(P) &= \mathcal{D}_\mu e_\nu{}^a - \mathcal{D}_\nu e_\mu{}^a - \tfrac{1}{2} \bar{\psi}_\mu \gamma^a \psi_\nu , \\ R_{\mu\nu}(Q) &= \mathcal{D}_\mu \psi_\nu - \mathcal{D}_\nu \psi_\mu . \end{aligned} \tag{2.7}$$

As usual these tensors can be found by considering the (anti)commutator of two fully covariant derivatives (Ricci identity). Likewise one may consider the cyclic sum of the triple commutator of covariant derivatives. This sum should vanish according to the Jacobi identity, which leads to the Bianchi identities

$$\sum_{(abc)} \left(D_a R_{bc}{}^{de}(M) - R_{ab}{}^f(P) R_{cf}{}^{de}(M)\right) = 0 , \tag{2.8a}$$

$$\sum_{(abc)} \left(D_a R_{bc}{}^d(P) + R_{ab}{}^{dc}(M) - R_{ab}{}^f(P) R_{cf}{}^d(P)\right) = 0 , \tag{2.8b}$$

where $\sum_{(abc)}$ denotes the cyclic sum over Lorentz indices a,b, and c. The derivatives D_μ are now covariant with respect to Lorentz and supersymmetry transformations. Note that we have converted world indices into Lorentz indices by contraction with the inverse of $e_\mu{}^a$. Hence we use $e_\mu{}^a$ as a vierbein field; the four covariant vectors $e_\mu{}^a$ specify the basis vectors of the linear tangent space at each point of a curved space-time manifold. This requires that $e_\mu{}^a$ is non-singular and has an inverse vierbein field $e_a{}^\mu$, defined by

$$e_a{}^\mu e_\mu{}^b = \delta_a{}^b \ , \quad e_\mu{}^a e_a{}^\nu = \delta_\mu{}^\nu \ . \tag{2.9}$$

In this context the group of spin rotations M is called the "structure group" or "tangent-space group", which rotates the local tangent space frame.

At this point we should emphasize that the Poincaré group has been treated entirely as an internal symmetry group with no relation to space-time. Indeed we distinguish independent general coordinate transformations, which describe reparametrizations of the base manifold spanned by the space-time coordinates

$$x^\mu \to x^\mu + \xi^\mu(x) \ . \tag{2.10}$$

Under general coordinate transformations gauge fields and curvatures transform as covariant tensors.

To see how one can possibly convert the Poincaré gauge transformations into a symmetry of space-time, we rewrite a general coordinate transformation on $e_\mu{}^a$ as follows:

$$\begin{aligned} \delta_{gct}(\xi)e_\mu{}^a &= -\partial_\mu \xi^\nu e_\nu{}^a - \xi^\nu \partial_\nu e_\mu{}^a \\ &= -\mathcal{D}_\mu(\xi^\nu e_\nu{}^a) - (\xi^\nu \omega_\nu{}^{ab})e_\mu{}^b - (\tfrac{1}{2}\xi^\nu \bar{\psi}_\nu)\gamma^a \psi_\mu + \xi^\nu R_{\mu\nu}{}^a(P) \ . \end{aligned} \tag{2.11}$$

The first two terms in (2.11) can be identified as special P, M and Q transformations with field-dependent parameters. Hence this result can be expressed as follows (cf. (2.6)):

$$\{\delta_{gct}(\xi) + \delta_P(\xi^\nu e_\nu{}^b) + \delta_M(\xi^\nu \omega_\nu{}^{bc}) + \delta_Q(\tfrac{1}{2}\xi^\nu \psi_\nu)\} \ e_\mu{}^a = \xi^\nu R_{\mu\nu}{}^a(P). \tag{2.12}$$

Therefore, modulo the term proportional to R(P), a general coordinate transformation simply coincides with the sum of a special P, M and Q transformation. This motivates us to introduce a constraint

$$R_{\mu\nu}{}^a(P) = 0 \ , \tag{2.13}$$

so that Poincaré and general coordinate transformations will no longer be independent. A constraint such as (2.13) is called a <u>conventional</u> constraint, because it expresses the spin connection $\omega_\mu{}^{ab}$ in terms of other fields. The solution of (2.13) can be expressed in the following form

$$\omega_\mu{}^{ab}(e,\psi) = \tfrac{1}{2} \, e_\mu{}^c \ (\hat{\Omega}_{ab}{}^c - \hat{\Omega}_{bc}{}^a - \hat{\Omega}_{ca}{}^b) \ , \tag{2.14}$$

The object of anholonomity $\Omega_{ab}{}^{c}$ measures the noncommutativity of the vierbein basis

$$\Omega_{ab}{}^{c} = e_a{}^{\mu} e_b{}^{\nu} (\partial_\mu e_\nu{}^{c} - \partial_\nu e_\mu{}^{c}) , \tag{2.15}$$

and we have used the notation $\hat{\Omega}$ to indicate that Ω has been made covariant with respect to supersymmetry

$$\hat{\Omega}_{ab}{}^{c} = \Omega_{ab}{}^{c} - \tfrac{1}{2} e_a{}^{\mu} e_b{}^{\nu} \bar{\psi}_\mu \gamma^c \psi_\nu . \tag{2.16}$$

Substitution of (2.14) into R(M) leads to the Riemann tensor with a minor modification to make it covariant with respect to supersymmetry. Its contraction leads to the corresponding generalization of the Ricci tensor

$$R_{\mu\nu}(e,\psi) = R_{\mu\rho}{}^{ab}(M)\, e_b{}^{\rho} e_{a\nu} , \tag{2.17}$$

which is symmetric by virtue of (2.13) and the Bianchi identity (2.8b).

A direct consequence of the conventional constraint (2.13) is that the number of gravitational degrees of freedom is reduced from 30 to only 6. Furthermore, since $\omega_\mu{}^{ab}$ is no longer an independent field, its transformations will not necessarily coincide with those of the super-Poincaré algebra. Indeed the constraint (2.13) is not invariant under P and Q transformations

$$\begin{aligned} \delta_P R_{\mu\nu}{}^{a}(P) &= - \xi_P{}^{b} R_{\mu\nu}{}^{ab}(M) , \\ \delta_Q R_{\mu\nu}{}^{a}(P) &= \bar{\epsilon}\gamma^a R_{\mu\nu}(Q) . \end{aligned} \tag{2.18}$$

Therefore the transformations of $\omega_\mu{}^{ab}$ deviate from their original form by terms proportional to curvature tensors. Consequently the transformations will no longer generate the super-Poincaré algebra in the presence of the constraint (2.13). It is important to realize that (2.12) is not uniformly valid, but only when acting on the vierbein field. Hence, by imposing (2.13) we have not achieved a uniform conversion of P transformations into general coordinate and other gauge transformations. Nevertheless, the P transformations are discarded after imposing the constraint, and in the algebra their role is effectively taken over by the general coordinate transformations. Strictly speaking, we no longer have a gauge theory of the super-Poincaré algebra; instead we have a mutilated version of this algebra which contains the general coordinate transformations in a nontrivial way.

3 GAUGE-EQUIVALENT FORMULATIONS; A SIMPLE EXAMPLE

We have already argued that the complexity of supergravity makes it advantageous to use formulations with the highest possible degree of gauge invariance. For that reason we prefer to work within the superconformal framework. As we shall see, the presence of extra invariances does not necessarily limit the possible applications of this formalism. At the end one can often remove the unwanted invariances by imposing a number of gauge conditions. In this way one constructs theories within the context of a higher symmetry, which are gauge equivalent to the theories with less symmetry whose construction one was originally aiming for. We stress that the gauge equivalence will only be used within the classical context. For the full quantum theory there may be subtleties at this point, in particular for conformally invariant theories.

In this section we will elucidate the use of gauge equivalent formulations by means of a simple example. We consider a theory of massive vector fields in the adjoint representation of SU(N)

$$L = \mathrm{Tr}\left(\tfrac{1}{4}G_{\mu\nu}{}^2(W) + \tfrac{1}{2}m^2W_\mu{}^2\right) . \tag{3.1}$$

We have used a notation where W_μ and $G_{\mu\nu}$ are N×N antihermitean traceless matrices. The tensor $G_{\mu\nu}$ has the familiar form

$$G_{\mu\nu} = \partial_\mu W_\nu - \partial_\nu W_\mu - [W_\mu, W_\nu] . \tag{3.2}$$

Because of the presence of the mass term the Lagrangian is not invariant under local SU(N) transformations, but it is still invariant under rigid transformations

$$W_\mu \to W_\mu{}' = VW_\mu V^{-1} , \tag{3.3}$$

with V an element of SU(N).

We now construct a gauge-equivalent version of this theory which has a local SU(N) invariance. This is done by introducing scalar fields $\Phi(x)$, which form space-time dependent elements of SU(N). Hence $\Phi(x)$ is a unitary matrix with unit determinant. To avoid new degrees of freedom, we require $\Phi(x)$ to transform under local SU(N) transformations. At the same time $\Phi(x)$ also transforms under the previously introduced rigid SU(N) transformations. The transformation rules are

$$\begin{aligned} &\Phi(x) \to \Phi'(x) = U(x)\Phi(x) , && \text{(local SU(N))} \\ &\Phi(x) \to \Phi'(x) = \Phi(x)V^{-1} , && \text{(rigid SU(N))} \end{aligned} \tag{3.4}$$

where U(x) and V characterize the local and rigid SU(N) transformations, respectively.

Covariant derivatives of Φ are defined in the standard way

$$D_\mu\Phi = (\partial_\mu - A_\mu)\Phi , \tag{3.5}$$

with A_μ the SU(N) gauge field written as an N×N antihermitean traceless matrix. These gauge fields are inert under the rigid transformations. Under local SU(N) they transform according to

$$A_\mu(x) \to A_\mu'(x) = U(x)A_\mu(x)U^{-1}(x) + \partial_\mu U(x)\, U^{-1}(x) \,. \qquad (3.6)$$

Again we wish to avoid new degrees of freedom. Therefore we impose the following gauge invariant condition on A_μ:

$$W_\mu = -\Phi^{-1}D_\mu\Phi = \Phi^{-1}A_\mu\Phi + \partial_\mu\Phi^{-1}\Phi \,. \qquad (3.7)$$

At this point we note that A_μ and W_μ are precisely related by a field-dependent gauge transformation characterized by Φ. This observation elucidates the underlying strategy: one uniformly expresses all gauge invariant fields transforming under rigid SU(N) as field-dependent gauge tranformations acting on similar fields but now transforming under local SU(N). In other words, the fact that the original fields are gauge invariant is recognized as the result of the field-dependent gauge transformations whose variation is such as to compensate for the gauge transformation of the new noninvariant fields. For this reason the fields Φ are called <u>compensating</u> fields.

Using the previous definitions we obtain an alternative form of the same theory

$$\begin{aligned} L &= \mathrm{Tr}\left(\tfrac{1}{4}G_{\mu\nu}{}^2(A) + \tfrac{1}{2}m^2\Phi^{-1}D_\mu\Phi\,\Phi^{-1}D_\mu\Phi\right) \\ &= \mathrm{Tr}\left(\tfrac{1}{4}G_{\mu\nu}{}^2(A) - \tfrac{1}{2}m^2D_\mu\Phi\,D_\mu\Phi^{-1}\right) \,. \end{aligned} \qquad (3.8)$$

This lagrangian is invariant both under local and rigid SU(N) transformations, specified by (3.4) and (3.6). The equivalence of (3.8) to (3.1) can be seen by absorbing the dependence on Φ into the definition of A_μ through a Φ-dependent gauge transformation. This is simply a reversion of the argument that led to (3.8). Alternatively one could make a choice of gauge. A suitable gauge condition is

$$\Phi(x) = 1 \,. \qquad (3.9)$$

With this choice we find

$$D_\mu\Phi = -A_\mu = -W_\mu \,, \qquad (3.10)$$

and the lagrangian (3.8) directly reduces to the form (3.1). Although the two theories are equivalent, there may be certain reasons to prefer one of these formulations. Clearly the first theory is based on an irreducible multiplet of $4(N^2-1)$ degrees of freedom represented by the fields W_μ. The second formulation has an SU(N) gauge invariance. This allows us to have two irreducible multiplets, represented by gauge fields A_μ and scalars Φ. The gauge fields have only $3(N^2-1)$ degrees of freedom, whereas the remaining N^2-1 are contained in the compensating scalars. Hence we have effectively decomposed the field W_μ into transversal and longitudinal components by introducing the extra gauge invariance. On the basis of the transversal components it is only possible to have a lagrangian of the standard form describing massless spin-1 degrees of freedom. For the massive case one has to introduce additional degrees of freedom in the form of compensating fields.

The reason why we prefer formulations of the second type in supergravity is that the higher degree of gauge invariance restricts the nonlinear complications present in these theories; at the same time we have multiplets with a smaller number of degrees of freedom. To obtain the theories without this high degree of symmetry one adds additional compensating multiplets. Examples of compensating fields in the context of gravity and supergravity will be discussed in sections 5 and 8.

4 GAUGE THEORY OF THE SUPERCONFORMAL ALGEBRA

The highest degree of invariance that one can consider in the context of supergravity is generated by the elements of the superconformal

Table 1: Generators and corresponding gauge fields of the superconformal group. The numbers in the right column denote the degrees of freedom represented by the gauge fields. The U(1) symmetry is absent in the case of N = 4.

superconformal gauge symmetries			gauge fields	
4	translations	P	$e_\mu{}^a$	45
6	Lorentz rotations	M	$\omega_\mu{}^{ab}$	
1	dilatation	D	b_μ	
4	conformal boosts	K	$f_\mu{}^a$	
4N	supersymmetries	Q	$\psi_\mu{}^i$	24N
4N	special supersymm.	S	$\phi_\mu{}^i$	
N^2-1	chiral SU(N)	C	$V_\mu{}^i{}_j$	$3(N^2-1)$
1	chiral U(1)	A	A_μ	3

algebra. The superconformal algebra is a generalization of the conformal algebra SU(2,2), and is denoted by SU(2,2 | N). The conformal algebra contains the generators of Lorentz transformations (M), translations (P), dilatations (D), and special conformal boosts (K). If one introduces supersymmetry (Q), one finds that a second type of supersymmetry is generated by the commutator of the generators of supersymmetry and the conformal boosts. This "special" supersymmetry will be denoted by S. To close the algebra one needs to include U(N) chiral transformations (except for the case of N=4 where SU(4) is sufficient). We have listed the generators of SU(2,2 | N) with their corresponding gauge fields in Table 1.

We now give the nonvanishing (anti)commutators of SU(2,2 | N) in schematic form. The conformal subalgebra is given by

$$\begin{aligned} &[M,M] \propto M \quad , && [M,P] \propto P \quad , \\ &[D,P] \propto P \quad , && [M,K] \propto K \quad , \\ &[D,K] \propto K \quad , && [K,P] \propto D + M , \end{aligned} \tag{4.1}$$

whereas the following (anti)commutators enter in its supersymmetric extension

$$\begin{aligned} &\{Q,Q\} \propto P \quad , && [K,Q] \propto S \quad , \\ &[P,S] \propto Q \quad , && \{S,S\} \propto K \quad , \\ &\{Q,S\} \propto D + M + (S)U(N) , && \\ &[M,Q] \propto Q \quad , && [M,S] \propto S \quad , \\ &[D,Q] \propto Q \quad , && [D,S] \propto S \quad , \\ &[U(N),Q] \propto Q \quad , && [U(N),S] \propto S \quad . \end{aligned} \tag{4.2}$$

As has been explained in section 2, the values of the structure constants are in direct correspondence to the transformation rules of the gauge fields. The transformations under Q and S supersymmetry, dilatations, conformal boosts and chiral U(1) are given below; the assignments with respect to the remaining symmetries follow directly from the index structure of the fields

$$\begin{aligned} \delta e_\mu{}^a &= -\Lambda_D e_\mu{}^a + (\bar\varepsilon^i \gamma^a \psi_{\mu i} + \text{h.c.}) \quad , \\ \delta \omega_\mu{}^{ab} &= \Lambda_K{}^{[a} e_\mu{}^{b]} - (\bar\varepsilon^i \sigma^{ab} \phi_{\mu i} + \text{h.c.}) + (\bar\psi_\mu{}^i \sigma^{ab} \eta_i + \text{h.c.}) \quad , \\ \delta b_\mu &= \partial_\mu \Lambda_D + \Lambda_K{}^a e_\mu{}^a + \tfrac{1}{2}(\bar\varepsilon^i \phi_{\mu i} + \text{h.c.}) - \tfrac{1}{2}(\bar\psi_\mu{}^i \eta_i + \text{h.c.}) , \\ \delta f_\mu{}^a &= \mathcal{D}_\mu \Lambda_K{}^a + \Lambda_D f_\mu{}^a + \tfrac{1}{2}(\bar\eta^i \gamma^a \phi_{\mu i} + \text{h.c.}) \quad , \\ \delta \psi_\mu{}^i &= 2\mathcal{D}_\mu \varepsilon^i - \tfrac{1}{2}\Lambda_D \psi_\mu{}^i - \gamma_\mu \eta^i - i\big((4-N)/(4N)\big)\Lambda_A \psi_\mu{}^i \quad , \end{aligned}$$

$$\delta\phi_\mu{}^i = 2\mathcal{D}_\mu\eta^i + \tfrac{1}{2}\Lambda_D\phi_\mu{}^i - 2f_\mu{}^a\gamma_a\varepsilon^i + \Lambda_K{}^a\gamma_a\psi_\mu{}^i + i\big((4-N)/(4N)\big)\Lambda_A\phi_\mu{}^i ,$$
$$\delta V_\mu{}^i{}_j = \bar\varepsilon^i\phi_{\mu j} - \bar\psi_\mu{}^i\eta_j - 1/N\,\delta^i{}_j(\bar\varepsilon^k\phi_{\mu k} - \bar\psi_\mu{}^k\eta_k) - \text{h.c.} ,$$
$$\delta A_\mu = i(\bar\varepsilon^i\phi_{\mu i} - \text{h.c.}) - i(\bar\psi_\mu{}^i\eta_i - \text{h.c.}) + \partial_\mu\Lambda_A . \qquad (4.3)$$

In (4.3) ε^i and η_i are the spinorial parameters of Q and S supersymmetry, respectively. We have used a chiral SU(N) notation in which ε^i and η_i have positive chirality. The opposite chirality components are denoted by ε_i and η^i. Similar conventions have been used for $\psi_\mu{}^i$ and $\phi_{\mu i}$. The transformation parameters of dilatations, conformal boosts and U(1) have been denoted by Λ_D, Λ_K, and Λ_A, respectively. Notice that the chiral charge of $\psi_\mu{}^i$ and $\phi_\mu{}^i$ vanishes for N = 4. The derivatives $\mathcal{D}_\mu$ are now covariant with respect to M,D and (S)U(N). From the transformations (4.3) it is straightforward to define the SU(2,2 | N) curvature tensors. They have been listed in Table 2.

We now proceed in analogy to the treatment of the gauge theory of the super-Poincaré algebra, which was discussed in section 2. Since the SU(2,2 | N) transformations have no relation to space-time at this point, we impose certain curvature constraints. A priori it is not known how many of such constraints are necessary. Guided by our knowledge of the specific cases of N = 1 and 2, we choose the maximal set of conventional constraints. Inspection of the explicit form of the curvature tensors shows that R(P), R(M), R(D) and R(Q) contain terms proportional to a connection field, multiplied by a vierbein field. Hence these connections, $\omega_\mu{}^{ab}$, $f_\mu{}^a$, and $\phi_\mu{}^i$, can be expressed in terms of other fields by imposing curvature constraints. For this purpose the following set of constraints suffices

$$\hat R_{\mu\nu}{}^a(P) = 0 ,$$
$$e^\nu{}_b\hat R_{\mu\nu}{}^{ab}(M) = 0, \qquad (4.4)$$
$$\gamma^\mu\hat R_{\mu\nu}{}^i(Q) = 0 .$$

At first sight it seems that one can also restrict R(D), but in the presence of the first constraint of (4.4) one can show that R(D) is no longer independent by virtue of an SU(2,2 | N) Bianchi identitity analogous to (2.8b). The constraints (4.4) fully determine $\omega_\mu{}^{ab}$, $f_\mu{}^a$, and $\phi_\mu{}^i$ in terms of the other SU(2,2 | N) gauge fields, so that their transformation rules will no longer coincide with the original rules given in (4.3). Consequently the curvatures will need additional terms to regain covariance. These terms have been implicitly included in (4.4) by replacing the original curvatures R by $\hat{R}$. We should mention that the detailed form of conventional constraints is not crucial, as long as they fully restrict the gauge fields in question.

The most immediate effect of the conventional constraints is that they restrict the number of degrees of freedom. Because the constraints (4.4) are K invariant, the K transformations of the restricted fields will not be affected and coincide with those listed in (4.3). The only unrestricted gauge field that transforms under K is the dilatational field b_μ. Therefore the K transformations of the restricted fields are entirely generated by their dependence on b_μ.

At this stage it is possible to impose a gauge condition for removing the invariance under K transformations. Since b_μ is the only field left which transforms under K, the choice of the gauge condition is obvious (note that b_μ may be viewed as the compensating field for K transformations):

$$b_\mu = 0 \ . \tag{4.5}$$

Apart from violating the K symmetry this condition also affects all other transformations of the (super)conformal algebra. For instance the invariance under dilatations is broken as well. However, we can define modified scale transformations by adding a special field-dependent K

Table 2: Curvatures of the graded SU(2,2 | N) algebra.

$$
\begin{aligned}
R_{\mu\nu}{}^{a}(P) &= \mathcal{D}_{[\mu}e_{\nu]}{}^{a} - \tfrac{1}{2}\bar{\phi}_{[\mu}{}^{i}\gamma^{a}\psi_{\nu]i} \ , \\
R_{\mu\nu}{}^{ab}(M) &= \partial_{[\mu}\omega_{\nu]}{}^{ab} - \omega_{[\mu}{}^{ac}\omega_{\nu]}{}^{cb} - f_{[\mu}{}^{[a}e_{\nu]}{}^{b]} \\
&\quad + \tfrac{1}{2}(\bar{\psi}_{[\mu}{}^{i}\sigma^{ab}\phi_{\nu]i} + \text{h.c.}) \ , \\
R_{\mu\nu}(D) &= \partial_{[\mu}b_{\nu]}{}^{a} - f_{[\mu}{}^{a}e_{\nu]}{}^{a} - \tfrac{1}{4}(\bar{\psi}_{[\mu}{}^{i}\phi_{\nu]i} + \text{h.c.}), \\
R_{\mu\nu}{}^{a}(K) &= \mathcal{D}_{[\mu}f_{\nu]}{}^{a} - \tfrac{1}{4}\bar{\phi}_{[\mu}{}^{i}\gamma^{a}\phi_{\nu]i} \ , \\
R_{\mu\nu}{}^{i}(Q) &= \mathcal{D}_{[\mu}\psi_{\nu]}{}^{i} - \tfrac{1}{2}\gamma_{[\mu}\phi_{\nu]}{}^{i} \ , \\
R_{\mu\nu}{}^{i}(S) &= \mathcal{D}_{[\mu}\phi_{\nu]}{}^{i} + f_{[\mu}{}^{a}\gamma_{a}\psi_{\nu]}{}^{i} \ , \\
R_{\mu\nu}(A) &= \partial_{[\mu}A_{\nu]}{}^{a} - \tfrac{1}{2}i(\bar{\psi}_{[\mu}{}^{i}\phi_{\nu]i} - \text{h.c.}) \ , \\
R_{\mu\nu}{}^{i}{}_{j}(V) &= \partial_{[\mu}V_{\nu]}{}^{i}{}_{j} - V_{[\mu}{}^{i}{}_{k}V_{\nu]}{}^{k}{}_{j} \\
&\quad - \tfrac{1}{2}(\bar{\psi}_{[\mu}{}^{i}\phi_{\nu]j} - \frac{1}{N}\delta^{i}{}_{j}\bar{\psi}_{[\mu}{}^{k}\phi_{\nu]k} - \text{h.c.}).
\end{aligned}
$$

transformation to D in order to compensate for violations induced by the condition (4.5). This is specified by a so-called decomposition rule, which holds uniformly on all fields

$$D'(\Lambda) = D(\Lambda) + K(-\partial_\mu \Lambda e^\mu{}_a) \; . \qquad (4.6)$$

Eq. (4.6) defines the new dilatations, denoted by D', in terms of the original transformations of the (super)conformal algebra. The K component in (4.6) is chosen such that the gauge condition (4.5) remains invariant under D'. Similar decomposition rules can be given for Q and S supersymmetry as well. The presence of these decomposition rules is typical for theories in which a symmetry has been broken down to a smaller one by means of a gauge choice. The two formulations are by definition gauge equivalent. Hence this is the situation that we have discussed in the previous chapter. The second formulation in which the gauge (4.5) has been imposed is obviously less transparent, and modified transformations such as D' have a more complex structure. This formulation is the standard one for discussions of conformal invariance. In these lectures we advocate formulations with a higher degree of symmetry, where the choice of gauge is postponed until the very end.

Let us now analyze the number of degrees of freedom in somewhat more detail. The gravitational sector is described by the fields $e_\mu{}^a$, $\omega_\mu{}^{ab}$, $f_\mu{}^a$, and b_μ, of which only $e_\mu{}^a$ and b_μ are independent. Hence we count 16 + 4 = 20 components, of which 6 + 4 + 4 + 1 = 15 correspond to the gauge degrees of freedom induced by Lorentz transformations, conformal boosts, general coordinate transformations, and dilatations. Therefore there are 5 gravitational degrees of freedom left, which is the minimal number for spin 2. Notice that we do not count dynamic degrees of freedom here. Massless particles of spin 2 have only two helicity states, but their corresponding fields should be able to generate the Lorentz algebra off-shell. Therefore for field degrees of freedom the relevant counting should be based on massive representations which implies 5 degrees of freedom for spin 2. This technique of "off-shell counting" also applies to supermultiplets as a whole, and as such it has played an important role in the discussion of auxiliary field formulations of supersymmetric theories.

A similar counting applies to the fermionic gauge fields of SU(2,2 | N). Because of the third constraint the fields $\psi_\mu{}^i$ are the only independent spinors, which have 16 components each. We should subtract the gauge degrees of freedom corresponding to Q and S supersymmetry. Therefore every field $\psi_\mu^{\ i}$ has 16 - 4 - 4 = 8 degrees of freedom. Again this is in agreement with off-shell counting arguments for spin 3/2. Massive spin 3/2 states have 4 degrees of freedom, but for fermions the number of field degrees of freedom is twice the number of physical states. These counting arguments show that the constraints (4.4) treat the gauge fields of various spins in a uniform fashion. Namely they remove the low-spin degrees of freedom as much as possible. This is only necessary for the spin 2 and 3/2 fields. For fields of lower spin such as the (S)U(N) gauge fields, the low spin degrees of freedom have already been removed by their corresponding gauge transformations.

We have already mentioned that the SU(2,2 | N) curvature constraints are of the conventional type. This causes the general coordinate transformations to become a nontrivial part of the gauge algebra, and to take the place of the P transformations of SU(2,2 | N); the latter are then ignored in subsequent considerations. In this way we have obtained a mutilated version of the original algebra, which is a supersymmetry algebra of the conventional type in the sense that the anticommutator of two supersymmetry transformations yields a general coordinate transformation. Consequently, off-shell representations of this algebra should contain equal numbers of fermionic and bosonic degrees of freedom. This simple counting argument allows us to show that the gauge theory that we have presented above does not yet constitute a complete field representation for the superconformal gauge algebra. Its gravitational sector is based on 5 degrees of freedom, and the remaining bosonic degrees of freedom are contained in the U(N) (or SU(N)) gauge fields. Hence, we have $5+3N^2$ (or $5+3N^2-3$ for N=4) bosonic components. On the other hand, the number of fermionic degrees of freedom is only 8N. Clearly the two numbers only match for N=1. Indeed N=1 conformal supergravity is based on the gauge fields presented in this section, but for higher N the theory is still incomplete. We return to this important problem in section 7.

5 CONFORMAL GRAVITY AND MATTER IN d DIMENSIONS

To get acqainted with conformally invariant field theories in the formulation presented above, we shall construct conformally invariant actions for matter fields of various spins. Since the conformal part of the algebra (4.2) holds true for any space-time dimension d we will leave d arbitrary in this section; the conformal group is isomorphic to SO(d,2) in that case, and the gravitational part of the transformations (4.3) remains the same. In the second part of this section we shall demonstrate how to construct Einstein gravity within the context of the conformal theory.

The first example concerns a real scalar field ϕ inert under conformal boosts

$$\delta_K\phi = 0 \quad . \tag{5.1}$$

Under dilatations the transformation of ϕ is characterized by a Weyl weight factor w, according to

$$\delta_D\phi = w\,\Lambda_D\,\phi \; . \tag{5.2}$$

As we have discussed previously, the P transformations are discarded; the conformal covariant derivative is thus given by

$$D_\mu\phi = (\partial_\mu - wb_\mu)\phi \; . \tag{5.3}$$

Because b_μ transforms under K transformations $D_a\phi$ is no longer inert,

$$\delta_K(D_a\phi) = - w\Lambda_{Ka}\phi \; . \tag{5.4}$$

Therefore the application of a second covariant derivative yields

$$D_\mu(D_a\phi) = (\partial_\mu - (w+1)b_\mu)D_a\phi - \omega_{\mu a}{}^b D_b\phi + wf_\mu{}^a\phi \; . \tag{5.5}$$

Note that the Weyl weight of $D_a\phi$ is w+1 because of the contraction with an inverse vierbein field. Using the various transformations of the gauge fields given in (4.3), one finds straighforwardly

$$\begin{aligned}
\delta_K(D_\mu D_a\phi) &= -(w+1)(\Lambda_{Ka}D_\mu\phi + \Lambda_{K\mu}D_a\phi) + \Lambda_K{}^b D_b\phi\, e_{\mu a} \quad , \\
\delta_K(D^a D_a\phi) &= (d-2-2w)\,\Lambda_K{}^a\, D_a\phi \quad , \\
\delta_D(D^a D_a\phi) &= (w+2)\Lambda_D\,(D^a D_a\phi) \quad .
\end{aligned} \tag{5.6}$$

Hence for $w=\frac{1}{2}(d-2)$ we construct a density

$$L = e\phi D^a D_a\phi \quad , \tag{5.7}$$

which is invariant under general coordinate transformations, Lorentz transformations, dilatations and conformal boosts. Note that the presence of the vierbein determinant e in (5.7) is crucial for invariance under both general coordinate transformations and dilatations.

If we assume the presence of the conventional constraints (4.4) b_μ is the only field transforming under conformal boosts. Therefore (5.7) does not depend on b_μ if $w=\frac{1}{2}(d-2)$. We may then write (5.7) as

$$L = e\;\phi\,\Box^{grav}\phi + \tfrac{1}{2}(d-2)ef_\mu{}^\mu\phi^2 \quad , \tag{5.8}$$

where $\Box^{grav}$ denotes the standard Lagrangian for scalar fields coupled to gravity, and $f_\mu{}^a$ is the solution of the second constraint with b_μ vanishing.

The construction of an invariant density for spin-½ fields proceeds in a similar fashion. We introduce a Dirac spinor ψ, which transforms under conformal boosts and dilatations according to

$$\delta_K \psi = 0 \quad , \quad \delta_D \psi = w \, \Lambda_D \, \psi \quad . \tag{5.9}$$

The covariant derivative

$$D_\mu \psi = (\partial_\mu - \tfrac{1}{2} \omega_\mu{}^{ab} \sigma_{ab} - w b_\mu)\psi \ , \tag{5.10}$$

transforms under K as follows:

$$\delta_K(D_\mu \psi) = (\tfrac{1}{2}\gamma_\mu \gamma_\nu \Lambda_K{}^\nu - (w+\tfrac{1}{2})\Lambda_{K\mu})\psi \quad . \tag{5.11}$$

Hence the Dirac operator on ψ transforms under conformal boosts and dilatations according to

$$\begin{aligned} \delta_K(\not{D}\psi) &= (\tfrac{1}{2}d - \tfrac{1}{2} - w)\, \not{\Lambda}_K \psi \ , \\ \delta_D(\not{D}\psi) &= (1 + w)\, \Lambda_D \, \not{D}\psi \quad . \end{aligned} \tag{5.12}$$

Therefore we obtain a density for $w = \tfrac{1}{2}(d-1)$

$$L = - e \, \bar{\psi} \not{D} \psi \ , \tag{5.13}$$

which is invariant under general coordinate transformations, Lorentz transformations, dilatations and conformal boosts. Again when we assume the validity of the constraints (4.4) the conformally invariant density (5.13) should be independent of the field b_μ. Note that (5.13) does not contain the gauge field $f_\mu{}^a$.

The next example is to consider the Maxwell action. We leave it to the reader to show that this action is only D invariant in 4 dimensions, whereas K invariance is manifest. Another application is the action of an antisymmetric tensor field, which is known to occur in superconformal actions. The corresponding lagrangian in d dimensions is

$$L = - \tfrac{1}{4} \, e \, T^{ab} D^c D_c T_{ab} + 2\,(d-2)^{-1} e \, T^{ab} D_a D^c T_{cb} \quad . \tag{5.14}$$

It is straightforward to show that (5.14) is K and D invariant if the Weyl weight of T is $w = \tfrac{1}{2}(d-2)$.

We will now demonstrate how a gravitational action without conformal invariance can be formulated in a conformally invariant way. Essentially this may be viewed as the spin-2 analogue of what happened in the example presented in section 3. To exhibit this in detail we shall construct Einstein gravity with a cosmological term. We start from the conformally invariant action (5.7) for a scalar field ϕ with Weyl weight $\frac{1}{2}(d-2)$, to which we add a conformally invariant self-coupling

$$L = - e\phi D^a D_a \phi + eg\phi^{2d/(d-2)} \quad . \tag{5.15}$$

We have chosen a minus sign for the kinetic term for reasons to be explained below. In the presence of the constraints conformal invariance implies that (5.15) is independent of the dilatational gauge field b_μ; furthermore we can solve the conformal gauge field $f_\mu{}^a$. Ignoring b_μ dependent terms for obvious reasons we find that $f_\mu{}^a$ can be expressed in terms of the Ricci tensor

$$f_\mu{}^a = \frac{1}{d-2} \left(R_{\mu a}(e) - \frac{1}{2(d-1)} e_{\mu a} R(e)\right) . \tag{5.16}$$

Therefore (5.15) can be written as follows

$$L = - e\phi \,\square^{grav} \phi \; - \frac{d-2}{4(d-1)} R(e)\phi^2 + eg\phi^{2d/(d-2)} \tag{5.17}$$

This lagrangian depends only on $e_\mu{}^a$ and ϕ, since all b_μ dependence has disappeared, and ω is a dependent quantity because of the constraint. The fields in (5.17) are therefore inert under K transformations, and the only nontrivial conformal symmetry is given by the local dilatations

$$e_\mu{}^a \to e^{-\Lambda_D} e_\mu{}^a \;, \quad \phi \to e^{\frac{1}{2}(d-2)\Lambda_D} \phi \; . \tag{5.18}$$

We shall now show that (5.17) is gauge equivalent to the Einstein lagrangian with a cosmological term (We should emphasize again that the gauge equivalence will only be used at the classical level. Quantum corrections will introduce a mass scale in conformally invariant theories, which spoils the conformal symmetry. Thereby the equivalence is lost at the same time). As we have argued in section 3 a convenient way to exhibit the gauge equivalence is by imposing a consistent set of gauge conditions. To break the invariance under conformal boosts the condition (4.5) is introduced, whereas the invariance under dilatations is broken by adjusting ϕ to a dimensionful constant. The Poincaré gauge is thus defined by

$$b_\mu = 0 \;, \quad \phi = \kappa^{-1} \; . \tag{5.19}$$

Note that the first condition has no consequences for the lagrangian. Using the second condition (5.17) becomes

$$L = - \frac{d-2}{4(d-1)} \kappa^{-2} R(e) + eg\kappa^{2d/(2-d)} \quad . \tag{5.20}$$

Hence the conformally invariant action (5.17) is gauge equivalent to the Einstein action with cosmological term. The reason for choosing the minus sign for the kinetic term in (5.15) was to obtain the correct sign for the Riemann scalar in (5.20). The constant κ is directly related to Newton's constant, whereas the size of the cosmological term is governed by the independent coupling constant g. For finite cosmological constant this theory is sometimes referred to as de Sitter gravity.

Notice that the action (5.20) is based on $\frac{1}{2}d(d-1)$ degrees of freedom: conformal gravity with conventional constraints is based on $\frac{1}{2}(d+1)(d-2)$ degrees of freedom, and an additional degree of freedom is provided by the compensating field ϕ. We should mention that it is possible in four dimensions to have a conformally invariant lagrangian for spin 2, which is quadratic in the curvature tensor R(M)

$$L \propto \varepsilon^{\mu\nu\rho\sigma}\varepsilon_{abcd}\, R_{\mu\nu}{}^{ab}(M)\, R_{\rho\sigma}{}^{cd}(M) \,. \tag{5.21}$$

Conformal invariance of (5.21) requires that the first constraint (4.4) is satisfied. Therefore the lagrangian (5.21) is quadratic in <u>second-order</u> derivatives. Assuming the second constraint (4.4), or alternatively by solving the algebraic field equation for $f_\mu{}^a$, (5.21) reduces to (modulo a total derivative)

$$L \propto e\left(R_{\mu\nu}{}^2(e) - 1/3\, R^2(e)\right) , \tag{5.22}$$

with $R_{\mu\nu}(e)$ the Ricci tensor, and R(e) its trace.

Precisely as in the example of section 3, the gauge condition is only a convenient way of demonstrating the gauge equivalence. In (5.17) one could also redefine the vierbein field such that it becomes scale invariant:

$$e_\mu{}^a = \kappa\phi^{2/(2-d)}(e_\mu{}^a)^{new} . \tag{5.23}$$

With this redefinition (5.17) is simply equal to (5.20), but now expressed in terms of the new vierbein field. The fact that (5.20) only depends on the vierbein, and not on b_μ and ϕ is a direct consequence of invariance under conformal boosts and dilatations.

6 CURRENTS

After having discussed some conformally invariant matter lagrangians we now consider the lowest-order coupling of the superconformal gauge fields to matter in a more general context. To that order we expand the superconformal gauge fields about their flat space values: $e_\mu{}^a = \delta_\mu{}^a$, and all other fields vanishing. Considering only the purely gravitational fields and the spinors of the superconformal theory, we have

$$L \approx L_{\text{matter}} + h_\mu{}^a\Theta_{\mu a} + \omega_\mu{}^{ab}S_\mu{}^{ab} + b_\mu T_\mu + f_\mu{}^a U_{\mu a} + \bar{\psi}_\mu J_\mu + \bar{\phi}_\mu J'_\mu , \quad (6.1)$$

where $h_\mu{}^a$ denotes the deviation of the vierbein field from its flat-space value. The first term in (6.1) denotes the matter lagrangian in flat space; for convenience we have suppressed the SU(N) indices for the spinors. The current that couples to the vierbein is called the energy-momentum tensor $\Theta_{\mu a}$. The currents in (6.1) satisfy a number of properties which are implied by the conjectured invariance of the full action under the superconformal transformations. For our purpose it suffices to include only the inhomogeneous terms in the superconformal transformation of the gravitational fields; we give the nonvanishing terms

$$\begin{aligned}
&\text{(i)} \quad \text{Coordinate transformations: } \delta h_\mu{}^a = -\partial_\mu \xi^a \quad ,\\
&\text{(ii)} \quad \text{Spin rotations M: } \delta\omega_\mu{}^{ab} = \partial_\mu \varepsilon^{ab} \;, \; \delta h_\mu{}^a = \varepsilon^a{}_\mu \;,\\
&\text{(iii)} \; \text{Dilatations D: } \delta b_\mu = \partial_\mu \Lambda_D \;, \; \delta h_\mu{}^a = -\delta_\mu{}^a \Lambda_D \quad , \qquad (6.2)\\
&\text{(iv)} \quad \text{Conformal boosts K: } \delta f_\mu{}^a = \partial_\mu \Lambda_K{}^a , \; \delta\omega_\mu{}^{ab} = \Lambda_K{}^{[a}\,\delta_\mu{}^{b]} , \; \delta b_\mu = \Lambda_{K\mu} \;,\\
&\text{(v)} \quad \text{Q supersymmetry: } \delta\psi_\mu{}^i = 2\partial_\mu \varepsilon^i \;,\\
&\text{(vi)} \quad \text{S supersymmetry: } \delta\phi_\mu{}^i = 2\partial_\mu \eta^i , \; \delta\psi_\mu{}^i = -\gamma_\mu \eta^i \;.
\end{aligned}$$

We now consider the variation of the action corresponding to (6.1) under one of these symmetries. Ignoring terms proportional to the gravitational field, we have

$$\begin{aligned}
\delta I = \int dx \, \Big\{ &\delta\psi(x) \frac{\delta I_{\text{matter}}}{\delta\psi(x)} + \delta h_\mu{}^a(x)\, \Theta_{\mu a}(x) + \delta\omega_\mu{}^{ab}(x)\, S_\mu{}^{ab}(x) \\
&+ \delta b_\mu(x)\, T_\mu(x) + \delta f_\mu{}^a(x)\, U_{\mu a}(x) \\
&+ \delta\bar{\psi}_\mu(x)\, J_\mu(x) + \delta\bar{\phi}_\mu(x)\, J'_\mu(x) \Big\} \;, \qquad (6.3)
\end{aligned}$$

where $\psi(x)$ generically denotes the matter fields. Requiring the full action to be invariant under the superconformal symmetries, and assuming that the matter fields satisfy their field equations in the absence of the gravitational fields, leads to

$$\begin{aligned}
\int dx \, \Big\{ &\delta h_\mu{}^a(x)\, \Theta_{\mu a}(x) + \delta\omega_\mu{}^{ab}(x)\, S_\mu{}^{ab}(x) + \delta b_\mu(x)\, T_\mu(x) \\
&+ \delta f_\mu{}^a(x)\, U_{\mu a}(x) + \delta\bar{\psi}_\mu(x)\, J_\mu(x) + \delta\bar{\phi}_\mu(x)\, J'_\mu(x) \Big\} = 0. \qquad (6.4)
\end{aligned}$$

After insertion of the transformation rules (6.2) with arbitrary space-time dependent parameters one easily deduces the relations

$$
\begin{aligned}
&\text{(i)} \quad \partial_\mu \Theta_{\mu\nu} = 0 && \text{(general coordinate transformations)},\\
&\text{(ii)} \quad \partial_\mu S_\mu{}^{ab} + \tfrac{1}{2}(\Theta_{ab} - \Theta_{ba}) = 0 && \text{(spin rotations)},\\
&\text{(iii)} \quad \partial_\mu T_\mu + \Theta_{\mu\mu} = 0 && \text{(dilatations)},\\
&\text{(iv)} \quad \partial_\mu U_\mu{}^a - 2S_\mu{}^{a\mu} - T_a = 0 && \text{(conformal boosts)},\\
&\text{(v)} \quad \partial_\mu J_\mu = 0 && \text{(Q supersymmetry)},\\
&\text{(vi)} \quad \partial_\mu J'_\mu - \gamma_\mu J_\mu = 0 && \text{(S supersymmetry)}.
\end{aligned}
\tag{6.5}
$$

Hence not all of these currents are conserved. However, it is possible to define conserved currents, which have an explicit dependence on the coordinates x^μ. Namely, by combining the equations of (6.5) we find four more conserved currents

$$
\begin{aligned}
&\text{(ii)}' \quad \partial_\mu\left(S_\mu{}^{ab} + \tfrac{1}{2}(\Theta_{\mu b}x_a - \Theta_{\mu a}x_b)\right) = 0 ,\\
&\text{(iii)}' \quad \partial_\mu\left(T_\mu + \Theta_{\mu\nu}x^\nu\right) = 0 ,\\
&\text{(iv)}' \quad \partial_\mu\left(U_\mu{}^a - 2S_\mu{}^{ab}x_b - T_\mu x^a + \Theta_{\mu b}(\tfrac{1}{2}x^2\delta^{ab} - x^a x^b)\right) = 0 ,\\
&\text{(vi)}' \quad \partial_\mu\left(J'_\mu - \gamma_\nu J_\mu x^\nu\right) = 0 .
\end{aligned}
\tag{6.6}
$$

The currents (ii)', (iii)' and (iv)' correspond to flat-space Lorentz transformations, dilatations and conformal boosts, characterized by

$$
\begin{aligned}
&\text{(ii)}': \quad \varepsilon^{ab}(x) = \varepsilon^{ab} , && \xi^\mu(x) = \varepsilon^{\mu\nu}x_\nu ,\\
&\text{(iii)}': \quad \Lambda_D(x) = \Lambda_D , && \xi^\mu(x) = -x^\mu\Lambda_D ,\\
&\text{(iv)}': \quad \Lambda_K{}^a(x) = \Lambda_K{}^a , && \varepsilon^{ab}(x) = x^a\Lambda_K{}^b - x^b\Lambda_K{}^a ,\\
&\qquad\quad \Lambda_D(x) = -\Lambda_K{}^a x_a , && \xi^\mu(x) = x^\mu x_a\Lambda_K{}^a - \tfrac{1}{2}x^2\Lambda_K{}^\mu ,
\end{aligned}
\tag{6.7}
$$

with ε^{ab}, Λ_D and $\Lambda_K{}^a$ space-time independent parameters. It is straightforward to show that the transformations (6.7) leave the flat-space vierbein $e_\mu{}^a = \delta_\mu{}^a$ invariant. The transformation corresponding to the current (vi)' in (6.6) is a combined Q- and S-supersymmetry transformation

$$
\text{(vi)}': \quad \eta(x) = \eta , \qquad \varepsilon(x) = \gamma^\mu x_\mu \eta . \tag{6.8}
$$

So far we have assumed that the gauge fields in (6.1) are mutually independent. We now reconsider (6.1) in the presence of the constraints (4.4), so that only $h_\mu{}^a$ and ψ_μ are independent. Consequently there are only two currents, which are modifications of $\Theta_{\mu a}$ and J_μ. They differ by so-called improvement terms which are generated by the dependence of $\omega_\mu{}^{ab}$ and $f_\mu{}^a$ on $h_\mu{}^a$, and of ϕ_μ on ψ_μ. Improvement terms are conserved because of their form: they can generally be written as the divergence of an antisymmetric tensor. That this is the case follows from the curl structure present in the explicit solutions for $\omega_\mu{}^{ab}$, $f_\mu{}^a$ and ϕ_μ. The improvement terms can then be expressed explicitly in terms of derivatives of the currents $S_\mu{}^{ab}$, $U_{\mu a}$ and J'_μ (notice that T_μ is not independent). Repeating the above analysis we find that the field b_μ should decouple from the theory in question. Namely, because of the constraints (4.4) b_μ couples to $S_\mu{}^{ab}$ and $U_\mu{}^a$ through the dependence of $\omega_\mu{}^{ab}$ and $f_\mu{}^a$ on b_μ; conformal invariance then implies that the sum of all contributions should vanish. Algebraically, this is a consequence of the divergence equation (iv) in (6.5). The remaining invariances imply for the two modified currents:

$$
\begin{aligned}
&\partial_\mu \Theta^{imp}_{\mu\nu} = 0 , && \text{(general coordinate transformations)},\\
&\Theta^{imp}_{\mu\nu} = \Theta^{imp}_{\nu\mu} , && \text{(Lorentz transformations)},\\
&\Theta^{imp}_{\mu\mu} = 0 , && \text{(dilatations)} ,\\
&\partial_\mu J^{imp}_\mu = 0 , && \text{(Q supersymmetry)} ,\\
&\gamma_\mu J^{imp}_\mu = 0 , && \text{(S supersymmetry)} .
\end{aligned}
\tag{6.9}
$$

In deriving (6.9) we have implicitly assumed that the transformations (6.2) have not been affected by the presence of the conventional constraints. One can easily verify that this is the case. The construction of modified currents which satisfy the conditions above is thus intrinsically related to the possibilities for choosing conventional constraints.

One can construct the analogue of (6.6) in terms of improved currents only. Namely we have

$$
\begin{aligned}
&\partial_\mu\big(\tfrac{1}{2}(\Theta^{imp}_{\mu b}x_a - \Theta^{imp}_{\mu a}x_b)\big) = 0 ,\\
&\partial_\mu\big(\Theta^{imp}_{\mu\nu}x^\nu\big) = 0 ,\\
&\partial_\mu\big(\Theta^{imp}_{\mu b}(\tfrac{1}{2}x^2\delta^{ab} - x^a x^b)\big) = 0 ,\\
&\partial_\mu\big(\gamma_\nu J^{imp}_\mu x^\nu\big) = 0 .
\end{aligned}
\tag{6.10}
$$

This leads to an elegant expression for the currents related to the transformations (6.7) and (6.8), which differ from the original ones defined in (6.6) by further improvement terms. At this point we make contact with more conventional treatments of these currents.

We may now also include the currents of the chiral (S)U(N) transformations, but their possible improvement terms are not relevant here. Let us recall that all these conserved currents are expressed in terms of matter fields that are subject to their equations of motion, discarding the interactions with the superconformal gauge fields. For our purpose it suffices to consider matter systems without self-interactions; in that case all currents are bilinear in the fields. In principle one can construct each of these currents for any given theory, provided that this theory is invariant under the corresponding transformations: translations, supersymmetry and chiral (S)U(N). However, the existence of suitable improvement terms is a priori not guaranteed.

Clearly the (improved) currents constructed so far do not constitute a multiplet under supersymmetry. This follows from the counting argument given in section 4, which is again applicable since the gauge fields and their corresponding currents contain the same numbers of degrees of freedom. However, in this case it is easy to find the complete multiplet of currents for any given theory. Namely, one can obtain the missing components by constructing successive variations of the known currents, up to the point where one encounters only derivatives of bilinears that have been found before. Since the matter fields themselves satisfy their field equations, it suffices to start from an on-shell formulation of the matter theory. Nevertheless the multiplet of currents is a genuine off-shell multiplet. It is also finite: after at most 4N supersymmetry variations one can only find derivatives of previously found bilinears, because of the anticommuting nature of the supersymmetry generators. (We have assumed here that the matter multiplet has no on-shell central charges. This must be the case for massless multiplets.)

The construction of the gravitational multiplet of currents should fail beyond N=4, where no matter multiplets exist. (Since N>4 multiplets contain spin-3/2 fields, these multiplets are an intrinsic part of supergravity). One may envisage the construction of the multiplet of currents for supergravity itself, but for a self-coupled gauge system the currents are never gauge invariant. Therefore the standard procedure leads to an infinite number of degrees of freedom, from which no further information can be deduced. In addition, it is difficult to interpret such a result, since this "multiplet of currents" is supposed to couple to the original fields, which have already been taken on-shell. Similar arguments apply for supersymmetric Yang-Mills theories with N>2, since no Yang-Mills matter multiplets exist beyond N=2.

Hence to construct the largest gravitational multiplet of currents one has to consider an N=4 supersymmetric matter theory. The only known candidate for this is the supersymmetric Yang-Mills theory, and for our purpose the abelian version suffices. It is based on a gauge field V_μ, a quartet of Majorana spinors ψ^i (the chirality of ψ^i is positive), and a Lorentz scalar ϕ^{ij} subject to an SU(4) covariant reality constraint

$$\phi^{ij} = (\phi_{ij})^* = \tfrac{1}{2}\,\varepsilon^{ijk\ell}\,\phi_{k\ell}\ . \tag{6.11}$$

The field strength corresponding to V_μ is denoted by $F_{\mu\nu}$, and we define (anti)selfdual components $F^{\pm} = \frac{1}{2}(F \pm \tilde{F})$. The fields transform under supersymmetry according to

$$\begin{aligned}
\delta V_\mu &= \bar{\varepsilon}^i\gamma_\mu\psi_i + \text{h.c.} \quad ,\\
\delta F^-_{\mu\nu} &= \bar{\varepsilon}_i\not{\partial}\sigma_{\mu\nu}\psi^i \quad ,\\
\delta\psi^i &= -\,\sigma.F^-\varepsilon^i - 2i\not{\partial}\phi^{ij}\,\varepsilon_j \quad ,\\
\delta\phi_{ij} &= i\bar{\varepsilon}_{[i}\psi_{j]} - i\varepsilon_{ijk\ell}\,\bar{\varepsilon}^k\psi^\ell\ .
\end{aligned} \tag{6.12}$$

The action, invariant under translations, supersymmetry and chiral SU(4), corresponds to the following lagrangian

$$L = -\tfrac{1}{4}\,F_{\mu\nu}{}^2 - \tfrac{1}{2}\,\bar{\psi}_i\overleftrightarrow{\not{\partial}}\psi^i - \tfrac{1}{2}\,\partial_\mu\phi^{ij}\,\partial_\mu\phi_{ij}\ . \tag{6.13}$$

The currents of translations, supersymmetry and chiral SU(4) transformations can be constructed in the standard way. Furthermore it is possible to find improvement terms for these currents, so that they satisfy (6.9). The explicit expressions, modulo terms proportional to the field equations, are

$$\begin{aligned}
\Theta_{\mu\nu} &= 2F^+_{\mu\rho}\,F^-_{\nu\rho} - \tfrac{1}{4}\,\bar{\psi}^i\gamma_{(\mu}\overleftrightarrow{\partial}_{\nu)}\psi_i + \tfrac{1}{2}\,\delta_{\mu\nu}\,|\,\partial_\rho\phi^{ij}\,|^2 \quad ,\\
&\quad -\ \partial_\mu\phi^{ij}\,\partial_\nu\phi_{ij} - 1/6\,(\delta_{\mu\nu}\Box - \partial_\mu\partial_\nu)\,|\,\phi^{ij}\,|^2 \quad ,\\
J_{\mu i} &= -\,\sigma.F^-\gamma_\mu\psi_i + 2i\,\phi_{ij}\overleftrightarrow{\partial}_\mu\psi^j + 4/3\,i\sigma_{\mu\lambda}\,\partial_\lambda(\phi_{ij}\psi^j),\\
v_\mu{}^i{}_j &= \phi^{ik}\overleftrightarrow{\partial}_\mu\phi_{kj} + \bar{\psi}^i\gamma_\mu\psi_j - \tfrac{1}{4}\,\delta^i{}_j\,\bar{\psi}^k\gamma_\mu\psi_k\ .
\end{aligned} \tag{6.14}$$

When we now apply further supersymmetry transformations (6.12), subject to the field equations, we determine the full supermultiplet of currents. The remaining components are

$$\begin{aligned}
d^{ij}{}_{k\ell} &= \phi^{ij}\phi_{k\ell} - 1/12\,\delta^{[i}{}_k\,\delta^{j]}{}_\ell\,\,|\,\phi\,|^2\ ,\\
\chi^{ij}{}_k &= \tfrac{1}{2}\,\varepsilon^{ijmn}\,(\phi_{mn}\,\psi_k + \phi_{kn}\psi_m),\\
t_{ab}{}^{ij} &= \bar{\psi}^i\sigma_{ab}\psi^j + 2i\phi^{ij}\,F^-_{ab} \quad ,
\end{aligned}$$

$$e_{ij} = \bar{\psi}_i \psi_j \quad ,$$
$$\lambda_i = \sigma . F^- \psi_i \ ,$$
$$c = (F^-_{ab})^2 \quad . \tag{6.15}$$

The multiplet of currents plays an important role in the construction of the field representation of N=4 superconformal gravity. This will be discussed in section 7. Its supersymmetry transformations can be found in the literature.

7 CONFORMAL SUPERGRAVITY

We have shown in section 4 that the gauge theory derived from SU(2,2 | N) does not lead to a complete field representation for conformal supergravity. Therefore one is forced to change strategy and to choose another starting point. The particular approach that we shall be discussing here is based on the multiplet of currents, where one envisages conformal supergravity coupled to some matter system; to first order in the superconformal fields this coupling is then described by certain currents. This corresponds to the situation that we have discussed in the previous section. The essential aspect of our approach is that knowledge of the multiplet of currents yields information about the multiplet of fields on which conformal supergravity is based, since every current should couple to a corresponding field.

The most obvious candidate for this construction is the current multiplet of the N=4 supersymmetric Yang-Mills theory. In view of the fact that this multiplet contains the energy-momentum tensor and supersymmetry current with the conformal improvement terms, one expects that the fields to which these currents couple contain the gauge fields of SU(2,2 | N) subject to the constraints (4.4) as a subset. Hence we start by assigning a field to each current component. Apart from the graviton, gravitino and chiral SU(4) gauge fields which couple to the conserved currents, we thus find a complex scalar C, a symmetric scalar E_{ij}, an antisymmetric tensor (both in Lorentz and SU(4) indices) $T_{ab}{}^{ij}$, a scalar $D^{ij}{}_{k\ell}$ in the 20-dimensional real representation of SU(4), and two spinors Λ_i and $\chi^{ij}{}_k$, the latter in a complex 20-dimensional SU(4) representation. Subsequently one writes a coupling of the form $\int d^4x$ (current)×(field) and requires invariance. Since the supersymmetry transformations of the currents are known, one thus determines the linearized transformations for the fields. This then defines a linearized multiplet on which the Q supersymmetry transformations close; altogether the multiplet contains 128 + 128 degrees of freedom. This is the smallest possible multiplet that one can have in N=4 as is revealed by off-shell counting: in order to realize a representation for arbitrary momenta the field components must correspond to the states of a massive multiplet of N=4 supersymmetry (we assume the absence of central charges, which is entirely appropriate in this context). Such representations can be classified according to the USp(8) group, and the smallest massive representation is the one with maximal spin 2 containing 128 + 128 states. The decomposition of this multiplet with the corresponding superconformal fields is exhibited in Table 3.

Subsequently one reinterprets the linearized transformation rules within the context of the superconformal algebra by writing derivatives on gauge fields in terms of the dependent fields $\omega_\mu{}^{ab}$, $f_\mu{}^a$ and $\phi_\mu{}^i$, or in terms of the SU(2,2 | N) curvatures given in Table 2. At that point one indeed recovers the gauge field transformations of SU(2,2 | N) (cf. (4.3)), but supplemented by the new fields and their transformations. The only field that is missing so far is the dilatational gauge field b_μ, which can be reinstated according to (4.3). The presence of b_μ does not violate the closure of the algebra, mainly because together with b_μ we introduce the transformations under conformal boosts. We now give the linearized Q transformations for the fields of N=4 conformal supergravity:

$$
\begin{aligned}
\delta C &= \bar{\varepsilon}^i \Lambda_i \quad , \\
\delta \Lambda_i &= 2\not{\partial} C \varepsilon_i + E_{ij}\varepsilon^j + \varepsilon_{ijk\ell}\,\sigma.T^{k\ell}\varepsilon^j \quad , \\
\delta E_{ij} &= \bar{\varepsilon}_{(i}\not{\partial}\Lambda_{j)} - \bar{\varepsilon}^k \chi^{mn}{}_{(i}\varepsilon_{j)kmn} \quad , \\
\delta T_{ab}{}^{ij} &= \bar{\varepsilon}^{[i} R_{ab}{}^{j]}(Q) + \bar{\varepsilon}^k \sigma_{ab}\chi^{ij}{}_k + \tfrac{1}{2}\varepsilon^{ijk\ell}\bar{\varepsilon}_k \not{\partial}\sigma_{ab}\Lambda_\ell \quad , \\
\delta \chi^{ij}{}_k &= -\sigma.T^{ij}\overleftarrow{\not{\partial}}\varepsilon_k - 1/3\ \delta^{[i}{}_k\sigma.T^{j]\ell}\overleftarrow{\not{\partial}}\varepsilon_\ell \\
&\quad -\sigma.R^{[i}{}_k(V)\varepsilon^{j]} - 1/3\ \delta^{[i}{}_k\sigma.R^{j]}{}_\ell(V)\varepsilon^\ell \\
&\quad -\tfrac{1}{2}\,\varepsilon^{ij\ell m}\not{\partial}E_{k\ell}\varepsilon_m + D^{ij}{}_{k\ell}\varepsilon^\ell \quad , \qquad (7.1) \\
\delta D^{ij}{}_{k\ell} &= -2\bar{\varepsilon}^{[i}\not{\partial}\chi^{j]}{}_{k\ell} + \delta^{[i}{}_{[k}\bar{\varepsilon}^m\not{\partial}\chi^{j]}{}_{m\ell]} + \text{h.c.} \quad , \\
\delta e_\mu{}^a &= \bar{\varepsilon}^i\gamma^a\psi_{\mu i} + \text{h.c.} \quad , \\
\delta \psi_\mu{}^i &= 2(\partial_\mu\varepsilon^i + \tfrac{1}{2}b_\mu\varepsilon^i - \tfrac{1}{2}\,\sigma.\omega_\mu\varepsilon^i - V_\mu{}^i{}_j\varepsilon^j) - \sigma.T^{ij}\gamma_\mu\varepsilon_j \quad , \\
\delta V_\mu{}^i{}_j &= \bar{\varepsilon}^i\phi_{\mu j} + \bar{\varepsilon}^k\gamma_\mu\chi^i{}_{kj} - \tfrac{1}{4}\delta^i{}_j\bar{\varepsilon}^k\phi_{\mu k} \quad , \\
\delta b_\mu &= \tfrac{1}{2}\,\bar{\varepsilon}^i\phi_{\mu i} + \text{h.c.} \quad .
\end{aligned}
$$

In (7.1) we have used the same chiral notation as in (4.3). The extra fields transform among themselves, or into curvature tensors. In these transformations $\omega_\mu{}^{ab}$, $f_\mu{}^a$ and $\phi_\mu{}^i$ are determined by the constraints (4.4).

It is encouraging to observe the close relation between (4.3) and (7.1), but on the other hand the results are rather complicated and involve a large number of degrees of freedom. The full nonlinear results are known as well, and have an even more complex structure. The derivation of the complete transformation rules and the corresponding algebra proceeds as follows. One first assigns the fields to representations of chiral SU(4) and dilatations. To find the linearized S-supersymmetry transformations one calculates the commutator of a conformal boost and a supersymmetry transformation, making an ansatz for the K transformations consistent with SU(4) and dilatations. In this case no such transformations are possible, but in the general case a unique result is obtained by imposing the anticommutators of Q and S and of two S transformations. Subsequently one makes the derivatives covariant with respect to all linearized transformations by including the appropriate gauge connections. The derivation of the nonlinear results and the corresponding gauge algebra then proceeds by induction: one calculates the commutator of two supersymmetry transformations on the basis of the linearized results to find new terms in the supersymmetry algebra linear in the fields. This is first done on the gauge fields. To impose the new algebra on all fields then requires the addition of terms of higher order in the fields to the transformation laws, which in turn introduce corresponding higher-order terms in the algebra. This iterative procedure is rather complicated, since

Table 3: USp(8) and SU(4) decompositions of the massive N=4 supermultiplet with the corresponding fields of N=4 conformal supergravity. Note that indices are raised or lowered by complex conjugation; for spinors upper and lower indices are related to their chiral components.

spin	USp(8)	SU(4)	fields
2	1	1	$e_\mu{}^a$
3/2	8	$4+4^c$	$\psi_\mu{}^i$, $\psi_{\mu i}$
1	27	$15+6+6^c$	$V_\mu{}^i{}_j$, $T_{ab}{}^{ij}$, T_{abij}
1/2	48	$20+20^c+4+4^c$	$\chi^{ij}{}_k$, $\chi_{ij}{}^k$, Λ^i, Λ_i
0	42	$20+10+10^c+1+1^c$	$D^{ij}{}_{k\ell}$, E^{ij}, E_{ij}, C, C*

many nonlinear terms are consistent with dilatational and SU(4) invariance, and usually all possible terms do indeed appear.

However, there is one aspect of those results which leads to certain simplifications, and which deserves further attention. If one follows a few steps in this program, one finds that the scalar C occurs in a nonpolynomial fashion; notice that this is possible because of the fact that C is inert under dilatations. However, some of the nonlinear terms that involve this field have a remarkably systematic structure. The transformations (7.1) are known to be consistent with a rigid chiral U(1) symmetry, and it turns out that the nonlinear Q-transformations contain precisely such a U(1) transformation with a field-dependent coefficient as a uniform component. Furthermore, all derivatives are augmented by $C\overleftrightarrow{\partial}_\mu C^*$ terms in such a way that this quantity can be interpreted as a new connection field that makes the derivative covariant with respect to local U(1) transformations. In other words $C\overleftrightarrow{\partial}_\mu C^*$ seems to play the role of an extra gauge field connected to chiral U(1) transformations. These facts can be viewed as an indication that the theory can be reformulated in a form which is manifestly symmetric under local chiral U(1) transformations.

The starting point for such a reformulation is derived from the observation that a complex scalar C is mathematically equivalent to a parametrization of the coset space SU(1,1)/U(1); SU(1,1) is the group of complex 2×2 matrices with unit determinant that leave the metric $\eta = \mathrm{diag}(1,-1)$ invariant. Therefore elements of SU(1,1) satisfy

$$U^{-1} = \eta U^{\dagger}\eta \ , \quad U\varepsilon U^{T} = \varepsilon \ , \tag{7.2}$$

where U^T is the transpose of U, and ε is the two-index Levi-Civita symbol $\varepsilon_{\alpha\beta}$. Consider now a doublet Φ_α, for which we generally use the notation

$$\Phi^{\alpha} = \eta^{\alpha\beta}\,(\Phi_\beta)^* = (\Phi_1{}^*,-\Phi_2{}^*) \ , \tag{7.3}$$

which satisfies an SU(1,1) invariant constraint

$$\Phi^{\alpha}\Phi_{\alpha} = |\Phi_1|^2 - |\Phi_2|^2 = 1 \ . \tag{7.4}$$

Because of (7.4) Φ is based on three degrees of freedom. Now assume that Φ transforms both under rigid SU(1,1) transformations and local U(1) transformations. Note that (7.4) is invariant under both. The extra local invariance allows us to remove one further degree of freedom by a choice of gauge; for instance, one may impose a reality condition on one of the components of Φ, such as

$$\Phi_1 = \Phi^1 \ . \tag{7.5}$$

In that case Φ is based on two degrees of freedom; hence modulo local U(1) gauge transformations Φ can be parametrized in terms of a single complex variable. A possible parametrization consistent with (7.5) is

$$\Phi_\alpha = (1 - |C|^2)^{-\frac{1}{2}} (1, C) \ , \tag{7.6}$$

where C is a complex field. To show that (7.6) corresponds to a representation of the SU(1,1)/U(1) coset space, we note that one can assign a group element of SU(1,1) to Φ according to

$$U(\Phi) = \begin{pmatrix} \Phi_1 & -\Phi^2 \\ \Phi_2 & \Phi^1 \end{pmatrix} \ . \tag{7.7}$$

The phase transformations on Φ correspond to multiplication from the right by elements of a U(1) subgroup of SU(1,1):

$$U(\Phi) \to \big(U(\Phi)\big)' = U(\Phi) \begin{pmatrix} \exp(-i\Lambda) & 0 \\ 0 & \exp(i\Lambda) \end{pmatrix} \tag{7.8}$$

Hence all gauge-equivalent degrees of freedom contained in $U(\Phi)$ fall in equivalence classes which are by definition the right cosets of SU(1,1)/U(1).

Assuming that the field C in (7.1) corresponds to a parametrization of SU(1,1)/U(1) some of the transformation rules take a unique form. Modulo an overall factor the only supersymmetry variation of Φ that is consistent with SU(1,1), chiral U(1)×SU(4) and dilatational invariance is of the form

$$\delta\Phi_\alpha = - \bar{\varepsilon}^i \Lambda_i \, \varepsilon_{\alpha\beta} \, \Phi^\beta \ . \tag{7.9}$$

Imposing the gauge condition (7.5) this result is indeed consistent with the linearized transformation of C, where C is defined by (7.6). To preserve the gauge condition the supersymmetry transformations are uniformly modified by the addition of a Λ-dependent U(1) transformation; this explains the previously mentioned U(1) component in the nonlinear transformation rules. The gauge field for the U(1) transformations is only defined modulo U(1) invariant terms. One definition is

$$a_\mu = - \tfrac{1}{2} \Phi^\alpha \overleftrightarrow{\partial}_\mu \Phi_\alpha - \tfrac{1}{4} \bar{\Lambda}^i \gamma_\mu \Lambda_i \ , \tag{7.10}$$

which contains a term $\frac{1}{2}C^* \overleftrightarrow{\partial}_\mu C$ in lowest order. The formulation with explicit rigid SU(1,1) and local U(1) invariance offers important advantages because it restricts the nonlinearities that may occur in the full transformation rules. Indeed this strategy is completely in accord with the central theme of these lectures, which was to emphasize the usefulness of gauge-equivalent formulations with the highest possible degree of gauge invariance. It has been known for some time that SU(1,1) invariance plays a role in Poincaré supergravity, but this symmetry was never linked to the superconformal sector of the theory.

8 N=2 POINCARÉ SUPERGRAVITY

We have already demonstrated in sections 3 and 5 how to construct lagrangians in theories with a high degree of gauge invariance that are gauge equivalent to theories with less invariance. In this section we will outline how this procedure works for extended supergravity. The starting point is conformal supergravity as discussed in the previous section, to which one adds one or several compensating supermultiplets in order to obtain lagrangians for extended Poincaré (or de Sitter) supergravity. The compensating supermultiplets must of course constitute representations of the full nonlinear superconformal algebra. The technique for extending representations of rigid supersymmetry to superconformal representations was briefly discussed in the previous section. It is important to realize that the appropriate components for the compensation mechanism may be contained in inequivalent configurations of compensating supermultiplets. This will then lead to inequivalent off-shell representations for Poincaré supergravity. One must verify that the addition of the compensating multiplets leads to a field representation for Poincaré supergravity which is most easily done by constructing a consistent set of

Fig. 1: Conformal decomposition of N=1 Poincaré supergravity theories.

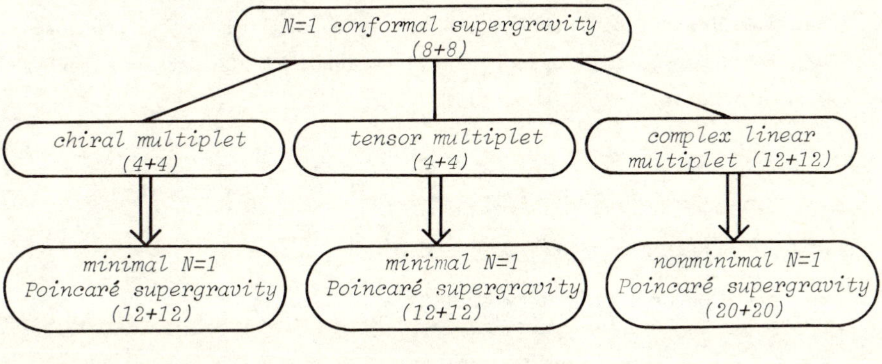

gauge conditions to break the gauge symmetries of the conformal sector, i.e. the invariance under K, D, S, and chiral (S)U(N). After the selection of the compensating supermultiplets one constructs a corresponding superconformal invariant. This invariant must then lead to a meaningful supergravity action, something that is not guaranteed a priori. If this is not the case one has to find a way of introducing more degrees of freedom into the theory.

Let us first review briefly how the known field representations of N=1 Poincaré supergravity decompose into the fields of conformal supergravity and those of the compensating multiplets. The situation is summarized in Fig. 1. The N=1 superconformal fields consist of 8+8 degrees of freedom. The two minimal formulations of Poincaré supergravity arise by introducing a chiral or a tensor multiplet, which both contain 4+4 degrees of freedom. In this way one obtains two inequivalent field representations both containing 12+12 field components. The nonminimal field representation of Poincaré supergravity is based on 20+20 degrees of freedom, decomposed into the components of conformal supergravity, and 12+12 components of a compensating complex linear multiplet.

The decomposition for N=2 supergravity is more complicated, and we will discuss it in somewhat more detail. A schematic outline is given in Fig. 2. The multiplet of conformal supergravity has 24+24 independent components. Its field content, which can be deduced from the N=4 results of the previous section, is represented by the vierbein $e_\mu{}^a$, the gravitinos $\psi_\mu{}^i$, the SU(2) and U(1) chiral gauge fields $V_\mu{}^i{}_j$ and A_μ, a spinor doublet χ^i, an antisymmetric tensor $T_{ab}{}^{ij}$, and a scalar D. As a first step one introduces an N=2 vector multiplet as a compensating field. This multiplet consists of an abelian gauge field $B_\mu{}^{ij}$, antisymmetric in SU(2) indices, a complex scalar field a, a doublet spinor ξ_i and a symmetric scalar S_{ij}, subject to a reality constraint. The gauge field $B_\mu{}^{ij}$ has a modified field strength, which is denoted by $t_{ab}{}^{ij}$. Its specific form is not important for our considerations. The Q- and S- supersymmetry transformations of this multiplet are

$$\begin{aligned}
\delta a &= \bar{\varepsilon}^i \xi_i \,, \\
\delta\xi_i &= 2\not{D} a \varepsilon_i - 2S_{ij}\varepsilon^j - \tfrac{1}{2}\sigma.t_{ij}\varepsilon^j + 2a\eta_i \,, \\
\delta S_{ij} &= -\tfrac{1}{2}\,\bar{\varepsilon}_{(i}\not{D}\xi_{j)} - \tfrac{1}{2}\varepsilon_{ik}\varepsilon_{j\ell}\,\bar{\varepsilon}^{(k}\not{D}\xi^{\ell)} \,, \\
\delta B_\mu{}^{ij} &= -\tfrac{1}{2}\sqrt{2}\,\varepsilon^{ij}\,(\varepsilon_{k\ell}\,\bar{\varepsilon}^k(\gamma_\mu\xi^\ell + 2a^*\psi_\mu{}^\ell) + \text{h.c.}) \,.
\end{aligned} \qquad (8.1)$$

The 32+32 degrees of freedom of the combined multiplets of N=2 conformal supergravity and the compensating vector multiplet constitute the so-called minimal field representation. It is minimal in the sense that it is the smallest multiplet that is gauge equivalent to N=2 Poincaré supergravity. This follows from the observation that the Poincaré gauge conditions

$$b_\mu = 0 \ , \quad a = \kappa^{-1} \ , \quad \xi_i = 0 \ , \tag{8.2}$$

break the invariance under conformal boosts, dilatations, chiral U(1) and S supersymmetry. It is easy to see from the transformations (8.1) that the third condition is required for consistency. Note that the chiral SU(2) transformations remain unaffected by (8.2). The dimensioned constant κ is again related to Newton's gravitational coupling constant, and is put equal to one henceforth.

The gauge conditions imply modifications of the transformation laws for the remaining invariances. These can be expressed in terms of uniform decomposition rules, which define the super-Poincaré transformations in terms of a linear combination of (field-dependent) supercon-

Fig. 2: Conformal decomposition of minimal N=2 Poincaré supergravity theories.

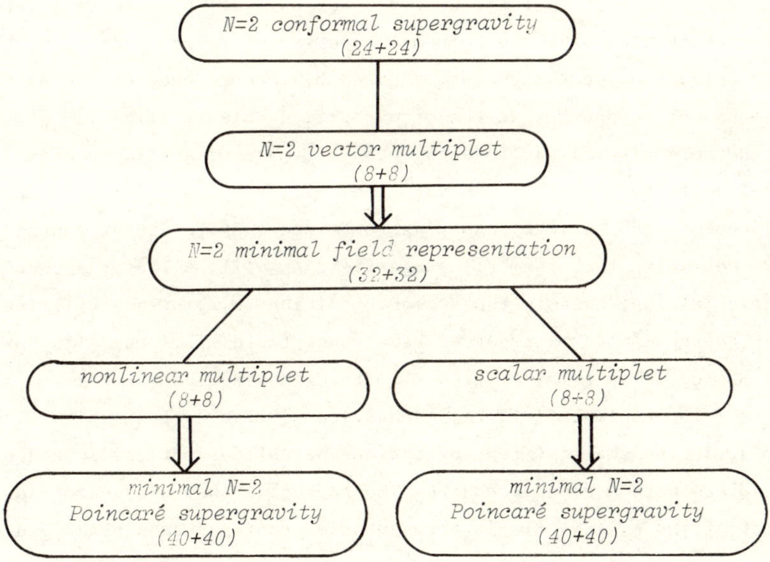

formal transformations. We have already seen an example of such a decomposition for the case of dilatations in section 4 (cf. (4.6)). We will now derive the decomposition rule for supersymmetry. Obviously the conditions (8.2) affect Q supersymmetry, but it is possible to compensate for this by adding a field-dependent conformal boost to Q supersymmetry, such that the condition $b_\mu = 0$ remains unaffected. Furthermore one adds a special field-dependent S-supersymmetry transformation, in order that the gauge $\xi_i = 0$ is maintained. Hence, Poincaré supersymmetry can be decomposed into three superconformal transformations.

$$\delta_{\text{Poincaré}}(\varepsilon_i) = \delta_Q(\varepsilon_i) + \delta_S(\eta_i) + \delta_K(\Lambda_K{}^a) . \tag{8.3}$$

The K transformation parameter $\Lambda_K{}^a$ takes a rather complicated form, but this component does not play an important role since most fields are inert under K. Therefore we concentrate on the S component in (8.3). To leave the condition $\xi_i = 0$ unaffected we must choose a linear combination of Q and S supersymmetry such that

$$\delta\xi_i = 2\not{D}a\varepsilon_i - 2S_{ij}\varepsilon^j - \tfrac{1}{2}\sigma.t_{ij}\varepsilon^j + 2a\eta_i = 0. \tag{8.4}$$

Using (8.1) and writing out the covariant derivative in (8.4) then leads to the solution for η_i

$$\eta_i = (S_{ij} + \tfrac{1}{4}\sigma.t_{ij})\varepsilon^j - i\not{A}\varepsilon_i . \tag{8.5}$$

This result fully demonstrates why Poincaré supergravity is more complicated than its conformal counterpart. Not only is it based on a larger field representation, but the Poincaré supersymmetry transformations have a more complex nonlinear structure induced by the presence of the extra field-dependent components in the decomposition rule (8.3). This is precisely the reason why it is advantageous to remain within the context of conformal supergravity until the very end. Only at that point one imposes a gauge choice such as (8.2), to pass to Poincaré or de Sitter supergravity. Note also that the U(1) gauge field A_μ will no longer appear in a systematic fashion; because the second condition (8.2) breaks U(1) invariance A_μ can no longer be regarded as a gauge field. This explains why A_μ does not occur in (8.5) as part of a covariant derivative.

The minimal field representation is also called minimal because it is the smallest representation that allows for the introduction of multiplets with a central charge. The reason is that the vector submultiplet of the minimal field representation contains an abelian gauge field $B_\mu{}^{ij}$ whose gauge invariance is distinct from the gauge symmetries of the superconformal multiplet. This abelian gauge symmetry occurs in the commutator of two Q transformations

$$[\delta_Q(\epsilon_1),\delta_Q(\epsilon_2)] = \ldots\ldots + \delta_Z(z) \ , \qquad (8.6)$$

where the dots indicate the standard superconformal transformations, and δ_Z is the abelian gauge transformation acting on $B_\mu{}^{ij}$

$$\delta_Z(z)B_\mu{}^{ij} = -\sqrt{2}\epsilon^{ij}\,\partial_\mu z \ . \qquad (8.7)$$

The transformation parameter z in (8.6) takes the form

$$z = 2\mathcal{A}^*\epsilon_{ij}\bar{\epsilon}_2{}^i\epsilon_1{}^j + \text{h.c.} \ . \qquad (8.8)$$

The extra term is not important for the minimal field representation, since δ_Z acts only on the gauge degree of freedom contained in $B_\mu{}^{ij}$. For that reason δ_Z commutes with all the gauge transformations of the superconformal algebra, and it is therefore a trivial example of a central charge transformation. However, there are multiplets of N=2 supersymmetry which contain a central charge in the algebra which acts nontrivially in the sense that it transforms field components into field components. In that case the transformation parameter z is proportional to

$$z = 2\epsilon_{ij}\,\bar{\epsilon}_2{}^i\epsilon_1{}^j + \text{h.c.} \ . \qquad (8.9)$$

without any dependence on the fields. In the context of local supersymmetry the central charge must correspond to a local gauge transformation, and therefore it requires a corresponding gauge field. Indeed $B_\mu{}^{ij}$ is an obvious candidate, but that requires that the central charge parameter is given by (8.8) rather than (8.9). Therefore one must first impose the local algebra with a field-dependent central charge (8.8) on the multiplet in question, which can be done by a straightforward iterative procedure. If one subsequently passes to the Poincaré gauge (8.2) where the field is constant, then the field-independent central charge (8.9) reemerges.

Hence the compensating vector multiplet of the minimal field representation plays a dual role. On the one hand it acts as a compensating multiplet, which leads to a field representation for Poincaré supergravity. However, it can also act as the gauge field multiplet of central charge transformations, in which case it simultaneously provides for the appropriate field dependence of the central charge transformation in the anticommutator of two Q transformations (cf. (8.8)). This dependence is crucial in order to discuss central charges within the superconformal framework. The central charge must commute with all superconformal symmetries, in particular with dilatations and chiral transformations; on the other hand the left-hand side of (8.6) is not inert under those transformations, and the field dependence in (8.8) is precisely such that it compensates for this lack of invariance.

The fields of the minimal field representation do not contain the appropriate degrees of freedom to compensate for the chiral SU(2) transformations. This is a first indication that 32+32 degrees of freedom are not sufficient to describe N=2 Poincaré supergravity. Indeed, if one constructs a superconformal invariant for the compensating vector multiplet it turns out that this invariant does not qualify as a suitable action, since in the Poincaré gauge (8.2) it contains a term linear in the field D (ignoring a vierbein determinant). Such an invariant does not lead to consistent equations of motion, and we have to introduce further compensating degrees of freedom. As is shown in Fig. 2 two possibilities have been investigated so far, both leading to Poincaré supergravity based on 40+40 degrees of freedom. Both formulations lead to equivalent equations of motion, but they are completely inequivalent in their off-shell aspects. In the first version one adds a so-called nonlinear multiplet, which we will introduce below; the second one uses a scalar multiplet as a compensator. The most crucial difference between the two concerns their behaviour under central charge transformations. The nonlinear multiplet is inert under the central charge; consequently the central charge acts trivially in the form of a gauge transformation (cf. (8.7)). There exists a minor modification of this where (8.7) corresponds to an SO(2) transformation, thus leading to Poincaré supergravity with gauged SO(2). On the other hand the central charge acts nontrivially on the scalar multiplet. Upon the Poincaré gauge this transformation is no longer a central charge in the strict sense, and only on-shell it coincides with the gauge transformation (8.7), or with an SO(2) transformation in the variant with gauged SO(2).

We refer to the literature for a more explicit exposition of N=2 Poincaré supergravity. In the remaining part of this section we shall further clarify the role played by some of the submultiplets in the variant of this theory that contains the nonlinear multiplet. Let us start by considering the N=2 vector multiplet, consisting of a complex scalar A, a doublet of Majorana spinors Ψ_i, a triplet of real scalars B_{ij}, and an antisymmetric tensor $F_{\mu\nu}$. Under rigid supersymmetry we have

$$\begin{aligned}
\delta A &= \bar{\varepsilon}^i \Psi_i \quad , \\
\delta \Psi_i &= 2\partial\!\!\!/ A \varepsilon_i + B_{ij}\varepsilon^j + \sigma.F\, \varepsilon_{ij}\varepsilon^j \quad , \\
\delta B_{ij} &= \bar{\varepsilon}_{(i}\partial\!\!\!/ \Psi_{j)} + \varepsilon_{ik}\varepsilon_{j\ell}\, \bar{\varepsilon}^k \partial\!\!\!/ \Psi^\ell \quad , \\
\delta F_{\mu\nu} &= \varepsilon^{ij}\, \bar{\varepsilon}_i \partial\!\!\!/ \sigma_{\mu\nu} \Psi_j + \text{h.c.} \quad .
\end{aligned} \tag{8.10}$$

The component B_{ij} and $F_{\mu\nu}$ are subject to constraints:

$$B^{ij} = (B_{ij})^* = \varepsilon^{ik}\,\varepsilon^{j\ell}\, B_{k\ell} \;, \quad \varepsilon^{\mu\nu\rho\sigma}\partial_\nu F_{\rho\sigma} = 0 \;, \tag{8.11}$$

and therefore the multiplet is based on 8+8 degrees of freedom. The second equation is a Bianchi identity, which implies that $F_{\mu\nu}$ can be expressed in terms of a vector potential in the standard way. The components A, Ψ_i, B_{ij}, $F_{\mu\nu}$, $-\varepsilon_{ij}\partial\!\!\!/\Psi^j$, $-2\,\Box A^*$ transform as the components of an N=2 <u>chiral multiplet</u>. The vector multiplet can thus be interpreted as a constrained chiral multiplet.

From the vector multiplet one may construct another multiplet, known as the <u>linear multiplet</u>. Its components are denoted by L_{ij}, φ^i, G and E_μ, and can be expressed in terms of those of the vector multiplet by

$$\begin{aligned}
L_{ij} &= B_{ij} \quad , \\
\varphi^i &= \partial\!\!\!/ \Psi^i \quad , \\
G &= -2\,\Box A^* \;, \\
E_\mu &= \partial_\nu F_{\nu\mu} \quad ,
\end{aligned} \tag{8.12}$$

Obviously this multiplet is also based on 8+8 degrees of freedom. The vector E_μ satisfies a constraint

$$\partial_\mu E_\mu = 0 \quad , \tag{8.13}$$

which ensures that E_μ can be expressed in terms of a tensor gauge field. The components (8.12) correspond precisely to the field equations of the N=2 supersymmetric Maxwell theory. Therefore, the N=2 Maxwell current is contained in a linear multiplet.

One may now repeat the above procedure and construct a vector multiplet from a linear one. Namely one chooses the G^* component as the first field of the vector multiplet, and determines the other components by considering successive supersymmetry variations. In this way one obtains an infinite sequence of vector and linear multiplets. However, we wish to argue that this is no longer possible when the vector and linear multiplets are realized in a superconformal context. The easiest way to show this is by establishing that these multiplets have a unique Weyl weight factor; in other words, the full superconformal algebra can only be imposed on these multiplets provided that their components scale appropriately under local dilatations. These weights are such that a vector multiplet cannot be obtained from a linear one in the way indicated above.

The Weyl weight of a supermultiplet is given by the weight of its first component, and the unique assignments of the vector and linear multiplets are w=1 and w=2, respectively. Hence under dilatations we have

$$\begin{aligned} A &\to A' = e^{\Lambda_D} A \quad , \\ L_{ij} &\to L_{ij}' = e^{2\Lambda_D} L_{ij} \quad . \end{aligned} \tag{8.14}$$

The weight factors of the remaining components follow by requiring covariance of the supersymmetry transformations; the generators of supersymmetry have weight ½, and therefore a supersymmetry variation leads to components with a weight factor increased by ½. Therefore, the Weyl weights corresponding to (8.14) of the field strengths of the vector and linear multiplet, F_{ab} and E_a, are w=2 and w=3, respectively. The vector and tensor gauge fields in terms of which the field strengths can be expressed,

$$\begin{aligned} F_{ab} &= e_a{}^{\mu} e_b{}^{\nu} (\partial_\mu V_\nu - \partial_\nu V_\mu) \quad , \\ E^a &= \tfrac{1}{2} i e^{-1} e_\mu{}^a \varepsilon^{\mu\nu\rho\sigma} \partial_\nu E_{\rho\sigma} \quad , \end{aligned} \tag{8.15}$$

have therefore vanishing weights (remember that the vierbein $e_\mu{}^a$ has weight -1). One can now argue that the assignments (8.14) are unique because the gauge fields must be inert under local scale transformations. The reason for the latter is that it is not possible to construct field strengths such as (8.15) that are simultaneously covariant under gauge transformations of the corresponding gauge fields and under local scale transformations, unless the gauge fields are inert under scale transformations. If the latter is not the case one could covariantize the derivatives in (8.15) with respect to dilatations, but this would then disturb the gauge invariance of V_μ and $E_{\mu\nu}$. The underlying reason for this incompatibility is that dilatations and gauge transformations commute, so that the gauge fields cannot transform under both.

Exploiting the minimal coupling inconsistency is only one of the many alternative ways of deriving (8.14). For instance, by applying the $\{Q,S\}$ anticommutator in A, one establishes that D and chiral U(1) transformations act on A with equal absolute strength. However, the chiral weight factor of A follows uniquely from the fact that V_μ and B_{ij} are inert under U(1) transformations in view of the fact that they are real. Similarly, one can establish the Weyl weight of L_{ij} by comparing the relative strengths of the SU(2) and D components in the $\{Q,S\}$ anticommutator acting on L_{ij}. According to the previous arguments it is still possible to define a linear multiplet from a vector multiplet according to (8.12), since B_{ij} and L_{ij} have the same behaviour under dilatations. To construct a vector multiplet from a linear one, seems impossible, unless one chooses a nonlinear relationship, such as

$$A = G^* L^{-1} - \bar{\varphi}_i \varphi_j L^{ij} L^{-3} \quad , \quad L = (L_{ij} L^{ij})^{\frac{1}{2}} \; . \tag{8.16}$$

Indeed, (8.16) leads to a multiplet with superconformal transformations that are very similar to those of the vector multiplet. But more explicit considerations show that (8.16) is in fact the first component of an unrestricted chiral multiplet based on 16+16 degrees of freedom. This is not the case for N=1, where a similar construction does generate a vector from a linear multiplet.

Hence we conclude that manipulations of multiplets within the context of rigid supersymmetry cannot automatically be extended to multiplets coupled to supergravity. In particular, it seems impossible to generate a vector from a linear multiplet. However, there exists another multiplet from which the vector multiplet can be constructed. This is the so-called nonlinear multiplet which we will now discuss. Its lowest-dimensional component is written as a 2×2 matrix $\Phi^i{}_\alpha$, which is restricted to elements of SU(2):

$$\Phi^i{}_\alpha \Phi^\alpha{}_j = \delta^i{}_j \; , \; \Phi^\alpha{}_i \Phi^i{}_\beta = \delta^\alpha{}_\beta \; , \quad \Phi^\alpha{}_i = \varepsilon^{\alpha\beta} \varepsilon_{ij} \Phi^j{}_\beta \; . \tag{8.17}$$

Note that $\Phi^i{}_\alpha$ transforms under chiral SU(2) from the left; from the right it transforms under another independent SU(2) group, which acts on the indices α, β, ..., and which is unrelated to the symmetries of conformal supergravity. Because of the nonlinear constraint (8.17) the nonlinear multiplet must have Weyl weight w=0. The three degrees of freedom contained in Φ transform under supersymmetry into an SU(2) doublet of Majorana spinors λ_i (λ_i has positive chirality). This requirement is sufficient to determine all supersymmetry transformations:

$$\begin{aligned} \delta\Phi^i{}_\alpha &= (2\bar{\varepsilon}^i \lambda_j - \delta^i{}_j \bar{\varepsilon}^k \lambda_k - \text{h.c.})\, \Phi^j{}_\alpha \; , \\ \delta\lambda^i &= -\tfrac{1}{2}\not{V}\varepsilon^i - \tfrac{1}{2}M^{ij}\varepsilon_j + \Phi^i{}_\alpha \not{\partial}\Phi^\alpha{}_j \varepsilon^j \\ &\quad - 2\lambda^i(\bar{\lambda}^j \varepsilon_j + \bar{\lambda}_j \varepsilon^j) + \gamma_a \varepsilon^i (\bar{\lambda}^j \gamma_a \lambda_j) + 2\sigma_{ab}\varepsilon_j\,(\bar{\lambda}^j \sigma_{ab} \lambda^i) \; , \end{aligned}$$

$$\delta M^{ij} = 4\,\bar\varepsilon^{[i}(\partial\!\!\!/\lambda^{j]} + \Phi^{j]}{}_\alpha\partial\!\!\!/\Phi^\alpha{}_k\lambda^k) - 2\bar\varepsilon^{[i}V\!\!\!/\lambda^{j]} - 2(\bar\varepsilon^k\lambda_k)\,M^{ij}\ , \qquad (8.18)$$
$$\delta V_a = 4\,\bar\varepsilon^i\sigma_{ab}\partial_b\lambda_i + 2\,\bar\varepsilon_i\gamma_a\Phi^i{}_\alpha\partial\!\!\!/\Phi^\alpha{}_j\lambda^j - \bar\varepsilon^i\gamma_a V\!\!\!/\lambda_i + \bar\varepsilon^i\gamma_a\lambda^j M_{ij} + \text{h.c.}.$$

We have introduced a complex antisymmetric scalar M^{ij} and a vector V_a. The latter satisfies a supersymmetric constraint

$$\partial\cdot V = \tfrac{1}{2}V_a{}^2 + \tfrac{1}{4}|M^{ij}|^2 - \partial_\mu\Phi^i{}_\alpha\,\partial_\mu\Phi^\alpha{}_i - 2\bar\lambda_i\overleftrightarrow{\partial\!\!\!/}\lambda^i\ , \qquad (8.19)$$

so that the nonlinear multiplet is also based on 8+8 degrees of freedom. Although the degrees of freedom of the linear and nonlinear multiplet are thus in close correspondence, the two multiplets are clearly inequivalent when coupled to supergravity, as is obvious from the different assignments with respect to dilatations and chiral SU(2). However, even within the context of rigid supersymmetry the nonlinear multiplet remains distinct from the linear one in view of its nonlinear transformations. Only when these transformations are linearized one makes contact with the linear multiplet. The extension of (8.18) to conformal supersymmetry has been given in the literature; its most conspicuous features are that λ^i transforms inhomogeneously under S supersymmetry, whereas V_a is not inert under conformal boosts:

$$\delta_S\lambda = \eta\ ,\ \delta_K V_a = 2\Lambda_{Ka}\ . \qquad (8.20)$$

The nonlinear multiplet acts as a "precurvature" for the vector multiplet, even when coupled to supergravity. The identification is as follows:

$$\begin{aligned}
A &= -\tfrac{1}{4}\,\varepsilon_{ij}M^{ij}\ ,\\
\Psi_i &= -2\varepsilon_{ij}\partial\!\!\!/\lambda^j - 2\varepsilon_{ij}\,\Phi^j{}_\alpha\partial\!\!\!/\Phi^\alpha{}_k\lambda^k + \varepsilon_{ij}\,V\!\!\!/\lambda^j + \tfrac{1}{2}\,\lambda_i\,\varepsilon_{jk}M^{jk}\ ,\\
B_{ij} &= \varepsilon_{ik}V_\mu\Phi^k{}_\alpha\partial_\mu\Phi^\alpha{}_j - \varepsilon_{ik}\partial_\mu(\Phi^k{}_\alpha\partial_\mu\Phi^\alpha{}_j) + \qquad (8.21)\\
&\quad + \bar\lambda_i\lambda_j\,\varepsilon_{mn}M^{mn} + \varepsilon_{ik}\varepsilon_{j\ell}\,\bar\lambda^k\lambda^\ell\,\varepsilon^{mn}M_{mn} + \ldots\ ,\\
F_{\mu\nu} &= \partial_\mu V_\nu - \partial_\nu V_\mu\ ,
\end{aligned}$$

where we have refrained from giving all terms in B_{ij} quadratic in λ.

We can now discuss the multiplet decomposition of N=2 Poincaré supergravity in the formulation based on the minimal field representation (i.e. conformal supergravity combined with the compensating vector multiplet (8.1)) fused with a nonlinear multiplet. The Poincaré gauge condition (8.2) is imposed together with a further condition to break the invariance under chiral SU(2). The inclusion of the nonlinear multiplet makes an SU(2) gauge condition possible:

$$\Phi^i{}_\alpha = \delta^i{}_\alpha \, . \tag{8.22}$$

Because the low-dimensional components of the compensating fields are eliminated by the gauge conditions the multiplet decomposition takes another form. The vector multiplet changes into a linear multiplet, with S_{ij} as its first component. Subsequent components are proportional to the superconformal fields. For instance, the spinor component is equal to $-\frac{1}{2}\not{D}\xi^i$, but since $\xi=0$ by (8.2) only its S covariantization remains, which leads to $\frac{1}{2}\gamma\cdot\phi^i$. A similar transition takes place for the nonlinear multiplet, which can now be interpreted as a vector multiplet, according to (8.21). Hence the superconformal decomposition of Poincaré supergravity in terms of the multiplet of conformal supergravity, the vector multiplet and the nonlinear multiplet, appears within the Poincaré context as a decomposition in the multiplet of conformal supergravity, a linear (tensor), and a vector multiplet, respectively. In addition another rearrangement of components takes place, which is related to the supersymmetric constraint (8.19). In the presence of conformal supergravity this constraint takes the following form:

$$D_a V_a - \tfrac{1}{2} V_a{}^2 - \tfrac{1}{4} |M^{ij}|^2 + D_a \Phi^i{}_\alpha \, D_a \Phi^\alpha{}_i + 2\left(\bar{\lambda}_i (\not{D}\lambda^i + 3/2\, \chi^i - \tfrac{1}{4}\sigma\cdot T^{ij}\lambda_j) + \text{h.c.}\right) = 3\, D, \tag{8.23}$$

where the field D on the right-hand side is one of the fields of conformal supergravity. It is now possible to eliminate D by using (8.23); hence, instead of assigning three degrees of freedom to V_a, and one to D, we may consider V_a as an unrestricted vector field. This then solves the problem of constructing the N=2 Poincaré supergravity lagrangian; as we have mentioned previously this lagrangian contains a term linear in D which leads to inconsistent equations of motion if D is regarded as an elementary field. Because of (8.23) D is now quadratic in other fields (modulo a total divergence), so that this problem no longer arises.

9 NOTES AND REFERENCES

• For earlier discussions of conformal invariance and further references, see

Weyl, H. (1918a). "Gravitation und Elektrizität", Sitzungsber. K. Preuss. Akad. Wiss. pp. 465-480 (reprinted in "The principle of relativity", Dover 1952); (1918b). "Reine Infinitesimal Geometrie", Math. Z. 2, pp. 384-411.

Wess, J. (1960). "The conformal invariance in quantum field theory", Nuovo Cim. 18, pp. 1086-1107.

Kastrup, H.A. (1962). "Zur physikalischen Deutung und darstellungstheoretischen Analyse der Konformen Transformationen von Raum und Zeit.", Ann. Physik 9, pp. 388-428.

Fulton, T., Rohrlich F. & Witten, L. (1962). "Conformal invariance in physics", Rev. Mod. Phys. 34, pp. 442-457.

Mack, G. & Salam, A. (1969). "Finite-component field representations of the conformal group", Ann. Phys. (N.Y.) 53, pp. 174-202.

Ivanov, E.A. & Ogievetsky, V.I. (1975). "Inverse Higgs effect in nonlinear realizations", Teor. Mat. Fiz. 25, pp. 1050-1059.

de Wit, B. (1981). "Conformal invariance in gravity and supergravity", in proc. 18th Winter School of Theoretical Physics, Karpacz, Gordon & Breach, to be published.

• Reviews and extensive references on supersymmetry, supergravity and superspace can be found in this volume, and in

Fayet, P. & Ferrara, S. (1977). "Supersymmetry", Phys. Rep. 32, pp. 249-334.

van Nieuwenhuizen, P. (1979). "Lectures in supergravity theory", in "Recent developments in gravitation", Summer Inst. Cargèse, 1978, eds. M. Lévy & S. Deser, Plenum Press, New York, pp. 519-548; (1981a). "Six lectures at the Cambridge Workshop on supergravity", in "Superspace and Supergravity", Nuffield Workshop Cambridge, 1980, eds. S.W. Hawking & M. Roček, Cambridge Univ. Press, pp. 9-70; (1981b). "Supergravity", Phys. Rep. 68, pp. 189-398.

Scherk, J. (1979). "Extended supersymmetry and extended supergravity theories", in "Recent developments in gravitation", Summer Inst. Cargèse, 1978, eds. M. Lévy & S. Deser, Plenum Press, New York, pp. 479-517.

Zumino, B. (1979). "Supergravity and Superspace", in "Recent developments in gravitation", Summer Inst. Cargèse, 1978, eds. M. Lévy & S. Deser, Plenum Press, New York, pp. 405-459.

Roček, M. (1981). "An introduction to superspace and supergravity", in "Superspace and Supergravity", Nuffield Workshop Cambridge, 1980, ed. S.W. Hawking and M. Roček, Cambridge Univ. Press. pp. 71-131.

Howe, P. (1981). "Supergravity in superspace", CERN preprint TH.3117.

• Formulations of supergravity as the gauge theory of the super-Poincaré algebra, the super-de Sitter algebras OSp(N | 4), or the superconformal SU(2,2 | N), algebras were presented in

MacDowell, S.W. & Mansouri, F. (1977). "Unified geometric theory of gravity and supergravity", Phys. Rev. Lett. 38, pp. 739-742, 1376.

Chamseddine, A.H. & West P.C. (1977). "Supergravity as a gauge theory of supersymmetry", Nucl. Phys. B129, p. 39-44.

Townsend, P.K. & Van Nieuwenhuizen, P. (1977). "Geometrical interpretation of extended supergravity", Phys. Lett. B67, p. 439-442.

Ferrara, S., Kaku, M., Townsend & Van Nieuwenhuizen, P. (1978). "Unified field theories with U(N) internal symmetries: gauging the superconformal group", Nucl. Phys. B129, pp. 125-134.

• The term "conventional" constraint has been introduced in the context of superspace by

Gates, S.J. & Siegel, W. (1980). "Understanding constraints in superspace formulations of supergravity", Nucl. Phys. B163, pp. 519-545.

Gates, S.J., Stelle K.S. & West P.C. (1980). "Algebraic origins of superspace constraints in supergravity", Nucl. Phys. B169, p. 347.

• The discussion in section 3 is related to Stueckelberg's treatment of massive spin-1 fields. See

Stueckelberg, E.C.G. (1938). "Die Wechselwirkungskräfte in der Elektrodynamik und in der Feldtheorie der Kernkräfte", Helv. Phys. Acta 11, pp. 225-244.

• Conformal supergravity theories have been constructed for N≤4 in

Kaku, M., Townsend, P.K. & van Nieuwenhuizen, P. (1978). "Properties of conformal supergravity", Phys. Rev. D17, pp. 3179-3187.

Townsend, P.K. & van Nieuwenhuizen, P. (1979). "Simplifications of conformal supergravity", Phys. Rev. D19, pp. 3166-3169.

de Wit, B. & van Holten, J.W. (1979). "Multiplets of linearized SO(2) supergravity", Nucl. Phys. B155, pp. 530-542.

de Wit, B., van Holten, J.W. & Van Proeyen, A. (1980a). "Transformation rules of N=2 supergravity multiplets", Nucl. Phys. B167, pp. 186-204 (E: B172, pp. 543-44).

Bergshoeff, E., de Roo, M. & de Wit, B. (1981). "Extended conformal supergravity", Nucl. Phys. B182, pp. 173-204.

• The last reference discusses the SU(1,1) invariance of N=4 conformal supergravity. For Poincaré supergravity this invariance was discovered in

Cremmer, E., Scherk, J & Ferrara, S. (1978). "SU(4) invariant supergravity theory", Phys. Lett. 74B, pp. 61-64.

• For the mathematical aspects of coset spaces such as SU(1,1)/U(1), see

Gilmore, R. (1974). "Lie groups, Lie algebras, and some of their applications", Wiley Interscience.

• Superspace formulations of conformal supergravity have been discussed in

van Nieuwenhuizen, P. & West, P.C. (1980). "From conformal supergravity in ordinary space to its superspace constraints", Nucl. Phys. B169, pp. 501-514.

Howe, P. (1981). "A superspace approach to extended conformal supergravity", Phys. Lett. 100B, pp. 389-392.

• Using the group manifold approach N=1 conformal supergravity has been rederived in

Castellani, L., Fré, P. & van Nieuwenhuizen, P. (1981). "A review of the group manifold approach and its applications to conformal supergravity", Ann. Phys. (N.Y.) 136, pp. 398-434.

• Conformal invariance in arbitrary space-time dimensions was studied by

Capper, D. & Duff, M. (1974). "Trace anomalies in dimensional regularization", Nuovo Cim. 23A, pp. 173-183.

Englert, F., Truffin, C. & Gastmans, R. (1976). "Conformal invariance in quantum gravity", Nucl. Phys. B117, pp. 407-432.

• Off-shell counting has been introduced in

de Wit, B. & Ferrara, S. (1979). "On higher-order invariants in extended supergravity", Phys. Lett. 81B, pp. 317-20.

For some applications, see de Wit & van Holten (1979) and

Sohnius, M.F., Stelle, K.S. & West, P.C. (1980). "Off-mass-shell formulation of extended supersymmetric gauge theories", Phys. Lett. 92B, pp. 123-127.

Siegel, W. (1981). "On-shell O(N) supergravity in superspace", Nucl. Phys. B177, pp. 325-332.

• For a more standard discussion of currents and improvement terms, see for instance Wess (1960), Mack & Salam (1969), and

Callan, C.G., Coleman, S. & Jackiw, R. (1970). "A new improved energy-momentum tensor", Ann. Phys. (N.Y.) 59, pp. 42-73.

Coleman, S. & Jackiw, R. (1971). "Why dilatation generators do not generate dilatations", Ann. Phys. (N.Y.) 67, pp. 552-598.

• The supersymmetry transformation (6.8) has been considered by

Wess, J. & Zumino, B. (1974). "Supergauge transformations in four dimensions", Nucl. Phys. B70, pp. 39-50.

• Multiplets of currents have been constructed for various theories in

Ferrara, S., & Zumino, B. (1975). "Transformation properties of the supercurrent", Nucl. Phys. B87, pp. 207-20.

Ogievetsky, V., & Sokatchev, E. (1977). "On a vector superfield generated by the supercurrent", Nucl. Phys. B124, pp. 309-316; (1978). Yad. Fiz. 28, p. 1631.

Sohnius, M. (1978). "The multiplet of currents for N=2 extended supergravity", Phys. Lett. 81B, pp. 8-10.

Bergshoeff, E., de Roo, M. & de Wit, B. (1981).

Howe, P. & Lindström, U. (1981). "The supercurrent in five dimensions", Phys. Lett. 103B, pp. 422-426.

Howe, P., Stelle, K.S. & Townsend P.K. (1981). "Supercurrents", Nucl. Phys. B192, pp. 332-352.

Bergshoeff, E., de Roo, M., de Wit, B. & van Nieuwenhuizen, P. (1982). "Ten-dimensional Maxwell-Einstein supergravity, its currents, and the issue of its auxiliary fields, Nucl. Phys. B195, pp. 97-136.

Bergshoeff, E. & de Roo, M. (1982). "The supercurrent in ten dimensions", Leiden preprint.

• The application of conformally invariant formulations to N=1 Poincaré supergravity and the various off-shell representations indicated in Fig. 1 are discussed in the following references:

Breitenlohner, P. (1977a). "A geometric interpretation of local supersymmetry", Phys. Lett. 67B, pp. 49-51; (1977b). "Some invariant lagrangians for local supersymmetry", Nucl. Phys. B124, pp. 500-510.

Ferrara, S. & van Nieuwenhuizen, P. (1978a). "The auxiliary fields of supergravity", Phys. Lett. 74B, pp. 333-335; (1978b). "Tensor calculus for supergravity", Phys. Lett. 76B, pp. 404-408.

Stelle, K.S. & West, P.C. (1978). "Auxiliary fields for supergravity", Phys. Lett. B74, pp. 330-332.

Kaku, M. & Townsend, P.K. (1978). "Poincaré supergravity as broken superconformal gravity", Phys. Lett. 76B, pp. 54-58.

Das, A., Kaku, M. & Townsend, P.K. (1978). "Unified approach to matter coupling in Weyl and Einstein supergravity", Phys. Rev. Lett. 40, pp. 1215-1218.

Ferrara, S., Grisaru, M. & van Nieuwenhuizen, P. (1978). "Poincaré and conformal supergravity models with closed algebras", Nucl.

Phys. B138, pp. 430-444.

Siegel, W. & Gates, S.J. (1979). "Superfield supergravity", Nucl. Phys. B147, pp. 77-104.

Sohnius, M.F. & West, P.C. (1981). "An alternative minimal off-shell version of N=1 supergravity", preprint ICTP 80-81/37.

de Wit, B. & Roček, M. (1981). "Improved tensor multiplets", preprint NIKHEF-H/81-28.

• A comprehensive study of N=2 supergravity based on the superconformal approach can be found in

de Wit, B., van Holten, J.W. & Van Proeyen, A. (1981). "Structure of N=2 supergravity", Nucl. Phys. B184, pp. 77-108.

For alternative treatments of N=2 supergravity, see

Fradkin, E.S., & Vasiliev, M.A. (1979a). "Minimal set of auxiliary fields and S-matrix for extended supergravity", Lett. Nuovo Cim. 25, pp. 79-87; (1979b). "Minimal set of auxiliary fields in SO(2)-extended supergravity", Phys. Lett. 85B, pp. 47-51.

Breitenlohner, P. & Sohnius, M.F. (1980). "Superfields, auxiliary fields and tensor calculus for N=2 extended supergravity", Nucl. Phys. B165, pp. 483-510; (1981). "An almost simple off-shell version of SU(2) Poincaré supergravity", Nucl. Phys. B178, pp. 151-176.

• Central charges in the supersymmetry algebra have been discussed by

Haag, R., Lopuszanski, J.T. & Sohnius, M.F. (1975). "All possible generators of supersymmetries of the S-matrix", Nucl. Phys. B88, pp. 257-274.

Sohnius, M.F., Stelle, K.S. & West, P.C. (1980). "Supersymmetric Yang-Mills theories", in Proc. Europhysics Study Conf. on "Unification of Fundamental Interactions", Erice, 1980, eds. J. Ellis, S. Ferrara, & P. van Nieuwenhuizen, Plenum Press, New York, pp. 187-244.

For superconformal central charges, see

de Wit, B., van Holten, J.W. & Van Proeyen, A. (1980b). "Central charges and conformal supergravity", Phys. Lett. 95B, pp. 51-55.

• For discussions of supermultiplets, see the lectures in this volume by S. Ferrara and J.G. Taylor, and, for instance,

Salam, A., & Strathdee, J. (1974). "Unitary representations of super-gauge symmetries", Nucl. Phys. B80, pp. 499-505; (1975). "SU(6) and supersymmetry", Nucl. Phys. B84, pp. 127-131.

Nahm, W. (1978). "Supersymmetries and their representations", Nucl. Phys. B135, pp. 149-166.

Taylor, J.G. (1980). "On representations of extended supersymmetry algebras on superfields", Nucl. Phys. B169, pp. 484-500.

Lopuszanski, J.T. & Wolf, M. (1980). "Central charges in the massive supersymmetric quantum theory of scalar spinor and scalar-spinor-vector fields", preprint MPI-PAE/PTh 41/80.

Ferrara, S., Savoy, C.A. & Zumino, B. (1981). "General massive multiplets in extended supersymmetry". Phys. Lett. 100B, pp. 393-398.

Rittenberg, V. & Sokatchev, E. (1981). "Decomposition of extended superfields into irreducible representations of supersymmetry", Bonn preprint Bonn-HE-81-5.

Siegel, W. & Gates, S.J. (1981). "Superprojectors", Nucl. Phys. B189, pp. 295-316.

• A defining condition for the linear multiplet is that its lowest dimensional component transforms according to the adjoint representation of SU(N), and that its supersymmetry variation contains no spinors in symmetric SU(N) representations. For N=4 this multiplet contains 384+384 degrees of freedom with highest spin 3. Its full decomposition was obtained by Howe, Stelle & Townsend (1981) and by Bergshoeff, de Roo, de Wit & Rocek (unpublished). Presumably a nonlinear multiplet can also be found for N=4; its w=0 component is restricted to an element of SU(4).

• The fact that representations are not always preserved in the coupling to supergravity was first found in

Fischler, M. (1979). "Globally supersymmetric multiplets without local extensions", Phys. Rev. D20, pp. 1842-1845.

Usually this is discussed within the context of superspace. See, for instance, Gates & Siegel (1980), and Gates, Stelle & West (1980). The preservation of representations is related to the question raised in section 8 whether certain relations between supermultiplets remain valid in the presence of supergravity. A nonlinear relation of this type, which was given in (8.16), has been discussed by de Wit & Roček (1981).

DIMENSIONAL REDUCTION IN FIELD THEORY AND HIDDEN SYMMETRIES IN EXTENDED SUPERGRAVITY

E. CREMMER
Laboratoire de Physique Théorique de l'Ecole Normale Supérieure
75005 PARIS - FRANCE

The idea of a 4 + N dimensional space-time is by no means a new one. In 1921, already, Kaluza suggested that gravitation and electromagnetism could be unified in a 5 dimensional theory of gravity. This idea has been revived several times, in particular, in connection with the possible unification of gravitation with gauge fields, or in the context of the fiber bundle approach to Yang-Mills theories trying to associate the extra coordinates with the group space. It has also been put forward in the context of dual models which are consistent only in a precise space-time dimension (D = 10 or D = 26 for the known models) (Scherk & Schwarz 1975 ; Cremmer & Scherk 1976). It is then possible to associate to the extra coordinates a compact space (product of N torus in this case) such that the ordinary 4-dimensional physics is a low energy approximation of a bigger theory. We call the 4-dimensional theory the dimensional reduction of the 4 + N ones. We shall concentrate in these lectures essentially on this aspect forgetting about the possible interpretation (or existence) of the extra dimensions. This applies to most of the attempts to unify gravity and Yang-Mills, the concept of low energy approximation being most of the time replaced by the requirement of specific symmetries of the solution of field equations.

The simplest dimensional reduction has been particularly fruitful for supersymmetric theories, especially extended supersymmetric Yang-Mills or extended supergravities. It has made connection between N = 4 Yang-Mills in 4 dimensions and N = 1 Yang-Mills in 10 dimensions, and N = 8 supergravity in 4 dimensions and N = 1 supergravity in 11 dimensions. Moreover, the dimensional reduction explains part of the hidden symmetries in extended supergravities.

In these lectures, we shall discuss first the dimensional reduction of theories which do not include gravitation and then proceed in the second part with the dimensional reduction of theories including gravitation. In particular, we shall describe the 11-dimensional supergravity and its reduction to 4 dimensions. The hunt for the hidden symmetries, global E_7 and local SU(8) of the N = 8 supergravity in 4 dimensions will be described in part III. These hidden symmetries shall provide geometrical meaning to scalar fields. This will be a property of all extended supergravities and will be discussed in part IV. Finally in part V we shall summarize what we know or would like to know about N = 8 supergravity at the classical and quantum level. We shall discuss the possible implications of these hidden symmetries.

I DIMENSIONAL REDUCTION WITHOUT GRAVITATION

The simplest example of dimensional reduction arises when we try to interpret a theory in 4 + N dimensions whose Lagrangian is Poincaré invariant. An interesting case is that of Yang-Mills theory in 4 + N dimensions which leads, via dimensional reduction, to a Yang-Mills + Higgs scalars coupled theory with specific couplings. However, this will lead to unified theory only in the case of supersymmetric theories which requires the study of the supersymmetry algebra in D dimensions. The dimensional reduction of supersymmetric Yang-Mills in 10 dimensions leads to the well-known N = 4 supersymmetric Yang-Mills in 4 dimensions.

1 Interpretation of extra dimensions

Let us start with a scalar theory in 4 + N dimensions whose Lagrangian $\mathcal{L}$ is Poincaré invariant (Cremmer & Scherk 1976 a)

$$S = \int d^{4+N}x \left[\frac{1}{2}\partial_M\phi\,\partial^M\phi - \frac{1}{2}\mu_0^2\phi^2 + \mathcal{V}(\phi)\right] \tag{1.1}$$

It is consistent with the metric $\eta_{NM} = (+-\ldots.-)$ to assume that the extra dimensions are circles of length $L_1 \ldots L_N$ or, denoting $x_M = (x_\mu, y_i)$ (i = 1...N), to assume that

$$\phi(x_\mu, y_i + L_i) = \phi(x_\mu, y_i) \tag{1.2}$$

This breaks "spontaneously" the Poincaré invariance of the <u>action S</u> : P_{4+N} to $P_4 \times U(1)^N$. $U(1)^N$ will be associated with the conservation of N "heaviness" numbers. We can now expand $\phi(x_\mu, y_i)$ in Fourier series

$$\phi(x_\mu, y_i) = \frac{1}{(L_1 \ldots L_N)^{1/2}} \sum_{\{n_i\}} \phi_{\{n_i\}}(x_\mu)\, \exp\left(2i\pi \sum_i \frac{y_i n_i}{L_i}\right) \tag{1.3}$$

with n_i integer. If ϕ is real, we have $\phi_{\{n_i\}} = \phi^*_{\{-n_i\}}$. We can integrate over y_i and obtain a 4-dimensional description of this theory

$$S = \int d_4x \left\{ \sum_{\{n_i\}} \left[\frac{1}{2} \partial_\mu \phi^*_{\{n_i\}} \partial^\mu \phi_{\{n_i\}} - \frac{1}{2} m^2_{\{n_i\}} \phi^*_{\{n_i\}} \phi_{\{n_i\}} + \text{"} V(\phi_{\{n_i\}}) \text{"} \prod_i \delta\left(\sum_\alpha n_i^\alpha\right) \right] \right\} \tag{1.4}$$

with $$m^2_{\{n_i\}} = \mu_0^2 + 4\pi^2 \sum_i \frac{n_i^2}{L_i^2} \tag{1.5}$$

"$\prod_i \delta(\sum_\alpha n_i^\alpha)$" symbolizes the conservation of the N heaviness numbers. S describes now an infinite number of interacting scalar particles.

The ultraviolet behaviour is the same as in 4 + N dimensions. This can be shown by using properties of Jacobi θ-functions. The infrared behaviour is the same as in 4 dimensions : if we start with 1 massless particle in 4 + N dimensions, we get 1 massless particle in 4 dimensions plus an infinite number of massive particles.

There are two limiting cases : if all the L_i's $\to \infty$: we recover the original theory in 4 + N dimensions. If all the L_i 's $\to 0$: only one state keeps a finite mass classically. This limit is associated with the process called "dimensional reduction". This remaining state, ϕ , is described by the "reduced action"

$$S_{(4)} = \int d_4x \left[\frac{1}{2} \partial_\mu \phi \partial^\mu \phi - \frac{1}{2} \mu_0^2 \phi^2 + V(\phi) \right] \tag{1.6}$$

where the coupling constants have been rescaled before the limit $L_i \to 0$ according to their canonical dimensions.

This is equivalent to retaining only the mode independent of y_i or to imposing the "symmetry conditions"

$$\frac{\partial \phi}{\partial y_i} = 0$$

This ansatz, together with the equations of motion in 4 + N dimensions, leads to equations which are derivable from a reduced Lagrangian L. L is identical with $\mathcal{L}$ where we have dropped the y_i dependence and we have done some canonical rescaling to the scalar fields and the coupling constants.

Remarks : we could have started with a field theory in curved space, for example, S x M_4 where S is a compact space of dimension N. If G is the symmetry group of S, we can expand a field $\phi(x_\mu, y_i)$ in "G harmonics" on S $Y_N(y_i)$

$$\phi(x_\mu, y_i) = \sum_N Y_N(y_i)\, \phi_N(x_\mu) \tag{1.7}$$

Φ_N will describe a scalar particle of mass $\mu_0^2 + \frac{C(N)}{L^2}$ where C(N) is the eigenvalue of the Laplace-Beltrami operator for S in the representation N, L being some length which characterizes S. If some of the C(N)'s are zero we can make a consistent truncation by letting $L_i \to 0$. The corresponding ansatz on Φ

$$\Phi(x_\mu, y_i) = \sum_{N_0} Y_{N_0}(y_i)\, \Phi_{N_0}(x_\mu) \qquad \text{with } C(N_0) = 0$$

is such that the y_i dependence factorizes partially from the 4 + N-dimensional equations of motion implying several 4-dimensional equations. The group G is then interpreted as an internal symmetry group, S can be the space of G itself or a coset space G/H (dim S = dim G - dim H)

2 Dimensional reduction of non abelian Yang-Mills theory

Let us consider the non abelian Yang-Mills theory in 4 + N dimensions described by the Lagrangian

$$S = -\frac{1}{4}\int d^{4+N}x \ \mathrm{Tr}\,(F_{MN}^2) \tag{1.8}$$

where η_{NM} the metric is (+-...-) and F_{MN} the field strength

$$F_{MN} = \partial_M A_N - \partial_N A_M + i g_0 [A_M, A_N] \tag{1.9}$$

A_M and F_{MN} are in the Lie algebra of the group G.

Let us now apply the process of dimensional reduction by implementing the condition

$$\partial A_M / \partial y_i = 0$$

The 4 + N components of A_M will split into two parts : a 4-dimensional vector A_μ and N scalars A_i. After canonical rescalings, we obtain the corresponding 4-dimensional action

$$S_4 = \int d_4x \ \mathrm{Tr}\left\{ -\frac{1}{4}F_{\mu\nu}^2 + \frac{1}{2}(D_\mu A_i)^2 + \frac{1}{4}g^2 [A_i, A_j]^2 \right\}$$

where $D_\mu A_i = \partial_\mu A_i + i g [A_\mu, A_i]$ (1.10)

Properties of the reduced theory :

(1) It contains both vector and scalar particles (in the adjoint representation of G)

(2) The global internal symmetry is O(N) instead of $U(1)^N$ if

we had kept all the modes of 4 + N dimensions. We have now the breaking

$$P_{4+N} \longrightarrow P_4 \otimes O(N) \tag{1.11}$$

(3) There is a specific Higgs coupling g^2 related to the gauge coupling g as well as a very specific structure of the potential

(4) The signature of the extra dimensions of space time (space like) is such that we obtain the right sign for the kinetic term of the scalar fields A_i

Important remark : This is not a unified theory of vector and scalar particles. There is no symmetry relating A_μ and A_i (No go theorem of Coleman-Mandula) which implies the relation between the Higgs coupling constant and the gauge coupling constant. Explicit calculations at the quantum level show that this relation is not preserved by radiative corrections.

This problem will be overcome if we start with supersymmetric theories which can relate scalar, vector and spinor particles.

3 Supersymmetry in D dimensions

Unified theories and less trivial examples of dimensional reduction are obtained if we start with a theory invariant under simple supersymmetry algebra in D dimensions

$$\{ Q_{\hat{\alpha}}, \bar{Q}_{\hat{\beta}} \} = 2 (\Gamma^M)_{\hat{\alpha}\hat{\beta}} P_M \tag{1.12}$$

where the D matrices Γ^M have dimensions $2^{[D/2]}$ and satisfy the Clifford algebra

$$\{ \Gamma^M, \Gamma^N \} = 2 \eta^{NM} \tag{1.13}$$

They have the hermiticity properties

$$(\Gamma^0)^\dagger = \Gamma^0 , \qquad (\Gamma^M)^\dagger = - \Gamma^M \qquad \text{for } M \neq 0$$

or $$(\Gamma^M)^\dagger = \Gamma^0 \Gamma^M \Gamma^0 \qquad \forall M$$

The possible properties of $Q_{\hat{\alpha}}$: Majorana, Weyl, Dirac or Majorana-Weyl will depend on the dimension D of space-time (Gliozzi et al. 1977)

D even :

(1) Since Γ^M and $(\Gamma^M)^*$ satisfy the same Clifford algebra, there exists a matrix B such that

$$(\Gamma^M)^* = -B\Gamma^M B^{-1} \tag{1.14}$$

we can fix the phases such that $BB^* = \epsilon I$ with $\epsilon = \pm 1$. A Majorana spinor will be defined as being its own antiparticle $B^{-1}\psi^* = \psi$ therefore it exists only if $\epsilon = +1$.

(2) Since Γ^M and $(\Gamma^M)^t$ satisfy also the same Clifford algebra, there exists a matrix C (charge conjugation) such that

$$(\Gamma^M)^t = -C\Gamma^M C^{-1} \tag{1.15}$$

Together with the hermiticity properties this implies (after a phase choice)

$$B^t = C\Gamma^0 \quad , \quad BB^\dagger = I \tag{1.16}$$

and consequently $\quad B = \epsilon B^t \quad , \quad C = -\epsilon C^t$

Defining $\Gamma^{(n)}$ as the antisymmetrized product of n matrices, we therefore have

$$(C\Gamma^{(n)})^t = \epsilon\,(-1)^{\frac{(n-1)(n-2)}{2}}\,C\Gamma^{(n)} \tag{1.17}$$

This allows us to count the number of antisymmetric matrices $2^{D/2} \times 2^{D/2}$ which should be $1/2\; 2^{D/2}(2^{D/2} - 1)$. This gives

$$\epsilon = -\sqrt{2}\cos\frac{\pi}{4}(D+1) \qquad \text{or} \qquad \begin{cases} \epsilon = +1 \text{ for } D = 2,4 \bmod 8 \\ \epsilon = -1 \text{ for } D = 0,6 \bmod 8 \end{cases} \tag{1.18}$$

For D = 2,4 mod 8, there exists a pure imaginary representation of Γ matrices. We can choose B = 1, C = Γ^0 , then a Majorana spinor is simply a real spinor.

For D = 0,6 mod 8, we can however define "Majorana spinors" when there is an internal symmetry (extended supersymmetry).

The matrix Γ^{D+1} of square 1 and anticommuting with the D Γ matrices is given by

$$\Gamma^{D+1} = (-1)^{\frac{D-2}{4}}\,\Gamma^0\Gamma^1 \ldots \Gamma^{D-1} \tag{1.19}$$

A Weyl spinor λ is defined by

$$\Gamma^{D+1}\lambda = \pm\lambda \tag{1.20}$$

We can have a Majorana-Weyl spinor if Γ^{D+1} is real, i.e. for D = 2 mod 8 (in particular D = 10)

<u>D odd</u> : D = d + 1 with d even

A Clifford algebra for d + 1 is obtained from the one in d

dimensions by adding to the Γ^M (M = 0, ... d - 1) the matrix

$$\Gamma^d = i(-1)^{\frac{d-2}{4}} \Gamma^0 \Gamma^1 \ldots \Gamma^{d-1} \quad (\equiv i \Gamma^{d+1}) \tag{1.21}$$

such that $(\Gamma^d)^2 = -1$.

Therefore, we shall have the possibility of Majorana spinors if Γ^d is pure imaginary when $\Gamma^0, \ldots \Gamma^{d-1}$ are pure imaginary or equivalently if Γ^{d+1} is real : there are Majorana spinors in d + 1 dimensions if there are Majorana-Weyl spinors in d dimensions, namely D = 3 mod 8

Note : these properties of the spinors depend only on the signature of space-time s-t (the metric being $(\underbrace{++\ldots+}_{s}, \underbrace{-\ldots-}_{t})$)

Dimensional reduction : Starting from a Clifford algebra in 4 + N dimensions, we can always define it (up to an equivalence) as a tensor product of γ matrices 4 x 4 by "internal" matrices $\tilde{\gamma}$ $2^{[N/2]} \times 2^{[N/2]}$ such that

$$\Gamma^\mu = \gamma^\mu \otimes (1 \text{ or } \Omega) \tag{1.22}$$

$$\Gamma^i = (1 \text{ or } \gamma^5) \otimes \tilde{\gamma}^i \tag{1.23}$$

Therefore, a Dirac spinor in 4 + N dimensions, through the ordinary dimensional reduction, is equivalent to a $2^{[N/2]}$ Dirac spinor in 4 dimensions or $2.2^{[N/2]}$ Majorana or Weyl spinors in 4 dimensions. There will be a reduction factor : 1/2 if we start with Majorana or Weyl spinors and 1/4 if we start with Majorana-Weyl spinors when they exist. These results are summarized in Table I giving the number of Majorana or Weyl spinors in 4 dimensions corresponding to a given spinor in D dimensions.

We see from Table I, that starting from a Majorana-Weyl spinor in 10 dimensions and reducing to 9, we should obtain a 4 component-spinor although there are no Majorana or Weyl spinors. This means that in 9 dimensions there exists another kind of spinors (pseudo-Majorana) which have half the number of the components of a Dirac spinor (Van Nieuwenhuizen

TABLE I

D	4	5	6	7	8	9	10	11	12
DIRAC	2	2	4	4	8	8	16	16	32
MAJ.	1	-	-	-	-	-	8	8	16
WEYL	1	-	2	-	4	-	8	-	16
M - W	-	-	-	-	-	-	4	-	-

1980)

4 Supersymmetric Yang-Mills in D = 10, N = 1 and Supersymmetric Yang-Mills in D = 4, N = 4

From the previous table, it can be seen that the maximal dimension in which we can have a supersymmetric Yang-Mills theory (with spin ≤ 1) is D = 10. For $D > 10$, the dimensional reduction would lead to particles with spin $\geq 3/2$ or equivalently with $N > 4$ supersymmetry.

Supersymmetric Yang-Mills in D = 10, N = 1 (Gliozzi et al. 1977 ; Brink et al. 1977). A vector field in 10 dimensions has 8 degrees of freedom as a Majorana-Weyl spinor field. Therefore it seems plausible that there could exist a supersymmetric Yang-Mills theory with only 1 vector field and a Majorana-Weyl spinor field. It is in fact possible and the Lagrangian is given by

$$S = \int d^{10}x \; \mathrm{Tr}\left[-\frac{1}{4} F_{MN} F^{MN} + \frac{i}{2} \bar{\lambda} \Gamma^M D_M \lambda \right] \qquad (1.24)$$

where

$$F_{MN} = \partial_M A_N - \partial_N A_M + i g_0 [A_M, A_N]$$

$$D_M \lambda = \partial_M \lambda + i g_0 [A_M, \lambda]$$

A_M and λ belong to the Lie algebra of a group G. The Γ_M's are 32 x 32 matrices. λ is a Majorana-Weyl spinor and satisfies $\lambda = \Gamma_{11} \lambda$

S is invariant under the following supersymmetry transformations

$$\delta A_M = i \bar{\epsilon} \Gamma_M \lambda \quad , \quad \delta \lambda = 2 F_{MN} \Gamma^{MN} \epsilon \qquad (1.25)$$

with $\epsilon = \Gamma_{11} \epsilon$

Let us note that the same action is supersymmetric for D = 6 if λ is a Weyl spinor and D = 4 if λ is a Majorana spinor.

Supersymmetric Yang-Mills in D = 4, N = 4 : It is obtained by dimensional reduction of the supersymmetric Yang-Mills theory in D = 10. Let us define 6 real 4 x 4 independent antisymmetric matrices $(\alpha^i)_{ab}$, $(\beta^i)_{ab}$ (i = 1,2,3) which satisfy the algebra of $O(4) \sim SU(2) \times SU(2)$

$$\{\alpha^i, \alpha^j\} = \{\beta^i, \beta^j\} = -2\delta^{ij}$$

$$[\alpha^i, \beta^j] = 0 \, , \; [\alpha^i, \alpha^j] = -2\epsilon^{ijk} \alpha^k \, , \; [\beta^i, \beta^j] = -2\epsilon^{ijk} \beta^k$$

$$\frac{1}{2}\varepsilon^{abcd}(\alpha^i)_{cd} = (\alpha^i)_{ab} \quad ; \quad \frac{1}{2}\varepsilon^{abcd}(\beta^i)_{cd} = -(\beta^i)_{ab} \qquad (1.26)$$

We can then write the Γ matrices in 10 dimensions and in the Majorana representation as

$$\Gamma^\mu = \gamma^\mu \otimes \begin{pmatrix} I_4 & 0 \\ 0 & -I_4 \end{pmatrix} \qquad \mu = 0,1,2,3$$

$$\Gamma^{3+i} = i\, I_4 \otimes \begin{pmatrix} 0 & \beta^3\alpha^i \\ \beta^3\alpha^i & 0 \end{pmatrix}$$

$$\Gamma^{6+i} = \gamma_5 \otimes \begin{pmatrix} \beta^i & 0 \\ 0 & \beta^3\beta^i\beta^3 \end{pmatrix} \qquad (1.27)$$

therefore

$$\Gamma^{11} = I_4 \otimes \begin{pmatrix} 0 & \beta^3 \\ -\beta^3 & 0 \end{pmatrix} \qquad (1.28)$$

A Majorana-Weyl spinor is then written as

$$\lambda = (\lambda_a , -\beta^3_{ab}\lambda_b) \qquad a,b = 1,2,3,4 \qquad (1.29)$$

λ_a being a 4 dimensional Majorana spinor.

As has been done previously, we split the 10 components of the vector fields $A_M \rightarrow (A_\mu, A_i, B_i)$. With the previous decomposition of the spinor field, this immediately gives the reduced Lagrangian

$$\begin{aligned} S_R = \int d_4x \, \mathrm{Tr}\Big\{ & -\frac{1}{2}F_{\mu\nu}F^{\mu\nu} + \frac{1}{2}(D_\mu A_i)^2 + \frac{1}{2}(D_\mu B_i)^2 \\ & + \frac{i}{2}\bar\lambda_a\gamma^\mu D_\mu\lambda_a + \frac{g}{2}\bar\lambda_a[\alpha^i_{ab}A_i + i\gamma_5\beta^i_{ab}B_i , \lambda_b] \\ & + \frac{g^2}{4}([A_i, A_j]^2 + [B_i, B_j]^2 + 2[A_i, B_j]^2) \Big\} \end{aligned} \qquad (1.30)$$

A_i and B_i are respectively scalar and pseudoscalar fields. The reduction of the transformation of supersymmetry $\epsilon = \mathrm{cste}$ is consistent with $\partial\epsilon/\partial y_i = 0$. Then we get 4 supersymmetries with parameters ϵ_a from

$$\epsilon = (\epsilon_a , -\beta^3_{ab}\epsilon_b)$$

S_R is then invariant under the N = 4 supersymmetry transformations

$$\delta A_\mu = 2i\,\bar\epsilon^a\gamma_\mu\lambda^a$$

$$\delta A_i = -2\bar{\epsilon}^a \alpha^i_{ab}\lambda^b \quad , \quad \delta B_i = 2i\bar{\epsilon}^a\gamma_5 \beta^i_{ab}\lambda^b$$

$$\delta\lambda^a = 2F_{\mu\nu}\gamma^{\mu\nu}\epsilon^a + 4i(D_\mu A_i \alpha^i_{ab} + i\gamma_5 D_\mu B_i \beta^i_{ab})\gamma_\mu \epsilon^b$$
$$-ig\,(2\epsilon_{ijk}[A^i, A^j]\alpha^k_{ab} + 2\epsilon_{ijk}[B^i, B^j]\beta^k_{ab}$$
$$-2i\gamma_5[A^i, B^j](\alpha^i\beta^j)_{ab})\epsilon^b \qquad (1.31)$$

It is also invariant as expected under a O(6)$\sim$ SU(4) global symmetry of parameters $\Lambda_{ij}=-\Lambda_{ji}$, $\Lambda'_{ij}=-\Lambda'_{ji}$ and $\tilde{\Lambda}_{ij}$

$$\delta A_\mu = 0$$
$$\delta A_i = \Lambda'_{ij}A_j - \tilde{\Lambda}_{ij}B_j \quad , \quad \delta B_i = \Lambda_{ij}B_j + \tilde{\Lambda}_{\ell i}A_\ell$$
$$\delta\lambda_a = -\tfrac{1}{4}[\varepsilon^{ijk}\beta^k_{ab}\Lambda_{ij} + \varepsilon^{ijk}\alpha^k_{ab}\Lambda'_{ij} + i\gamma_5(\alpha^i\beta^j)_{ab}\tilde{\Lambda}_{ij}]\lambda_b \qquad (1.32)$$

The ordinary formulation of N = 4 Yang-Mills, as it is described by the representation of N = 4 supersymmetry, is made with the scalar fields A_{ab} and B_{ab} respectively self-dual and antiself-dual, defined by

$$A_{ab} = \alpha^i_{ab}A_i \quad , \quad B_{ab} = \beta^i_{ab}B_i \qquad (1.33)$$

Since this multiplet (A_μ ,A_i, B_i, λ_a) is an irreducible representation of N = 4 supersymmetry, we now really have a unified theory of vector, scalar and spinor particles. All relations between coupling constants are dictated by the N = 4 supersymmetry.

This conformal invariant theory has remarkable renormalizability properties. It has been shown that the β function is zero up to three-loop order and that there exists, at the one loop order, a gauge in which the theory is finite (Grisaru 1981). It has been conjectured that β = 0 to all orders and that the theory is finite. This could be linked with the fact that the N = 4 Yang-Mills supermultiplet is CPT self-conjugate and that the global symmetry is SU(4) and not U(4).

II DIMENSIONAL REDUCTION WITH GRAVITATION

When we consider theories in higher dimensions which include gravity (or equivalently which are invariant under local transformations of coordinates) their interpretation in 4 dimensions leads to the concept of spontaneous compactification of space-time. The 4-dimensional theories obtained by dimensional reduction have new features. As previously, really unified theories are obtained if we start now with supergravity theories in higher dimensions (for example D = 11).

1 Spontaneous compactification of space-time

Let us start with a theory with gravitation in 4 + N dimensions. As previously, if we want to interpret it in 4 dimensions, we need the background 4 + N space-time to be a product of a compact space of dimension N by an ordinary 4 dimensional space-time. But now this requirement should follow from the field equations (for the metric g_{MN} and eventually the other fields) : This is called spontaneous compactification of space-time (Cremmer & Scherk 1977a ; Cremmer et al. 1977 b ; Luciani 1978 ; Palla 1979).

Then the theory is expanded around this particular solution using "harmonic" functions for the invariance group G of this compact space of N dimensions. The problem of showing the stability of such a solution is in general difficult. If there are several possible classical solutions, the choice between them is not obvious since the energy is not well defined. We could even have solutions for various decompositions of the 4 + N space-time : boundary conditions could eventually choose between them. We could also ask that only quantum corrections provide such a spontaneous compactification, but we would need the 4 + N dimensional theory to have a meaning at the quantum level (finite theory ?).

It has been shown that such solutions exist. In particular, in the system Einstein + Yang-Mills, it has been shown that we can require that the 4-dimensional space-time should be flat and the internal space a sphere S_N. If we can make the limit $L \rightarrow 0$, ($L \sim$ size of the compact space) and keep some fields with a finite (or zero) 4-dimensional mass, we can truncate the theory in a consistent way and deduce a theory in 4 dimensions with a finite number of fields. This is equivalent to retaining solutions which have a specific property of symmetry.

The most simple dimensional reduction consists in compactifying on a product of torus (this is always consistent with the equations of motion and is, as previously, dictated only by boundary conditions) and letting the size of the torus go to zero. This is equivalent to assuming that the fields do not depend on the extra-coordinates. This is the ordinary dimensional reduction. To 1 degree of freedom in 4 + N dimensions corresponds 1 degree of freedom in 4 dimensions. This should not be the case for other compactified spaces like a sphere for example.

2 Dimensional reduction

As we have already seen, the solutions of 4 + N dimensional equations of motion which satisfy $\partial\phi/\partial y_i = 0$ can be derived from the Lagrangian $\int d_4x \sqrt{g(x_\mu)}\; L(\phi(x_\mu))$ where L is identical with $\mathcal{L}$ up to some rescaling of the coupling constants and the fields in order to give them the canonical dimensions of a theory in 4 dimensions. These rescaling factors disappear after integration over $d^N y$. The world indices M will be split as previously into 4-dimensional world indices and internal indices.

In 4 + N dimensions, the theory has the complete reparametrization invariance under the local coordinates transformations

$$\begin{aligned} \delta x^M &= -\xi^M(x) \\ \delta\phi &= \xi^M \partial_M \phi \\ \delta A_M &= \xi^N \partial_N A_M + \partial_M \xi^N A_N \end{aligned} \qquad (2.1)$$

In 4 dimensions, after dimensional reduction, the remaining invariances will be those consistent with the condition

$$\partial_i \phi = \partial_i A_N = 0$$

$$* \; \delta\phi = \xi^\mu \partial_\mu \phi + \underline{\xi^i \partial_i \phi}$$

$$\partial_i(\delta\phi) = 0 \Rightarrow \partial_i \xi^\mu = 0$$

$$* \; \delta A_\mu = \xi^\nu \partial_\nu A_\mu + \partial_\mu \xi^\nu A_\nu + \partial_\mu \xi^i A_i + \underline{\xi^i \partial_i A_\mu}$$

$$\partial_i(\delta A_\mu) = 0 \Rightarrow \partial_j \partial_\mu \xi^i = 0$$

$$* \; \delta A_i = \xi^\nu \partial_\nu A_i + \underline{\partial_i \xi^\nu A_\nu} + \underline{\xi^j \partial_j A_i} + \partial_i \xi^j A_j$$

$$\partial_k(\delta A_i) = 0 \Rightarrow \partial_k \partial_i \xi^j = 0 \qquad (2.2)$$

The results are summarized in

$$\xi^\mu = \xi^\mu(x_\nu)\,,\quad \xi^i = a^i{}_j x^j + \xi^i(x_\nu)\,,\quad a^i{}_j = \text{Cste} \qquad (2.3)$$

The rescalings we had to make on the fields restrict to transformations which preserve the volume element $d^N y$ i.e. $a^i{}_i = 0$. These transformations correspond to the following symmetries in 4 dimensions

$$\begin{cases} \delta x^\mu = -\xi^\mu(x) & \text{Reparametrization invariance in 4 dimensions} \\ \delta x^i = -\xi^i(x) & U(1)^N \text{ local invariance} \\ \delta x^i = -a^i{}_j x^j & SL(N,R) \text{ global invariance} \end{cases}$$

3 Dimensional reduction of pure gravitation

(a) Let us apply the previous discussion to the reduction of the pure gravitation (Cho & Freund 1975 ; Cho & Jang 1975 ; Cremmer & Julia 1978, 1979 ; Scherk & Schwarz 1979 b). We start with the Einstein-Cartan formulation of gravitation described by the action (we have chosen K = 1)

$$S = -\frac{1}{4}\int d^{4+N}x \; e\, R(\omega, e) \tag{2.4}$$

where $e_M{}^A$ is the vielbein field, e = det $e_M{}^A$, ω_{MAB} is the connection for the local Lorenzt group SO(N + 3, 1)

$$R(\omega,e) = e^{MA} e^{NB} \left(\partial_M \omega_{NAB} - \partial_N \omega_{MAB} + \omega_{MAC}\, \omega_N{}^C{}_B - \omega_{NAC}\, \omega_M{}^C{}_B\right) \tag{2.5}$$

ω is an independent field which does not propagate. We can solve its equation of motion and obtain $\omega = \omega(e)$. The invariances are the reparametrization in 4 + N dimensions and the local SO(N + 3,1) Lorentz invariance. Let us now perform the dimensional reduction. Writing $e_M{}^A$ as

$$e_M{}^A = \begin{pmatrix} e_\mu{}^\alpha & e_\mu{}^a \\ e_m{}^\alpha & e_m{}^a \end{pmatrix} \qquad \begin{array}{l} \alpha \text{ indices for } SO(3,1) \\ a \text{ indices for } SO(N) \end{array}$$

Of course, we require $\partial_i e_M{}^A = 0$. Moreover, we break the local SO(3+N,1) invariance into SO(3,1) x SO(N) by imposing the condition : $e_m{}^\alpha = 0$. This condition does not restrict the invariances derived from the reparametrization invariance discussed in the previous section.

$$\begin{aligned} \delta e_m{}^\alpha &= \xi^\mu \partial_\mu e_m{}^\alpha + a_m{}^i e_i{}^\alpha = 0 \\ \delta e_\mu{}^\alpha &= \xi^\nu \partial_\nu e_\mu{}^\alpha + \partial_\mu \xi^\nu e_\nu{}^\alpha \\ \delta e_m{}^a &= \xi^\nu \partial_\nu e_m{}^a + a_m{}^i e_i{}^a \\ \delta e_\mu{}^a &= \xi^\nu \partial_\nu e_\mu{}^a + \partial_\mu \xi^i e_i{}^a + \partial_\mu \xi^\nu e_\nu{}^a \end{aligned} \tag{2.6}$$

Defining $B_\mu{}^i = e_a{}^i e_\mu{}^a \quad (e_a{}^i e_j{}^a = \delta^i_j)$, we obtain

$$\delta B_\mu^i = \xi^\nu \partial_\nu B_\mu^i + \partial_\mu \xi^\nu B_\nu^i + \partial_\mu \xi^i - a^i{}_j B_\mu^j \tag{2.7}$$

Therefore :

- $B_\mu{}^i$ are N vectors, gauge fields for $U(1)^N$ in the representation $\bar{N}$ of SL(N,R)

- $e_m{}^a$ are N^2 scalar fields, in the N representation of SL(N,R) (index m) and in the N representation of SO(N) (index a)

- $e_\mu{}^\alpha$ is the ordinary vierbein

We can define two tensor metrics invariant in the local transformations SO(3,1) x SO(N)

$$g_{\mu\nu} = \eta_{\alpha\beta}\, e_\mu{}^\alpha e_\nu{}^\beta \qquad \eta_{\alpha\beta} = (+---)$$

$$g_{ij} = \eta_{ab}\, e_i{}^a e_j{}^b \qquad \eta_{ab} = (--\dots-) \tag{2.8}$$

The complete metric tensor g_{MN} is then written in the Kaluza-Klein parametrization

$$g_{MN} = \begin{pmatrix} g_{\mu\nu} + B_\mu^k g_{k\ell} B_\nu^\ell & B_\mu^k g_{ik} \\ B_\nu^k g_{kj} & g_{ij} \end{pmatrix} \tag{2.9}$$

It has, in particular, the property

$$\det g_{MN} = \det g_{\mu\nu} \det g_{ij} \quad (\equiv g\,\Delta) \tag{2.10}$$

Because of the $U(1)^N$ gauge invariance, $B_\mu{}^k$ must appear in the reduced Lagrangian only through $G_{\mu\nu}{}^k$. The simplest way to obtain gauge invariant objects in 4 dimensions is to start with flat tensors in 4 + N dimensions which are scalar under reparametrization. Defining the anholonomy coefficients Ω_{ABC} by

$$[\partial_A, \partial_B] \equiv [e^M{}_A \partial_M, e^N{}_B \partial_N] = \Omega_{AB}{}^C \partial_C \tag{2.11}$$

we can rewrite S, after integration by parts, as

$$S = \frac{1}{16}\int d^{4+N}x \sqrt{g}\,[\Omega_{ABC}^2 - 2\,\Omega_{ABC}\,\Omega^{CAB} - 4\,\Omega_{CA}{}^A\,\Omega^C{}_B{}^B] \tag{2.12}$$

After dimensional reduction in the SO(3 + N,1) gauge $e_m{}^\alpha = 0$, the only non-vanishing coefficients are $\Omega_{\alpha\beta\gamma}$,

$$\Omega_{\beta\alpha c} = e_{ic}\, e^{\mu}{}_{\alpha}\, e^{\nu}{}_{\beta}\, G^{i}_{\mu\nu}$$

$$\Omega_{\alpha bc} = -\Omega_{b\alpha c} = -e^{i}{}_{b}\, e^{\mu}{}_{\alpha}\, \partial_{\mu} e_{ic}$$

Then, we immediately get the reduced action in 4 dimensions

$$S = -\frac{1}{4}\int d_4x \sqrt{g}\sqrt{\Delta}\Big\{ R - \frac{1}{4} g_{ij}\, G^{i}_{\mu\nu}\, G^{j}_{\rho\sigma}\, g^{\mu\rho} g^{\nu\sigma} + \frac{1}{4} g^{\rho\sigma}(g^{ik}g^{jl} - g^{il}g^{jk})\partial_{\rho} g_{ik}\partial_{\sigma} g_{jl}\Big\} \quad (2.13)$$

We can eliminate $\Delta^{1/2} = |\det g_{ij}|^{1/2}$ in front of R by a Weyl rescaling

$$e_{4\mu}{}^{\alpha} = e_{\mu}{}^{\alpha}\,\Delta^{1/4}$$

We finally obtain

$$S = -\frac{1}{4}\int d_4x \sqrt{g_4}\Big\{ R_4 - \frac{1}{4}\sqrt{\Delta}\, g_{ij}\, g_4^{\mu\rho} g_4^{\nu\sigma}\, G^{i}_{\mu\nu}\, G^{j}_{\rho\sigma} - \frac{1}{8} g_4^{\mu\nu}\partial_{\mu}\mathrm{Log}\,\Delta\,\partial_{\nu}\mathrm{Log}\,\Delta + \frac{1}{4} g_4^{\mu\nu}\partial_{\mu} g_{ij}\partial_{\nu} g^{ij}\Big\} \quad (2.14)$$

This describes 1 graviton, N gauge vector fields and N(N-1)/2 massless scalar fields. The invariances are the reparametrization in 4 dimensions, local $U(1)^N$ and global SL(N,R) as well as local SO(N) and SO(3,1) Lorentz hidden in the metrics g_{ij} and $g_{\mu\nu}$. As previously, it is not a unified theory of scalar, vector and tensor particles since no symmetry relates $g_{\mu\nu}$, g_{ij} and $B_{\mu}{}^{i}$. Let us remark that we can extend the SL(N,R) symmetry to GL(N,R) by adding the following scale transformations which preserve S

$$g_{ij} \to \lambda\, g_{ij}\ , \ (\Delta \to \lambda^{N}\Delta)\ ,\ B_{\mu}{}^{i} \to \lambda^{-\frac{1}{2}-\frac{N}{4}} B_{\mu}{}^{i} \quad (2.15)$$

(b) Structure of the scalar fields

Defining $\bar{g}_{ij}$ such that $\det \bar{g}_{ij} = 1$ or $g_{ij} = \bar{g}_{ij}\Delta^{1/N}$, the part of the Lagrangian which describes the self interaction of the scalar fields (together with their couplings to gravity) is written as

$$\mathcal{L}_S = \frac{1}{16}\int d_4x\, V_4\, g_4^{\mu\nu}\Big[\Big(\frac{1}{2}+\frac{1}{N}\Big)\partial_{\mu}\log\Delta\,\partial_{\nu}\log\Delta - \partial_{\mu}\bar{g}_{ij}\,\partial_{\nu}\bar{g}^{ij}\Big] \quad (2.16)$$

$\bar{g}_{ij}$ is a symmetric matrix of determinant equal to 1. It is an element of the coset space SL(N,R)/SO(N). This structure is better seen when we reintroduce a N-bein $\bar{e}_i^{\ a}$ such that

$$\bar{g}_{ij} = \bar{e}_i^{\ a} \eta_{ab} \bar{e}_j^{\ b} \tag{2.17}$$

$\bar{e}_i^{\ a}$ is an element of SL(N,R) defined only up to a SO(N) local transformation. If we forget for a moment about the local SO(N) transformations, a possible Lagrangian for $\bar{e}_i^{\ a}$ invariant under SL(N,R) is

$$L \sim - \partial_\mu \bar{e}_i^{\ a} \partial^\mu \bar{e}^i_{\ a} \qquad (\bar{e}^i_{\ a} \text{ inverse matrix of } \bar{e}_i^{\ a})$$

$$\sim \mathrm{Tr}[(\bar{e}^{-1}\partial_\mu \bar{e})^2] \qquad \text{where } \bar{e}^{-1}\partial_\mu \bar{e} \text{ is now an element of the Lie algebra of SL(N,R)}$$

It is invariant even under SL(N,R) x SL(N,R). Since SL(N,R) is non-compact, $\mathrm{Tr}[(\bar{e}^{-1}\partial_\mu \bar{e})^2]$ has both positive and negative terms. Therefore, L cannot be positive definite. This non-positivity problem can be solved as usual by introducing a local gauge invariance : SO(N) in this case. We now start from the gauge invariant Lagrangian

$$L \sim \mathrm{Tr}[(\bar{e}^{-1} D_\mu \bar{e})^2]$$

$$\text{with } D_\mu \bar{e}_i^{\ a} = \partial_\mu \bar{e}_i^{\ a} - \bar{e}_i^{\ b} \Omega_{\mu a}^{\ \ b} \tag{2.18}$$

$\Omega_{\mu a}^{\ \ b}$ is a gauge field for the local SO(N) invariance, it belongs to the Lie algebra of SO(N), i.e. is antisymmetric in a and b. However, we do not introduce any kinetic terms for Ω_μ so that the Lagrangian being quadratic in Ω_μ, we can solve its equations of motion

$$L \sim \mathrm{Tr}[(\bar{e}^{-1}\partial_\mu \bar{e} - \Omega_\mu)^2] \tag{2.19}$$

We can decompose $\bar{e}^{-1}\partial_\mu \bar{e}$ into two parts parallel and perpendicular to SO(N) with respect to the Killing metric i.e. in antisymmetric and symmetric part in a and b

$$\bar{e}_a^{\ i} \partial_\mu \bar{e}_i^{\ b} = (\bar{e}_a^{\ i} \partial_\mu \bar{e}_i^{\ b})_{/\!/} + (\bar{e}_a^{\ i} \partial_\mu \bar{e}_i^{\ b})_{\perp} \tag{2.20}$$

The equation for Ω_μ is solved immediately by

$$\Omega_{\mu a}^{\ \ b} = (\bar{e}_a^{\ i} \partial_\mu \bar{e}_i^{\ b})_{/\!/} \quad \left(= \tfrac{1}{2}(\bar{e}_a^{\ i} \partial_\mu \bar{e}_i^{\ b} - a \leftrightarrow b)\right) \tag{2.21}$$

so that after insertion in the Lagrangian, it becomes

$$L \sim \mathrm{Tr}[(\bar{e}^{-1}\partial_\mu \bar{e})_\perp^2] \qquad (2.22)$$

Since SO(N) is the maximal compact subgroup of SL(N,R), L is now positive definite. Although Ω_μ has disappeared, L is still gauge invariant. It is invariant under the following transformation on $\bar{e}$

$$\bar{e}_i{}^a \longrightarrow S_i{}^j \, \bar{e}_j{}^b \, \mathcal{O}_b{}^a(x) \qquad (2.23)$$

$S_i{}^j$ being a constant matrix of SL(N,R) and $\mathcal{O}_b{}^a(x)$ a local matrix of SO(N). With a little algebra, we can show that this Lagrangian is in fact only a function of the gauge invariant quantity $\bar{g}_{ij}$ and can be rewritten as

$$L \sim - \mathrm{Tr}(\partial_\mu \bar{g}\, \partial^\mu \bar{g}^{-1}) \qquad (2.24)$$

If we write $\bar{e}$ as exp w, where w is an element of the Lie algebra of SL(N,R), SL(N,R) and O(N) do not act linearly on w. The maximal subgroup which acts linearly on w is a diagonal subgroup isomorphic to SO(N)

$$\bar{e} \rightarrow \mathcal{O}\, \bar{e}\, \mathcal{O}^{-1}$$

This structure is a particular case of what is called G/H σ-model where now the scalar fields are defined on the coset space G/H, for example, the CP^{N-1} models are SU(N)/U(N-1) σ- models. We can use a non compact group G provided H is the maximal subgroup. Although the gauge invariance H could seem artificial, the N-bein formalism is necessary when we couple the theory to fermions. Moreover, we know from the properties of the 2-dimensional CP^{N-1} model, that it could eventually lead to non-trivial quantum effects : this local symmetry could become dynamical.

4 Dimensional reduction of coupled matter

Let us now consider some matter in 4 + D dimensions coupled to gravity : scalar matter and vector matter. This last one will exhibit some particular problems which will occur in a more complicated way later on in the dimensional reduction of 11-dimensional supergravity.

(a) Scalar matter

Let us start with the scalar Lagrangian

$$L_S = \int d^{4+N}x \, \sqrt{g}\, g^{MN} \partial_M \phi \, \partial_N \phi \qquad (2.25)$$

Assuming as usual $\partial_i \phi = 0$ and with the previous formula for g_{MN} we get immediately (taking into account the previous Weyl rescaling)

$$L_S^{(4)} = \int d_4 x \; V_4 \; g_4^{\mu\nu} \partial_\mu \phi \, \partial_\nu \phi \qquad (2.26)$$

(b) Vector matter

For simplicity, we shall consider only an abelian gauge field described by the Lagrangian

$$L_V = \int d^{4+N}x \sqrt{g}\, g^{MN} g^{PQ} F_{MP} F_{NQ} \tag{2.27}$$

L_V is invariant under the following transformations of A_M (together with corresponding transformations of g_{MN})

$$\delta A_M = \xi^N \partial_N A_M + \partial_M \xi^N A_N + \partial_M \Lambda \tag{2.28}$$

After dimensional reduction with $\partial_i \Lambda = 0$, we get

$$\delta A_\mu = \xi^\nu \partial_\nu A_\mu + \partial_\mu \xi^\nu A_\nu + \partial_\mu \xi^i A_i + \partial_\mu \Lambda$$

$$\delta A_i = \xi^\nu \partial_\nu A_i + a_i{}^j A_j \tag{2.29}$$

As already noticed, A_μ transforms not only under its own gauge group (Λ) but also under the $U(1)^N$ gauge group associated with the vector fields $B_\mu{}^i$. However, we can define a new vector field A'_μ which is invariant under this $U(1)^N$ gauge group. For this, we apply the general method which is to start with a flat tensor in 4 + N dimensions A_A

$$A_A = e_A{}^M A_M \tag{2.30}$$

and reduce it in 4 dimensions. We get

$$A_\alpha = e_\alpha{}^\mu A_\mu + e^i{}_\alpha A_i = e_\alpha{}^\mu (A_\mu - B_\mu{}^i A_i) \equiv e_\alpha{}^\mu A'_\mu$$

$$A_a = e_a{}^i A_i + \cancel{e_a{}^\mu A_\mu} = e^i{}_a A_i \tag{2.31}$$

The reduction of the Lagrangian is obtained by starting with the flat tensor

$$F_{AB} = e_A{}^M e_B{}^N F_{MN} = e_A{}^M \partial_M A_B - e_B{}^M \partial_M A_A - \Omega_{AB}{}^C A_C \tag{2.32}$$

for which we get

$$F_{\alpha\beta} = e_\alpha{}^\mu e_\beta{}^\nu (\partial_\mu A'_\nu - \partial_\nu A'_\mu + G^i_{\mu\nu} A_i)$$

$$F_{\alpha a} = e_a{}^i e_\alpha{}^\mu \partial_\mu A_i$$

$$F_{ab} = 0 \tag{2.33}$$

The Lagrangian in 4 dimensions is then immediately obtained after the previously defined Weyl rescaling noting that $\sqrt{g}\, g^{\mu\rho} g^{\nu\sigma}$ is a Weyl rescaling invariant

$$L_V^{(4)} = \int d_4 x \, V_4 \left[\sqrt{\Delta} \, g_4^{\mu\rho} g_4^{\nu\sigma} (F'_{\mu\nu} + G^i_{\mu\nu} A_i)(F'_{\rho\sigma} + G^j_{\rho\sigma} A_j) + g_4^{\mu\nu} g^{ij} \partial_\mu A_i \partial_\nu A_j \right] \quad (2.34)$$

As previously, the SL(N,R) invariance can be extended to GL(N,R) by adding to the previous scale transformations, the following ones

$$A'_\mu \to \lambda^{-N/4} A'_\mu \quad , \quad A_i \to \lambda^{1/2} A_i \quad (2.35)$$

This example has shown us how to define pure gauge fields in 4 dimensions.

5 Supergravity in 11 dimensions (Cremmer et al. 1978 b)

As we have seen, the maximal dimension in which a supersymmetric theory leads, after reduction in 4 dimensions, to $N \leq 8$ supersymmetric theory is D = 11. Therefore, supergravity in 11 dimensions should correspond to maximal N = 8 supergravity in 4 dimensions. The onshell massless states in D dimensions are classified by O(D - 2). A graviton g_{MN} has 1/2 (D -2) (D - 1) - 1 or 44 degrees of freedom in 11 dimensions. A gravitino ψ_M with a Majorana condition has $1/2 \; 2^{[D/2]} (D - 3)$ or 128 degrees of freedom in 11 dimensions. Therefore, 84 bosonic degrees of freedom are missing : it corresponds in fact to the representation A_{ijk} antisymmetric of O(9). We can associate to it the covariant tensor A_{MNP} with the gauge invariance

$$\delta A_{MNP} = 3 \, \partial_{[M} \xi_{NP]} \qquad \xi_{NP} = - \xi_{PN} \quad (2.36)$$

If this guess is correct, only gauge fields appear in this theory and then its construction should be relatively simple. This is true and we can construct a locally supersymmetric Lagrangian function of the 3 fields $e_M^{\ A}$, ψ_M and A_{MNP}. Because of gauge invariance A_{MNP} appears mainly through its field strength $F_{MNPQ} = 4 \, \partial_{[M} A_{NPQ]}$. The Lagrangian is given by

$$\begin{aligned} L = \int d^{11}x \Big\{ & -\frac{e}{4} R(\omega) - \frac{ie}{2} \bar{\psi}_M \Gamma^{MNP} D_N\!\left(\frac{\omega + \hat{\omega}}{2}\right) \psi_P \\ & - \frac{e}{48} F_{MNPQ} F^{MNPQ} + \frac{2}{(144)^2} \varepsilon^{IJKLMNOPQRS} F_{IJKL} F_{MNOP} A_{QRS} \quad (2.37) \\ & + \frac{e}{192} \left(\bar{\psi}_R \Gamma^{RSMNPQ} \psi_S + 12 \bar{\psi}^M \Gamma^{PQ} \psi^N \right) \left(F_{MNPQ} + \hat{F}_{MNPQ} \right) \Big\} \end{aligned}$$

with
$$\omega_{MAB} = \hat{\omega}_{MAB} - \frac{i}{4}\bar{\Psi}_P \Gamma_{MAB}{}^{PQ}\Psi_Q$$
$$\hat{\omega}_{MAB} = \overset{\circ}{\omega}_{MAB}(e) + \frac{i}{2}(\bar{\Psi}_M\Gamma_B\Psi_A - \bar{\Psi}_M\Gamma_A\Psi_B + \bar{\Psi}_B\Gamma_M\Psi_A)$$
Replacing $\hat{\omega}$ by $\omega + \ldots$, ω is a solution of its own equation of motion when it is considered as an independent variable (1st order formalism)
$$\hat{F}_{MNPQ} = F_{MNPQ} - 3\bar{\Psi}_{[M}\Gamma_{NP}\Psi_{Q]}$$
The term ε FFA is gauge invariant up to a total derivative. L is invariant under the following supersymmetry transformations
$$\delta e_M{}^A = -i\bar{\epsilon}\Gamma^A\Psi_M$$
$$\delta A_{MNP} = \frac{3}{2}\bar{\epsilon}\Gamma_{[MN}\Psi_{P]} \tag{2.38}$$
$$\delta\Psi_M = (\partial_M + \frac{1}{4}\hat{\omega}_{MAB}\Gamma^{AB} + \frac{i}{144}(\Gamma^{NOPQ}{}_M - 8\delta_M^N\Gamma^{OPQ})\hat{F}_{NOPQ})\epsilon$$

$\hat{\omega}$ and $\hat{F}$ are supercovariant objects (their variations by supersymmetry do not contain $\partial\epsilon$ terms).

The bosonic invariances are :

(1) reparametrization invariance $\xi^M(x)$

(2) local Lorentz invariance SO(10,1) $\Lambda_A{}^B(x)$

(3) local abelian invariance for A_{MNP} $\xi_{NP}(x)$

We can show that the algebra closes on-shell as usual. The geometrical interpretation of A_{MNP} as a gauge field is still unclear : $e_M{}^A$ "gauge" the translations P_M, Ψ_M gauge the supersymmetries , what do A_{MNP} gauge ?

Let us finally note that for the counting of degrees of freedom we could have used a field A_{MNOPQR} with gauge invariance $\delta A_{MNOPQR} = 6\partial_{[M}\xi_{NOPQR]}$. However, although we can construct a free supersymmetric theory with this field, it is impossible to construct an interactive theory (Nicolai et al. 1981)

6 Dimensional reduction of 11-dimensional supergravity in 4 dimensions

We shall perform now the ordinary dimensional reduction of the 11-dimensional supergravity (Cremmer & Julia 1978, 1979). We shall also perform duality transformations which change a pseudovector field into a vector field as well as a two-rank antisymmetric tensor field into a scalar field. This will make the physical content of the theory more similar to the ordinary content of N = 8 supergravity. However, the hunt for the hidden symmetries will be discussed only in the next part.

The field content in 4 dimensions is summarized in Table II and has to be compared with the field content of N = 8 supergravity :

$g_{\mu\nu}$	$\psi_\mu{}^A$	$A_\mu^{[AB]}$	$\chi^{[ABC]}$	$\phi^{[ABCD]} = \frac{1}{24}\varepsilon^{ABCDEFGH}\phi_{EFGH}$
1	8	28	56	70

TABLE II

11-D Field	4-D Field	Number	d° of freedom	Total d° of freedom
	$g_{\mu\nu}$	1	2	2
$g_{MN} \rightarrow$	$g_{\mu i}$	7	2	14
	g_{ij}	28	1	28
				44
	$A_{\mu\nu\rho}$	1	0	0
	$A_{\mu\nu i}$	7	1	7
$A_{MNP} \rightarrow$	$A_{\mu ij}$	21	2	42
	A_{ijk}	35	1	35
				84
	$\psi_\mu{}^A$	8	2	16
$\psi_M \rightarrow$	$\psi_i{}^A$	56	2	112
	(A = 1...8)			128

(a) Diagonalization of gauge transformations

In 11 dimensions for A_{MNP} we have the invariance under the following reparametrization and gauge transformations

$$\delta A_{MNP} = 3\,\partial_{[M}\xi^{Q} A_{NP]Q} + \xi^{Q}\partial_{Q} A_{MNP} + 3\,\partial_{[M}\zeta_{NP]} \tag{2.39}$$

Restricting to $\partial_i \zeta_{NP} = 0$ and previous conditions on ξ^M, we get

$$\begin{aligned} \delta A_{ijk} &= \xi^{\mu}\partial_{\mu} A_{ijk} + 3\, a_{[i}{}^{\ell} A_{jk]\ell} \\ \delta A_{\mu ij} &= \xi^{\nu}\partial_{\nu} A_{\mu ij} + \partial_{\mu}\xi^{\nu} A_{\nu ij} + \partial_{\mu}\xi^{k} A_{kij} - 2 a_{[i}{}^{k} A_{\mu j]k} + \partial_{\mu}\zeta_{ij} \end{aligned} \tag{2.40}$$

As usual, $A_{\mu ij}$, $A_{\mu\nu i}$ and $A_{\mu\nu\rho}$ are invariant under the $U(1)^7$ gauge transformations. We can as previously define invariant tensors starting from $A_{ABC} = e_A{}^M e_B{}^N e_C{}^P A_{MNP}$ or in 4 dimensions

$$\begin{aligned} A'_{ijk} &= A_{ijk} \\ A'_{\mu ij} &= e_{\mu}{}^{\alpha} e_i{}^a e_j{}^b A_{\alpha ab} = A_{\mu ij} - B_{\mu}{}^{k} A_{kij} \\ A'_{\mu\nu i} &= e_{\mu}{}^{\alpha} e_{\nu}{}^{\beta} e_i{}^a A_{\alpha\beta a} = A_{\mu\nu i} + 2 B_{[\mu}{}^{j} A_{\nu]ji} + B_{\mu}{}^{i} B_{\nu}{}^{j} A_{jki} \end{aligned} \tag{2.41}$$

Under the gauge transformations ξ^i and ζ_{ij} we have

$$\begin{aligned} &\delta A'_{ijk} = 0 \quad ; \quad \delta A'_{\mu ij} = \partial_{\mu}\zeta_{ij} \\ &\delta A'_{\mu\nu i} = 2\,\partial_{[\mu}\zeta_{\nu]i} + 2 B_{[\mu}{}^{j}\partial_{\nu]}\zeta_{ji} \end{aligned} \tag{2.42}$$

therefore $A'_{\mu\nu i}$ is no longer invariant under the ζ_{ij} gauge transformations. There is in fact no way to define a $A''_{\mu\nu i}$ which should be a gauge field for $\zeta_{\mu i}$ transformations and invariant under the ξ^i and ζ_{ij} gauge transformations. The same problem arises also for $A'_{\mu\nu\rho}$. We can nevertheless define completely gauge invariant tensors starting from F_{ABCD}

$$\begin{aligned} F_{\alpha bcd} &= e_{\alpha}{}^{\mu} e_b{}^i e_c{}^j e_d{}^k \partial_{\mu} A_{ijk} \\ F_{\alpha\beta cd} &= e_{\alpha}{}^{\mu} e_{\beta}{}^{\nu} e_c{}^i e_d{}^j F^4_{\mu\nu ij} \\ F_{\alpha\beta\gamma d} &= e_{\alpha}{}^{\mu} e_{\beta}{}^{\nu} e_{\gamma}{}^{\rho} e_d{}^i F^4_{\mu\nu\rho i} \\ F_{\alpha\beta\gamma\delta} &= e_{\alpha}{}^{\mu} e_{\beta}{}^{\nu} e_{\gamma}{}^{\rho} e_{\delta}{}^{\sigma} F^4_{\mu\nu\rho\sigma} \end{aligned} \tag{2.43}$$

This defines in particular

$$F^4_{\mu\nu ij} = \partial_\mu A'_{\nu ij} - \partial_\nu A'_{\mu ij} + G^k_{\mu\nu} A_{ijk}$$

$$F^4_{\mu\nu\rho i} = 3\, \partial_{[\mu} A'_{\nu\rho]i} + 3\, G^k_{[\mu\nu} A'_{\rho]ik} \tag{2.44}$$

(b) Duality transformations

Duality transformations allow transformations of a magnetic field into an electric field or an antisymmetric tensor $A_{\mu\nu}$ into a scalar field ϕ. Let us start with the following Lagrangian

$$\mathcal{L} = \mathcal{L}(F_{\mu\nu}, F_{\mu\nu\rho}, F_{\mu\nu\rho\sigma}) \tag{2.45}$$

where $F_{\mu\nu}$, $F_{\mu\nu\rho}$ and $F_{\mu\nu\rho\sigma}$ are, respectively, the field strengths of the fields A_μ , $A_{\mu\nu}$ and $A_{\mu\nu\rho}$ which do not appear explicitly in $\mathcal{L}$. Up to topological subtleties or harmonic solutions, these field strengths are characterized by the constraints

$$\partial_{[\mu} F_{\nu\rho]} = 0 \quad , \quad \partial_{[\mu} F_{\nu\rho\sigma]} = 0 \quad , \qquad \text{no constraints for } F_{\mu\nu\rho\sigma} \tag{2.46}$$

We can therefore use a first order formalism by implementing these constraints via Lagrange multipliers B_μ and ϕ i.e. by adding to $\mathcal{L}$

$$\begin{aligned} \Delta\mathcal{L} &= \varepsilon^{\mu\nu\rho\sigma} (B_\mu \partial_\nu F_{\rho\sigma} + \phi\, \partial_\mu F_{\nu\rho\sigma}) \\ &= -\varepsilon^{\mu\nu\rho\sigma} (\partial_\nu B_\mu F_{\rho\sigma} + \partial_\mu\phi\, F_{\nu\rho\sigma}) \end{aligned} \tag{2.47}$$

In $\mathcal{L} + \Delta\mathcal{L}$ we can now integrate on the fields $F_{\mu\nu}$, $F_{\mu\nu\rho}$ and $F_{\mu\nu\rho\sigma}$ which appear algebraically and we get a new 2nd order Lagrangian $\mathcal{L}'$

$$\mathcal{L}' = \mathcal{L}'(G_{\mu\nu}, \partial_\mu \phi)$$

$$\text{with} \quad G_{\mu\nu} = \partial_\mu B_\nu - \partial_\nu B_\mu \tag{2.48}$$

In the present case, we shall make the following transformation

$$A_{\mu ij} \rightarrow B_\mu^{ij} \; , \quad A_{\mu\nu i} \longrightarrow \phi^i \quad , \qquad A_{\mu\nu\rho} \quad \text{eliminated.}$$

Because of the previous discussion on the various gauge invariances, $\mathcal{L}$ is a function of $F^4_{\mu\nu ij}$, $F^4_{\mu\nu\rho i}$, $F^4_{\mu\nu\rho\sigma}$ and A_{ijk} . Taking into account the definition of F^4_i), we must add to $\mathcal{L}$ (after integration by parts)

$$\Delta\mathcal{L} = -\varepsilon^{\mu\nu\rho\sigma}\Big[\tfrac{1}{12}\partial_\mu\phi^i F^4_{\nu\rho\sigma i} + \tfrac{1}{8}\phi^i G^k_{\nu\rho}F^4_{\mu\sigma ik} + \tfrac{1}{8}G^{ij}_{\mu\nu}\left(F^4_{\rho\sigma ij} - G^k_{\rho\sigma}A_{ijk}\right)\Big] \tag{2.49}$$

Remarks : 1. At the quantum level ; there is no exact equivalence between $A_{\mu\nu}$ and ϕ . For example, they contribute in a different way to topological counterterms at the one-loop level.

2. Taking into account the harmonic solutions allows one to introduce some extra-parameters in the theory. This is easily seen in the case of the elimination of $F_{\mu\nu\rho\sigma}$ (Duff 1981 ; Aurilla et al. 1981). If we start from the Lagrangian $\mathcal{L} = -\frac{e}{48}F^2_{\mu\nu\rho\sigma}$

The variations with respect to $F_{\mu\nu\rho\sigma}$ and $A_{\mu\nu\rho}$ lead respectively to

$$\delta\mathcal{L}/\delta F_{\mu\nu\rho\sigma} = 0 \qquad F_{\mu\nu\rho\sigma} = 0 \tag{I}$$

$$\delta\mathcal{L}/\delta A_{\mu\nu\rho} = 0 \qquad \partial^\mu F_{\mu\nu\rho\sigma} = 0 \quad \text{or} \quad F_{\mu\nu\rho\sigma} = a\,\varepsilon_{\mu\nu\rho\sigma} \tag{II}$$

where a is an arbitrary constant. (2.50)

It is easy to see that (II) can be derived from the following Lagrangian

$$\mathcal{L}' = -\frac{e}{48}F^2_{\mu\nu\rho\sigma} + a\,\varepsilon^{\mu\nu\rho\sigma}F_{\mu\nu\rho\sigma} \tag{2.51}$$

when $F_{\mu\nu\rho\sigma}$ is a curl, the extra term is a total derivative. For the solution $F_{\mu\nu\rho\sigma} = a\,\varepsilon_{\mu\nu\rho\sigma}$, $\mathcal{L}$ has the value $e a^2$, it corresponds to a cosmological constant. This mechanism can be implemented in the reduction of 11-dimensional supergravity (Aurilla et al. 1981)

(c) Bosonic Lagrangian

We then get the complete bosonic Lagrangian in terms of the fields $g_{\mu\nu}$, g_{ij} , B^i_μ , B^{ij}_μ , Φ^i and A_{ijk} (i = 1...7) after the Weyl rescaling

$$\mathcal{L}_B = -\frac{e}{4}R + L_S + L_V$$

with

$$L_S = -\frac{e}{16}g^{\mu\nu}\partial_\mu g^{ij}\partial_\nu g_{ij} + \frac{e}{32}g^{\mu\nu}\partial_\mu\log\Delta\,\partial_\nu\log\Delta - \frac{e}{12}g^{\mu\nu}\partial_\mu A_{ijk}\partial_\nu A_{lmn}g^{il}g^{jm}g^{kn}$$
$$-\frac{e}{8\Delta}g_{iq}g^{\mu\nu}\left(\partial_\mu\phi^i - \tfrac{1}{3}\sqrt{\Delta}\,{}^*A^{ijkl}\partial_\mu A_{jkl}\right)\left(\partial_\nu\phi^q - \tfrac{1}{3}\sqrt{\Delta}\,{}^*A^{qrst}\partial_\nu A_{rst}\right) \tag{2.52}$$

where ${}^*A^{ijkl} = \frac{1}{6\sqrt{\Delta}}\varepsilon^{ijklmno}A_{mno}$

$$L_V = \frac{e}{16}\sqrt{\Delta}\, g^{\mu\rho} g^{\nu\sigma} g_{ij}\, G^i_{\mu\nu} G^j_{\rho\sigma} + \frac{1}{48}\varepsilon^{\mu\nu\rho\sigma}\, {}^*A^{ijk\ell} A_{ijp} A_{k\ell q} G^p_{\mu\nu} G^q_{\rho\sigma}$$
$$+ \frac{1}{8}\varepsilon^{\mu\nu\rho\sigma} G^k_{\mu\nu} G^{ij}_{\rho\sigma} A_{ijk} - \frac{e}{8\sqrt{\Delta}}\, Y^{ij} \mathcal{M}_{ij,pq} Y^{pq} \qquad (2.53)$$

where $Y^{ij}_{\mu\nu} = G^{ij}_{\mu\nu} + \frac{1}{2}\sqrt{\Delta}\, {}^*A^{ijk\ell} A_{k\ell m} G^m_{\mu\nu} + \frac{1}{2}(\phi^i G^j_{\mu\nu} - \phi^j G^i_{\mu\nu})$

$$\mathcal{M}_{ij,k\ell} = \left[\tfrac{1}{2}(g^{ik}g^{j\ell} - g^{i\ell}g^{jk}) - j\, {}^*A^{ijk\ell}\right]^{-1}$$

For the indices $\mu\nu$, i stands for $\frac{1}{2}(g^{\mu\rho}g^{\nu\sigma} - g^{\mu\sigma}g^{\nu\rho})$

and j stands for $\frac{1}{2e}\varepsilon^{\mu\nu\rho\sigma}$

(d) Reduction of the spinor fields

The 11 Γ matrices 32 x32 of the Clifford algebra can be written as

$$\Gamma^\alpha = \gamma^\alpha \otimes \mathbb{1}^A{}_B \qquad \alpha = 0,1,2,3$$

$$\Gamma^a = \gamma_5 \otimes (\Gamma^a)^A{}_B \qquad a = 1,\ldots 7$$

$$A = 1\ldots 8$$

The $(\Gamma^a)^A{}_B$ form a Clifford algebra of SO(7) and are real and antisymmetric 8 x 8 matrices.

As for $A_{\mu\, ij}$ we can define ψ_μ and ψ_i fields which are invariant under $U(1)^7$ gauge transformations. We also make some Weyl rescaling on these spinor fields associated with a rediagonalization of the kinetic term for ψ_μ and χ_a

$$\psi_\mu = e^\alpha_{\mu(4)}\, \Delta^{-1/8} (\psi_M e^M{}_\alpha - \tfrac{1}{2}\gamma_5\gamma_\alpha \Gamma^a \psi_M e^M{}_a) \qquad (2.54)$$

$$\chi_a = \Delta^{-1/8} \psi_M e^M{}_a \qquad (2.55)$$

Then, from the kinetic term in 11 dimensions for ψ_M, we get

$$L^{kin}_F = -\frac{ie}{2}\bar{\psi}_\mu \gamma^{\mu\nu\rho} \partial_\nu \psi_\rho - \frac{ie}{2}\bar{\chi}^A{}_a (\tfrac{1}{2}\Gamma^a\Gamma^b + \eta^{ab})_A{}^B \gamma^\mu \partial_\mu \chi_{bB} \qquad (2.56)$$

+ scalar couplings

A final diagonalization of χ^A_a 's kinetic terms is made by the redefinition $\lambda_{ABC} = \frac{3}{2}\Gamma^a_{[AB}\chi_{aC]}$

which leads to

$$L^{kin}_{\frac{1}{2}} = \frac{ie}{12} \bar{\lambda}_{ABC} \gamma^{\mu} \partial_{\mu} \lambda_{ABC} \tag{2.57}$$

Some further chiral transformations on ψ_{μ}^{A} and λ_{ABC} are made.

$$\psi_{\mu}^{A} = (i\gamma_5)^{-1/2} \psi^{A}_{\mu\,new} \quad ; \quad \lambda_{ABC} = (i\gamma_5)^{-1/2} \lambda_{ABC\,new} \tag{2.58}$$

(e) Spinor couplings

In order to obtain a simple form for the spinor couplings, a lot of algebraic properties of $(\Gamma^{a})_{A}{}^{B}$ matrices are required (Cremmer & Julia 1979), for instance

$$\Gamma^{[ab}_{AB} \chi^{c]}_{B} = -\frac{1}{3\sqrt{2}} \Gamma^{[ab}_{[AB} \Gamma^{c]}_{CD]} \lambda_{BCD} \tag{2.59}$$

The use of λ_{ABC} is crucial in this tremendous simplification and we get finally

$$\begin{aligned} L_F = \ & \frac{1}{2} \varepsilon^{\mu\nu\rho\sigma} \bar{\psi}^{A}_{\mu} \gamma_{\sigma} \gamma_5 (\partial_{\nu} \delta_{A}^{B} - Q^{o}_{\nu A}{}^{B}) \psi_{\rho}^{B} \\ & + \frac{ie}{12} \bar{\lambda}_{ABC} \gamma^{\mu} (\partial_{\mu} \delta_{A}^{D} - 3 Q^{o}_{\mu A}{}^{D}) \lambda_{DBC} \\ & - \frac{e}{3\sqrt{2}} \bar{\psi}_{\mu D} \gamma^{\nu} \gamma^{\mu} \bar{P}^{o\,ABCD}_{\nu} \lambda_{ABC} \\ & + \frac{e}{4\sqrt{2}} \bar{\psi}^{A}_{\mu} \gamma^{[\nu} F^{o}_{AB} \gamma^{\mu]} \psi_{\nu}^{B} - \frac{ie}{8} \bar{\psi}^{C}_{\mu} F^{o}_{AB} \gamma^{\mu} \lambda_{ABC} \\ & - \frac{e}{288\sqrt{2}} \eta \varepsilon^{ABCDEFGH} \bar{\lambda}_{ABC} F^{o}_{DE} \lambda_{FGH} \end{aligned} \tag{2.60}$$

+ quartic fermionic terms

where η is a self duality coefficient ($\eta = \pm 1$) depending on the explicit representation of the Γ 's matrix of SO(7) defined by

$$\Gamma^{a}_{[AB} \Gamma^{b}_{CD]} = \frac{\eta}{4!} \varepsilon_{ABCDEFGH} \Gamma^{a}_{[EF} \Gamma^{b}_{GH]}$$

$Q^{o}_{\nu A}{}^{B}$, $\bar{P}^{o\,ABCD}_{\nu}$ and F^{o}_{AB} are given in terms of e^{i}_{a}, ϕ^{i}, A_{ijk} , $G^{i}_{\mu\nu}$ and $G^{ij}_{\mu\nu}$ by

$$\begin{aligned} Q^{o}_{\nu A}{}^{B} = -\frac{1}{4} \Big[& e^{i}_{a} \partial_{\nu} e_{ib} \Gamma^{ab} + e_{ia} \Big(\frac{\partial_{\nu} \phi^{i}}{\sqrt{\Delta}} - \frac{1}{3} {}^{*}A^{ijkl} \partial_{\nu} A_{jkl} \Big) \Gamma^{a} \\ & - \frac{i}{3} \gamma_5 e^{i}_{a} e^{j}_{b} e^{k}_{c} \partial_{\nu} A_{ijk} \Gamma^{abc} \Big]_{A}{}^{B} \end{aligned} \tag{2.61}$$

$$\bar{P}^{o\ ABCD}_{\nu} = \frac{1}{8}\{e^{i}_{a}\partial_\nu e_{ib}\,\Gamma^{ac}_{[AB}(\Gamma^{b}{}_{c})_{CD]} + e_{ia}\left(\frac{\partial_\nu \phi^{i}}{\sqrt{\Delta}} - \frac{1}{3}{}^{*}A^{ijke}\partial_\nu A_{jke}\right)\Gamma^{ac}_{[AB}(\Gamma_{c})_{CD]}$$
$$+2i\gamma_5 e^{i}_{a} e^{j}_{b} e^{k}_{c}\partial_\nu A_{ijk}\,\Gamma^{ab}_{[AB}\Gamma^{c}_{CD]}\} \qquad (2.62)$$

$P^{o}_{\nu ABCD}$ satisfies the self-duality relation. ($\bar{P}$ designates the complex conjugate)

$$\bar{P}^{o\ ABCD}_{\nu} = \frac{\eta}{4!}\,\varepsilon^{ABCDEFGH}P^{o}_{\nu EFGH} \qquad (2.63)$$

$$F^{o}_{AB} = \frac{1}{2\sqrt{2}}\gamma^{\rho\sigma}\{-G^{i}_{\rho\sigma}e_{ia}\Gamma^{a}\Delta^{1/4} + \Delta^{-1/4}\bar{M}_{ij,pq}e^{i}_{a}e^{j}_{b}\Gamma^{ab}\times$$
$$\times(G^{pq}_{\rho\sigma} + \phi^{p}G^{q}_{\rho\sigma} + \frac{1}{8}\sqrt{\Delta}\,{}^{*}A^{pqmn}A_{mnl}G^{l}_{\rho\sigma})\}_{AB} \qquad (2.64)$$

We can note at this stage that the Noether coupling to $\bar{\Psi}_\mu\lambda : P^{o}_{\nu\ ABCD}$ is nothing else than the square root of the scalar "kinetic term" L_S as it is usually in supergravity.

(f) Supersymmetry transformation laws

In the reduction process, we assume that ϵ does not depend on the extra-coordinates : that is necessary for the compatibility with the independence of the fields on these coordinates. The 32 components then split into 8 ϵ^{A} 4-component spinors in 4 dimensions.

In order to determine the supersymmetry transformation laws in 4 dimensions, we have to keep in mind all transformations on the fields we have made during the reduction and also redefine ϵ^{A} by chiral and Weyl transformations. In order to preserve the gauge condition $e_m{}^{\alpha} = 0$, it is necessary to add a compensating SO(10,1) gauge transformation $\Omega^{\alpha}{}_{a}$. We can always redefine δ_S by combining with local SO(3,1) or local SO(7) transformations. This will simplify δ_S and allow one to keep the canonical form for $\delta e_\mu{}^{\alpha}$

$$\delta e_\mu{}^{\alpha} = -i\,\bar{\epsilon}^{A}\gamma^{\alpha}\psi_{\mu A} \qquad (2.65)$$

When we perform duality transformations, the new fields are not functions of the original fields (except on-shell and in a non-local way). We derive their transformation laws such that the first order Lagrangian is supersymmetric following Van Nieuwenhuizen's method. Let us show in a simple example how this works. Let us start from a symmetric theory with a Lagrangian $\mathcal{L}(F_{\mu\nu\rho}, \psi)$ where $F_{\mu\nu\rho} = 3\,\partial_{[\mu}A_{\nu\rho]}$. It is invariant under some transformations $\delta A_{\mu\nu}$ and $\delta\psi$. Let us now consider the first order Lagrangian

$$\mathcal{L}' = \mathcal{L}(F_{\mu\nu\rho}, \psi) + \varepsilon^{\mu\nu\rho\sigma}\partial_\sigma F_{\mu\nu\rho}\,\phi \tag{2.66}$$

Under $\delta\psi$ and $\delta F_{\mu\nu\rho} = 3\,\partial_{[\mu}\delta A_{\nu\rho]}$, $\delta\mathcal{L}$ is zero only if $F_{\mu\nu\rho}$ is a curl, therefore it can be written as

$$\delta\mathcal{L} = \varepsilon^{\mu\nu\rho\sigma}\partial_\sigma F_{\mu\nu\rho}\, S \tag{2.67}$$

Then it is obvious from

$$\delta\mathcal{L}' = \varepsilon^{\mu\nu\rho\sigma}\partial_\sigma F_{\mu\nu\rho}\, S + \varepsilon^{\mu\nu\rho\sigma}\partial_\sigma F_{\mu\nu\rho}\,\delta\phi \tag{2.68}$$

$$(\partial_{[\sigma}\delta F_{\mu\nu\rho]} = 0)$$

that by choosing $\delta\phi = -S$ the Lagrangian $\mathcal{L}'$ is now invariant under $\delta\psi$, $\delta F_{\mu\nu\rho} = 3\,\partial_{[\mu}\delta A_{\nu\rho]}$, and $\delta\phi = -S$ and we can proceed to the elimination of $F_{\mu\nu\rho}$. Although this method is simple in principle, it can be very complicated in practice to determine S.

The next step for this theory will be the hunt for hidden symmetries which will be discussed in the next part.

7 Concluding remarks on dimensional reduction

(a) The dimensional reduction suggests hidden symmetries. One example which will be developed in length is of course the N = 8 supergravity. But previously already, the possibility of reducing the 10 dimensional supergravity has suggested that there could exist a formulation of N = 4 supergravity with an SU(4)∼SO(6) invariance of the Lagrangian, instead of only an invariance of the equation of motion. This theory has been constructed directly and shown to be equivalent to the other formulation of N = 4 supergravity as far as the equations of motion are concerned. This has allowed, at the same time, the discovery of the first hidden symmetry in supergravity : SU(1,1) for N = 4 supergravity (Cremmer et al. 1978 a).

(b) There exists a method called "dimensional reduction by Legendre transform" which can partially solve the problem of finding the auxiliary fields (Sohnius et al. 1981 ; West 1981).

(c) There exists a modified and more general version of dimensional reduction (Cho & Freund 1975 ; Scherk & Schwarz 1979 b). It still associates to 1 degree of freedom in 4 + N dimensions, 1 degree of freedom in 4 dimensions. The gravitational sector instead of generating $U(1)^N$ gauge group, generates a non-abelian gauge group with N generators.

If we require the absence of a cosmological constant and the positivity of scalar potential, we obtain what has been called a "flat group". Starting from a massless theory, it allows one to introduce mass parameters and to generate a potential term for the scalar fields. In the case of a supersymmetric theory (without gravitation), it ends up with a theory which has soft supersymmetry breaking in general. In the case of a supergravity theory (local supersymmetry) (Scherk & Schwarz 1979 ; Cremmer et al.1979) we end up with a theory invariant under some new local transformations like supersymmetry but the local algebra has changed and in general it is not possible to extract a global algebra from it.

III HIDDEN SYMMETRIES OF N = 8 SUPERGRAVITY IN 4 DIMENSIONS

From dimensional reduction of the 11-dimensional supergravity we have obtained a N = 8 supergravity theory described in terms of the fields $g_{\mu\nu}$, $\Psi_\mu^{\ A}$, $B_\mu^{\ i}$, $B_\mu^{\ ij}$, λ_{ABC} (or $\chi^A_{\ a}$), g^{ij}, ϕ^i which form representations of SO(7). The natural question arises : is it the same as the N = 8 supergravity expected and described in terms of irreducible representation of SO(8) ? A first remark is that the N = 7 supergravity is expected to be the same as the N = 8, the fields being described by irreducible representation of SO(7) in particular the scalar fields which are in the irreducible 35 representation of SO(7), but the scalar fields coming from the reduction g^{ij}, ϕ^i are a 28 + 7 representation of SO(7). This problem will be solved by the discovery of the hidden symmetries of N = 8 supergravity and by a special property of SO(8) : the triality which allows two different embeddings of SO(7) in SO(8).

1 General ideas for the hunt for hidden symmetries

The main idea is to try to generalize what we have learned from dimensional reduction. The scalar fields coming from the tensor metric are described by a coset space SL(7,R)/SO(7). Therefore we would like to describe all scalars and pseudoscalars by a coset space G/H. It describes, as we have seen, an element $\mathcal{V} \in G$ defined up to a local transformation of H (Coleman et al. 1969 ; Callan et al. 1969). The theory will be invariant under the change $\mathcal{V} \to g\mathcal{V}h(x)$ $g \in G$, $h(x) \in H$ together with some transformations on the other fields. Writing $\mathcal{V} = \exp w$ $\quad w \in \mathcal{L}ie\,(G)$
the maximal linear subgroup acting on w will be H_d defined by

$$\mathcal{V} \to h^{-1}\mathcal{V}h \qquad h \in H \tag{3.1}$$

For the N = 8 supergravity, the maximal linear group acting on the 70 scalar fields ϕ_{ABCD} is expected to be SU(8). Therefore, we should have dim G = dim SU(8) + 70 = 133. This is consistent with G = E_7. G must be non-compact since it contains the non-compact group SL(7,R) derived from dimensional reduction. In fact, there exists a non-compact version $E_{7(+7)}$ (the normal form) whose maximum compact subgroup is SU(8). This will ensure the positivity of the Lagrangian for the scalar fields.

From previous results for $N \leqslant 4$ (Ferrara et al. 1977 ; Cremmer et al. 1977 a ; Cremmer et al. 1977 b) we know that H_D is realized only on the equations of motion when vector fields are present. In fact, for vector fields G and H_D exchange the equations of motion $\partial^\mu H_{\mu\nu} = 0$ and the Bianchi identities $\partial^\mu \tilde{G}_{\mu\nu} = 0$ (see also Gaillard & Zumino 1981). Therefore, if there are N vector fields, 2 N should be an irreducible representation of H_D and therefore of G. In fact, the fundamental representation of E_7 has dimension 56 consistent with 28 vector fields.

From dimensional reduction we have learned also that vector fields are singlet for the gauge group SO(7) and that the spinor fields are singlet for the global group SL(7,R). We then expect to describe the supergravity by : 1 graviton singlet for E_7 and SU(8) ; 8 gravitinos singlet for E_7 and in representation 8 for SU(8) ; 28 vector fields whose field strengths $G_{\mu\nu}$ and $\tilde{H}_{\mu\nu} \sim \partial\mathcal{L}/\partial G_{\mu\nu}$ are singlet for SU(8) and in the 56 representation of E_7, 56 spin 1/2 fields singlet for E_7 and in the 56 representation of SU(8), the 70 scalar fields being described by a 56 x 56 matrix of E_7, which transforms as 56 for E_7 and 28 complex for SU(8).

Therefore we shall show that there exists a E_7/SU(8) formulation of N = 8 supergravity described by the pattern shown in Table III.

TABLE III

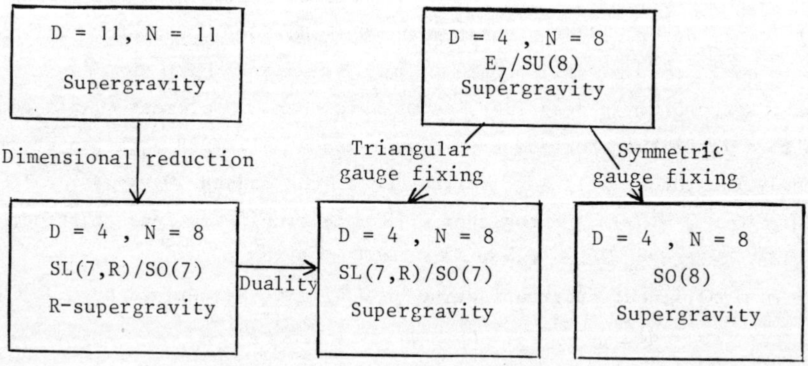

The R-supergravity is the one obtained directly from 11 dimensions where some of the scalar fields are described by antisymmetric tensors. The SO(8) supergravity is the ordinary formulation of N = 8 supergravity on which partial results were obtained (de Wit & Freedman 1977 ; de Wit 1979). When we fix the gauge of SU(8), then the group E_7 acts on all the fields, in particular on the spinor fields via transformations of SU(8) which are function of the scalar fields.

2 Restoration of SL(8,R) symmetry

Before defining E_7 and showing that it is a symmetry of the theory, we shall perform an intermediate step at the bosonic level and exhibit the SL(8,R) symmetry and local SO(8) symmetry which unify B_μ^i with B_μ^{ij}, and g^{ij} with ϕ^i.

In the absence of pseudoscalar A_{ijk} the bosonic Lagrangian is written

$$\mathcal{L}_B^{(+)} = -\frac{e}{16}\partial_\mu g^{ij}\partial^\mu g_{ij} - \frac{e}{8\Delta}\partial_\mu\phi^i\partial^\mu\phi^j g_{ij} + \frac{e}{32}\frac{\partial_\mu\Delta\,\partial^\mu\Delta}{\Delta^2}$$
$$+\frac{e}{16}\sqrt{\Delta}\,g_{ij}G^i_{\mu\nu}G^{\mu\nu j} - \frac{e}{8\sqrt{\Delta}}g_{ip}g_{jq}\Big[G^{ij}_{\mu\nu}+\tfrac{1}{2}(\phi^i G^j_{\mu\nu}-\phi^j G^i_{\mu\nu})\Big]\Big[G^{pq}_{\rho\sigma}+\tfrac{1}{2}(\phi^p G^q_{\mu\nu}-\phi^q G^p_{\mu\nu})\Big]g^{\mu\rho}g^{\nu\sigma} \tag{3.2}$$

Defining $S^{i'j'} = \Delta^{-3/4}\begin{pmatrix}\Delta g^{ij}-\phi^i\phi^j & , \phi^j\\ \phi^i & , -1\end{pmatrix}$ with det S = 1

$$G^{i'j'}_{\mu\nu} = \left(G^{ij}_{\mu\nu}, -\tfrac{1}{2}G^i_{\mu\nu}\right) \quad ; \quad \left(G^{i8}_{\mu\nu} = -\tfrac{1}{2}G^i_{\mu\nu}\right)$$

$\mathcal{L}_B^{(+)}$ takes now the very simple form

$$\mathcal{L}_B^{(+)} = -\frac{e}{16}\partial_\mu S^{i'j'}\partial^\mu S_{i'j'} - \frac{e}{16}S_{i'p'}S_{j'q'}G^{i'j'}_{\mu\nu}G^{p'q'}_{\rho\sigma}g^{\mu\rho}g^{\nu\sigma} \tag{3.3}$$

As discussed previously, we can reintroduce a 8-bein $v_{i'}{}^{a'}$ defined up to local SO(8) transformations $S_{i'j'} = v_{i'}{}^{a'}v_{j'}{}^{b'}\eta_{a'b'}$
A specific choice of $v_{i'}{}^{a'}$ which breaks SO(8) local invariance down to SO(7) is

$$v_{i'}{}^{a'} = \Delta^{-1/8}\begin{pmatrix} e_i{}^a & , 0\\ \phi^j e_j{}^a & , \sqrt{\Delta}\end{pmatrix} \qquad (\det v = 1) \tag{3.4}$$

Defining $A_{abc} = e^i{}_a e^j{}_b e^k{}_c A_{ijk}$ and $*A^{abcd} = \frac{1}{6}\varepsilon^{abcdefg}A_{efg}$, we can rewrite the vector part of the Lagrangian as

$$L_V = -\frac{1}{8}e\,v_{i'}{}^{c'}v_{j'}{}^{b'}v_{k'}{}^{c'}v_{\ell'}{}^{d'}G^{i'j'}_{\mu\nu}G^{k'\ell'}_{\rho\sigma}\mathcal{N}^{\mu\nu,\rho\sigma}_{a'b',c'd'} \tag{3.5}$$

$\mathcal{N}$ is a 28 x 28 matrix function of A_{abc} only, that we shall not write explicitly (see Cremmer & Julia 1979). The complete bosonic Lagrangian will simplify when we shall take into account the E_7 symmetry.

3 Definition of E_7(+7)

We shall define E_7 by the infinitesimal transformations in the 56 dimensional fundamental representation. This 56-space is spanned by 2 antisymmetric tensors x^{ij} and y_{ij} ($i, j = 1...8$)

$$\delta x^{ij} = \Lambda^i{}_k x^{kj} + \Lambda^j{}_k x^{ik} + \frac{1}{24}\varepsilon^{ijklmnop}\Sigma_{mnop}\, y_{kl}$$
$$\delta y_{ij} = \Lambda_i{}^k y_{kj} + \Lambda_j{}^k y_{ik} + \Sigma_{ijkl}\, x^{kl} \qquad (3.6)$$

with $\Lambda_i{}^k = -\Lambda^i{}_k$; $\Lambda^i{}_i = 0$, Σ_{ijkl} totally antisymmetric.

$\Sigma = 0$ corresponds to the subgroup SL(8,R).

There are two invariants for E_7, one bilinear invariant which shows that E_7 is a subgroup of Sp(56) with the symplectic metric Ω

$$I = x_1^{ij} y_{2ij} - x_2^{ij} y_{1ij} \qquad (3.7)$$

Another one is quartic and characterizes E_7

$$J = x^{ij} y_{jk} x^{kl} y_{li} - \frac{1}{4} x^{ij} y_{ij} x^{kl} y_{kl}$$
$$+ \frac{1}{96}\left[\varepsilon^{ijklmnop} y_{ij} y_{kl} y_{mn} y_{op} + \varepsilon_{ijklmnop} x^{ij} x^{kl} x^{mn} x^{op}\right] \qquad (3.8)$$

An element X of E_7 being also an element of Sp(56) satisfies

$$X^t \Omega X = \Omega \qquad X^{-1} = -\Omega X^t \Omega$$
$$\Omega^2 = -\mathbb{1} \qquad \Omega = \begin{pmatrix} 0 & -\mathbb{1} \\ \mathbb{1} & 0 \end{pmatrix} \qquad (3.9)$$

The subgroup SU(8) of E_7 is characterized by

$$[X, \Omega] = 0 \quad \text{or} \quad X^t X = \mathbb{1} \qquad (3.10)$$

4 E_7 formulation for scalars and vectors

Forgetting for the present time the fermions, the vectors' equations of motion can be written as $\partial_\mu (e \tilde{H}^{\mu\nu}_{i'j'}) = 0$ where $e\tilde{H}^{\mu\nu}_{i'j'} = \partial L/\partial G^{i'j'}_{\mu\nu}$ (since $B^{i'j'}_\mu$ appears only through $G^{i'j'}_{\mu\nu}$) . The Bianchi identity is written as $\partial_\mu (e\, \tilde{G}^{\mu\nu}_{i'j'}) = 0$.

We therefore expect that $\mathcal{F}_{\mu\nu} = (G^{i'j'}_{\mu\nu}, H_{\mu\nu i'j'})$ could be a vector of the

56 representation of E_7. Since $H_{\mu\nu i'j'}$ is a function of $G_{\mu\nu}^{ij}$ and of the scalar fields, there must exist a relation between $\mathcal{F}$ and the matrix describing the scalar fields $\mathcal{V}$ (or the metric $\mathcal{R} = \mathcal{V}^t\mathcal{V}$ invariant under the expected local SU(8)). The simplest covariant relation is $\Omega\tilde{\mathcal{F}} = \mathcal{R}\mathcal{F}$ Since we know $\mathcal{F}$, this defines the matrix $\mathcal{R}$ in terms of the original scalar fields $S^{i'j'}$ and A_{abc}. We find

$$\mathcal{R} = \mathcal{V}_+^t \mathcal{P}_- \mathcal{V}_+$$

where
$$\mathcal{V}_+ = \begin{pmatrix} v^{[a'}{}_{[i'} v^{b']}{}_{j']} , & 0 \\ 0 & , v_{[a'}{}^{[i'} v_{b']}{}^{j']} \end{pmatrix} \tag{3.11}$$

and $\mathcal{P}_-$ is function of A_{abc} only and is computed from $\mathcal{N}_{a'b'c'd'}$

Defining
$$A_{a'b'c'd'} = 4\, A_{[a'b'c'}\,\delta^8_{d']}$$
$$*A^{a'b'c'd'} = \frac{1}{24}\,\varepsilon^{a'b'c'd'e'f'g'h'} A_{e'f'g'h'} \tag{3.12}$$

($^*A$ is zero if one of its indices is equal to 8) and making the convention that all contractions are made on antisymmetric pairs of indices to define the product of 28 x 28 matrices, we find

$$\mathcal{P}_- = \mathcal{P}_-^t = \begin{pmatrix} X , & Z \\ Z^t , & Y \end{pmatrix} \tag{3.13}$$

where $X = X^t$, $Y = Y^t$ and Z are 28 x 28 matrices defined by

$$X = \mathbb{1} + A^2 + \tfrac{1}{2}\,{}^*\!A A + \tfrac{1}{2} A\,{}^*\!A + \tfrac{1}{4} A({}^*\!A)^2 A$$
$$Z = A + {}^*\!A + \tfrac{1}{2} A^2\,{}^*\!A + \tfrac{1}{2} A({}^*\!A)^2 + \tfrac{1}{6} A\,{}^*\!A A + \tfrac{1}{12} A\,{}^*\!A A^2\,{}^*\!A \tag{3.14}$$
$$Y = \mathbb{1} + ({}^*\!A)^2 + \tfrac{1}{2} A\,{}^*\!A + \tfrac{1}{2}\,{}^*\!A A + \tfrac{1}{4}\,{}^*\!A A^2\,{}^*\!A$$

We must prove that $\mathcal{R}$ is a matrix of E_7. Since $\mathcal{V}_+^t$ and $\mathcal{V}_+$ are in E_7, it is sufficient that $\mathcal{P}_-$ should be in E_7. For this we show that $\mathcal{P} = \mathcal{V}_-^t\mathcal{V}_-$ with

$$\mathcal{V}_- = \exp\begin{pmatrix} 0 & , {}^*\!A^{a'b'c'd'} \\ A_{a'b'c'd'} , & 0 \end{pmatrix} \equiv \exp V \tag{3.15}$$

In this form it is obvious that $\mathcal{V}_- \in E_7$. To prove $\mathcal{P}_- = \mathcal{V}_-^t\mathcal{V}_-$ it is necessary to notice that A and $^*A$ as defined previously satisfy $^*AA^*A = 0$ so that V is nilpotent $V^4 = 0$. Therefore the expression of the exponential has only a finite number of terms and is polynomial in A. Defining $\mathcal{V} = \mathcal{V}_-\mathcal{V}_+$, $\mathcal{F}_{\mu\nu}$ now satisfies

$$\mathcal{V}\mathcal{F}_{\mu\nu} = \Omega\,\mathcal{V}\tilde{\mathcal{F}}_{\mu\nu} \tag{3.16}$$

If this matrix $\mathcal{R}$ describes the scalar field, we must have

$$L_S = -\gamma e\,\mathrm{Tr}(\partial_\mu \mathcal{R}\,\partial_\nu \mathcal{R}^{-1})\,g^{\mu\nu} \tag{3.17}$$

Computing this explicitly in terms of $\mathcal{V}_i{}^{a'}$, A_{ijk} we check that it is true and find γ = 1/192. This can be written in a form which exhibits the local SU(8) gauge invariance using the vielbein $\mathcal{V}$

$$L_S = \frac{e}{48}\,\mathrm{Tr}([D_\mu \mathcal{V}\,\mathcal{V}^{-1}]^2) \tag{3.18}$$

The complete bosonic Lagrangian is now written in the simple form

$$L_B = -\frac{e}{4}R - \frac{e}{192}g^{\mu\nu}\mathrm{Tr}(\partial_\mu \mathcal{R}\,\partial_\nu \mathcal{R}^{-1}) + \frac{1}{16}\varepsilon^{\mu\nu\rho\sigma}G_{\mu\nu}^{ij}H_{\rho\sigma ij} \tag{3.19}$$

where $H_{\rho\sigma ij}$ is defined in terms of $G_{\mu\nu}^{ij}$ and $\mathcal{R}$ by the relation

$$\Omega\tilde{\mathcal{F}} = \mathcal{R}\mathcal{F} \quad \text{and} \quad \mathcal{F} = \begin{pmatrix} G_{\mu\nu}^{ij} \\ H_{\mu\nu ij} \end{pmatrix}$$

$$\tilde{H}^{\mu\nu}{}_{ij} = \frac{1}{2e}\varepsilon^{\mu\nu\rho\sigma}H_{\rho\sigma ij} = \frac{1}{e}\,\partial\mathcal{L}_B/\partial G_{\mu\nu}^{ij} \tag{3.20}$$

$\mathcal{R}$ being a symmetric matrix 56 x 56 of E_7.

Let us now prove the complete invariance of the equations of motion derived from this Lagrangian which has the general form

$$L_B = L_{E_7} + \frac{1}{16}\varepsilon^{\mu\nu\rho\sigma}G_{\mu\nu}^{ij}H_{\rho\sigma ij}(B_\mu^{ij}, g_{\mu\nu}, \mathcal{R}) \tag{3.21}$$

where L_{E_7} is invariant under E_7.

(1) The equations of motion of B_μ^{ij} together with the Bianchi identities for $G_{\mu\nu}$ are

$$\partial_\mu(e\tilde{H}^{\mu\nu}{}_{ij}) = 0 \quad , \quad \partial_\mu(e\tilde{G}^{\mu\nu ij}) = 0 \tag{3.22}$$

which can be written $\partial_\mu(e\tilde{\mathcal{F}}^{\mu\nu}) = 0$ covariant for E_7.

(2) the other fields belong to a precise representation of E_7. Let us call them ϕ

$$\frac{\delta\mathcal{L}}{\delta\phi} = \frac{\delta L_{E_7}}{\delta\phi} + \frac{1}{16}\varepsilon^{\mu\nu\rho\sigma}G_{\mu\nu}^{ij}\frac{\delta H_{\rho\sigma ij}}{\delta\phi}$$

$$= \frac{\delta L_{E_7}}{\delta\phi} + \frac{1}{16}\varepsilon^{\mu\nu\rho\sigma}\left(G_{\mu\nu}^{ij}\frac{\delta H_{\rho\sigma ij}}{\delta\phi} - \frac{\delta G_{\mu\nu}^{ij}}{\delta\phi}H_{\rho\sigma ij}\right) \tag{3.23}$$

The second term has the form $\varepsilon^{\mu\nu\rho\sigma}\mathcal{F}_{\mu\nu}^t\,\Omega(\delta\mathcal{F}_{\rho\sigma}/\delta\phi)$ and consequently transforms under E_7 in the inverse way to ϕ since the invariant of E_7 is written $\mathcal{J}^t \Omega \mathcal{G}$. Therefore

$\delta \mathcal{L}/\delta \phi = 0$ is covariant under E_7.

This completely proves the invariance of this bosonic theory under E_7, the local SU(8) (as the local Lorentz invariance SO(3,1)) being hidden in the metric $\mathcal{R}$ (as $g_{\mu\nu}$).

5 SO(7), SO(8), SU(8), E_7(+7)

Before discussing the coupling to fermions, it is necessary to reformulate E_7 in a basis which features the subgroup SU(8). This basis linked to the basic space of the Clifford algebra of SO(7) will show also the two ways of embedding SO(7) into SO(8).

(a) Clifford algebra of SO(7) and Lie algebra of SO(8)

Let us start with the 7 $(\Gamma^a)_A{}^B$ matrices (a= 1...7, A = 1...8) which form the Clifford algebra of SO(7)

$$\{\Gamma^a, \Gamma^b\} = 2\,\eta^{ab}\,\mathbb{1} \tag{3.24}$$

The $(\Gamma^a)_A{}^B$ are real and antisymmetric. From them we define the antisymmetrized product of Γ matrices. Γ^a , Γ^{ab} are antisymmetric matrices and Γ^{abc} are symmetric matrices. They form together a complete basis for the 8 x 8 matrices. The other matrices are related to them via the "duality" relation

$$\Gamma^{abcdefg} = \varepsilon^{abcdefg}\,\mathbb{1} \quad , \quad \Gamma^{abcd} = -\frac{1}{6}\,\varepsilon^{abcdefg}\,\Gamma_{efg} \tag{3.25}$$

. The Γ^a and Γ^{ab} satisfy the commutation relations

$$\left\{\begin{aligned} [\Gamma^{ab}, \Gamma^{cd}] &= -8\,\eta^{[a}{}_{[c}\,\Gamma^{b]}{}_{d]} \\ [\Gamma^{ab}, \Gamma^{c}] &= -2\,(\eta^{ac}\Gamma^b - \eta^{bc}\Gamma^a) \\ [\Gamma^{a}, \Gamma^{b}] &= 2\,\Gamma^{ab} \end{aligned}\right. \tag{3.26}$$

Therefore, defining $\Gamma^{a8} = -\Gamma^a$ $(\eta^{88} = -1)$ and noting $a' = (a, 8)$ we get

$$[\Gamma^{a'b'}, \Gamma^{c'd'}] = -8\,\eta^{[a'}{}_{[c'}\,\Gamma^{b']}{}_{d']} \tag{3.27}$$

This is the Lie algebra of SO(8) : a' labels the 8 vector representation of SO(8) and A labels the 8 spinor representation of SO(8). To the transformation $\delta x^{a'} = \Lambda^{a'}{}_{b'} x^{b'}$ ($\Lambda^{a'}{}_{b'}$ is antisymmetric) of SO(8) corresponds in the spinor representation

$$\delta x^A = \frac{1}{4}(\Gamma^{a'b'})_A{}^B \Lambda_{a'b'}\,x^B \equiv \Lambda^A{}_B\,x^B \tag{3.28}$$

$\Lambda^A{}_B$ is again an antisymmetric 8 x 8 matrix (but different from $\Lambda_{a'b'}$). However, the two 8 representation of SO(8) are not equivalent : there is no transformation $x^A = Z^A{}_{a'} x^{b'}$ which would imply the relation between $\Lambda^{a'b'}$ and $\Lambda_A{}^B$. But for the 28 antisymmetric representation of SO(8) spanned by $x^{a'b'} = -x^{b'a'}$ the SO(8) transformation $\Lambda_{a'}{}^{b'}$ on $x^{a'b'}$ generates the SO(8) "vector" representation of parameter $\Lambda_A{}^B$ on $x^{AB} = \frac{1}{4}(\Gamma^{a'b'})_A{}^B x^{a'b'}$.

(b) SU(8) basis for E_7

The basis which features the SU(8) subgroup is defined from x^{ij} and y_{ij} by $Z_{AB} = \frac{1}{4}(\Gamma^{ij})_A{}^B (x^{ij} + i\, y_{ij})/\sqrt{2} = -Z_{BA}$

Using a lot of properties of Γ matrices we can show that E_7 is defined in this basis by the infinitesimal transformations

$$\delta Z_{AB} = \Lambda_A{}^C Z_{CB} + \Lambda_B{}^C Z_{AC} + \Sigma_{ABCD} \bar{Z}^{CD} \tag{3.29}$$

where $\bar{Z}^{CD}$ is the complex conjugate of Z_{CD} ; $\Lambda_A{}^B$ is a traceless antihermitian 8 x 8 matrix, Σ_{ABCD} is totally antisymmetric and satisfies the self-duality condition

$$\Sigma_{ABCD} = \frac{\eta}{4!} \varepsilon_{ABCDEFGH} \bar{\Sigma}^{EFGH} \qquad (\eta = \pm 1) \tag{3.30}$$

The bilinear invariant is now written in term of the symplectic matrix Ω'

$$Z_{1AB} \bar{Z}_2^{AB} - Z_{2AB} \bar{Z}_1^{AB} \qquad \Omega' = \begin{pmatrix} i, & 0 \\ 0, & -i \end{pmatrix} \tag{3.31}$$

The subgroup SU(8) corresponds to $\Sigma = 0$ or to matrices of the group E_7 which satisfy $[X', \Omega'] = 0$ or equivalently since

$X'^{-1} = -\Omega' X'^{\dagger} \Omega'$ for $X' \in E_7$, to $X' X'^{\dagger} = 1\!\!1$

(c) Reformulation of the previous results in this basis

We make the corresponding change on the matrix describing the scalar fields $\mathcal{V}'$

$$\mathcal{V}' = \begin{pmatrix} U_{AB}{}^{MN} & , V_{ABMN} \\ \bar{V}^{AB\,MN} & , \bar{U}^{AB}{}_{MN} \end{pmatrix} \tag{3.32}$$

$\partial_\mu \mathcal{V}' \mathcal{V}'^{-1}$ is an element of the Lie algebra of E_7 and can therefore be written

$$\partial_\mu \mathcal{V}' \mathcal{V}'^{-1} = \begin{pmatrix} 2\, Q_{\mu[A}{}^{[C} \delta_{B]}{}^{D]} & , P_{\mu ABCD} \\ \bar{P}_\mu{}^{ABCD} & , 2\, \bar{Q}_\mu{}^{[A}{}_{[C} \delta^{B]}{}_{D]} \end{pmatrix} \tag{3.33}$$

with $P_{\mu ABCD} = \frac{\eta}{24} \varepsilon_{ABCDEFGH} \bar{P}_\mu{}^{EFGH}$

The splitting between $(\partial_\mu \mathcal{V}'\mathcal{V}'^{-1})_{\parallel}$ and $(\partial_\mu \mathcal{V}'\mathcal{V}'^{-1})_{\perp}$ with respect to SU(8) is now obvious

$$(\partial_\mu \mathcal{V}'\mathcal{V}'^{-1})_{\parallel} = 2\begin{pmatrix} Q_{\mu[A}{}^{[C}\,\delta_{B]}{}^{D]} & , & 0 \\ 0 & , & \bar{Q}_\mu{}^{[A}{}_{[C}\,\delta^{B]}{}_{D]} \end{pmatrix} \tag{3.34}$$

$$(\partial_\mu \mathcal{V}'\mathcal{V}'^{-1})_{\perp} = \begin{pmatrix} 0 & , & P_{\mu ABCD} \\ \bar{P}_\mu{}^{ABCD} & , & 0 \end{pmatrix} = D_\mu \mathcal{V}'\mathcal{V}'^{-1} \tag{3.35}$$

Under the transformations

$$\mathcal{V}' \longrightarrow S(x)\mathcal{V}'G \qquad S(x) \in SU(8)\ ,\ G \in E_7 \tag{3.36}$$

$Q_{\mu A}{}^{B}$ and $P_{\mu ABCD}$ are invariant under E_7. $Q_{\mu A}{}^{B}$ transforms as a gauge field for the local SU(8) and $P_{\mu\, ABCD}$ transforms as a covariant tensor for this local SU(8).

For the vector fields we redefine $B_\mu^{MN} = \frac{1}{4}(\Gamma^{ij})_M{}^N B_\mu^{ij}$ and correspondingly $G_{\mu\nu}^{MN}$ and $H_{\mu\nu\, MN}$. $\mathcal{F}$ will become

$$\mathcal{F}'_{\mu\nu} = \frac{1}{\sqrt{2}}\begin{pmatrix} G_{\mu\nu}^{MN} + i\,H_{\mu\nu MN} \\ G_{\mu\nu}^{MN} - i\,H_{\mu\nu MN} \end{pmatrix} = \begin{pmatrix} \mathcal{F}_{\mu\nu MN} \\ \bar{\mathcal{F}}_{\mu\nu}{}^{MN} \end{pmatrix} \tag{3.37}$$

The condition on $\mathcal{F}'$ $\qquad \mathcal{V}'\mathcal{F}'_{\mu\nu} = \Omega'\mathcal{V}'\tilde{\mathcal{F}}'_{\mu\nu}$

is now written if we define $\mathcal{V}'\mathcal{F}' = \begin{pmatrix} \mathcal{F}_{AB} \\ \bar{\mathcal{F}}^{AB} \end{pmatrix}$ as

$$\mathcal{F}_{\mu\nu\, AB} = i\,\tilde{\mathcal{F}}_{\mu\nu\, AB} \tag{3.38}$$

The Lagrangian for the bosonic fields is then written

$$L_B = -\frac{e}{4}R + \frac{e}{24}P_{\mu ABCD}\bar{P}_\nu{}^{ABCD}g^{\mu\nu} + \frac{1}{16}\varepsilon^{\mu\nu\rho\sigma}G_{\mu\nu}^{MN}H_{\rho\sigma MN} \tag{3.39}$$

6 Couplings to the fermions

(a) Let us define the transformations of SU(8) on Majorana spinors : For $\Lambda_A{}^B = \underset{\text{real antisymmetric}}{\Lambda'_A{}^B} + i\,\underset{\text{real symmetric}}{\Lambda''_A{}^B} \in \mathcal{L}ie\,(SU(8))$

$$\delta\lambda_A = (\Lambda'_A{}^B + i\gamma_5\Lambda''_A{}^B)\,\lambda_B \tag{3.40}$$

This preserves the Majorana property of λ^A. It is equivalent to using Weyl spinors

$$\lambda_A^{(R)} = \frac{1+\gamma_5}{2}\lambda_A \quad , \quad \lambda_{(L)}^A = \frac{1-\gamma_5}{2}\lambda_A$$

$$\delta\lambda_A^{(R)} = \Lambda_A{}^B \lambda_B^{(R)} \quad , \quad \delta\lambda^A_{(L)} = \Lambda^A{}_B \lambda^B_{(L)} \quad , \quad \Lambda^A{}_B = \overline{\Lambda_A{}^B} \tag{3.41}$$

(b) We have seen in the previous part that from dimensional reduction the fermionic couplings can be written in terms of the 3 objects $Q^{\circ}_{\mu A}{}^B$, $P^{\circ}_{\mu ABCD}$ and $\mathcal{F}^{\circ}_{AB}$ which are functions of the fields g^{ij}, ϕ^i, A_{ijk}, $B_\mu{}^{ij}$ and $B_\mu{}^i$. Let us now compute $Q_{\mu A}{}^B$, $P_{\mu ABCD}$ defined in section 5, using the scalar matrix $\mathcal{V}$ (or $\mathcal{V}'$) defined in section 4. We find (up to the replacement : i by $i\gamma_5$)

$$Q_{\nu A}{}^B = Q^{\circ}_{\nu A}{}^B \quad , \quad P_{\mu ABCD} = P^{\circ}_{\mu ABCD} \tag{3.42}$$

This shows the SU(8) local invariance for the fermionic terms.(Note the factor 3 between $\psi_{\mu A}$ and λ_{ABC} for the gauge coupling to $Q_{\nu A}{}^B$.)

The coupling to the vector fields appears only through

$$\mathcal{F}^{\circ}_{AB} = \gamma^{\mu\nu} \mathcal{F}^{\circ}_{\mu\nu AB} \tag{3.43}$$

(with always the replacement of $i \to i\gamma_5$). Since $\gamma_{\mu\nu}$ satisfies

$$\tilde{\gamma}_{\mu\nu} = -i\gamma_5\gamma_{\mu\nu} \tag{3.44}$$

only the part of $\mathcal{F}^{\circ\mu\nu}_{AB}$ which satisfies $\mathcal{F}^{\circ\mu\nu}_{AB} = i\tilde{\mathcal{F}}^{\circ\mu\nu}_{AB}$ is coupled to fermions. In fact computing $\mathcal{F}_{AB\mu\nu}$ defined in section 5 with $\mathcal{V}'(\mathcal{V})$ defined in section 4, we find

$$\mathcal{F}_{\mu\nu AB} = \mathcal{F}^{\circ}_{\mu\nu AB} \tag{3.45}$$

(c) We have to take into account that now the real vector for E_7 is no longer $\mathcal{F}$ but $\mathcal{F}^F_{\mu\nu} = (G^{MN}_{\mu\nu}, H^F_{\mu\nu MN})$, where $H^F_{\mu\nu MN}$ contains now fermionic terms. $\mathcal{F}^F_{\mu\nu AB}$ no longer satisfies the constraint $\mathcal{F}^F_{\mu\nu AB} = i\tilde{\mathcal{F}}^F_{\mu\nu AB}$ but the modified one $\hat{\mathcal{F}}_{\mu\nu AB} = i\tilde{\hat{\mathcal{F}}}_{\mu\nu AB}$ with

$$\hat{\mathcal{F}}_{\mu\nu AB} = \mathcal{F}^F_{\mu\nu AB} + 2\sqrt{2}\Big\{\overline{\psi}^{(L)}_{[\mu A}\psi^{(R)}_{\nu]B} - \frac{i}{\sqrt{2}}\overline{\psi}^{(R)C}_{[\mu}\gamma_{\nu]}\lambda^{(R)}_{ABC} + \frac{\sqrt{2}}{288}\varepsilon_{ABCDEFGH}\overline{\lambda}^{CDE}_{(R)}\gamma_{\mu\nu}\lambda^{FGH}_{(L)}\Big\} \tag{3.46}$$

(d) We have not worked completely all quartic terms although they are completely determined by dimensional reduction. We have conjectured that they can be reabsorbed in the Lagrangian by the minimal replacement

$$P_{\mu ABCD} \to \frac{1}{2}(P_{\mu ABCD} + \hat{P}_{\mu ABCD}) \quad , \quad \mathcal{F}_{\mu\nu AB} \to \frac{1}{2}(\mathcal{F}_{\mu\nu AB} + \hat{\mathcal{F}}_{\mu\nu AB}) \tag{3.47}$$

$\omega_{\mu\alpha\beta}$ being given by its equation of motion.
We have checked that all the ψ_μ^4 terms are correct, the dependence of the quartic terms in scalar fields is correct and that the reduction from N = 8 to N = 4 gives the correct quartic terms.

This Lagrangian having the same structure as the general one discussed in section 4, the proof of the E_7 invariance of the theory is exactly the same.

7 The E_7/SU(8) supergravity

We can now write the complete Lagrangian

$$\begin{aligned}\mathcal{L} = & -\frac{e}{4} R(\omega, e) + \frac{1}{2} \varepsilon^{\mu\nu\rho\sigma} \bar{\psi}_{\mu A} \gamma_\sigma \gamma_5 (\delta_A^B D_\nu(\omega) - Q_{\nu A}{}^B) \psi_{\rho B} \\ & + \frac{e}{8} G_{\mu\nu}^{MN} \tilde{H}^{\mu\nu(F)}_{MN} (B, \phi, \psi, \lambda) + \frac{ie}{12} \bar{\lambda}_{ABC} \gamma^\mu (\delta_A^D D_\mu(\omega) - 3 Q_{\mu A}{}^D) \lambda_{BCD} \\ & + \frac{e}{24} P_{\mu ABCD} \bar{P}_\nu^{ABCD} g^{\mu\nu} + \frac{e}{6\sqrt{2}} \bar{\psi}_{\mu A} \gamma^\nu \gamma^\mu (\bar{P}_\nu^{ABCD} + \hat{\bar{P}}_\nu^{ABCD}) \lambda_{BCD} \\ & + \frac{e}{8\sqrt{2}} \left\{ \bar{\psi}_{\mu A} \gamma^{[\nu} \hat{\mathcal{F}}_{AB} \gamma^{\mu]} \psi_{\nu B} - \frac{i}{\sqrt{2}} \bar{\psi}_{\mu C} \hat{\mathcal{F}}_{AB} \gamma^\mu \lambda_{ABC} - \frac{1}{72} \varepsilon^{ABCDEFGH} \lambda_{ABC} \hat{\mathcal{F}}_{DE} \lambda_{FGH} \right\}\end{aligned}$$

with

$$\begin{aligned}\hat{P}_{\mu ABCD} &= P_{\mu ABCD} + 2\sqrt{2} \left[\bar{\psi}^{(L)}_{\mu [A} \lambda^{(R)}_{BCD]} + \frac{1}{24} \varepsilon_{ABCDEFGH} \bar{\psi}^E_{\mu (R)} \lambda^{FGH}_{(L)} \right] \\ \hat{\mathcal{F}}_{AB} &= \gamma^{\mu\nu} \hat{\mathcal{F}}_{\mu\nu AB}\end{aligned} \tag{3.48}$$

$$\begin{aligned}\hat{\mathcal{F}}_{\mu\nu AB} = \mathcal{F}^F_{\mu\nu AB} + \sqrt{2} \Big\{ & \bar{\psi}^{(R)}_{[\mu A} \psi^{(L)}_{\nu B]} - \frac{i}{\sqrt{2}} \bar{\psi}^{(L)C}_{[\mu} \gamma_{\nu]} \lambda^{(R)}_{ABC} \\ & + \frac{1}{288} \varepsilon_{ABCDEFGH} \bar{\lambda}^{CDE}_{(L)} \gamma_{\mu\nu} \lambda^{FGH}_{(R)} \Big\}\end{aligned} \tag{3.49}$$

$\hat{\mathcal{F}}_{\mu\nu AB}$ (and therefore $\mathcal{F}^F_{\mu\nu AB}$) is defined by the constraints

$$\hat{\mathcal{F}}_{\mu\nu AB} = i \, \hat{\tilde{\mathcal{F}}}_{\mu\nu AB} \tag{3.50}$$

We note that in $\mathcal{L}$, one half of the coupling to $\mathcal{F}_{\mu\nu AB}$ has been absorbed in the term $G\tilde{H}^{(F)}$ so that $H^{(F)}$ satisfies as it should

$$e H^{(F)}_{\mu\nu MN} = - \partial \mathcal{L} / \partial \tilde{G}_{\mu\nu MN} \tag{3.51}$$

Let us summarize the invariances of the theory.

(a) It is invariant under reparametrization in 4 dimensions and local Lorentz SO(3,1) transformations.

(b) It is invariant under local SU(8) transformations acting on spinor fields and on scalar fields.

(c) The equations of motion are invariant under global $E_{7(+7)}$ transformations acting on field strengths of the vector fields and on scalar fields.

(d) It is invariant under the following supersymmetry transformations

$$\delta_s e_\mu{}^\alpha = -i\bar{\epsilon}_A \gamma^\alpha \psi_{\mu A} \tag{3.52}$$

$$\delta_s \mathcal{V}' \mathcal{V}'^{-1} = -2\sqrt{2}\begin{pmatrix} 0 & x_{ABCD} \\ \bar{x}^{ABCD} & 0 \end{pmatrix} \quad \text{with}$$

$$x_{ABCD} = \bar{\epsilon}^{(L)}_{[A}\lambda^{(R)}_{BCD]} + \frac{y}{24}\varepsilon_{ABCDEFGH}\bar{\epsilon}^E_{(R)}\lambda^{FGH}_{(L)} \tag{3.53}$$

$\delta_s B_\mu^{MN}$ is derived from

$$\delta_s \begin{pmatrix} B_\mu^{MN} + iC_{\mu MN} \\ B_\mu^{MN} - iC_{\mu MN} \end{pmatrix} = -2\sqrt{2}\mathcal{V}'^{-1}\begin{pmatrix} \bar{\epsilon}^{(L)}_{[A}\psi^{(R)}_{\mu B]} - i\frac{\sqrt{2}}{4}\bar{\epsilon}^C_{(R)}\gamma_\mu\lambda^{(R)}_{ABC} \\ \bar{\epsilon}^{[A}_{(R)}\psi^{B]}_{\mu(L)} - \frac{i\sqrt{2}}{4}\bar{\epsilon}^{(L)}_C\gamma_\mu\lambda^{ABC}_{(L)} \end{pmatrix} \tag{3.54}$$

$C_{\mu MN}$ is the dual potential defined only on-shell

$$\delta_s \psi^{(R)}_{\mu A} = (D_\mu(\omega)\delta_A^B - \mathcal{Q}_{\mu A}{}^B)\epsilon^{(R)}_B - \frac{i}{4\sqrt{2}}\hat{\mathcal{F}}_{AB}\gamma_\mu\epsilon^B_{(L)}$$
$$+ \frac{i}{4}\bar{\lambda}^{(L)}_{ABC}\gamma^\alpha\lambda^{DBC}_{(L)}\gamma_\alpha\gamma_\mu\epsilon^{(R)}_D - \frac{1}{\sqrt{2}}\bar{\psi}^B_{\mu(R)}\gamma^\alpha\lambda^{(R)}_{ABC}\epsilon^C_{(L)} \tag{3.55}$$

$$\delta_s \lambda^{(R)}_{ABC} = -i\sqrt{2}\,\hat{P}_{\mu ABCD}\gamma^\mu\epsilon^D_{(L)} + \frac{3}{4}\hat{\mathcal{F}}_{[AB}\epsilon^{(R)}_{C]} \tag{3.56}$$

We check directly that $\hat{P}_{\mu ABCD}$, $\hat{\mathcal{F}}_{AB}$ and $\mathcal{Q}_{\mu A}{}^B$ are supercovariant.

In fact, having guessed the symmetries of the theory (SU(8) local x E_7 global) we could have written down the Lagrangian and the supersymmetry transformations up to a few numerical coefficients which could have been determined by checking directly the supersymmetry. We shall show how this works for N = 8 supergravity in 5 dimensions in the next part.

8 The symmetric gauge

The gauge (or equivalently the parametrization) which puts the Lagrangian in the SO(8) symmetric form usually used in supergravity is the so-called symmetric gauge. Moreover, it is the only gauge in which there are SU(8) linear transformations acting on the parameters of the vielbein $\mathcal{V}'$ describing the scalar fields (and not only the metric $\mathcal{R}$). By an SU(8) gauge transformation we can impose the condition

$$\mathcal{V}' = \mathcal{V}'^{+}$$

In this gauge, $\mathcal{V}'$ is generated by the part of $E_{7(+7)}$ perpendicular to SU(8) $\mathcal{V}' = \exp X$

$$X = \begin{pmatrix} 0 & w_{ABCD} \\ \bar{w}^{ABCD} & 0 \end{pmatrix} \qquad w_{ABCD} = \frac{\eta}{24}\,\varepsilon_{ABCDEFGH}\,\bar{w}^{EFGH} \tag{3.57}$$

$\mathcal{V}'$ is then written

$$\mathcal{V}' = \begin{pmatrix} ch\sqrt{w\bar{w}} \;, & w\,\dfrac{sh\sqrt{\bar{w}w}}{\sqrt{\bar{w}w}} \\ \bar{w}\,\dfrac{sh\sqrt{w\bar{w}}}{\sqrt{w\bar{w}}} \;, & ch\sqrt{\bar{w}w} \end{pmatrix} \tag{3.58}$$

(w and $\bar{w}$ are considered as 28 x 28 matrices)

At this point, it is useful to introduce the so-called inhomogeneous coordinates of E7/SU(8) y. (This discussion is roughly independent of the coset space considered G/H ; we only need H to be a maximal compact subgroup of G or equivalently G/H to be a symmetric space.)

$$y_{AB,CD} = \left(w\,\frac{th\sqrt{\bar{w}w}}{\sqrt{\bar{w}w}} \right)_{AB,CD} \tag{3.59}$$

This variable y is bounded by $\|y\| < 1$. Then we get

$$\mathcal{V}' = \begin{pmatrix} \dfrac{1}{\sqrt{1-y\bar{y}}} \;, & y\,\dfrac{1}{\sqrt{1-\bar{y}y}} \\ \bar{y}\,\dfrac{1}{\sqrt{1-y\bar{y}}} \;, & \dfrac{1}{\sqrt{1-\bar{y}y}} \end{pmatrix} \tag{3.60}$$

The action of $E_{7\,(+7)}$ on y is simple and gives the y 's their names.

$$\mathcal{V}(y)\begin{pmatrix} A & B \\ C & D \end{pmatrix} = \begin{pmatrix} U(y) & 0 \\ 0 & \overset{*}{U}(y) \end{pmatrix} \mathcal{V}(y') \tag{3.61}$$

constant matrix of E_7 U matrix of SU(8)

with $\quad y' = (A + yC)^{-1}\,(B + yD)$

The SU(8) local transformation U is determined by the condition that $\mathcal{V}(y')$ is symmetric and is reabsorbed by gauge transformations.

Let us note that apart from its definition, there is no simple way to characterize y_{ABCD}, in particular it is not antisymmetric in ABCD. It is easy to compute $\partial_\mu \mathcal{V}'\,\mathcal{V}'^{-1}$ and get $P_{\mu ABCD}$ and $Q_{\mu A}{}^{B}$. In particular

$$P_{\mu\,ABCD} = \left(\frac{1}{\sqrt{1-y\bar{y}}}\;\partial_\mu y\,\frac{1}{\sqrt{1-\bar{y}y}} \right)_{AB,CD} \tag{3.62}$$

so that the kinetic term of the scalar fields is written in the simple way

$$P_{\mu ABCD}\,\bar{P}^{\mu\,ABCD} = \mathrm{Tr}\left(\frac{1}{1-y\bar{y}}\,\partial_\mu y\,\frac{1}{1-\bar{y}y}\,\partial_\mu \bar{y} \right) \tag{3.63}$$

In this gauge the constraint $\mathcal{F}_{\mu\nu AB} = i\,\tilde{\mathcal{F}}_{\mu\nu AB}$ is solved by

$H_{\mu\nu AB} = \mathcal{N}_{AB,CD}\,\tilde{G}^{CD}_{\mu\nu}$ (the notation is now covariant only for SO(8)

and no longer for SU(8)) with

$$\mathcal{N}_{AB,CD}=\left(\frac{\mathbb{1}+y}{\mathbb{1}-y}\right)_{AB,CD} \qquad \text{where} \qquad \mathbb{1}_{AB,CD}=\tfrac{1}{2}(\delta_{AC}\delta_{BD}-\delta_{AD}\delta_{BC}) \tag{3.64}$$

(we recall that i stands for $\frac{1}{2}(g_{\mu\rho}g_{\nu\sigma}-g_{\mu\sigma}g_{\nu\rho})$ and j for $\frac{1}{2e}\varepsilon_{\mu\nu\rho\sigma}$)

Then we get

$$F_{\mu\nu\,AB}=\sqrt{2}\left(\sqrt{1-\bar{y}y}\;\frac{1}{1-y}\right)_{AB,CD} G^{CD}_{\mu\nu} \tag{3.65}$$

The supersymmetry transformations have to be modified because the specific gauge transformations we have made, which put $\mathcal{V}'$ in the symmetric gauge, depend on the scalar fields.

IV HIDDEN SYMMETRIES FOR SUPERGRAVITIES IN D DIMENSIONS WITH N SPINORIAL CHARGES

We have found that for N = 8 supergravity in 4 dimensions, there exists a non-compact global invariance realized non-linearly and a hidden local invariance of the theory. These properties are true for all supergravities. Let us give the general idea to find these symmetries. We shall assume that the scalars are always described by a coset space G/H. H which is isomorphic to the maximal group linearly realized on the physical states is the group of invariance of the algebra of supersymmetry : for instance SU(N) or U(N) in 4 dimensions, USp(2N) in 5 dimensions. The dimension of G is just equal to the sum of the dimension of H and the number of scalar fields. In order for the scalar to have the right signature, H must be the maximal compact subgroup of G, this fixes the signature of G. We expect that except the scalar fields, all bosonic fields are singlet for the local group, G being realized eventually on-shell, the fermionic fields are singlet for the non-compact group G, the scalar fields transform under both groups and can be considered as "vielbein". All these counting arguments fix quite uniquely the group G. We shall have also a guide line by decreasing N or increasing D.

1 Extended supergravities in 4 dimensions

We start from N = 8 and make some consistent truncation which leads to N $<$ 8 supergravity theories. We have to take care of CPT invariance. In this way we recover in particular the global invariance SU(1,1) of N = 4 supergravity. It can be formulated as SU(1,1) x SU(4) global invariance of the equations of motion and U(4) local invariance of the Lagrangian.

For $N \leq 3$, since there is no scalar fields the local invariance and the global invariance are both isomorphic to U(N) and can be reduced to the known global U(N) invariance by a field redefinition.

TABLE IV

N	8	7	6	5	4
Spin 2	1	1	1	1	1
Spin 3/2	8	7+1	6	5	4
Spin 1	28	21+7	15+1	10	6
Spin 1/2	56	35+21	20+6	10+1	4
Spin 0	70	35+35	15+15	5+5	1+1
Global group rank	$E_7(+7)$ 7	$E_7(+7)$ 7	SO*(12) 6	SU(5,1) 5	SU(4) x SU(1,1) 4
Local group rank	SU(8) 7	SU(8) 7	U(6) 6	U(5) 5	U(4) 4

2 Maximal extended supergravities in D dimensions

The maximal supergravities in D dimensions ($D \leq 11$) are all obtained by dimensional reduction of D = 11, N = 1 supergravity. In order to exhibit the maximal symmetries (we have at least SL(11 - D,R)global x SO(11 - D) local)we have to perform duality transformations on tensor fields, for instance

$$
\begin{array}{lll}
D = 7 & A_{\mu\nu\rho} \rightarrow B_{\mu\nu} & \\
D = 6 & A_{\mu\nu\rho} \rightarrow B_{\mu} & \\
D = 5 & A_{\mu\nu\rho} \rightarrow \phi & , A_{\mu\nu} \rightarrow B_{\mu} \\
D = 4 & A_{\mu\nu\rho} \rightarrow \text{NOTHING} & , A_{\mu\nu} \rightarrow \phi
\end{array}
\qquad (4.1)
$$

The global transformations G_D can eventually be realized on field strength of tensor fields and not on the fields themselves. In this case G_D is a symmetry of the equations of motion and not of the Lagrangian for instance D = 8 for $A_{\mu\nu\rho}$, D = 6 for $A_{\mu\nu}$, D = 4 for A_{μ} . If D is odd, G_D is always a symmetry of the Lagrangian. The content of the maximal supergravities as well as their symmetries are summarized in Table V (Cremmer 1980 ; see also Morel & Thierry-Mieg 1981 ; Schwarz 1980)

TABLE V

D=9 GL(2, R)global ⊗ SO(2)local

$1\ e_\mu^r$, $2\ \Psi_\mu$, $1A_{\mu\nu\rho}$, $2A_{\mu\nu}$, $3A_\mu$, $4\ \chi$, 3 scalars

D=8 $E_{3(+3)}$=SL(3,R) x SL(2,R)global ⊗ [SO(3) x SO(2)] local

$1\ e_\mu^r$, $2\ \Psi_\mu$, $1A_{\mu\nu\rho}$ (wavy underline), $3A_{\mu\nu}$, $6A_\mu$, $6\ \chi$, 7 scalars

D=7 $E_{4(+4)}$=SL(5,R)global ⊗ SO(5)local

$1\ e_\mu^r$, $4\ \Psi_\mu$, $5A_{\mu\nu}$, $10A_\mu$, $16\ \chi$, 14 scalars

D=6 $E_{5(+5)}$=SO(5, 5)global ⊗ SO(5) x SO(5) local

$1\ e_\mu^r$, $4\ \Psi_\mu$, $5A_{\mu\nu}$ (wavy underline), $16A_\mu$, $20\ \chi$, 25 scalars

D=5 $E_{6(+6)}$global ⊗ USp(8)local

$1\ e_\mu^r$, $8\ \Psi_\mu$, $27A_\mu$, $48\ \chi$, 42 scalars

D=4 $E_{7(+7)}$global ⊗ SU(8)local

$1\ e_\mu^r$, $8\ \Psi_\mu$, $28A_\mu$ (wavy underline), $56\ \chi$, 70 scalars

D=3 $E_{8(+8)}$global ⊗ SO(16)local

$1\ e_\mu^r$, $16\Psi_\mu$, $128\ \chi$, 128 scalars

The global symmetries are realized on field strengths for the underlying fields. Let us note that in 3 dimensions there are no degrees of freedom for the graviton and the gravitino.

In all dimensions, we have of course the same number of degrees of freedom i.e. 128 bosonic states and 128 fermionic states. For D = 3...8, the global group is E_{11-D}. It has been suggested by Julia (1981 a) that there could exist an inverse process of dimensional reduction : the group desintegration. It is based on the observation that in D dimension the local group x little spin group is always a maximal subgroup of SO(16). A similar statement holds for the global invariances

$$E_8 \supset E_{11-D} \otimes SL(D-2) \quad , SO(16) \supset H_D \otimes SO(D-2) \qquad (4.2)$$

The scalars are described by the coset E_{11-D}/H_D and the on-shell graviton is described by SL(D - 2)/SO(D-2).

In two dimensions, we expect that both G_D and H_D will have an infinite number of generators, in particular G_D should be the affine group $E_8^{(1)}$ ($\sim E_9$) associated to E_8 (Julia 1981 b).

3 Supergravities in 5 dimensions

In 5 dimensions, the symmetries are invariances of the Lagrangian, so that the structure should be simpler. The knowledge of these theories in 5 dimensions not only allows a better understanding of the 4 dimensional theories, but also leads to new things such as supersymmetry breaking or off-shell formulation in 4 dimensions via various dimensional reduction procedure.

(a) Notations

The metric will be (+ - - - -). As has been explained in part I, we choose as Clifford algebra $\{\gamma_r, \gamma_s\} = 2\eta_{rs}$ the one using the 4-dimensional γ matrices, γ_0, γ_i (i = 1,2,3) pure imaginary and $\gamma_4 = i\gamma_5$ real with the relation $\gamma_{rstuv} = \varepsilon_{rstuv}$. There are no Majorana spinors in 5 dimensions. Instead of using Dirac spinors, we can double the spinors and use new "reality" conditions. We have only 2N supersymmetry in 5 dimensions. They are classified by USp(2N) (as the massive multiplet in 4 dimensions with central charge $\sim$ 5th dimensions). Defining, Ω^{ab}, the antisymmetric symplectic matrix of USp(2N) (a = 1...N), and using it to lower or raise indices, we define the new "reality" conditions

Bosons $\quad A_\mu^{\ ab} = A^{*}_{\mu\ ab}$

Fermions $\quad \psi_\mu^{\ a} = \gamma_5 \psi^{*}_{\mu a}$

The algebra of supersymmetry is then written

$$\{\bar{Q}^a_\alpha, Q^b_\beta\} = \Omega^{ab} \gamma^\mu_{\alpha\beta} P_\mu \qquad (4.3)$$

(b) Physical content of supergravities

The representation of USp(2N) appearing in supergravity are traceless antisymmetric tensors for instance A_μ^{ab}, χ^{abc}, with

$$\Omega_{ab} A_\mu^{\ ab} = 0 \quad , \quad \Omega_{ab} \chi^{abc} = 0 \qquad (4.4)$$

The representations of supergravity are obtained from the lowest spin representation of 2N supersymmetry by multiplication by appropriate angular momentum (Ferrara & Zumino 1979), (SO(3) classifies the massless states in 5 dimensions) and are given in Table VI.

TABLE VI

s	2	3/2	1	1/2	0	group
N=8	1	8	27	48	42	USp(8)
N=6	1	6	14+1	14'+6	14	USp(6)
	$(J=\frac{1}{2})$	⊗ [1	6	14	14']	
N=4	1	4	5+1	4	1	USp(4)
	(J=1)	⊗	[1	4	5]	
N=2	1	2	1			USp(2)
		$(J=\frac{3}{2})$	⊗	[1 ,	2]	

By counting arguments we can conjecture what the global and the local invariances of the Lagrangian should be. This gives the following results

N=8	$E_{6(+6)}$ global ⊗ USp(8) local
N=6	SU*(6) global ⊗ USp(6) local
N=4	USp(4) x R global ⊗ USp(4) local
N=2	USp(2) global ⊗ USp(2) local

(c) N = 8 supergravity in 5 dimensions

We shall show how from the knowledge of the global and local symmetries $E_{6(+6)}$ and USp(8) it is possible to construct the Lagrangian for N = 8 supergravity in 5 dimensions directly (Cremmer et al. 1978 ; Cremmer 1981).

Let us briefly describe $E_{6(+6)}$ by its infinitesimal transformations in the fundamental representation of dimension 27. They act on the vector space spanned by $z^{\alpha\beta} = -z^{\alpha\beta}$ $(= z^{*}_{\alpha\beta})$ $\Omega_{\alpha\beta} z^{\alpha\beta} = 0$ $(\alpha, \beta = 1 \ldots 8)$ and are given by

$$\delta z^{\alpha\beta} = \Lambda^{\alpha}{}_{\gamma} z^{\gamma\beta} + \Lambda^{\beta}{}_{\gamma} z^{\alpha\gamma} + \Sigma^{\alpha\beta\gamma\delta} z_{\gamma\delta} \qquad (4.5)$$

There are 78 generators, 36 compact ones generating the maximum compact subgroup Sp(8) of parameters $\Lambda^{\alpha}{}_{\gamma}$ antihermitian such that $\Lambda_{\alpha\gamma}$ is symmetric, and 42 non-compact ones with parameters $\Sigma^{\alpha\beta\gamma\delta}$ antisymmetric and traceless.

E_6 has no bilinear invariant but has a trilinear invariant

$$J = Z^{\alpha\beta}\Omega_{\beta\gamma}Z^{\gamma\delta}\Omega_{\delta\epsilon}Z^{\epsilon\lambda}\Omega_{\lambda\alpha} \qquad (4.6)$$

These properties will be sufficient to write the general structure of the theory. The field content is : 1 graviton e_μ^r, 8 gravitinos ψ_μ^a, 27 vector fields $A_\mu^{\alpha\beta}$ ($A_\mu^{\alpha\beta} = -A_\mu^{\beta\alpha} = (A_{\mu\,\alpha\beta})^*$, $\Omega_{\alpha\beta}A_\mu^{\alpha\beta} = 0$), 48 spin 1/2 fields χ^{abc} (antisymmetric, pseudoreal and traceless) and 42 scalar fields described by a matrix 27 x 27 of E_6 $\mathcal{V}_{\alpha\beta}{}^{ab}$ up to a local transformation of USp(8).

The scalar fields being described by an element $\mathcal{V}_{\alpha\beta}{}^{ab}$ of the coset space E_6/USp(8), their self interaction is described by the associated non-linear σ-model with Lagrangian

$$\mathcal{L} \simeq D_\mu \mathcal{V}_{\alpha\beta}{}^{ab}\, D^\mu (\mathcal{V}^{-1})_{ab}{}^{\alpha\beta} \sim -\mathrm{Tr}(\mathcal{V}^{-1}D_\mu\mathcal{V})^2 \qquad (4.7)$$

D_μ being the covariant derivative with respect to USp(8) using the connection $Q_{\mu\,a}{}^b$ defined by

$$\mathcal{V}_{cd}{}^{\alpha\beta}\,\partial_\mu \mathcal{V}_{\alpha\beta}{}^{ab} = 2\,Q_{\mu[c}{}^{[a}\,\delta_{d]}{}^{b]} + P_\mu{}^{ab}{}_{cd} \qquad (4.8)$$

$\in$ Lie (E_6) $\qquad$ $\in$ Lie (USp(8)) $\qquad$ $\notin$ Lie (USp(8))

The Lagrangian is then written

$$\mathcal{L} \simeq |P_{\mu\,abcd}|^2 \qquad (4.9)$$

$P_{\mu\,abcd}$ as well as $Q_{\mu\,a}{}^b$ are invariant under E_6. We can also describe the scalar fields by a metric for this coset space which is covariant for E_6 and invariant for USp(8) $g_{\alpha\beta,\gamma\delta}$

$$g_{\alpha\beta,\gamma\delta} = \mathcal{V}_{\alpha\beta}{}^{ab}\,\Omega_{ac}\,\Omega_{bd}\,\mathcal{V}_{\gamma\delta}{}^{cd} \qquad (4.10)$$

the Lagrangian can be written as

$$\mathcal{L} \sim \partial_\mu g_{\alpha\beta,\gamma\delta}\,\partial^\mu (g^{-1})^{\alpha\beta,\gamma\delta} \qquad (4.11)$$

However, g cannot be used for describing the couplings to fermions (analogy with $g_{\mu\nu}$ and $e_\mu{}^r$)

Since there is no quadratic invariant for the vector fields, in order to construct an E_6 invariant kinetic term for the vector fields we must use the metric $g_{\alpha\beta,\gamma\delta}$

$$\mathcal{L}_{V^2} \sim \sqrt{g}\, \mathcal{G}_{\alpha\beta,\gamma\delta}\, F_{\mu\nu}^{\alpha\beta} F_{\rho\sigma}^{\gamma\delta} g^{\mu\rho} g^{\nu\sigma} \tag{4.12}$$

As in 11 dimensions, there exists a trilinear gauge invariant coupling (up to a total derivative). Since there exists a trilinear invariant for E_6 we do not need the scalar metric $\mathcal{G}$ (nor the tensor metric $g_{\mu\nu}$)

$$\mathcal{L}_{V^3} \sim \varepsilon^{\mu\nu\rho\sigma\lambda}\, \Omega_{\alpha\beta} F_{\mu\nu}^{\beta\gamma}\, \Omega_{\gamma\delta} F_{\rho\sigma}^{\delta\epsilon}\, \Omega_{\epsilon\eta} A_{\lambda}^{\eta\alpha} \tag{4.13}$$

Since the fermions are scalar for E_6, the coupling of the bosons to fermions must appear through E_6 invariant bosonic expressions : $P_{\mu\, abcd}$, $Q_{\mu a}{}^{b}$ (in covariant derivative $D_\mu \psi_\nu{}^a$ and $D_\mu \chi_{abc}$) or $F_{\mu\nu}{}^{ab} = \mathcal{V}_{\alpha\beta}{}^{ab} F_{\mu\nu}^{\alpha\beta}$. The supersymmetry transformation laws $\delta\phi$ are assumed to be covariant with respect to USp(8) and E_6. At this step $\mathcal{L}$ and $\delta\phi$ are determined up to numerical coefficients and quartic fermionic terms. In particular all the non-polynomial structure of the scalar fields is known. Supersymmetry of $\mathcal{L}$ and closure of the algebra are used to get rid of the remaining arbitrariness. We then get the Lagrangian

$$\begin{aligned} e^{-1}\mathcal{L} = & -\frac{1}{4}R - \frac{i}{2}\bar{\psi}_\mu{}^a \gamma^{\mu\nu\rho} D_\nu \psi_{\rho a} - \frac{1}{8} g^{\mu\rho} g^{\nu\sigma} \mathcal{G}_{\alpha\beta,\gamma\delta} F_{\mu\nu}^{\alpha\beta} F_{\rho\sigma}^{\gamma\delta} \\ & + \frac{i}{12}\bar{\chi}^{abc}\gamma^\mu D_\mu \chi_{abc} - \frac{1}{24} g^{\mu\nu} D_\mu \mathcal{V}_{\alpha\beta}{}^{ab} D_\nu (\mathcal{V}^{-1})^{\alpha\beta}{}_{ab} \\ & - \frac{e^{-1}}{12}\varepsilon^{\mu\nu\rho\sigma\lambda} (F_{\mu\nu})^\alpha{}_\beta (F_{\rho\sigma})^\beta{}_\gamma (A_\lambda)^\gamma{}_\alpha + \frac{i}{3\sqrt{2}} P_\rho{}^{abcd} \bar{\psi}_{\mu a}\gamma^\rho\gamma^\mu \chi_{bcd} \\ & + \frac{i}{4}\mathcal{V}_{\alpha\beta}{}^{ab} F_{\mu\nu}^{\alpha\beta}\Big[\bar{\psi}_a^\rho \gamma_\rho \gamma^{\mu\nu}\gamma_\sigma \psi_b^\sigma + \frac{1}{\sqrt{2}}\bar{\psi}_\rho^c \gamma^{\mu\nu}\gamma^\rho \chi_{abc} + \frac{1}{2}\bar{\chi}_{acd}\gamma^{\mu\nu}\chi_b{}^{cd}\Big] \end{aligned} \tag{4.14}$$

+ quartic terms

We shall not write the quartic terms (see Cremmer 1981). It is invariant under global $E_{6(+6)}$, local USp(8) and the following supersymmetry transformations (up to trilinear fermionic terms)

$$\delta e_\mu{}^r = -i\,\bar{\varepsilon}^a \gamma^r \psi_{\mu a} \tag{4.15}$$

$$(\mathcal{V}^{-1})^{\alpha\beta}{}_{cd}\, \delta \mathcal{V}_{\alpha\beta,ab} = -2i\sqrt{2}\,\big(\bar{\varepsilon}_{[a}\chi_{bcd]} + \frac{3}{4}\Omega_{[ab}\bar{\varepsilon}_e \chi^e{}_{cd]}\big) \tag{4.16}$$

$$\delta A_\mu{}^{\alpha\beta} = 2i\, \mathcal{V}^{\alpha\beta}{}_{ab}\,\big(\bar{\varepsilon}^a \psi_\mu^b + \frac{1}{2}\bar{\varepsilon}_c \gamma_\mu \chi^{abc}\big) \tag{4.17}$$

$$\delta\psi_{\mu a} = (D_\mu(\hat\omega)\delta_a^b + \varphi_{\mu a}{}^b)\epsilon_b - \frac{1}{6}\hat F^{\alpha\beta}_{\rho\sigma} v_{\alpha\beta ab}(\gamma^{\rho\sigma}\gamma_\mu + 2\gamma^\rho\delta^\sigma_\mu)\epsilon^b + \cdots \tag{4.18}$$

$$\delta\chi_{abc} = \sqrt{2}\,\hat P_{\mu abcd}\gamma^\mu\epsilon^d - \frac{3}{2\sqrt{2}}\gamma^{\rho\sigma}\hat F^{\alpha\beta}_{\rho\sigma}\left(v_{\alpha\beta[ab}\epsilon_{c]} + \frac{1}{3}\Omega_{[ab} v_{\alpha\beta c]d}\epsilon^d\right) + \cdots \tag{4.19}$$

(d) N = 2 supergravity in 5 dimensions

By consistent truncation we obtain all supergravities in 5 dimensions and their symmetries, in particular the N = 2 supergravity whose Lagrangian is

$$e^{-1}\mathcal{L} = -\frac{1}{4}R(\omega) - \frac{i}{2}\bar\psi_\mu{}^a\gamma^{\mu\nu\rho}D_\nu\left(\frac{\omega+\hat\omega}{2}\right)\psi_{\rho a} - \frac{1}{4}F_{\mu\nu}F_{\rho\sigma}g^{\mu\rho}g^{\nu\sigma}$$
$$+\frac{e^{-1}}{6\sqrt{3}}\varepsilon^{\mu\nu\rho\sigma\lambda}F_{\mu\nu}F_{\rho\sigma}A_\lambda - \frac{i\sqrt{3}}{16}(F_{\mu\nu}+\hat F_{\mu\nu})\bar\psi^{\rho c}\gamma_{[\rho}\gamma^{\mu\nu}\gamma_{\sigma]}\psi^\sigma_c \tag{4.20}$$

$\omega_{\mu rs}$ is defined by its own equation of motion if $\hat\omega_{\mu rs} = \omega_{\mu rs} + \frac{i}{4}\bar\psi^{\rho a}\gamma_{\mu rs\rho\sigma}\psi^\sigma_a$

$$\hat F_{\mu\nu} = F_{\mu\nu} + \frac{\sqrt{3}}{4}\bar\psi_\mu{}^c\psi_{\nu c} \tag{4.21}$$

It is invariant under the following supersymmetry transformations

$$\delta e_\mu^r = -i\,\bar\epsilon^c\gamma^r\psi_{\mu c} \tag{4.22}$$

$$\delta\psi_{\mu a} = \left[D_\mu(\hat\omega) + \frac{1}{4\sqrt{3}}\hat F_{\rho\sigma}(\gamma^{\rho\sigma}\gamma_\mu + 2\gamma^\rho\delta^\sigma_\mu)\right]\epsilon_a \tag{4.23}$$

$$\delta A_\mu = -\frac{\sqrt{3}}{4}\bar\epsilon^c\psi_{\mu c} \tag{4.24}$$

The structure of the theory is completely analogous to the 11-dimensional supergravity from which we can derive all supergravities by dimensional reduction and consistent truncation.

V POSSIBLE IMPLICATIONS OF THE SYMMETRIES : CONJECTURES FOR N = 8 SUPERGRAVITY IN 4 DIMENSIONS

In this part, we shall summarize what we know and what we would like to know about N = 8 supergravity at the classical or at the quantum level. This will be developed in more detail in other contributions to the School or to the Workshop. We shall also state the two conjectures for N = 8 supergravity and their implication about the possible relevance of N = 8 supergravity to particle physics.

1 Other problems for N = 8 supergravity

(a) We have seen that the existence of the global E_7 and the local SU(8) symmetries has allowed a "geometrical" description of the scalar fields as describing a coset space E_7/SU(8). A natural question arises : is there a possible "geometrical" interpretation of the spin 1/2 fields ? Although it has not been yet found, it seems that a natural answer should be yes. This should likely provide a tremendous simplification of the quartic fermionic terms and give more light on the supersymmetric structure and eventually on auxiliary fields.

(b) Off-shell formulation

There exists superspace formulation of N = 8 supergravity consistent with the symmetries E_7 and SU(8) but it is an on-shell formulation in the sense that the Bianchi identities on torsion, curvature... imply the equations of motion. The situation is the same for the N = 1 supergravity in 11 dimensions where all geometrical quantities can be written in terms of a single superfield $F_{rstu}(x, \theta)$ satisfying an equation (which implies the equations of motion) (Cremmer & Ferrara 1980 ; Brink & Howe 1980).

At the linearized level of N = 8 supergravity, some attempts have been made to obtain auxiliary fields using the dimensional reduction by Legendre transform starting with N = 8 linearized supergravity in 5 dimensions (Cremmer et al. 1980). This leads to fields which satisfy differential constraints (for gauge fields) which cannot be solved exactly (at least without introducing ghosts or higher derivative terms in the resulting Lagrangian). These constraints can be imposed by Lagrange multiplier, but we lose the closure of the algebra (on the Lagrange multipliers particularly). Moreover, this formulation even if it can be extended to the non-linearized case, will not be consistent with the symmetries E_7 and SU(8) (probably only with E_6 and USp(8)).

(c) Supersymmetry breaking

Using a generalized dimensional reduction of Scherk and Schwarz (1979 b), it is possible starting from N = 8 supergravity in 5 dimensions to derive a N = 8 supergravity in 4 dimensions with 4 mass parameters (Cremmer et al. 1979). This new theory is still invariant under some local like-supersymmetry transformations. These new transformations are spontaneously broken and we cannot extract a global algebra from them. However, at one loop, this theory seems still finite (Sezgin & Van Nieuwenhuizen 1981).

(d) Gauging of O(8)

The possibility of gauging O(N) in N-supergravity in 4 dimensions is suggested by the fact that the vector fields of the supergravity multiplet are in general in the adjoint representation of O(N) (true for

N = 1...5 and 8, for N = 6 there is an extra singlet, N = 7 is identical to N = 8). This has been done completely for N = 1...5 and partially for N = 8 (de Wit & Nicolai 1981). This allows the introduction of a new coupling constant : the dimensionless gauge coupling. This requires a huge cosmological constant or a scalar potential which is unbounded from below. This gauging of O(N) can also be viewed as a gauging of the super De Sitter algebra instead of the super Poincaré algebra, the coupling constant being related to the De Sitter radius. Both have the same local algebra . The introduction of the gauge coupling breaks the global invariance and even its subgroup U(N).

The possibility of gauging O(N) has provided the first attempt to superunification, unifying in a same multiplet the vector gauge fields and the graviton (and gravitinos). But, this attempt has not been successful to reproduce the present low energy phenomenology. Let us sketch the arguments for such a statement. First of all, if we want to avoid particles with spin$>$2 and/or several spin 2 massless particles, since we have no consistent interacting theories for them, we must have $N \leq 8$. Assuming that we can gauge SO(8) and that we can solve the problems due to unbounded potential from below, we immediately see that $SO(8) \not\supset SU(3)_c \times SU(2) \times U(1)$. Despite this problem Gell Mann (1977) has tried to go on noting that $SO(8) \supset SU(3)_c \times U(1)$ and has made an analysis based on the vector-like description $8 \to 3 + \bar{3} + 1 + 1$ with respective charge (-1/3, 1/3, 1, 0). This has shown that at least the muon and its neutrino, the τ and ν_τ , $W^\pm$ are missing.

2 Quantum corrections

This will be discussed in detail by P. Van Nieuwenhuizen, M. Duff and R. Kallosh in this book. As gravity, supergravity cannot be renormalizable in the ordinary sense because the coupling constant K has a dimension. This implies that the counterterms have not the same structure as the original Lagrangian and prevent the absorption of infinities in renormalization constants. Therefore, such a theory can only be finite (or meaningless in perturbation theory).

(a) 1-loop counterterm on shell

For pure gravity with cosmological constant Λ whose action is

$$S = -\frac{1}{2\kappa^2}\int d_4x \sqrt{g}\,(R-2\Lambda) \tag{5.1}$$

After use of the equation of motion (or redefinition of the background field) the counterterm can be written

$$\Delta S = -\frac{1}{D-4}\,(A\chi + B\delta) \tag{5.2}$$

with
$$\delta = -\frac{\kappa^2\Lambda}{12\pi^2}\,S \tag{5.3}$$

$$\chi = \frac{1}{32\pi^2}\int d_4x\,\underbrace{\sqrt{g}\,(R_{\mu\nu\rho\sigma}^{\;2} - 4R_{\mu\nu}^{\;2} + R^2)}_{\text{total divergence}} = \text{integer} \tag{5.4}$$

A and B are constants $A = 106/45$, $B = -87/10$.
Therefore, pure gravity is 1-loop finite if $\Lambda = 0$ (no cosmological constant) and $\chi = 0$ (the topological structure of space-time is trivial). For supergravity, S, δ and χ are extended to supersymmetric invariants. It is found that

. if $N \geqslant 5$, then $B = 0$ (this implies that $\beta(g) = 0$ if g is the O(N) gauge coupling constant ; this has been checked directly).

. A is an integer for $N \geqslant 3$

. For N = 8 if we use the field description of dimensional reduction (63 scalars $+7A_{\mu\nu} + 1A_{\mu\nu\rho}$) then A = 0.
This applies also to N = 4 if we use 1 scalar + $1A_{\mu\nu}$. Although these theories have not been constructed for N = 5 and 6 we can find such field descriptions using $A_{\mu\nu}$ and $A_{\mu\nu\rho}$ fields which make A = 0.

. There are no boundary contributions to the 1-loop counterterms for N = 8.

N = 8 supergravity is, up to now, the only theory including gravity which is completely finite at 1-loop.

(b) <u>N-loop counterterms</u>

There exists no supersymmetric extension of the 2-loop counterterms of pure gravity. At the linearized level there exist 3-loop counterterms (which could eventually violate E_7 symmetry for N = 8 supergravity ?). Finally there exists a 8-loop counterterm which respects all the symmetries of N = 8 supergravity.

(c) <u>Conjecture N° 1</u>

<u>N = 8 supergravity exists !</u> (it is finite)

Although there exist counterterms, the example of N = 4 supersymmetric Yang-Mills is encouraging. This could be linked to the fact that the multiplet of N = 8 supergravity as well as N = 4 super Yang-Mills are CPT self-conjugate. However, let us mention that it has been conjectured that all $N \geqslant 3$ supersymmetric theories should be finite. This conjecture implies that we cannot adjust parameters through the renormalization procedure. K is not really a parameter but only a mass scale.

Then supergravity N = 8 (ungauged) has no parameter.

3 Implications of the local symmetry SU(8)

The existence of a local symmetry whose gauge fields are composite fields as in the CP^N models (D'Adda et al. 1978 ; Witten 1979) (or SU(P + N)/U(P) x SU(N))in 2 dimensions (where they are renormalizable) leads to the following conjecture

Conjecture N°2

At the quantum level, the local symmetry SU(8) becomes dynamical. In particular the SU(8) gauge fields can propagate (acquire a kinetic term) and have a Yang-Mills type of coupling with gauge constant g which should be computable (conjecture N°1).

Since SU(8) is big enough to accommodate a grand unified group (like SU(5) for instance), this eliminates the problem found with the gauged SO(8) supergravity and leads to a completely different concept of superunification.

Supergravity is fundamental. It should be viewed as a preon type of theory whose spectrum we should compute.

By supersymmetry we can conjecture that not only the SU(8) gauge fields become dynamical but also other fields and that they could form together a supersymmetric multiplet.

Some conjectures on this multiplet have been made by Ellis et al. (1980 a,b). Namely they choose it as

$$\left(\tfrac{3}{2}\right)^A , \left(1\right)^A_{\ B} ; \left(\tfrac{1}{2}\right)^A_{\ [BC]} \cdots\cdots\cdots \left(-\tfrac{5}{2}\right)^A \quad + \text{ CPT conjugate}$$

Making some drastic simplifications (and far from being justified) they can imagine a scenario leading to a breaking of supersymmetry and of SU(8) at the Planck mass directly into SU(5). The massless states they keep are those of a grand unified theory SU(5), a maximal principle leading to 3 families of fermions $(\overline{5} + 10)_L$. All particles are originally in the same multiplet. Some speculations have also been made concerning the restoration of symmetries i.e. the E_7 symmetries of the level of the bound states. Since E_7 is non-compact, the only unitary representation have infinite dimension.

More recently, under more conservative assumptions especially in the spin 1/2 sector, Derendinger et al.(1981) have shown that there exist some combinations of supermultiplets which have an anomaly free sector of spin 1/2 : $\left[\text{real SU(8) or SU(8) families } (8 + \overline{28} + 56)_L\right]$ or [real SU(5) or SU(5) families $(\bar{5}+10)_L$]. The first choice (GUTSU(8)) is not compatible with the requirement that these multiplets should be composite states of N = 8 supergravity. The second choice (GUTSU(5)) implies that there should be an even number of SU(5) families and requires a large number of multiplets ($\sim$ 15).

Although these scenarios are far from being justified, this shows that provided the two conjectures are true, the physical spectrum of N = 8 supergravity is probably rich enough to accommodate present low energy phenomenology ($\lesssim 10^{15}$ GeV). To go further, we should have a better understanding of the structure of the N = 8 supergravity, especially the structure of the "multiplet" which contains the SU(8) gauge fields and the behaviour of the various members of this multiplet under the global E_7 and the local SU(8). We should also have solved at least partially the problem of auxiliary fields which should provide a "linearization" of the fermionic terms of the Lagrangian necessary to begin to study the dynamics, in particular the appearance of a kinetic term and Yang-Mills coupling for the SU(8) gauge fields. We are also facing now a new problem : what is a finite theory ? In particular, it should be useful to know if there exist some properties which could replace the concept of asymptotic freedom for renormalizable theories.

REFERENCES

Additional references can be found in the review papers by Fayet & Ferrara (1977), Van Nieuwenhuizen (1980) and Scherk (1979)

Aurilia, A., Nicolai, H. & Townsend, P.K. (1981). Spontaneous breaking of supersymmetry and the cosmological constant. In Superspace and Supergravity, ed. S.W. Hawking & M. Rocek, pp. 403-412. Cambridge : Cambridge University Press

Brink, L, & Howe, P; (1979). The N = 8 Supergravity in Superspace, Phys. Lett. 88B, pp. 268-272

Brink, L. & Howe, P. (1980). Eleven dimensional supergravities on the mass-shell in superspace, Phys. Lett 91B, pp. 384-386

Brink, L. , Scherk, J. & Schwarz, J.H. (1977). Supersymmetric Yang-Mills theories, Nucl. Phys. B121, pp. 77-92

Callan, C., Coleman, S., Wess, J. & Zumino, B. (1969). The structure of phenomenological Lagrangians II, Phys. Rev. 177, pp. 2247-2250

Cho, Y.M. & Freund, P.G.O. (1975). Non-Abelian gauge fields as Nambu-Goldstone fields, Phys. Rev. D12, pp. 1711-1720

Cho, Y.M. & Jang, P.S. (1975). Unified geometry of internal space with space-time, Phys. Rev. D12, pp. 3789-3792

Coleman, S., Wess, J. & Zumino, B. (1969). The structure of phenomenological Lagrangians I, Phys. Rev. 177, pp. 2239-2247

Cremmer, E. (1980). The N = 8 supergravity In Unification of the fundamental particle interactions, ed. S. Ferrara, J. Ellis & P. Van

Nieuwenhuizen, pp.137-155. New York : Plenum Publishing Corporation.
Cremmer, E.(1981). Supergravities in 5 dimensions. In Superspace and Supergravity, ed. S.W. Hawking & M. Rocek, pp. 267-282. Cambridge : Cambridge University Press
Cremmer, E. & Ferrara, S. (1980). Formulation of 11-dimensional supergravity in superspace, Phys. Lett. 91B, pp. 61-66
Cremmer, E. & Julia, B. (1978). The N = 8 supergravity theory I The Lagrangian, Phys. Lett 80B, pp. 48-51
Cremmer, E. & Julia, B. (1979). The SO(8) supergravity, Nucl. Phys. B159, pp.141-212
Cremmer, E. & Scherk, J. (1976). Dual models in four dimensions with internal symmetries, Nucl. Phys. B103, pp. 399-425
Cremmer, E. & Scherk, J. (1977 a). Spontaneous compactification of extra-space dimensions, Nucl. Phys. B118, pp. 61-75
Cremmer, E. & Scherk, J. (1977 b). Algebraic simplifications in supergravity theories, Nucl. Phys. B127, pp. 259-268
Cremmer, E. ; Ferrara, S. & Scherk, J. (1977 a). U(N) invariance in extended supergravity, Phys. Lett. 68B, pp. 234-238
Cremmer, E.; Horvath, Z. ; Palla, L. & Scherk, J. (1977 b). Grand unified schemes and spontaneous compactification, Nucl. Phys. B127, pp. 57-65
Cremmer, E. ; Ferrara, S. & Scherk, J. (1978 a). SU(4) invariant supergravity theory, Phys. Lett. 74B, pp. 61-64
Cremmer, E. ; Julia, B & Scherk, J. (1978 b). Supergravity theory in 11 dimensions, Phys. Lett. 76B, pp. 409-412
Cremmer, E. ; Scherk, J; & Schwarz, J.H. (1979). Spontaneous broken N = 8 supergravity, Phys. Lett. 84B, pp. 83-86
Cremmer, E. ; Ferrara, S. ; Stelle, K. & West, P.C. (1980). Off shell N = 8 supersymmetry with central charges, Phys. Lett. 94B, pp. 349-354
D'Adda, D. ; Divecchia, P. & Lüscher, M. (1978). A 1/N expandable series of non-linear σ-models with instantons, Nucl. Phys. B146, pp. 63-76
Deredinger, J.P. ; Ferrara, S. & Savoy, C.A. (1981). Flavor and superunification, Preprint CERN TH 3052
De Wit, B. (1979). Properties of SO(8) supergravity, Nucl. Phys. B158, pp. 189-212
De Wit, B. & Freedman, D.Z. (1977). On SO(8) extended supergravity, Nucl. Phys. B130, pp. 105-113
De Wit, B. & Nicolai, H. (1981). Extended supergravity with local SO(5) invariance, Nucl. Phys. B188, pp. 98-108
Duff, M. (1981). Antisymmetric tensors and supergravity. In Superspace and Supergravity, ed. S.W. Hawking & M. Rocek, pp. 381-402. Cambridge : Cambridge University Press
Ellis, J. ; Gaillard, M.K. & Zumino, B. (1980 a) A grand unified theory obtained from broken supergravity, Phys. Lett. 94B, pp. 343-348
Ellis, J. ; Gaillard, M.K. ; Maiani, L. & Zumino, B. (1980 b). In Unification of the Fundamental Particle Interactions, ed. S. Ferrara, J. Ellis & P. Van Nieuwenhuizen, pp. 69-88
Fayet, P. & Ferrara, S. (1977). Supersymmetry, Phys. Report 32C, pp. 250-334
Ferrara, S. & Zumino, B. (1979). The mass matrix of N = 8 supergravity, Phys. Lett. 86B, pp. 279-282
Ferrara, S. ; Scherk, J. & Zumino, B. (1977). Algebraic properties of

extended supergravity theories, Nucl. Phys. B121, PP. 393-402
Gaillard, M.K. & Zumino, B. (1981). Duality rotations for interacting fields, Preprint LAPP,TH 37/CERN TH 3078
Gell-Mann, M. (1977). Talk at the Washington Meeting of the A.P.S.
Gliozzi, F. ; Olive, D. & Scherk, J. (1977). Supersymmetry, supergravity theories and the dual spinor model, Nucl. Phys. B122, pp. 253-290
Grisaru, M. (1981). Contribution to this school
Julia, B. (1981 a). Group desintegrations. In Superspace and Supergravity ed. S.W. Hawking & M. Rocek, pp. 331-350. Cambridge : Cambridge University Press
Julia, B. (1981 b). Infinite Lie algebra in physics, Preprint LPTENS 81/14 Invited talk at the Johns Hopkins Workshop on Particle Theory
Luciani, J.F. (1978). Space-time geometry and symmetry breaking, Nucl. Phys. B135, pp. 111-130
Morel, B. & Thierry-Mieg (1981). Superalgebras in exceptional gravity. In Superspace and supergravity, ed. S.W. Hawking & M. Rocek, pp. 351-362. Cambridge : Cambridge University Press
Nicolai, H. ; Townsend, P.K. & Van Nieuwenhuizen, P. (1981). Comments on 11-dimensional supergravity, Nuovo Cimento Lett.
Palla, L. (1979). Spontaneous compactification. In Proceedings of the 19th International Conference in High Energy Physics, ed. S. Homma, M. Kawaguchi & H. Miyazawa, pp. 629-631. Tokyo : Physical Society of Japan
Scherk, J. (1979). Extended supersymmetry and extended supergravity. In Recent Developments in Gravitation, ed. M. Levy & S. Deser, pp. 479-517. New York : Plenum Publishing Corporation
Scherk, J. & Schwarz, J.H. (1975). Dual field theory of quarks and gluons, Phys. Lett. 57B, pp. 463-466
Scherk, J. & Schwarz, J.H. (1979 a) Spontaneous breaking of supersymmetry through dimensional reduction, Phys. Lett. 82B, pp. 60-64
Scherk, J. & Schwarz, J.H. (1979 b). How to get masses from extra-dimensions Nucl. Phys. B153, pp. 61-88
Schwarz, J.H. (1980). N = 8 supergravity in various dimensions and the implications for four dimensions, Phys. Lett. 95B, pp. 219-221
Sezgin, E. & Van Nieuwenhuizen, P. (1981). Renormalization properties of spontaneously broken N = 8 supergravity, Nucl. Phys. B, in press
Sohnius, M. ; Stelle, K. & West, P.C. (1981). Dimensional reduction by Legendre transformation generates off-shell supersymmetric Yang-Mills theories, Nucl. Phys. B173, pp. 127-153
Van Nieuwenhuizen, P. (1980). Two lectures on Supergravity and Phenomenology at the Bad-Honnef Summer Institute, Preprint IT-SB-80-84
West, P.C. (1981). Contribution to this school
Witten, E. (1979). Instantons, the quark model and the 1/N expansion, Nucl. Phys. B149, pp. 285-320
Zumino, B. (1981). Supergravity and grand unification. In Superspace and supergravity, ed. S. Hawking & M. Rocek, pp. 423-434. Cambridge : Cambridge University Press

THE STRUCTURE OF SUPERGRAVITY IN SUPERSPACE

R. Grimm
Institut für theoretische Physik, Universität Karlsruhe, Karlsruhe, West Germany

Abstract. The first part of my lectures will contain a condensed review of the essential definitions and equations of differential geometry in flexible superspace. A detailed and more complete description of the subject as well as its application to the formulation of supergravity (Freedman, van Nieuwenhuizen & Ferrara, 1976, Deser & Zumino 1976, Stelle & West 1978, Ferrara & van Nieuwenhuizen 1978) in superspace may be found in the original literature (Arnowitt & Nath 1975, Arnowitt, Nath & Zumino 1975, Akulov, Volkov & Soroka 1975, Zumino 1975, Wess & Zumino 1977, Grimm, Wess & Zumino 1978, Wess & Zumino 1978a, Wess & Zumino 1978b, Wess 1977, Zumino 1978, Dragon 1979, Grimm, Wess & Zumino 1979)[1] In the second part the structure of N = 2 extended supergravity (Ferrara & van Nieuwenhuizen 1976, Freedman 1977, Freedman & Das 1977, Ferrara, Scherk & Zumino 1977a, Ferrara, Scherk & Zumino 1977b, Fradkin & Vasiliev 1979, contributions to 'Supergravity' by Wess, deWit & Breitenlohner 1979, Wess 1979, Brink, Gell-Mann, Raymond & Schwarz 1978, Volkov & Akulov 1973, Salam & Strathdee 1976, Ferrara, Wess & Zumino 1974) in superspace will be exemplified by means of an investigation of the linearized structure equations for the supertorsion (Wess 1979), subject to a proper set of constraints.

DIFFERENTIAL GEOMETRY[2] OF FLEXIBLE SUPERSPACE

Flexible (curved) superspace is a generalization of ordinary curved space time. The coordinates z^M of a point in flexible superspace consist of its coordinates x^m $(m = 0,1,2,3)$ in space time and certain anticommuting variables θ^μ, $\theta_{\dot\mu}$ $(\mu, \dot\mu = 1,2)$ which are elements of a Grassmann algebra. In extended superspace,

$$z^M \sim (x^m,\ \theta^\mu_M,\ \bar\theta^M_{\dot\mu}) \quad , \tag{1}$$

the Grassmann variables carry an additional index M = 1, ...,N corresponding to the number N of supercharges. The supercoordinates z^M are graded according to the bosonic (x^m) or fermionic $(\theta^\mu_M, \bar\theta^M_{\dot\mu})$ nature of their components:

$$z^M z^N = (-1)^{d(M)\, d(N)} z^N z^M \tag{2}$$

$d(M)$ equals zero if M is a bosonic (even) index and one if M is a fermionic (odd) index:

$$d(M) = \begin{cases} 0 & M = m \\ 1 & M = \begin{cases} \mu, M \\ M, \dot{\mu} \end{cases} \end{cases} \tag{3}$$

We introduce derivatives $\partial_M = \frac{\partial}{\partial z^M}$ and differentials dz^M subject to the following commutation rules with supercoordinates and among themselves:

$$\begin{aligned} \partial_M z^N &= \delta_M^N + (-1)^{d(M)\, d(N)} z^N \partial_M \\ \partial_M \partial_N &= (-1)^{d(M)\, d(N)} \partial_N \partial_M \\ dz^M dz^N &= -(-1)^{d(M)\, d(N)} dz^N dz^M \\ dz^M z^N &= (-1)^{d(M)\, d(N)} z^N dz^M \end{aligned} \tag{4}$$

Supercoordinate transformations are defined in analogy to general co-ordinate transformations of ordinary curved space time as follows:

$$\begin{aligned} z^M &\to z'^M(z) \\ dz^M &\to dz^N \frac{\partial z'^M}{\partial z^N} \\ \partial_M &\to \frac{\partial z^N}{\partial z'^M} \partial_N \end{aligned} \tag{5}$$

and their infinitesimal form is

$$z'^M(z) = z^M - \xi^M(z) \tag{6}$$

with supercoordinate dependent infinitesimal parameters $\xi^M(z)$. Super-scalars $f(z)$, covariant supervectors $u_M(z)$ and contravariant supervectors

$v^M(z)$ are defined by their transformation properties:

$$
\begin{aligned}
\delta f(z) &= \xi^M \, \partial_M \, f(z) \\
\delta u_M(z) &= \xi^L \, \partial_L \, u_M(z) + (\partial_M \, \xi^L \,) \, u_L(z) \\
\delta v^M(z) &= \xi^L \, \partial_L \, v^M(z) - v^L(z) \, \partial_L \, \xi^M
\end{aligned}
\tag{7}
$$

The transformation law for a scalar density D(z) is

$$
\delta D(z) = (-1)^{d(M)} \, \partial_M \, (\xi^M \, D(z)) \tag{8}
$$

In order to define local linear frames we introduce the "supervielbein" $E_M{}^A(z)$ and its inverse $E_A{}^M(z)$.

$$
E_A{}^M(z) \, E_M{}^B(z) = \delta_A^B \tag{9}
$$

As a matrix the supervielbein has even and odd parts corresponding to the bosonic or fermionic nature of its components:

$$
E_M{}^A(z) \sim \begin{pmatrix}
E_m{}^a(z) & E_m{}^{\alpha}_{A}(z) & E_m{}^{A}_{\dot\alpha}(z) \\
E^{M}_{\mu}{}^{a}(z) & E^{M}_{\mu}{}^{\alpha}_{A}(z) & E^{M}_{\mu}{}^{A}_{\dot\alpha}(z) \\
E^{\dot\mu}_{M}{}^{a}(z) & E^{\dot\mu}_{M}{}^{\alpha}_{A}(z) & E^{\dot\mu}_{M}{}^{A}_{\dot\alpha}(z)
\end{pmatrix} \tag{10}
$$

The delta symbol δ_A^B is taken to be

$$
\delta_B^B \sim \begin{pmatrix}
\delta_a^b & 0 & 0 \\
0 & \delta_\alpha^\beta \, \delta_B^A & 0 \\
0 & 0 & \delta_{\dot\beta}^{\dot\alpha} \, \delta_A^B
\end{pmatrix} \tag{11}
$$

At a given point of superspace the supervielbein is used to express covariant or contravariant tangent space vectors $u_A \sim (u_a, \; u_{\alpha}^{A}, \; u_A^{\dot\alpha})$ or $v^A \sim (v^a, \; v_A^{\alpha}, \; v_{\dot\alpha}^{A})$ in terms of the corresponding world vectors u_M and v^M.

$$v^A = v^M E_M{}^A, \qquad u_A = E_A{}^M u_M \tag{12}$$

To distinguish tangent space and world indices we always take tangent space indices from the beginning and world indices from the middle of the alphabet.

The group of linear transformations which acts on the tangent spaces transforms covariant and contravariant vectors like

$$v'^A = v^B G_B{}^A(z), \qquad u'_A = G^{-1}_A{}^B u_B \tag{13}$$

Taken together, infinitesimal general supercoordinate transformations

$$z'^M(z) = z^M - \xi^M(z)$$

and infinitesimal tangent space transformations

$$G_B{}^A(z) = \delta_B^A + X_B{}^A(z)$$

change the form of the supervielbein and its inverse as follows:

$$E'_M{}^A = E_M{}^A + \delta E_M{}^A$$

$$\delta E_M{}^A = \xi^L \partial_L E_M{}^A + (\partial_M \xi^L) E_L{}^A + E_M{}^B X_B{}^A \tag{14}$$

$$\delta E_A{}^M = \xi^L \partial_L E_A{}^M - E_A{}^L \partial_L \xi^M - X_A{}^B E_B{}^M$$

Next we will define covariant derivatives in terms of the superconnection $\phi_{MB}{}^A(z)$ which takes its values in the Lie algebra of the structure group and transforms inhomogeneously:

$$\begin{aligned}\delta\phi_{MB}{}^A &= \xi^L \partial_L \phi_{MB}{}^A + (\partial_M \xi^L) \phi_{LB}{}^A \\ &\quad + \phi_{MB}{}^C X_C{}^A - X_B{}^C \phi_{MC}{}^A - \partial_M X_B{}^A\end{aligned} \tag{15}$$

(Here we have assumed $d(B) = d(C)$ for $X_B{}^C$, otherwise the fourth term on the r.h.s. of equ. (15) would be multiplied by $(-1)^{d(M)d(B+C)}$.) The covariant derivatives of tangent space vectors u_A and v^A then become

$$\mathcal{D}_M v^A = \partial_M v^A + (-1)^{d(M)\,d(B)}\, v^B\, \phi_{MB}{}^A$$
$$\mathcal{D}_M u_A = \partial_M u_A - \phi_{MA}{}^B\, u_B \qquad (16)$$

In order to avoid explicit reference to local coordinate systems we define differential forms and exterior deviatives in superspace.

A super p form σ^A,

$$\sigma^A = dz^{M_1} \ldots dz^{M_p}\, f_{M_p \ldots M_1}(z) \qquad (17)$$

has coefficient functions $f_{M_p \ldots M_1}(z)$ which reflect the graded symmetry properties of the product of differentials $dz^{M_1} \ldots dz^{M_p}$ in their indices. If, for instance, σ were a two form, $\sigma = dz^M\, dz^N\, f_{NM}$, we have $f_{NM} = -\,(-1)^{d(M)\,d(N)}\, f_{NM}$. The exterior derivative d of a p form σ^A,

$$d\sigma^A = dz^{M_1} \ldots dz^{M_p}\, dz^L\, \partial_L\, f_{M_p \ldots M_1}(z) \qquad (18)$$

is a p + 1 form and using (4) it is easy to see that

$$dd = 0 \qquad (19)$$

Locally, the reverse is also true. If σ is any p form and $d\sigma = 0$ then σ is the exterior derivative of a p - 1 form τ, $\sigma = d\tau$. The product of a p form σ^A and a q form τ^B is a p + q form and has the properties

$$\sigma^A\, \tau^B = (-1)^{p \cdot q}\, (-1)^{d(A)\,d(B)}\, \tau^B\, \sigma^A \qquad (20)$$

and

$$d(\sigma^A\, \tau^B) = \sigma^A\, d\tau^B + (-1)^q\, (d\sigma^A)\, \tau^B \qquad (21)$$

The fundamental differential forms of superspace differential geometry are the supervielbein and connection one forms E^A and $\phi_B{}^A$.

$$E^A = dz^M\, E_M{}^A(z) \qquad (22)$$

$$\phi_B{}^A = dz^M\, \phi_{MB}{}^A(z) \qquad (23)$$

The effects of a structure group transformation on E^A and $\phi_B{}^A$ are

$$E'^A = E^B \, G_B{}^A \tag{24}$$

$$\phi'_B{}^A = G^{-1}{}_B{}^D \, \phi_D{}^C \, G_C{}^A - G^{-1}{}_B{}^C \, dG_C{}^A \tag{25}$$

Using (20) and (21) one can see that for any p form σ^A or τ_A the covariant differentials

$$D\sigma^A = d\sigma^A + \sigma^B \, \phi_B{}^A \tag{26}$$

$$D\tau_A = d\tau_A - (-1)^p \, \phi_A{}^B \, \tau_B \tag{27}$$

are p + 1 forms and transform homogeneously.

The exterior derivatives of E^A and $\phi_B{}^A$ are related to supertorsion and supercurvature by means of the first and second structure equations

$$T^A = dE^A + E^B \, \phi_B{}^A \tag{28}$$

$$R_B{}^A = d\phi_B{}^A + \phi_B{}^C \, \phi_C{}^A \tag{29}$$

T^A and $R_B{}^A$ are two forms. In the same way as the superconnection, $R_B{}^A$ is Lie algebra valued. Applying exterior differentiation again to (28) and (29) gives rise to the Bianchi identities

$$DT^A = E^B \, R_B{}^A \tag{30}$$

$$DR_B{}^A = 0 \tag{31}$$

or equivalently,

$$dT^A + T^B \, \phi_B{}^A - E^B \, R_B{}^A = 0 \tag{32}$$

$$dR_B{}^A - \phi_B{}^C \, R_C{}^A + R_B{}^C \, \phi_C{}^A = 0 \tag{33}$$

When we factor out the covariant differentials E^A and use the explicit form of supertorsion and supercurvature,

$$T^A = \frac{1}{2} dz^M dz^N T_{NM}{}^A = \frac{1}{2} E^B E^C T_{CB}{}^A \tag{34}$$

$$R_B{}^A = \frac{1}{2} dz^M dz^N R_{NMB}{}^A = \frac{1}{2} E^C E^{\mathcal{D}} R_{\mathcal{D}CB}{}^A \tag{35}$$

we obtain for the Bianchi identities the following expressions:

$$\oint_{\mathcal{D}CB} (R_{\mathcal{D}CB}{}^A - \mathcal{D}_{\mathcal{D}} T_{CB}{}^A - T_{\mathcal{D}C}{}^F T_{FB}{}^A) = 0 \tag{36}$$

and

$$\oint_{E\mathcal{D}C} (\mathcal{D}_E R_{\mathcal{D}CB}{}^A + T_{E\mathcal{D}}{}^F R_{FCB}{}^A) = 0 \tag{37}$$

Performing the graded cyclic sum indices are interchanged according to the rule

$$\mathcal{D}C = - (-1)^{d(\mathcal{D})\, d(C)}\, C\mathcal{D} \tag{38}$$

Having defined torsion and curvature of flexible superspace, it is easy to work out the generalized commutation relation of two supercovariant derivatives,

$$(\mathcal{D}_C , \mathcal{D}_B) = \mathcal{D}_C \mathcal{D}_B - (-1)^{d(C)\, d(B)} \mathcal{D}_B \mathcal{D}_C \tag{39}$$

This may, for example, be done by applying $D \sim E^A \mathcal{D}_A$ as defined in (26) and (27) twice. For simplicity, take $u_A(z)$ to be a 0 form, then

$$\begin{aligned} DD\, u_A &= DE^C \mathcal{D}_C u_A = \\ &= E^C E^B \mathcal{D}_B \mathcal{D}_C u_A + T^C \mathcal{D}_C u_A \end{aligned} \tag{40}$$

On the other hand, using the definitions explicitly, we obtain

$$\begin{aligned} DDu_A &= d(du_A - \phi_A{}^B u_B) + \phi_A{}^B (du_B - \phi_B{}^C u_C) = \\ &= - (d\phi_A{}^B) u_B - \phi_A{}^C \phi_C{}^B u_B = - R_A{}^B u_B. \end{aligned} \tag{41}$$

Comparing (40) and (41) yields

$$(\mathcal{D}_C, \mathcal{D}_B) u_A = - T_{CB}{}^F \mathcal{D}_F u_A - R_{CBA}{}^F u_F \tag{42}$$

The geometric structure we have defined so far is still quite general. In order to describe pure N = 1 supergravity one has to specify the structure group and impose certain constraints on the torsion (Wess & Zumino 1977). The structure group is taken to be the Lorentz group such that the same Lorentz transformation acts on the components v^a, v^α and $v_{\dot{\alpha}}$ of a tangent space vector v^A. This means that the only nonvanishing components of $R_B{}^A$ are $R_b{}^a$, $R_\beta{}^\alpha$ and $R^{\dot{\beta}}{}_{\dot{\alpha}}$ with the following properties and relations among them:

$$\begin{aligned} R_{ba} &= - R_{ab} \\ R_\beta{}^\beta &= R^{\dot{\beta}}{}_{\dot{\beta}} = 0 \end{aligned} \tag{43}$$

$$\begin{aligned} R_\beta{}^\alpha &= - \tfrac{1}{4} (\sigma^b \bar{\sigma}_a)_\beta{}^\alpha \, R_b{}^a \\ R^{\dot{\beta}}{}_{\dot{\alpha}} &= - \tfrac{1}{4} (\bar{\sigma}^b \sigma_a)^{\dot{\beta}}{}_{\dot{\alpha}} \, R_b{}^a \end{aligned} \tag{44}$$

or

$$R_b{}^a = \tfrac{1}{2} (\bar{\sigma}_b \sigma^a)^{\dot{\alpha}}{}_{\dot{\beta}} \, R^{\dot{\beta}}{}_{\dot{\alpha}} + \tfrac{1}{2} (\sigma_b \bar{\sigma}^a)_\alpha{}^\beta \, R_\beta{}^\alpha \tag{45}$$

In spinor notation(3) each vector index is replaced by a pair of dotted and undotted spinor indices,

$$X_a \rightarrow X_{\alpha\dot{\alpha}} = \sigma^a{}_{\alpha\dot{\alpha}} X_a \tag{46}$$

and (45) becomes

$$R_{\beta\dot{\beta}\,\alpha\dot{\alpha}} = - 2\varepsilon_{\beta\alpha} R_{\underbrace{\dot{\beta}\dot{\alpha}}} + 2\varepsilon_{\dot{\beta}\dot{\alpha}} R_{\underbrace{\beta\alpha}} \tag{47}$$

$$R_{\beta\dot{\beta}\,\alpha\dot{\alpha}} = \sigma^b{}_{\beta\dot{\beta}} \, \sigma^a{}_{\alpha\dot{\alpha}} \, R_{ba} \tag{48}$$

This choice of structure group ensures that the resulting superspace contains rigid (flat) superspace (Volkov & Akulov 1973, Salam & Strathdee

1976, Ferrara, Wess & Zumino 1974) with nonvanishing torsion $T_{\gamma}{}^{\dot{\gamma} a} = -2i\,(\sigma^a \varepsilon)_{\gamma}{}^{\dot{\beta}}$ in a natural way (Wess & Zumino 1977). It is also interesting to observe that due to the choice of structure group it is possible to express all components of supercurvature in terms of components of supertorsion and their covariant derivatives by means of the Bianchi identities (36). Also, with this result, (37) becomes an identity. (This remains true for extended superspace when the structure group is the direct product of the Lorentz group and some group of internal transformations (Dragon 1979)). The torsion constraints for N = 1 supergravity are:

$$
\begin{aligned}
&T_{\gamma\beta}{}^{A} = 0, \qquad T^{\dot{\gamma}\dot{\beta} A} = 0, \\
&T_{\gamma}{}^{\dot{\beta}\alpha} = 0, \qquad T_{\gamma}{}^{\dot{\beta}}{}_{\dot{\alpha}} = 0, \qquad\qquad (49)\\
&T_{Cb}{}^{a} = 0, \qquad T_{\gamma}{}^{\dot{\beta} a} = -\,2i(\sigma^a\varepsilon)_{\gamma}{}^{\dot{\beta}} \;,
\end{aligned}
$$

the components $T_{Cb}{}^{\alpha}$ and $T_{Cb\dot{\alpha}}$ are left arbitrary.

Solving the Bianchi identities one finds that all covariant quantities of the geometry can be expressed in terms of the superfields R, R^*, $G_{\alpha\dot{\alpha}} = (G_{\alpha\dot{\alpha}})^*$, $W_{\underline{\alpha\beta\gamma}}$ and $\bar{W}_{\underline{\dot{\alpha}\dot{\beta}\dot{\gamma}}} = (W_{\underline{\alpha\beta\gamma}})^*$ and that the following set of relations among them is equivalent to the Bianchi identities (Grimm, Wess & Zumino 1979)

$$
\begin{aligned}
&\mathcal{D}_{\dot{\alpha}} R = 0, \qquad \mathcal{D}_{\alpha} R^* = 0 \\
&\mathcal{D}_{\alpha} \bar{W}_{\underline{\dot{\alpha}\dot{\beta}\dot{\gamma}}} = 0, \qquad \mathcal{D}_{\dot{\alpha}} W_{\underline{\alpha\beta\gamma}} = 0 \\
&\mathcal{D}^{\alpha} G_{\alpha\dot{\alpha}} = \mathcal{D}_{\dot{\alpha}} R^*, \qquad \mathcal{D}^{\dot{\alpha}} G_{\alpha\dot{\alpha}} = \mathcal{D}_{\alpha} R \qquad\qquad (50)\\
&\mathcal{D}^{\alpha} W_{\underline{\alpha\beta\gamma}} = -\frac{i}{2}\left(\mathcal{D}_{\beta\dot{\alpha}} G_{\gamma}{}^{\dot{\alpha}} + \mathcal{D}_{\gamma\dot{\alpha}} G_{\beta}{}^{\dot{\alpha}}\right) \\
&\mathcal{D}^{\dot{\alpha}} \bar{W}_{\underline{\dot{\alpha}\dot{\beta}\dot{\gamma}}} = -\frac{i}{2}\left(\mathcal{D}_{\alpha\dot{\beta}} G^{\alpha}{}_{\dot{\gamma}} + \mathcal{D}_{\alpha\dot{\gamma}} G^{\alpha}{}_{\dot{\beta}}\right)
\end{aligned}
$$

The torsion components $T_{\gamma e\dot{\alpha}}$, $T^{\dot{\gamma}}{}_{e}{}^{\alpha}$, $T_{\gamma e}{}^{\alpha}$ and $T^{\dot{\gamma}}{}_{e\dot{\alpha}}$, for example, have the following form:

$$T_{\gamma e\dot{\alpha}} = -\, i\, \sigma_{e\gamma\dot{\alpha}}\, R^{*}$$

$$T^{\dot{\gamma}\alpha}{}_{e} = -\, i\, \bar{\sigma}_{e}{}^{\dot{\gamma}\alpha}\, R$$

$$T_{\gamma e}{}^{\alpha} = i\, (\delta_{\gamma}^{\alpha}\, \delta_{e}^{a} - \sigma_{e}{}^{a}{}_{\gamma}{}^{\alpha})\, G_{a} \tag{51}$$

$$T^{\dot{\gamma}}{}_{e\dot{\alpha}} = -\, i\, (\delta_{\dot{\alpha}}^{\dot{\gamma}}\, \delta_{e}^{a} - \bar{\sigma}_{e}{}^{a\dot{\gamma}}{}_{\dot{\alpha}})\, G_{a}$$

whereas $W_{\alpha\beta\gamma}$ and $W_{\dot{\alpha}\dot{\beta}\dot{\gamma}}$ appear in $T_{cb}{}^{\alpha}$ and $T_{cb\dot{\alpha}}$ respectively.

A supertranslation in flexible superspace is defined as a special combination of a general supercoordinate and a structure group transformation (Wess & Zumino 1978b). First of all, we note that in terms of parameters $\xi^{A}(z)$,

$$\xi^{A}(z) = \xi^{M}\, E_{M}{}^{A} \tag{52}$$

the transformation law (14) of the supervielbein becomes

$$\delta E_{M}{}^{A} = \mathcal{D}_{M}\xi^{A} + E_{M}{}^{B}\xi^{C}T_{CB}{}^{A} + E_{M}{}^{B}(X_{B}{}^{A} - \xi^{C}\,\phi_{CB}{}^{A}) \tag{53}$$

In deriving this expression we have used the explicit form of supertorsion,

$$T_{NM}{}^{A} = \mathcal{D}_{N}\, E_{M}{}^{A} - (-1)^{d(N)d(M)}\, \mathcal{D}_{M}\, E_{N}{}^{A} \tag{54}$$

It is easy to see that a combination of structure group and ξ^{A}-transformations, such that

$$X_{B}{}^{A} = \xi^{C}\, \phi_{CB}{}^{A} \tag{55}$$

changes the form of the supervielbein according to

$$\delta E_{M}{}^{A} = \mathcal{D}_{M}\xi^{A} + E_{M}{}^{B}\xi^{C}\, T_{CB}{}^{A} \tag{56}$$

A covariant supervector v^{A} transforms as

$$\delta v^{A} = \xi^{B}\mathcal{D}_{B}\, v^{A} \tag{57}$$

The commutator of two supertranslations is then simply related to the commutation relations (42) for covariant derivatives.

It is well known (Wess & Zumino 1978b) how the results of the component formalism of supergravity (Stelle & West 1978, Ferrara & van Nieuwenhuizen 1978) follow from superspace. Let me only recall that the component fields of the supergravity multiplet are identified as the $\theta,\bar{\theta}$-independent values of the superfields R, R^* and $G_{\alpha\dot{\alpha}}$ and of the supervielbein, in a special gauge in superspace:

$$E_M{}^A(x,0,0) \sim \begin{pmatrix} e_m{}^a(x) & \frac{1}{2}\psi_m{}^\alpha(x) & \frac{1}{2}\bar{\psi}_{m\dot{\alpha}}(x) \\ 0 & \delta^\alpha_\mu & 0 \\ 0 & 0 & \delta^{\dot{\mu}}_{\dot{\alpha}} \end{pmatrix} \tag{58}$$

The lowest components of $T_{cb}{}^\alpha$ and $T_{\underline{cb}\dot{\alpha}}$ are related to the field strength of the Rarita Schwinger field $\psi_m{}^\alpha$, $\bar{\psi}_{m\dot{\alpha}}$.

Supersymmetry transformations have parameters $\xi^A(x,0,0) = (0,\zeta^\alpha(x), \bar{\zeta}_{\dot{\alpha}}(x))$ and the transformations rules are given by (56) and (57). The requirement that the gauge conditions in superspace in (58) remain unchanged fixes the terms linear in θ and $\bar{\theta}$ of the parameter ξ^a:

$$\xi^a{}_{/\sim\theta,\bar{\theta}} = 2i\,\theta^\alpha\,(\sigma^a\varepsilon)_\alpha{}^{\dot{\alpha}}\;\bar{\zeta}_{\dot{\alpha}}(x) + 2i\,\bar{\theta}_{\dot{\alpha}}(\bar{\sigma}^a\varepsilon)^{\dot{\alpha}}{}_\alpha\zeta^\alpha(x) \tag{59}$$

For a detailed discussion and for the construction of superspace actions, density superfields as well as the coupling of other supermultiplets to supergravity in superspace I would like to refer to refs. (Wess & Zumino, 1978a, Wess & Zumino, 1978b, Zumino 1978).

THE LINEARIZED TORSION EQUATIONS FOR N = 2 EXTENDED SUPERGRAVITY.

As an illustration of the methods of superspace differential geometry we will now investigate the linearized structure equations of N = 2 extended supergravity (Wess 1979) in detail. For this purpose we expand the supervielbein $E_M{}^A$ and its inverse around the rigid superspace vielbein $E^{(o)}{}_M{}^A$ and its inverse $E^{(o)}{}_A{}^M$.

$$E_M{}^A = E^{(o)}{}_M{}^A + E^{(o)}{}_M{}^B H_B{}^A \tag{2.1}$$

The rigid superspace vielbein is given by

$$E^{(o)}{}_M{}^A \sim \begin{pmatrix} \delta_m^a & 0 & 0 \\ -i\,\bar{\theta}^M_{\dot{\mu}}(\sigma^a\varepsilon)_\mu{}^{\dot{\mu}} & \delta^\alpha_\mu \delta^M_A & 0 \\ -i\,\theta^\mu_M(\alpha^a\varepsilon)_\mu{}^{\dot{\mu}} & 0 & \delta^{\dot{\mu}}_{\dot{\alpha}}\,\delta^A_M \end{pmatrix} \tag{2.2}$$

and the torsion of rigid superspace is

$$T^{(o)}{}_{\gamma B}{}^{C\dot{\beta}a} = -\,2i\,\delta^C_B\,(\sigma^a\varepsilon)_\gamma{}^{\dot{\beta}} \tag{2.3}$$

Inserting (2.1) in the torsion equations (28) we obtain for the linearized equations

$$\begin{aligned} T_{CB}{}^A = {} & T^{(o)A}_{CB} + D_C H_B{}^A - (-1)^{d(B)d(C)} D_B H_C{}^A \\ & + T^{(o)}{}_{CB}{}^{\mathcal{D}} H_{\mathcal{D}}{}^A - H_C{}^{\mathcal{D}} T^{(o)}{}_{\mathcal{D}B}{}^A + (-1)^{d(B)d(C)} H_B{}^{\mathcal{D}} T^{(o)}{}_{\mathcal{D}C}{}^A \\ & + \phi_{CB}{}^A - (-1)^{d(B)d(C)} \phi_{BC}{}^A \quad , \end{aligned} \tag{2.4}$$

$$D_B = E^{(o)}{}_B{}^M \partial_M \tag{2.5}$$

Corresponding to the components of superindices $A \sim (a_1\ A^\alpha_1\ \dot{\alpha}^A)$ we decompose the supertorsion tensor and obtain a set of 18 equations

(T.1) $\quad T^{\dot{\gamma}\dot{\beta}\alpha}_{CBA} = D^{\dot{\gamma}}_D H^{\dot{\beta}\alpha}_{BA} + D^{\dot{\beta}}_B H^{\dot{\gamma}\alpha}_{CA}$

(T.2) $\quad T^{CBA}_{\gamma\beta\dot{\alpha}} = D^C_\gamma H^{BA}_{\beta\dot{\alpha}} + D^B_\beta H^{CA}_{\gamma\dot{\alpha}}$

(T.3) $\quad T^{CBa}_{\gamma\beta} = D^C_\gamma H^{Ba}_\beta + D^B_\beta H^{Ca}_\gamma - H^{CF}_{\gamma\dot{\phi}}\, T^{(o)\dot{\phi}Ba}_{F\beta} - H^{BF}_{\beta\dot{\phi}}\, T^{(o)\dot{\phi}Ca}_{F\gamma}$

(T.4) $$T_{CB}^{\dot\gamma\dot\beta a} = D_C^{\dot\gamma} H_B^{\dot\beta a} + D_B^{\dot\beta} H_C^{\dot\gamma a} - H_{CF}^{\dot\gamma\phi} T^{(o)F\dot\beta a}_{\phi B} - H_{BF}^{\dot\beta\phi} T^{(o)F\dot\gamma a}_{\phi C}$$

(T.5) $$T^{c\dot\beta a}_{\gamma B} = T^{(o)c\dot\beta a}_{\gamma B} + T^{(o)c\dot\beta f}_{\gamma B} H_f^{\ a} + D_\gamma^c H^{\dot\beta a}_B + D_B^{\dot\beta} H^{ca}_\gamma$$

$$- H^{c\phi}_{\gamma F} T^{(o)F\dot\beta a}_{\phi B} - H^{\dot\beta F}_{B\dot\phi} T^{(o)\dot\phi ca}_{F\gamma} .$$

(T.6) $$T_{cb}^{\ \ a} = \partial_c H_b^{\ a} - \partial_b H_c^{\ a} + \phi_{cb}^{\ \ a} - \phi_{bc}^{\ \ a}$$

(T.7) $$T^{c\ a}_{\gamma b} = D^c_\gamma H_b^{\ a} - \partial_b H^{ca}_\gamma + H_{b\dot\phi}^{\ F} T^{(o)\dot\phi ca}_{F\gamma} + \phi^{c\ a}_{\gamma b}$$

(T.8) $$T^{\dot\gamma\ a}_{cb} = D^{\dot\gamma}_c H_b^{\ a} - \partial_b H^{\dot\gamma a}_c + H_b^{\ \phi}{}_F T^{(o)F\dot\gamma a}_{\phi c} + \phi^{\dot\gamma\ a}_{cb}$$

(T.9) $$T^{\dot\gamma B\alpha}_{c\beta A} = D^{\dot\gamma}_c H^{B\alpha}_{\beta A} + D^B_\beta H^{\dot\gamma\alpha}_{cA} + T^{(o)\dot\gamma Bf}_{c\beta} H_{fA}^{\ \alpha} + \phi^{\dot\gamma B\alpha}_{c\beta A}$$

(T.10) $$T^{\dot\gamma\dot\beta A}_{CB\dot\alpha} = D^{\dot\gamma}_C H^{\dot\beta A}_{B\dot\alpha} + D^{\dot\beta}_B H^{\dot\gamma A}_{C\dot\alpha} + \phi^{\dot\gamma\dot\beta A}_{CB\dot\alpha} + \phi^{\dot\beta\dot\gamma A}_{BC\dot\alpha}$$

(T.11) $$T^{c\dot\beta A}_{\gamma B\dot\alpha} = D^C_\gamma H^{\dot\beta A}_{B\dot\alpha} + D^{\dot\beta}_B H^{CA}_{\gamma\dot\alpha} + T^{(o)C\dot\beta f}_{\gamma B} H_{f\dot\alpha}^{\ A} + \phi^{c\dot\beta A}_{\gamma B\dot\alpha}$$

(T.12) $$T^{CB\alpha}_{\gamma\beta A} = D^C_\gamma H^{B\alpha}_{\beta A} + D^B_\beta H^{C\alpha}_{\gamma A} + \phi^{CB\alpha}_{\gamma\beta A} + \phi^{BC\alpha}_{\beta\gamma A}$$

(T.13) $$T^{c\ A}_{\gamma b\dot\alpha} = D^C_\gamma H_{b\dot\alpha}^{\ A} - \partial_b H^{CA}_{\gamma\dot\alpha}$$

(T.14) $$T^{\dot\gamma\ \alpha}_{cbA} = D^{\dot\gamma}_c H_{bA}^{\ \alpha} - \partial_b H^{\dot\gamma\alpha}_{CA}$$

(T.15) $$T^{c\ \alpha}_{\gamma bA} = D^C_\gamma H_{bA}^{\ \alpha} = \partial_b G^{C\alpha}_{\gamma A} - \phi_{b\gamma A}^{\ C\alpha}$$

(T.16) $$T^{\dot\gamma}{}_{cb\dot\alpha}{}^{A} = D^{\dot\gamma}_{c}\, H_{b\dot\alpha}{}^{A} - \partial_b\, H^{\dot\gamma A}_{c\dot\alpha} - \phi_{bc\dot\alpha}{}^{\dot\gamma A}$$

(T.17) $$T_{cb}{}^{\alpha}{}_{A} = \partial_c\, H_{b}{}^{\alpha}{}_{A} - \partial_b\, H_{c}{}^{\alpha}{}_{A}$$

(T.18) $$T_{cb\dot\alpha}{}^{A} = \partial_c\, H_{b\dot\alpha}{}^{A} - \partial_b\, H_{c\dot\alpha}{}^{A}$$

We will impose certain constraints on the torsion in order to reduce the number of independent degrees of freedom as much as possible. The constraints will be chosen such that no equations of motion for component fields arise and a meaningful and consistent theory for one spin two field and its supersymmetric partners emerges. Recently, for N = 1 and N = 2 supergravity, methods to construct and explain constraints or to check their consistency were developed (Gates Jr., Stelle & West 1980, van Nieuwenhuizen & West 1980). We will not go through the details of these investigations but merely postulate a set of constraints and study their consequences for the linearized theory. The constraints will be imposed step by step in the course of our analysis of structure equations (2.4).

Let me begin with

(C.1) $$T^{\dot\gamma\dot\beta\alpha}{}_{cBA} = 0, \quad T^{cBA}{}_{\gamma\beta\dot\alpha} = 0.$$

This means that (T.1) and (T.2) can be solved by introducing potentials

$$H^{\dot\beta\alpha}_{BA} = D^{\dot\beta}_{B}\, \Lambda^{\alpha}_{A} \tag{2.6}$$

$$H^{BA}_{\beta\dot\alpha} = D^{B}_{\beta}\, \Sigma^{A}_{\dot\alpha} \tag{2.7}$$

It is evident that the transformations

$$\Lambda^{A}_{\alpha} \to \Lambda^{A}_{\alpha} + \lambda^{A}_{\alpha}, \quad D^{\dot\beta}_{B}\, \lambda^{A}_{\alpha} = 0. \tag{2.8}$$

$$\Sigma^{\dot\alpha}_{A} \to \Sigma^{\dot\alpha}_{A} + \sigma^{\dot\alpha}_{A}, \quad D^{B}_{\beta}\, \sigma^{\dot\alpha}{}_{A} = 0. \tag{2.9}$$

of the potentials do not change $H_{BA}^{\beta\dot{\alpha}}$, $H_{\beta\dot{\alpha}}^{BA}$.

Next, we impose

$$T_{\gamma\beta}^{CB\ a} = 0, \qquad T_{CB}^{\dot{\gamma}\dot{\beta}\ a} = 0. \tag{C.2}$$

Using (2.6) and (2.7) we obtain from (T.3) and (T.4)

$$0 = D_{\gamma}^{C}(H_{\beta}^{Ba} - \Sigma_{\cdot\ \phi}^{F}\ T_{F\beta}^{(o)\dot{\phi}Ba}) + D_{\beta}^{B}(H_{\gamma}^{ca} - \Sigma_{\cdot\ \phi}^{F}\ T_{F\gamma}^{(o)\dot{\phi}ca}) \tag{2.10}$$

$$0 = D_{C}^{\dot{\gamma}}\ (H_{B}^{\dot{\beta}a} - \Lambda_{F}^{\phi}\ T_{\phi B}^{(o)F\dot{\beta}a}) + D_{B}^{\dot{\beta}}\ (H_{C}^{\dot{\gamma}a} - \Lambda_{F}^{\phi}\ T_{\phi C}^{(o)F\dot{\gamma}a}) \tag{2.11}$$

Again, equations (2.10) and (2.11) are solved by potentials:

$$H_{\beta}^{Ba} - \Sigma_{\cdot\ \phi}^{F}\ T_{F\beta}^{(o)\dot{\phi}Ba} = D_{\beta}^{B}\ Y^{a} \tag{2.12}$$

$$H_{B}^{\dot{\beta}a} - \Lambda_{F}^{\phi}\ T_{\phi B}^{(o)F\dot{\beta}a} = D_{B}^{\dot{\beta}}\ X^{a} \tag{2.13}$$

The transformations, which leave H_{β}^{Ba} and $H_{B}^{\dot{\beta}a}$ invariant are now:

$$Y^{a} \rightarrow Y^{a} + y^{a} \tag{2.14}$$

$$X^{a} \rightarrow X^{a} + x^{a} \tag{2.15}$$

$$D_{\beta}^{B}\ y^{a} + \sigma_{\cdot\ \phi}^{F}\ T_{F\beta}^{(o)\dot{\phi}Ba} = 0 \tag{2.16}$$

$$D_{B}^{\dot{\beta}}\ x^{a} + \lambda_{F}^{\phi}\ T_{\phi B}^{(o)F\dot{\beta}a} = 0 \tag{2.17}$$

Observe, that the transformation law of the linearized vielbein with respect to supertranslations,

$$\delta H_{\mathcal{B}}{}^{\mathcal{A}} = \mathcal{D}_{\mathcal{B}}\xi^{\mathcal{A}} + \xi^{\mathcal{C}}\ T_{\mathcal{CB}}{}^{\mathcal{A}} \tag{2.18}$$

implies the following transformations for the potentials introduced above:

$$\delta\Lambda_A^{\alpha} = \xi_A^{\alpha}, \qquad \delta\Sigma_{\dot{\alpha}}^{A} = \xi_{\dot{\alpha}}^{A} \ ,$$
$$\delta Y^a = \delta X^a = \xi^a \ . \tag{2.19}$$

It is convenient and very useful for the discussion of the remaining torsion equations to define the following quantities invariant with respect to supertranslations:

$$H_{\gamma B}^{C\beta} - D_{\gamma}^{C}\,\Lambda_B^{\beta} = \mathcal{H}_{\gamma B}^{C\beta}$$

$$H_{C\dot{\beta}}^{\dot{\gamma}B} - D_C^{\dot{\gamma}}\,\Sigma_{\dot{\beta}}^{B} = \mathcal{H}_{C\dot{\beta}}^{\dot{\gamma}\,B}$$

$$X^a - Y^a = Z^a, \qquad H_c^{\ b} - \partial_c Y^b = \mathcal{H}_c^{\ b} \tag{2.20}$$

$$H_{b\dot{\gamma}}^{\ c} - \partial_b\,\Sigma_{\dot{\gamma}}^{c} = \mathcal{H}_{b\dot{\gamma}}^{\ c}$$

$$H_{b\ c}^{\ \gamma} - \partial_b\,\Lambda_c^{\gamma} = \mathcal{H}_{b\ c}^{\ \gamma}$$

We postulate now

(C.3) $$T_{\gamma B}^{c\dot{\beta}a} = T^{(o)c\dot{\beta}a}_{\gamma B} = -\,2i\ \delta_B^c\ (\sigma^a\varepsilon)_{\gamma}^{\ \dot{\beta}}$$

and use the results obtained before as well as (2.20) to write (T.5) in the form:

$$D_{\gamma}^{c}\, D_B^{\dot{\beta}}\, Z^a - 2i\ \delta_B^c\ (\sigma^f\varepsilon)_{\gamma}^{\ \dot{\beta}}\ \mathcal{H}_f^{\ a}$$
$$+\ 2i(\sigma^a\varepsilon)_{\beta}^{\ \dot{\beta}}\ \mathcal{H}_{\gamma B}^{C\beta} + 2i\ (\sigma^a\varepsilon)_{\gamma}^{\ \dot{\gamma}}\ \mathcal{H}_{B\dot{\gamma}}^{\dot{\beta}C} = 0 \tag{2.21}$$

In spinor notation, with

$$Z_a \rightarrow Z_{\alpha\dot{\alpha}} = \sigma^a_{\alpha\dot{\alpha}}\, Z_a \tag{2.22}$$

and

$$\sigma^a_{\alpha\dot{\alpha}}\,\sigma_{a\,\beta\dot{\beta}} = -\,2\,\varepsilon_{\alpha\beta}\,\varepsilon_{\dot{\alpha}\dot{\beta}} \tag{2.23}$$

it becomes

$$D^A_\alpha D_{\dot\beta B} Z_{\alpha\dot\alpha} + 2i\, \delta^A_B H_{\beta\dot\beta\,\alpha\dot\alpha}$$
$$+ 4i\, \varepsilon_{\beta\alpha} H_{\dot\beta B \dot\alpha}{}^{A} - 4i\, \varepsilon_{\dot\beta\dot\alpha} H^A_{\beta\ \alpha B} = 0 \tag{2.24}$$

Taking the traceless part of this equation with respect to the indices A and B and symmetrizing with respect to $\beta\alpha$ and $\dot\beta\dot\alpha$ gives rise to the condition

$$\sum_{P(\beta\alpha,\dot\beta\dot\alpha)} (\delta^C_A \delta^B_D - \tfrac{1}{2} \delta^B_A \delta^C_D)\, D^A_\beta D_{\dot\beta B} Z_{\alpha\dot\alpha} = 0 \tag{2.25}$$

This condition considerably reduces the number of independent degrees of freedom in the θ-expansion of $Z_{\alpha\dot\alpha}$ without leading to x-space equations, as we will see later on. The remainder of (2.24) expresses certain combinations of H_{ba}, $H^{B\alpha}_{\beta A}$ and $H^{\beta A}_{B\dot\alpha}$ in terms of Z^a:

$$H_{\beta\dot\beta\,\alpha\dot\alpha} - \varepsilon_{\dot\beta\dot\alpha} H^B_{\beta\ \alpha B} + \varepsilon_{\beta\alpha} H_{\dot\beta B\dot\alpha}{}^{B} = \tfrac{i}{4} D^B_\beta D_{\dot\beta B} Z_{\alpha\dot\alpha} \tag{2.26}$$

$$\varepsilon_{\beta\alpha} \tilde H_{\dot\beta B\dot\alpha}{}^{A} - \varepsilon_{\dot\beta\dot\alpha} \tilde H^A_{\beta\ \alpha B} = \tfrac{i}{4} (\delta^A_C \delta^D_B - \tfrac{1}{2} \delta^A_B \delta^D_C)\, D^C_\beta D_{\dot\beta D} Z_{\alpha\dot\alpha} \tag{2.27}$$

The tilde means

$$\tilde H_{\dot\beta B}{}^{A}{}_{\dot\alpha} = (\delta^D_B \delta^A_C - \tfrac{1}{2} \delta^A_B \delta^D_C)\, H_{\dot\beta D}{}^{C}{}_{\dot\alpha}$$
$$\tilde H^A_{\beta\ \alpha B} = (\delta^D_B \delta^A_C - \tfrac{1}{2} \delta^A_B \delta^D_C)\, H^C_{\beta\ \alpha D} \tag{2.28}$$

The combination

$$\tfrac{1}{2} \tilde R^A{}_B = \tilde H^{A\alpha}_{\alpha B} - \tilde H^{\dot\alpha A}_{B\dot\alpha} \tag{2.29}$$

remains undetermined, as well as $H^{B\beta}_{\beta B}$, $H^{\dot\beta B}_{B\dot\beta}$.

In the next step we impose the constraints

(C.4) $\qquad T_{cb}{}^{a} = 0$

(C.5) $T_{\gamma b}^{c\ a} = 0, \qquad T_{cb}^{\dot{\gamma}\ a} = 0$,

which will allow us (among other things) to solve for the superconnections $\phi_{cb}{}^{a}$, $\phi_{\gamma b}^{c\ a}$ and $\phi_{cb}^{\dot{\gamma}\ a}$. Since the connection takes its values in the Lie algebra of the Lorentz group it has the following properties:

$$\phi_{ba} = -\phi_{ab}$$

$$\phi_{\beta\dot{\beta}\ \alpha\dot{\alpha}} = \sigma^{b}_{\beta\dot{\beta}}\ \sigma^{a}_{\alpha\dot{\alpha}}\ \phi_{ba}$$

$$\phi_{\beta\dot{\beta}\ \alpha\dot{\alpha}} = -2\varepsilon_{\dot{\beta}\dot{\alpha}}\ \phi_{\underbrace{\beta\alpha}} + 2\varepsilon_{\beta\alpha}\ \phi_{\underbrace{\dot{\beta}\dot{\alpha}}} \qquad (2.30)$$

$$\phi_{\underbrace{\beta\alpha}} = \phi_{\underbrace{\alpha\beta}}, \qquad \phi_{\underbrace{\dot{\alpha}\dot{\beta}}} = \phi_{\underbrace{\dot{\beta}\dot{\alpha}}} \qquad (2.31)$$

First of all, as a consequence of (C.4), equ. (T.6) can be solved for $\phi_{c\ ba}$:

$$\phi_{c\ ba} + \frac{1}{2}\partial_c(H_{ba} - H_{ab})$$

$$= \frac{1}{2}\partial_b\ (H_{ac} + H_{ca}) - \frac{1}{2}\partial_c\ (H_{ba} + H_{cb}) \qquad (2.32)$$

Introducing the definitions

$$\phi_{c\ ba} = \phi_{cba} + \frac{1}{2}\partial_c\ (H_{ba} - H_{ab}) \qquad (2.33)$$

$$h_{ba} = \frac{1}{2}(H_{ba} + H_{ab}) \qquad (2.34)$$

this equation acquires a particularly simple form

$$\phi_{c\ ba} = \partial_b\ h_{ac} - \partial_a\ h_{bc} \qquad (2.35)$$

In spinor notation the symmetric combination h_{ba} is decomposed as follows

$$h_{\beta\dot{\beta}\ \alpha\dot{\alpha}} = \sigma^{b}_{\beta\dot{\beta}}\ \sigma^{a}_{\alpha\dot{\alpha}}\ h_{ba}$$

$$h_{\beta\dot{\beta}\ \alpha\dot{\alpha}} = \frac{1}{2}\ h_{\underbrace{\beta\alpha}\ \underbrace{\dot{\beta}\dot{\alpha}}} + \frac{1}{2}\ \varepsilon_{\beta\alpha}\ \varepsilon_{\dot{\beta}\dot{\alpha}}\ h \qquad (2.36)$$

and from (2.26) we know that

$$h_{\underset{\smile}{\beta\alpha}\,\underset{\smile}{\dot\beta\dot\alpha}} = \frac{i}{8} \sum_{P(\beta\alpha,\dot\beta\dot\alpha)} D_\beta^B D_{\dot\beta B} Z_{\alpha\dot\alpha} \tag{2.37}$$

$$h = \frac{i}{8} D_\phi^B D_{\dot\phi B} Z^{\phi\dot\phi} - H^{B\phi}_{\phi B} - H^{\dot\phi B}_{B\dot\phi} \tag{2.38}$$

The advantage of these definitions will become evident when we analyze the next equations, (T.7) and (T.8). From (T.7) we obtain

$$\phi^{c\ a}_{\gamma b} + D^c_\gamma H_b^{\ a} - 2i\,(\sigma^a \varepsilon)_\gamma^{\ \dot\gamma} H_{b\dot\gamma}^{\ c} = 0 \tag{2.39}$$

Similar to (2.33) we define

$$\phi^c_{\gamma\ ba} = \phi^c_{\gamma\ ba} + \frac{1}{2} D^c_\gamma (H_{ba} - H_{ab}) \tag{2.40}$$

Now eq. (2.39) becomes

$$\phi^c_{\gamma\ ba} + D^c_\gamma h_{ba} - 2i\,(\sigma_a \varepsilon)_\gamma^{\ \dot\gamma} H_{b\dot\gamma}^{\ c} = 0 \tag{2.41}$$

and in spinor notation it is

$$\begin{aligned} &- 2\varepsilon_{\beta\alpha}\, \phi^c_{\gamma\ \underset{\smile}{\dot\beta\dot\alpha}} + 2\varepsilon_{\dot\beta\dot\alpha}\, \phi^c_{\gamma\ \underset{\smile}{\beta\alpha}} \\ &+ D^c_\gamma h_{\beta\dot\beta\,\alpha\dot\alpha} + 4i\,\varepsilon_{\alpha\gamma} H_{\beta\dot\beta\,\dot\alpha}^{\ \ A} = 0 \end{aligned} \tag{2.42}$$

with the obvious definitions

$$\begin{aligned} H_{\beta\dot\beta\,\dot\alpha}^{\ \ A} &= \sigma^b_{\beta\dot\beta} H_{b\dot\alpha}^{\ A} \\ \phi^c_{\gamma\ \beta\dot\beta\,\alpha\dot\alpha} &= - 2\varepsilon_{\beta\alpha}\, \phi^c_{\gamma\ \underset{\smile}{\dot\beta\dot\alpha}} + 2\varepsilon_{\dot\beta\dot\alpha}\, \phi^c_{\gamma\ \underset{\smile}{\beta\alpha}} \end{aligned} \tag{2.43}$$

Symmetrizing (2.42) with respect to a $(\alpha\dot\alpha)$ and b $(\beta\dot\beta)$ eliminates $\phi^c_{\gamma\ \underset{\smile}{\beta\alpha}}$ and $\phi^c_{\gamma\ \underset{\smile}{\dot\beta\dot\alpha}}$ from the equation and we are left with

$$2D^c_\gamma h_{\beta\dot\beta\,\alpha\dot\alpha} + 4i\,\varepsilon_{\alpha\gamma} H_{\beta\dot\beta\,\dot\alpha}^{\ \ c} + 4i\,\varepsilon_{\beta\gamma} H_{\alpha\dot\alpha\,\dot\beta}^{\ \ c} = 0 \tag{2.44}$$

The totally symmetric part of the equation,

$$\oint_{\alpha\beta\gamma} D_{\gamma c}\, h_{\underbrace{\beta\alpha}\,\underbrace{\dot\beta\dot\alpha}} = 0 \tag{2.45}$$

can be shown to be satisfied by using (2.31) and (2.25), the remainder of (2.44) is an equation for $H_{\beta\dot\beta\,\dot\alpha}{}^{A}$. We first symmetrize in β and γ and contract the resulting equation with $\varepsilon^{\gamma\alpha}$ to obtain

$$H_{\beta\dot\beta\,\dot\alpha}{}^{A} = -\frac{i}{12} D^{\alpha A}\, h_{\underbrace{\beta\alpha}\,\underbrace{\dot\beta\dot\alpha}} - \frac{i}{4}\varepsilon_{\dot\beta\dot\alpha} D_\beta^A h \tag{2.46}$$

We then go back to (2.42) and express $\phi_\gamma{}^c{}_{\underbrace{\beta\alpha}}$ and $\phi_\gamma{}^c{}_{\underbrace{\dot\beta\dot\alpha}}$ in terms of $h_{\beta\dot\beta\,\alpha\dot\alpha}$:

$$\phi_\gamma{}^c{}_{\underbrace{\beta\alpha}} = \frac{1}{4}\left(\varepsilon_{\gamma\beta} D_\alpha^c + \varepsilon_{\gamma\alpha} D_\beta^c\right) h \tag{2.47}$$

$$\phi_\gamma{}^c{}_{\underbrace{\dot\beta\dot\alpha}} = -\frac{1}{12} D^{\phi c}\, h_{\underbrace{\phi\gamma}\,\underbrace{\dot\beta\dot\alpha}} \tag{2.48}$$

Equation (T.8)

$$\phi_c{}^{\dot\gamma}{}_b{}^a + D_c^{\dot\gamma}\left(H_b{}^a - \partial_b Z^a\right) - 2i\,(\sigma^a\varepsilon)_\gamma{}^{\dot\gamma}\, H_b{}^\gamma{}_c = 0 \tag{2.49}$$

is treated completely analogous to (T.7). It is however convenient to use the slightly modified definitions

$$\hat H_b{}^a = H_b{}^a - \partial_b Z^a$$

$$\hat\phi_c{}^{\dot\gamma}{}_{ba} = \phi_c{}^{\dot\gamma}{}_{ba} + \frac{1}{2} D_c^{\dot\gamma}\left(\hat H_{ba} - \hat H_{ab}\right)$$

$$\hat h_{ba} = \frac{1}{2}\left(\hat H_{ba} + \hat H_{ab}\right) \tag{2.50}$$

$$\hat h_{\beta\dot\beta\,\alpha\dot\alpha} = \frac{1}{2}\hat h_{\underbrace{\beta\alpha}\,\underbrace{\dot\beta\dot\alpha}} + \frac{1}{2}\varepsilon_{\beta\alpha}\varepsilon_{\dot\beta\dot\alpha}\hat h,$$

$$\hat h_{\underbrace{\beta\alpha}\,\underbrace{\dot\beta\dot\alpha}} = -\frac{i}{8}\sum_{P(\beta\alpha,\dot\beta\dot\alpha)} D_{\dot\beta B} D_\beta^B Z_{\alpha\dot\alpha}$$

$$\hat h = -\frac{i}{8} D_{\dot\phi B} D_\phi^B Z^{\phi\dot\phi} - H_{\phi B}{}^{B\phi} - H_{B\dot\phi}{}^{\dot\phi B} \tag{2.51}$$

As a consequence of (T.8) we are able to express $H_{\beta\dot{\beta}}{}^{\alpha}{}_{A} = \sigma^{b}_{\beta\dot{\beta}} H_{b}{}^{\alpha}{}_{A}$, $\hat{\phi}_{\dot{\gamma}c\,\underbrace{\beta\alpha}}$ and $\hat{\phi}_{\dot{\gamma}c\,\underbrace{\dot{\beta}\dot{\alpha}}}$ in terms of $\hat{h}_{\beta\dot{\beta}\,\alpha\dot{\alpha}}$:

$$H_{\beta\dot{\beta}\alpha A} = \frac{i}{12} D^{\dot{\alpha}}_{A}\, \hat{h}_{\underbrace{\beta\alpha}\,\underbrace{\dot{\beta}\dot{\alpha}}} + \frac{i}{4} \varepsilon_{\beta\alpha} D_{\dot{\beta}A}\, \hat{h} \tag{2.52}$$

$$\phi_{\dot{\gamma}c\,\underbrace{\beta\alpha}} = \frac{1}{12} D^{\dot{\phi}}_{c}\, \hat{h}_{\underbrace{\beta\alpha}\,\underbrace{\dot{\phi}\dot{\gamma}}} \tag{2.53}$$

$$\hat{\phi}_{\dot{\gamma}c\,\underbrace{\dot{\beta}\dot{\alpha}}} = -\frac{1}{4} (\varepsilon_{\dot{\gamma}\dot{\beta}} D_{\dot{\alpha}c} + \varepsilon_{\dot{\gamma}\dot{\alpha}} D_{\dot{\beta}c})\, \hat{h} \tag{2.54}$$

Let me summarize the results obtained so far. We have postulated a set of constraints (C.1-5) and have analyzed the linearized torsion equations (T.1-8) subject to these restrictions. We found that the components of the linearized vielbein and the connections $\phi_{\phi b}{}^{a}$ can be expressed in terms of the quantities Z^{a}, $H^{B\alpha}_{\beta B}$, $H^{\dot{\beta}B}_{B\dot{\alpha}}$ and $\frac{1}{2}\tilde{R}^{A}{}_{B} = \tilde{H}^{A\phi}_{\phi B} - \tilde{H}^{\dot{\phi}A}_{B\dot{\phi}}$. $\tilde{R}^{A}{}_{B}$ is traceless and Z^{a} is subject to the condition (2.25).

For sake of simplicity we assume the structure group to be the Lorentz group only (Wess, 1979),

$$\phi^{B\alpha}_{\beta A} = \delta^{B}_{A}\, \phi_{\beta}{}^{\alpha}, \qquad \phi^{\dot{\beta}A}_{B\dot{\alpha}} = \delta^{A}_{B}\, \delta^{\dot{\beta}}_{\dot{\alpha}} \tag{2.55}$$

and proceed with the investigation of equations (T.9) and (T.11) for the spinorial torsion components $T^{\dot{\gamma}B\alpha}_{C\beta A}$, $T^{C\dot{\beta}A}_{\gamma B\dot{\alpha}}$,

$$T^{\dot{\gamma}B\alpha}_{C\beta A} = \phi^{\dot{\gamma}B\alpha}_{C\beta A} - 2i\, \delta^{B}_{C} (\sigma^{d}\varepsilon)_{\beta}{}^{\dot{\gamma}} H_{d}{}^{\alpha}{}_{A} + D^{\dot{\gamma}}_{C} H^{B\alpha}_{\beta A} \tag{2.56}$$

and

$$T^{C\dot{\beta}A}_{\gamma B\dot{\alpha}} = \phi^{C\dot{\beta}A}_{\gamma B\dot{\alpha}} - 2i\, \delta^{C}_{B} (\sigma^{d}\varepsilon)_{\gamma}{}^{\dot{\beta}} H_{d}{}^{A}{}_{\dot{\alpha}} + D^{C}_{\gamma} H^{\dot{\beta}A}_{B\dot{\alpha}} \tag{2.57}$$

As a result of the proceeding analysis one can show that

$$\sum_{P(\alpha\beta)} T_{C\ \beta\ \alpha A}^{\dot{\gamma}\ B} = 0 \qquad (2.58)$$

$$\sum_{P(\dot{\alpha}\dot{\beta})} T_{\gamma\ \dot{\beta}B\ \dot{\alpha}}^{C\ \ \ \ A} = 0$$

In addition, we require

$$\text{(C.6)} \qquad T_{C\ \beta B}^{\dot{\gamma}\ B\beta} = 0, \qquad T_{\gamma B\dot{\beta}}^{C\beta\dot{B}} = 0$$

As a consequence of these constraints and equations (2.56) and (2.57) the supervielbein components $H_{\beta B}^{B\beta}$ and $H_{B\dot{\beta}}^{\dot{\beta}B}$ can be expressed in terms of Z^a and the chiral superfields S and T ($\sim S^*$) as follows:

$$H_{B\dot{\beta}}^{\dot{\beta}B} = -\frac{i}{8} D_{\dot{\beta}B} D_{\beta}^{B} Z^{\beta\dot{\beta}} + S \qquad (2.59)$$

$$D_{\dot{\gamma}}^{C} S = 0$$

and

$$H_{\beta B}^{B\beta} = \frac{i}{8} D_{\beta}^{B} D_{\dot{\beta}B} Z^{\beta\dot{\beta}} + T \qquad (2.60)$$

$$D_{\gamma}^{C} T = 0$$

Furthermore, as a consequence of (2.59) and (2.60) we obtain

$$h = \frac{i}{8} D_{\dot{\phi}B} D_{\phi}^{B} Z^{\phi\dot{\phi}} - (S + T) \qquad (2.61)$$

$$\hat{h} = -\frac{i}{8} D_{\phi}^{B} D_{\dot{\phi}B} Z^{\phi\dot{\phi}} - (S + T) \qquad (2.62)$$

and

$$\tilde{H}_{B\dot{\beta}}^{\dot{\beta}A} + \tilde{H}_{\beta B}^{A\beta} = \frac{i}{8} (\delta_{C}^{A} \delta_{B}^{D} - \frac{1}{2} \delta_{B}^{A} \delta_{C}^{D}) D_{\phi}^{C} D_{\dot{\phi}D} Z^{\phi\dot{\phi}} \qquad (2.63)$$

We find that $T_{C\beta A}^{\dot{\gamma}B\beta}$ and $T_{\gamma B\dot{\beta}}^{C\beta\dot{A}}$ can be described as functions of Z^a, $\tilde{R}^{A}{}_{B}$, S and T only,

$$T^{\gamma B\dot\beta}_{C\beta A} = D^{\dot\gamma}_{C}\tilde{H}^{B\beta}_{\beta A} + (\delta^{B}_{C} D^{\dot\gamma}_{A} - \frac{1}{2}\delta^{B}_{A} D^{\dot\gamma}_{C})\hat{h} \tag{2.64}$$

$$T^{C\dot\beta A}_{\gamma B\dot\beta} = D^{C}_{\gamma}\tilde{H}^{\dot\beta A}_{B\dot\beta} + (\delta^{C}_{B} D^{A}_{\gamma} - \frac{1}{2}\delta^{A}_{B} D^{C}_{\gamma})h \tag{2.65}$$

and from (T.10) and (T.12) we learn that the torsion components $T^{\dot\gamma\beta A}_{CB\dot\alpha}$ and $T^{CB\alpha}_{\gamma\beta A}$ are given by $T^{\dot\gamma A\beta}_{C\beta B}$ and $T^{C\dot\beta A}_{\gamma B\dot\beta}$ respectively.

$$T^{\dot\gamma\beta A}_{CB\dot\alpha} = -\frac{1}{2}\delta^{\dot\beta}_{\dot\alpha} T^{\dot\gamma A\phi}_{C\phi B} - \frac{1}{2}\delta^{\dot\gamma}_{\dot\alpha} T^{\dot\beta A\phi}_{B\phi C} \tag{2.66}$$

$$T^{CB\alpha}_{\gamma\beta A} = -\frac{1}{2}\delta^{\alpha}_{\beta} T^{C\dot\phi B}_{\gamma A\dot\phi} - \frac{1}{2}\delta^{\alpha}_{\gamma} T^{B\dot\phi C}_{\beta A\dot\phi} \tag{2.67}$$

Finally, we impose the constraints

(C.7) $$\oint_{ABC} g_{BD} T^{\dot\gamma D\beta}_{C\beta A} = 0, \qquad \oint_{ABC} g_{AD}\, g_{CE}\, T^{E\dot\beta D}_{\gamma B\dot\beta} = 0$$

and

(C.8) $$D^{C}_{\dot\gamma} T^{\dot\gamma B\beta}_{C\beta A} + D^{\gamma}_{C} T^{C\dot\beta B}_{\gamma A\dot\beta} = 0$$

They relate the superfields $\tilde{R}^{B}{}_{A}$, S and T to $Z_{\alpha\dot\alpha}$ in the following way:

$$\oint_{ABC} D_{\gamma C}\,(\tilde{R}_{\underset{\smile}{BA}} - \frac{i}{8}\sum_{P(BA)} D_{\beta B}\, D_{\dot\beta A}\, Z^{\beta\dot\beta}) = 0 \tag{2.68}$$

$$\oint_{ABC} D_{\dot\gamma C}\,(\tilde{R}_{\underset{\smile}{BA}} + \frac{i}{8}\sum_{P(BA)} D_{\beta B}\, D_{\dot\beta A}\, Z^{\beta\dot\beta}) = 0 \tag{2.69}$$

$$D_{\dot\phi B}\, D^{\dot\phi}_{A}\, T + D^{\phi}_{B}\, D_{\phi A}\, S$$

$$+ \frac{i}{8}(D_{\dot\beta B}\, D^{\beta}_{A}\, D^{F}_{\phi}\, D_{\dot\phi F} - D^{\beta}_{B}\, D_{\beta A}\, D_{\dot\phi F}\, D^{F}_{\phi})\, Z^{\phi\dot\phi} = 0 \tag{2.70}$$

Let me summarize the results. We have restricted the geometry of N = 2 extended superspace by the requirement for the structure group to be the Lorentz group only and by torsion constraints (C.1-8) and analyzed the consequences of these restrictions in studying the linearized structure equations for the respective supertorsion components (T.1-12).

We were able to show that the whole geometry can be described in terms of superfields Z^a, S, T and $\tilde{R}_A{}^B$ subject to conditions (2.25), (2.59), (2.60) and (2.68-70). It still remains to be shown that this set of conditions does not imply equations of motion for component fields. Indeed, expanding in powers of the anti-commuting variables and solving the equations explicitly one can see that the multiplet consists of 40 bosonic and 40 fermionic degrees of freedom and that no equations of motion emerge. Furthermore the correct equations of motion are given by (Dragon, Grimm & Wess).

$$T_{C\phi A}{}^{\dot{\gamma} C\phi} = 0, \qquad T_{\gamma C\dot{\phi}}{}^{\dot{C}\phi A} = 0 . \tag{2.71}$$

Finally I would like to add

(1) For other approaches to supergravity, components as well as superspace, pure N = 1 supergravity as well as extended supergravities, I would like to refer to the other lectures of this school and to proceedings of the following conferences and schools (and references quoted there):

Conference on gauge theories and modern field theory, Northeastern University, Boston, 1975, eds. R. Arnowitt and P. Nath (M.I.T. Press).

VIII G.I.F.T. International Seminar on Theoretical Physics, Salamanca, 1977, Lecture Notes in Physics (Springer-Verlag) vol. 77.

NATO Advanced Study Institute on Recent Developments in Gravitation, Cargèse, 1978, Plenum Press.

Supergravity Workshop at Stony Brook, 1979, eds. P. van Nieuwenhuizen and D.Z. Freedman, NHPC.

Conference on Unification of the fundamental particle interactions, Erice, 1980, eds. S. Ferrara, J. Ellis and P. van Nieuwenhuizen, Plenum.

Nuffield supergravity workshop, 1980, Cambridge University Press, S. Hawking and M. Roček eds.

and to:

P. van Nieuwenhuizen, Physics Report, to appear.

I believe this contains an easily accessible and fairly complete list of references.

(2) For an introduction to differential geometry and differential forms see, for example,

S. Kobayashi, K. Nomizu, Foundations of Differential Geometry Interscience tracts in pure and applied mathematics, vol. 15, 1963.

H. Flanders, Differential Forms, Academic Press, 1963.

(3) Notations

Summation convention for superindices

$$v^A u_A = v^a u_a + v^{\alpha}_{A} u^{A}_{\alpha} + v^{A}_{\dot{\alpha}} u^{\dot{\alpha}}_{A}$$

$$\varepsilon^{\alpha\beta} = -\varepsilon^{\beta\alpha}, \quad \varepsilon^{12} = 1, \quad \varepsilon_{12} = -1$$

$$\varepsilon^{\dot{\alpha}\dot{\beta}} \equiv g^{AB} \equiv \varepsilon^{\alpha\beta}.$$

$$\alpha_1 \dot{\alpha} = 1,2, \qquad SL(2C) \qquad A = 1,2, \quad SU(2)$$

$$\eta^{ab} = \mathrm{diag}\,(-1,1,1,1) \qquad a = 0,1,2,3$$

ε, g and η are used to raise and lower the respective indices as usual.

ε_{abcd}, totally antisymmetric

$$\varepsilon_{0123} = +1.$$

Spinor notation

$$V_{\alpha\dot{\alpha}} = \sigma^a_{\alpha\dot{\alpha}} V_a$$

$$\sigma^0 \sim \begin{pmatrix} 1 & 0 \\ 0 & 1 \end{pmatrix}, \quad \sigma^1 \sim \begin{pmatrix} 0 & 1 \\ 1 & 0 \end{pmatrix}, \quad \sigma^2 \sim \begin{pmatrix} 0 & -i \\ i & 0 \end{pmatrix}, \quad \sigma^3 \sim \begin{pmatrix} 1 & 0 \\ 0 & -1 \end{pmatrix}$$

$$\bar{\sigma}_a^{\dot{\alpha}\alpha} = - \varepsilon^{\alpha\beta} \sigma_{a\beta\dot{\beta}} \varepsilon^{\dot{\beta}\dot{\alpha}} = - (\varepsilon\sigma_a\varepsilon)^{\alpha\dot{\alpha}}$$

$$\mathrm{tr}(\sigma^a \bar{\sigma}_b) = - 2 \delta_b^a, \quad \sigma^a_{\alpha\dot{\alpha}} \bar{\sigma}_a^{\dot{\beta}\beta} = - 2 \delta_\alpha^\beta \delta_{\dot{\alpha}}^{\dot{\beta}}$$

$$\sigma^{ab}{}_\alpha{}^\beta = \frac{1}{4} (\sigma^a \bar{\sigma}^b - \sigma^b \bar{\sigma}^a)_\alpha{}^\beta, \quad \sigma^{ab}{}_\alpha{}^\alpha = 0$$

$$\bar{\sigma}^{ab\dot{\alpha}}{}_{\dot{\beta}} = \frac{1}{4} (\bar{\sigma}^a \sigma^b - \bar{\sigma}^b \sigma^a)^{\dot{\alpha}}{}_{\dot{\beta}}, \quad \bar{\sigma}^{ab\dot{\alpha}}{}_{\dot{\alpha}} = 0$$

$$\sigma_{ab} = \frac{i}{2} \varepsilon_{abcd} \sigma^{cd}, \quad \bar{\sigma}_{ab} = - \frac{i}{2} \varepsilon_{abcd} \bar{\sigma}^{cd}$$

$$(\sigma^a \bar{\sigma}^b)_\alpha{}^\beta = - \eta^{ab} \delta_\alpha^\beta + 2 \sigma^{ab}{}_\alpha{}^\beta$$

$$(\bar{\sigma}^a \sigma^b)^{\dot{\alpha}}{}_{\dot{\beta}} = - \eta^{ab} \delta^{\dot{\alpha}}_{\dot{\beta}} + 2 \bar{\sigma}^{ab\dot{\alpha}}{}_{\dot{\beta}}$$

See also refers. (Wess & Zumino 1977, Grimm, Wess & Zumino 1978, Wess & Zumino 1978a, Wess & Zumino 1978b, Wess 1977, Zumino 1978), and

B.L. van der Wearden, Group Theory and Quantum Mechanics, Springer-Verlag, 1932 and 1974.

R. Penrose, Ann. Phys. 10(1960) 171.

M. Carmeli, Group Theory and General Relativity, McGraw-Hill, 1977.

I would like to acknowledge the hospitality extended to me by the International Centre for Theoretical Physics at Miramare during this Spring School and Workshop on Supergravity and by the Aspen Center for Physics where part of the lecture notes was completed. I would also like to thank Nancy Jacobson for typing the manuscript.

REFERENCES

Akulov, V.P., Volkov, D.V. & Soroka, V.A. (1975) JETP Lett. 22 396 (English, 181).

Arnowitt, R. & Nath. P. (1975) Phys. Lett. 56B 177.

Arnowitt, R. Nath, P. & Zumino, B. (1975) Phys. Lett. 56B 81.

Brink, L. Gell-Mann, M. Raymond, P. & Schwarz, J.H. (1978), Phys. Lett. 74B, 336.

Contributions to
'Supergravity', Proceedings of the supergravity workshop at Stony Brook, (1979), es. van Nieuwenhuizen, P. & Freedman, D.Z.
NHPC,
by
Wess, J. p.103
deWit, B. p.113
Breitenlohner, P. p.123
and references cited there.

Deser, S. & Zumino, B. (1976), Phys. Lett. 628 335.

Dragon, N. (1979), Z. Phys. C2 29.

Dragon, N. Grimm, R. & Wess, J. unpublished.

Ferrara, S. Scherk, J. & Zumino, B. (1977a), Phys. Lett. 66B, 35.

Ferrara, S. Scherk, J. & Zumino, B. (1977b) Nucl. Phys. B121 393.

Ferrara, S. & van Nieuwenhuizen, P. (1976), Phys. Rev. Lett. 37 1669.

Ferrara, S. & van Nieuwenhuizen, P. (1978) Phys. Lett. 74B 333.

Ferrara, S. Wess, J. & Zumino, B. (1974), Phys. Lett. 51B 239.

Fradkin, E.S. & Vasiliev, M.A. (1979), Lett. Nuovo Cim. 25 79, Phys. Lett. 85B 47.

Freedman, D.Z. (1977) Phys. Rev. D15 1173.

Freedman, D.Z. & Das, A. (1977) Nucl. Phys. B120 221.

Freedman, D.Z. van Nieuwenhuizen, P. & Ferrara, S. (1976), Phys. Rev. D13 3214; Freedman, D.Z. & van Nieuwenhuizen, P. (1976), Phys. Rev. D14 912.

Gates, S.J. Jr., Stelle, K.S. & West, P.C. (1980), Nucl. Phys. B 169 341.

Grimm, R. Wess, J. & Zumino, B. (1978), Phys. Lett. 73B 415.

Grimm, R. Wess, J. & Zumino, B. (1979), Nucl. Phys. B152 255.

Salam, A. & Strathdee, J. (1976), Nucl. Phys. B76 477.

Stelle, K.S. & West, P.C. (1978), Phys. Lett. 74B 330.

van Nieuwenhuizen, P. & West, P.C. (1980), Nucl. Phys. B169 501.

Volkov, D.V. & Akulov, V.P. (1973), Phys. Lett. 40B 109.

Wess, J. (1977), Proceedings of the VIII, G.I.F.T. International Seminars on Theoretical Physics, Salamanca. Lecture Notes in Physics, (Springer-Verlag) vol. 77, p.81.

Wess, J. (1979), Boulder lectures, Karlsruhe preprint.

Wess, J. & Zumino, B. (1977), Phys. Lett. 66B 361.

Wess, J. & Zumino, B. (1978a) Phys. Lett. 74B 51.

Wess, J. & Zumino, B. (1978b), Phys. Lett. 79B 394.

Zumino, B. (1975), Proceedings of the Conference on gauge theories and modern field theory, Northeastern University, Boston, eds. Arnowitt, R. & Nath, P. (MIT Press) p.255.

Zumino, B. (1978), Proceedings of the NATO Advanced Study Institute on Recent Developments in Gravitation, Cargèse (Penum Press) p. 405.

COUNTERTERMS IN EXTENDED SUPERGRAVITIES

R. Kallosh
Lebedev Physical Institute 117924 Moscow, USSR

1 INTRODUCTION

In the solving of the renormalizability problem there exist at least two stages:

i) to investigate all superinvariants (with any number of derivatives in theories with dimensional coupling constant) which could serve as counterterms in the theory.

ii) to calculate the coefficients in front of these invariants (i.e., to perform the calculations of multiloop quantum corrections). It is possible that in some cases we will find zero (as e.g., in the first three loops in the N = 4 super-Yang-Mills theory). This means that one should try to find some hidden symmetry or some other more deep reasons for the absence of divergences.

At present in supergravity the point i) is mainly investigated. The S-matrix calculations (point ii)) were performed only in the first loop for N = 1, ..., 8 supergravities by van Nieuwenhuizen, Grisaru, Vermaseren and Fishler beginning from 1976. The one-loop calculations of conformal anomalies were also performed.

When analysing counterterms, the coefficients in front of which are independent of the gauge condition for quantum fields, in the first approximation we consider only superinvariants which do not vanish when classical equations of motion hold, i.e., on mass shell. The presence of such invariants in the theory is an indication of the existence of gauge-independent divergences. However the absence of such invariants is in general not sufficient to prove that the theory is finite, a more deep investigation concerning e.g., non-local divergences is necessary.

The corresponding problem in pure gravity is to find all general covariant actions in the Einstein space $R_{\mu\nu} = R = 0$. There is an infinite number of such actions, constructed from curvature tensor $R_{\mu\nu\lambda\varepsilon}$.

There exists the following correspondence between superinvariants with higher derivatives and number of loops for which some superinvariant may serve as a counterterm. Note that we are analyzing only logarithmic (poles in dimensional regularization or reduction) divergences.

We shall consider only superfields without explicit dependence on gravitational coupling constant k^2. All dependence on k^2 is absorbed by rescaling of $s = 0, {}^1/_2, 1, {}^3/_2$ fields. In this case all bosons have dimension 0, all fermions have dimension ${}^1/_2$. Lagrangian of all extended supergravities in terms of these fields has a simple dependence on k^2

$$L = \frac{1}{k^2} L^1(\Phi, \chi, \ldots, g_{\mu\nu}) \tag{1.1}$$

It follows that any propagator has k^2-dependence $\sim k^2$, any vertex has a k^2-dependence $\sim \frac{1}{k^2}$. One should also use the fact that for any multiloop diagram the following relation holds:

$$\ell = n - m + 1, \tag{1.2}$$

where ℓ is the number of loops, n is the number of propagators and m is the number of vertices. Therefore any superinvariant constructed from ordinary fields or superfields described above has a simple dependence on k^2 of the form

$$k^{2(n-m)} = k^{2(\ell-1)}, \tag{1.3}$$

where ℓ is the number of loops.

2 LINEARIZED ON-SHELL SUPERFIELDS

Linearized on-shell superfields in extended supergravities have been described in papers (Kallosh 1981, Howe & Lindström 1981, Siegel 1981). These superfields contain the linearized curvature tensor and therefore they are the building blocks for constructing the superinvariants.

It is useful to present these superfields explicitly since

they contain only physical fields (or field strengths) of spin $s = 0, \ldots, 2$.

In the N = 1 supergravity there is a chiral three-component spinor superfield (Ferrara & Zumino 1978) (we use throughout the two-component spinor notations, a, ȧ being spinor indices, i, j being internal indices)

$$W_{abc} = \Psi_{abc} + \theta^{d} R_{abcd} , \qquad (2.1)$$

where Ψ_{abc} is the spin $^3/_2$ field strength. In eq. (2.1) R_{abcd} is the left curvature spinor, connected with the curvature tensor by means of matrices $\sigma^{\mu\nu}_{ab}$. The conjugated superfield

$$\bar{W}_{\dot{a}\dot{b}\dot{c}} = \bar{\Psi}_{\dot{a}\dot{b}\dot{c}} + \bar{\theta}^{\dot{d}} \bar{R}_{\dot{a}\dot{b}\dot{c}\dot{d}} \qquad (2.2)$$

contains the right spinor strengths of fields $s = {}^3/_2$ and $s = 2$.

In the supergravity N = 2 there is a chiral two-component spinor superfield

$$W_{ab} = F_{ab} + \theta^{c}_{i} \Psi^{i}_{abc} + \frac{1}{2!} \theta^{d}_{i} \theta^{c}_{j} \varepsilon^{ij} R_{abcd} \qquad (2.3)$$

where F_{ab} is the spin 1 field strength. The conjugated superfield $\bar{W}_{\dot{a}\dot{b}}$ contains all right spinor strengths for $s = 1, {}^3/_2, 2$.

In the supergravity N = 3 the lowest component of the chiral spinor superfield is the spin $^1/_2$ field χ_a.

$$W_{a} = \chi_{a} + \theta^{b}_{i} F^{i}_{ab} + \frac{1}{2!} \theta^{b}_{i} \theta^{c}_{j} \varepsilon^{ijk} \Psi_{kabc} +$$

$$+ \frac{1}{3!} \theta^{b}_{i} \theta^{c}_{j} \theta^{d}_{k} \varepsilon^{ijk} R_{abcd} \qquad (2.4)$$

The conjugated superfield $\bar{W}_{\dot{a}}$ begins with the right chirality spinor $\bar{\chi}_{\dot{a}}$.

Note that in the theories N = 1, 2, 3 the superfields $W_{a_1 \ldots a_{4-N}}$ satisfy besides the chirality condition also the "transversality" condition on shell.

$$D^{a_1}_{i} W_{a_1 \ldots a_{4-N}} = 0. \qquad (2.5)$$

In the N = 4 supergravity there is a chiral scalar superfield

$$W = \Phi + \theta_{ia}\,\chi^i_a + \frac{1}{2!}\,\theta^a_i\,\theta^b_j\,F^{ij}_{ab} + \frac{1}{3!}\,\theta^a_i\,\theta^b_j\,\theta^c_k\,\varepsilon^{ijk\ell}\,\times$$

$$\times\,\Psi_{\ell\,abc} + \frac{1}{4!}\,\theta^a_i\,\theta^b_j\,\theta^c_k\,\theta^d_\ell\,\varepsilon^{ijk\ell}\,R_{abcd} \qquad (2.6)$$

where $\Phi = A + iB$, A is the scalar field and B is the pseudoscalar field. The conjugated superfield $\bar{W}$ begins with $\bar{\phi} = A - iB$.

Thus the N = 1, ..., 4 supergravities are described at the linearized level by the 4 - N component spinor superfields

$$W_{a_1 \ldots a_{4-N}},\quad \bar{W}_{\dot{a}_1 \ldots \dot{a}_{4-N}}$$

$$D_{\dot{a}i}\,W_{a_1 \ldots a_{4-N}} = D^i_a\,\bar{W}_{\dot{a}_1 \ldots \dot{a}_{4-N}} = 0 \qquad (2.7)$$

$$i = 1, \ldots, N \qquad N \leqslant 4$$

The supergravity N = 8 at the linear level is described (Kallosh 1981) on shell by the superfield $W_{ijk\ell}$, i = 1, ..., 8, which is self-dual in the internal space.

$$W_{i_1 i_2 i_3 i_4} = \frac{1}{24}\,\varepsilon_{i_1 i_2 i_3 i_4 j_1 j_2 j_3 j_4}\,\bar{W}^{j_1 j_2 j_3 j_4} \qquad (2.8)$$

with the following properties

$$D_{\dot{a}(i}\,W_{jk\ell m)} = 0 \qquad (2.9)$$

$$D^i_a\,W_{jk\ell m} = \frac{1}{N-3}\,\delta^i_{[j}\,D^n_a\,W_{k\ell m]n} \qquad (2.10)$$

In general this superfield depends on all 32 components θ_i, $\bar{\theta}^i$, i = 1, ... 8. But due to eqs. (2.9), (2.10) a "proper" basis could be found where this superfield depends only on 16 components of θ_i, $\bar{\theta}^i$.

$$W_{i_1 i_2 i_3 i_4} = W_{i_1 i_2 i_3 i_4}\,(X_{a\dot{a}} + i\sum_I \theta_{i_k}\,\sigma_{a\dot{a}}\,\bar{\theta}^{i_k} - \qquad (2.11)$$

$$- i\sum_J \bar{\theta}^{j_k}\,\sigma_{a\dot{a}}\,\theta_{j_k}\,,\ \theta_{i_k}\,,\ \bar{\theta}^{j_k}) = \frac{1}{24}\,\varepsilon_{i_1 i_2 i_3 i_4 j_1 j_2 j_3 j_4}\,\bar{W}^{j_1 j_2 j_3 j_4}\,,$$

$$I = i_1, i_2, i_3, i_4 \ ,$$

$$J = j_1, j_2, j_3, j_4 \ .$$

This means e.g., that the superfield W_{1234} depends only on $\theta_1, \theta_2, \theta_3, \theta_4$ and $\bar{\theta}^5, \bar{\theta}^6, \bar{\theta}^7, \bar{\theta}^8$ in the basis $x_{a\dot{a}} + i \sum_{i=1}^{4} \theta_{\dot{i}} \sigma_{a\dot{a}} \bar{\theta}^{\dot{i}} - \sum_{j=5}^{8} \bar{\theta}^{j} \sigma_{a\dot{a}} \theta_{\dot{j}}$. Unfortunately the definition of this basis depends on the choice of the internal space components of the superfield $W_{\dot{i}jk\ell}$.

In the "proper" basis the linear on-shell superfield in the N = 8 supergravity has the form:

$$\begin{aligned} W_{i_1 i_2 i_3 i_4} &= \phi_{i_1 i_2 i_3 i_4} + \theta_{[i_1} \chi_{i_2 i_3 i_4]} + \frac{1}{24} \varepsilon_{i_1 i_2 i_3 i_4 j_1 j_2 j_3 j_4} \times \\ &\times \bar{\theta}^{j_1} \bar{\chi}^{j_2 j_3 j_4} + \theta_{i_1} \partial_{a\dot{a}} \bar{\theta}^{j_1} \phi_{j_1 i_2 i_3 i_4} + \frac{1}{2} \theta^a_{[i_1} \theta^b_{i_2} F_{ab\, i_3 i_4]} + \\ &+ \ldots + \frac{1}{24} (\theta^a_{\dot{i}_1} \theta^b_{\dot{i}_2} \theta^c_{\dot{i}_3} \theta^d_{\dot{i}_4} R_{abcd} + \frac{1}{24} \varepsilon_{i_1 i_2 i_3 i_4 j_1 j_2 j_3 j_4} \times \\ &\times \bar{\theta}^{j_1}_{\dot{a}} \bar{\theta}^{j_2}_{\dot{b}} \bar{\theta}^{j_3}_{\dot{c}} \bar{\theta}^{j_4}_{\dot{d}} \bar{R}^{\dot{a}\dot{b}\dot{c}\dot{d}}) + \ldots + \frac{1}{8!} \theta^a_{\dot{i}_1} \theta^b_{\dot{i}_2} \theta^c_{\dot{i}_3} \theta^d_{\dot{i}_4} \times \\ &\times \bar{\theta}^{j_1}_{\dot{a}} \bar{\theta}^{j_2}_{\dot{b}} \bar{\theta}^{j_3}_{\dot{c}} \bar{\theta}^{j_4}_{\dot{d}} \partial_{a\dot{a}} \partial_{b\dot{b}} \partial_{c\dot{c}} \partial_{d\dot{d}} \phi_{j_1 j_2 j_3 j_4} \ . \end{aligned} \tag{2.12}$$

This superfield possesses many interesting properties. First of all the left and right spinors (χ_a, $\bar{\chi}_{\dot{a}}$, ... R_{abcd}, $\bar{R}_{\dot{a}\dot{b}\dot{c}\dot{d}}$) are contained in the same superfield (self-conjugated multiplet) as distinct from the superfields presented above in the N = 1, ..., 4 supergravities. In this theories χ_a, ... R_{abcd} are contained in the left chiral superfield and $\chi_{\dot{a}}$ ... $\bar{R}_{\dot{a}\dot{b}\dot{c}\dot{d}}$ are in the conjugate right superfield. This property is important for investigations of self-duality (see §5).

Another interesting property of this superfield is the following. The last non-trivial component of this superfield is proportional to θ^8 though _a priori_ one could expect θ^{32}. Such a strong decrease of the number of components is connected with the constraints (2.9), (2.10) as well as with the mass shell condition. In the case of small N we have had also the decrease of the number of components due to chirality and transversality or due to mass shell condition. For example in the N = 4 supergravity the maximal component is proportional to

θ^4 though we could expect off-shell θ^8 for chiral superfield and θ^{16} otherwise. Thus in both examples the conditions (2.9), (2.10) or the chirality condition reduce the number of components twice. The mass shell condition also reduces the remaining number of components twice.

Let us consider how the self-dual superfield $W_{ijk\ell}$ of the theory N = 8 becomes the chiral superfield W of the N = 4 supergravity when we come from N = 8 to N = 4. Let in eq. (2.12) the internal indices i,j,... take only the values 1,2,3,4. In this case $W_{1234} = \varepsilon_{1234} W$. The "proper" basis in eq. (2.11) becomes the chiral basis $x_{a\dot{a}} + i \sum_{i=1}^{4} \theta_i \sigma_{a\dot{a}} \bar{\theta}^i$ and the full self-dual superfield in (2.12) becomes exactly the chiral superfield W (2.6) of the N = 4 supergravity.

3 FIRST AND SECOND LOOPS IN EXTENDED SUPERGRAVITIES (Kallosh 1981). LINEARIZED SUPERTOPOLOGICAL INVARIANTS

The problem of one- and two-loop renormalizability of the extended supergravities N = 2, ..., 8 was not completely investigated until the last year. However the hope existed that increasing of N could only improve the situation. As it is well known the theory N = 1 is finite in the first two loops (Grisaru, van Nieuwenhuizen & Vermaseren 1976, Grisaru 1977, Deser, Kay & Stelle 1977). In this theory there is only one superinvariant generalizing $R^2_{\mu\nu\alpha\beta}$ and vanishing on shell. In the second loop only the term generalizing $R^3_{\mu\nu\alpha\beta}$ could appear but it has no superpartner. In the third loop there exists a candidate to counterterms (Deser, Kay & Stelle 1977).

We will perform the analysis of superinvariant counterterms in the first two loops using the linearized superfields presented above. We shall use the fact that the dimension of dx is equal to -1, the dimension of $d\theta$ is equal to $+ ^1/_2$ and the dimension of gravitational constant k is equal to -1. The dimension of the superfields will be given below. As was already explained, the superinvariant corresponding to ℓ loops has the dependence on gravitational constant as $k^{2(\ell-1)}$. We consider one- and two-loop counterterms in the N = 2 supergravity. In the component approach the one-loop counterterms were analyzed and the coefficients in front of them were calculated in (Grisaru, van Nieuwenhuizen & Vermaseren 1976). In this paper the counterterm $R_{\mu\nu\alpha\beta} F^{\mu\alpha} F^{\nu\beta}$ ($R_{\mu\nu\alpha\beta}$ is the curvature tensor, $F_{\mu\nu}$ is the vector field strength) was also discussed and it was assumed that this term is forbidden by dual invariance. The coefficient in front of $T_{\mu\nu}^{\ 2} = F_{\mu\mu'} F_{\mu'\nu} - \frac{1}{4} g_{\mu\nu} F_{\mu'\nu'}^{\ 2}$, was also calculated and it was found to be zero.

In the superfield approach the N = 2 supergravity on shell is described at linear level by the chiral two-component spinor superfield W_{ab} of dimension 1, see eq., (2.3). In the one-loop approximation it is possible to construct only one invariant of zero order in k^2

$$S_{N=2}^{\ell=1} = \int d^4x\, d^4\theta\, W_{ab}^{\ 2} + \text{h.c.} = \int d^4x\, (R_{abcd}^2 + R_{\dot{a}\dot{b}\dot{c}\dot{d}}^{\ 2} + \ldots, \tag{3.1}$$

which vanishes on-shell. The invariant $R_{\mu\nu\alpha\beta} F^{\mu\alpha} F^{\nu\beta}$ as we can see from superfield formalism, is forbidden not by duality but simply by supersymmetry which forbids also the abovementioned term $T_{\mu\nu}^{\ 2}$. In the two-loop approximation there is one non-trivial linearized invariant

$$S_{N=2}^{\ell=2} = k^2 \int d^4x\, d^4\theta\, (W_{ab} W_{ab})^2 + \text{h.c.} =$$

$$= k^2 \int d^4x\, (F_{ab}^{\ 2} R_{cde\ell}^2 + \bar{F}_{\dot{a}\dot{b}}^2 \bar{R}_{cde\ell}^2) + \ldots \tag{3.2}$$

This invariant does not contain the pure gravitational part. It is not known now if it has a non-linear generalization. Note that in eq. (3.2) there is an integral only over the left superspace from the left chiral superfields. We could analyse it from the point of view of combined chiral-dual invariance (van Nieuwenhuizen & Vermaseren 1976). In terms of four-component spinors θ_α, $[\gamma^\mu,\gamma^\nu]_{\alpha\beta} W_{\mu\nu} = W_{\alpha\beta}$ the combined chiral dual transformation is merely a γ^5-transformation $\theta_\alpha \to a^{+\frac{1}{2}\gamma^5_{\alpha\beta}} \theta_\beta$, $W_{\alpha\beta} \to a^{\gamma^5_{\alpha\gamma}} W_{\gamma\beta}$, i.e., the γ^5 transformation of the superspace and of the superfields. The chiral-dual invariance on-shell at the linear level in terms of superfields means that under the global transformation

$$d\theta_{bi} \to a^{-\frac{1}{2}} d\theta_{bi}, \qquad W_{ab} \to a\, W_{ab}$$

$$d\bar{\theta}_{\dot{a}}^{\ i} \to a^{\frac{1}{2}} d\bar{\theta}_{\dot{a}}^{\ i} \qquad \bar{W}_{\dot{a}\dot{b}} \to a^{-1} \bar{W}_{\dot{a}\dot{b}}$$

the counterterms should not depend on a. The two-loop invariant (3.2) is forbidden by the abovementioned chiral-dual invariance. However we stress that the linearized supersymmetry allows such an invariant and therefore the problem of renormalizability of the N = 2 supergravity in two loops is based on less reliable (because of possible anomalies) chiral-dual invariance.

As is known, at the tree-loop level in the N = 2 supergravity the supersymmetric invariant exists (Deser & Kay 1980).

In the N = 3 supergravity on-shell there is a linearized chiral spinor superfield W_a of dimension $^1/_2$, see eq. (2.4). It is not possible to construct from this superfield the counterterms for the S-matrix in the first and second loops, but the three-loop counterterm exists (Kallosh 1981, Howe & Lindström 1981).

In the theory N = 4 on-shell there is a linearized chiral scalar superfield W of dimension 0, see eq. (2.6). In the one-loop approximation the superinvariant

$$S^{\ell=1}_{N=4} = \int d^4x\, d^8\theta\, W^4 + \text{h.c.} = \int d^4x\, \phi^2\, R^2_{abcd} + \ldots \qquad (3.3)$$

can be constructed. This invariant is forbidden by the chiral-dual invariance and also it is not clear if it exists at the non-linear level. But as in the case of the second loop for the N = 2 supergravity, the linearized supersymmetry allows such a counterterm.

Now we consider the most interesting case of the N = 8 supergravity, which is described at the linear level by the self-dual scalar zero-dimensional superfield $W_{ijk\ell}$. Using the abovementioned properties of this superfield, in particular, the fact that W_{1234} depends on 16 Grassman variables, it is not possible to construct the one- and two-loop S-matrix counterterms. Note that the dimension of $d^4x\, d^{16}\theta$ is equal to + 4, the superfield $W_{ijk\ell}$ has zero dimensions, any derivative has positive dimension and it is necessary to have at least k^4 to compensate the dimension of the volume of integration, i.e., to have the third or higher loop counterterm.

Very recently Howe, Stelle and Townsend have proposed the method of constructing the superactions. In this method the Lagrangian may not be the last component of the superfield. We guess that the superinvariants which are not connected with the last component of some superfield are full divergences. The first example is (3.3), the second example is the N = 4 super Yang-Mills linearized action (Howe, Stelle & Townsend preprint) which vanishes on-shell up to the total divergence. In the N = 8 supergravity in the first loop we can construct the following superinvariants on-shell using this method.

$$\binom{\chi}{\tau} = \int d^4x\, D^{[ijk\ell]}_{abcd}\, D^{[mnpq]}_{abcd}\, W_{ijk\ell}\, W_{mupq} \pm \text{h.c.} =$$
$$= \int d^4x\, R^2_{abcd} \pm \text{h.c.} \qquad (3.4)$$

where

$$D^{[ijk\ell]}_{abcd} \equiv D^i_a\, D^j_b\, D^k_c D^{\ell]}_d \quad .$$

Eq. (3.4) presents the linearized version of <u>supertopological invariants</u> χ and τ which are pure divergences on shell. For N = 4 the corresponding terms are

$$\int d^4x\, d^8\theta\, W^2 \pm \text{h.c.} \qquad (3.5)$$

The invariant (3.4) can be represented also in another form respecting E_7 symmetry of the N = 8 supergravity

$$\int d^4x\, D^{x\ell}\, D^{mn}\, D^{pq}\, W^a_{kmp}\, W^a_{\ell np} \pm \text{h.c.}, \qquad (3.6)$$

where

$$D^{(k\ell)} \equiv D^{(k}_a\, D^{\ell)}_a \quad . \qquad (3.7)$$

For the N = 3 supergravity we have from (3.7)

$$\int d^4x\, d^6\theta\, W^2_a \pm \text{h.c.}$$

It is not clear whether the non-linear generalization of the supertopological invariants presented above exist, on the contrary there are some indications that they do not exist from N = 3 and higher.

We stress that in the N = 8 supergravity there are no superinvariants in the first two loops in the topologically trivial backgrounds, which means that the S-matrix is finite in the first two loops. This statement is based also on the fact that even using the method of constructing the superactions from (Howe, Stelle & Townsend preprint) it is not possible to find in the N = 8 theory something like

$$\int d^4x\,(d^8\theta)^{ij\ldots}\,W^4_{ij\ldots}\,.$$

The conclusion is that the S-matrix in the first two loops in the N = 1 and N = 8 supergravities is finite and this follows only from supersymmetry. In the intermediate theories N = 2, N = 4 in the second and first loops correspondingly there exist linearized superinvariants (3.1), (3.2). It would be interesting to find the non-linear generalization of these invariants. It seems that it is possible to find them using the tensor calculus for conformal supergravities N = 2, N = 4 (deWit, private communication, Bergshoeff, de Roo & deWit preprint). In the case that these counterterms will be constructed non-linearly, the calculations of the corresponding divergences in the 2-d loop in the N = 2 supergravity and in the 1-st loop in the N = 4 supergravity will be necessary to check whether the combined chiral-dual invariance has anomalies in these theories.

4 THE THREE-LOOP COUNTERTERMS

In the supergravity N = 1 the three-loop invariant was discovered in (Deser, Kay & Stelle 1977), in the theory N = 2 in (Deser, & Kay 1980). In both papers the invariants were found in component approach at the linear level on shell. The linearized three-loop counterterms were found in (deWit, private communcation, Bergshoeff, Roo & deWit preprint) in the superfield approach. For the theories N = 2,3,4 they were found in (Kallosh 1981, Howe & Lindström 1981). In general for the N = 1, ..., 4 supergravities the three-loop invariants could be represented as

$$S^{\ell=3}_{N\leq 4} = k^4 \int d^4x\, d^{4N}\theta\,(W_{a_1\ldots a_{4-N}}\,\bar{W}_{\dot{a}_1\ldots\dot{a}_{4-N}})^2 =$$

$$= k^4 \int d^4x\,(R_{abcd}\,\bar{R}_{\dot{a}\dot{b}\dot{c}\dot{d}})^2 + \ldots \qquad (4.1)$$

where the superfields $W_{a_1\ldots a_{4-N}}$ are defined in equs. (2.1) - (2.4), (2.7). This counterterm is the superpartner of the square of the Bell-Robinson tensor, represented in the r.h.s. of eq. (4.1).

Before considering the case N > 4 and the non-linear theory we are going to transform eq. (4.1) in the theory N = 4 to the form where it depends not on scalar superfields but on spinor superfields $D^i_a W = \lambda^i_a$, which have a non-linear extension. We shall transform the equation

$$S^{\ell=3}_{N=4} = k^4 \int d^4x\, d^{16}\theta\, (W\bar{W})^2 \tag{4.2}$$

to the form

$$\int d^4x\, (d^{12}\theta)^{ij}_{k\ell}\; W^i_a\, W^j_a\, \bar{W}_{k\dot{a}}\, \bar{W}_{\ell\dot{a}} \quad . \tag{4.3}$$

We shall use the following property of the invariant (linearized) volume

$$\int d^4x\, d^{16}\theta \sim \int d^4x\, \mathcal{D}^{jj_1}_{ii_1}\, P^{mm_1\ ii_1}_{nn_1\ jj_1} \times \bar{D}_{m\dot{a}}\, \bar{D}_{m_1\dot{a}}\, D^n_a\, D^{n_1}_a , \tag{4.4}$$

where

$$\mathcal{D}^{jj_1}_{ii_1} = \bar{Q}^j_{\dot{a}\dot{b}\dot{c}}\; \bar{Q}^{i_1\dot{a}\dot{b}\dot{c}}\; Q_{iabc}\; Q_{i_1}^{\ abc} , \tag{4.5}$$

$$Q_{iabc} \equiv \varepsilon_{ijk\ell}\, d\theta^j_a\, d\theta^k_b\, d\theta^\ell_c , \tag{4.6}$$

and the projector P is defined as follows

$$P^{mm_1\ ii_1}_{nn_1\ jj_1} = \delta^i_n\, \delta^{i_1}_{n_1}\, \delta^{(m}_j\, \delta^{m_1)}_{j_1} - \frac{2}{3}\, \delta^{(i}_{(n}\, \delta^m_j\, \delta^{i_1)}_{j_1}\, \delta^{m_1}_{n_1)}$$
$$+ \frac{1}{15}\, \delta^{(m}_n\, \delta^{m_1)}_{n_1}\, \delta^i_{(j}\, \delta^{i_1}_{j_1)} \tag{4.7}$$

and possesses the property

$$\delta^j_i\, P^{(mm_1)(ii_1)}_{(nn_1)(jj_1)} = \ldots = 0. \tag{4.8}$$

According to (4.8) the action of four derivatives $D^n_a D^{n_1}_a \times \bar{D}_{m\dot{a}} \bar{D}_{m_1\dot{a}}$ on $W^2 \bar{W}^2$ produces only $W^n_a W^{n_1}_a \bar{W}_{m\dot{a}} \bar{W}_{m_1\dot{a}}$, and terms like $\delta^m_n\, \partial_{a\dot{a}} W W^{n_1}_a \bar{W}_{\dot{a}m_1} \bar{W}$ do not appear. As a result the original action (4.2) is proportional with some non-zero coefficient to the following action

$$\tilde{S}^{\ell=3}_{N=4} = k^4 \int d^4x\, \mathcal{D}^{jj_1}_{ii_1}\, P^{mm_1}_{nn_1}\, {}^{ii_1}_{jj_1}\, W^n_a\, W^{n_1}_a\, \bar{W}^{\dot{a}}_m\, \bar{W}^{\dot{a}}_{m_1} \equiv$$

$$\equiv k^4 \int d^4x\, \mathcal{D}^{jj_1}_{ii_1}\, (W^i_a\, W^{i_1}_a\, \bar{W}^{\dot{a}}_j\, \bar{W}^{\dot{a}}_{j_1})_{t_2=0} \quad . \tag{4.9}$$

It is interesting that the direct proof that (4.9) is a superinvariant was obtained in (Howe, Stelle & Townsend preprint) without using the fact that it is proportional to the original superinvariant (4.2).

In the N = 8 supergravity it is easy to construct the three-loop superinvariant using the properties of the superfield $W_{ijk\ell}$ introduced above. Originally this counterterm was constructed in a form which was not explicitly SU(8)-invariant. As was already explained, the superfield W_{1234} depends on 16 Grassman variables in the "proper" basis. Therefore we have the 3-loop superinvariant in the form:

$$S^{\ell=3}_{N=8} = k^4 \int d^4x\, d^2\theta_1 \ldots d^2\theta_4\, d^2\bar{\theta}^5 \ldots d^2\bar{\theta}^8\, W^4_{1234} =$$

$$= k^4 \int d^4x\, (R_{abcd}\, \bar{R}_{\dot{a}\dot{b}\dot{c}\dot{d}})^2 + \ldots \tag{4.10}$$

It is interesting to compare our eq. (4.10) with the recently obtained 3-loop counterterm in (Howe, Stelle & Townsend).

$$S^{\ell=3}_{N=8} = k^4 \int d^4x\, D^{[i_1 \ldots i_4],[j_1 \ldots j_4]}\, \bar{D}^{[k_1 \ldots k_4],[\ell_1 \ldots \ell_4]} \times$$

$$\times (W_{i_1 \ldots i_4}\, W_{j_1 \ldots j_4}\, W_{k_1 \ldots k_4} \cdot W_{\ell_1 \ldots \ell_4})_{232848} \tag{4.11}$$

where the kernel is taken in the representation <u>232848</u>. The invariance of the action in the form (4.11) is proved in (Howe, Stelle & Townsend) by using the constraints on the kernel which follow from the constraints on the superfield $W_{ijk\ell}$. Expression (4.10) is an explicit superinvariant. It can be also proved (Kallosh 1981) that it is SU(8) invariant after all θ-integrations are performed. It follows from the r.h.s. of eq. (4.10) which contains the SU(8)-invariant pure gravitational part $(R_{abcd}\, \bar{R}_{\dot{a}\dot{b}\dot{c}\dot{d}})^2$. All other terms in the r.h.s. of eq. (4.10) are also SU(8)-invariant, being the superpartners of $(R\bar{R})^2$. This was also verified by direct calculations. Both expressions (4.10) and (4.11) are proved to be superinvariant, both contain the same spin 2 part. This means that they coincide.

5 NON-LINEAR THEORY. SUPER-SELF-DUALITY

Now we shall study the problem which counterterms could be constructed in exact non-linear theory. To solve this problem we shall use the results of the paper of Brink & Howe (1978), where all the components of torsion and curvature in the N = 8 supergravity on-shell are expressed by means of Bianchi identities through the spinor superfield $W^{a}_{\cdot ijk}$, which is one of the components of the torsion. We shall use the notations of (Brink & Howe 1978) where in tangent space all indices: M N P Q, spinor and internal: A B C D, spinor: abcd...$\dot{a}\dot{b}\dot{c}\dot{d}$, internal SU(N): ijk...t, vector uv.. . The corresponding indices in curved space are denoted by $\underline{M}$ $\underline{N}$; $\underline{A}$ $\underline{B}$; αβ..., $\dot{\alpha}\dot{\beta}$...; ξη...;μν .

$$T^{a}_{\dot{b}\dot{c}\ ijk} = \varepsilon_{\dot{b}\dot{c}}\ W^{a}_{ijk}\ (x,\theta_i,\bar{\theta}^i). \tag{5.1}$$

The standard geometrical invariant in extended supergravities can be represented in the following form

$$S^{\ell}_{N} = (k^2)^{\ell-1} \int dV\ L(R_{MNPQ},\ T^{P}_{MN}) \quad , \tag{5.2}$$

where L is some scalar, constructed from curvature and torsion tensors in the tangent space. The integration in (5.2) goes over the whole supermanifold with invariant volume dV = $d^4x\ d^{4N}\theta$ Bere E, where Berezinian E is a superdeterminant of the vielbein, and ℓ is the number of loops. Dimensional considerations show that

$$\dim L = 4 - 2N + 2(\ell - 1). \tag{5.3}$$

Taking into account that L contains only torsion and curvature tensors (dim R, T $\geqslant$ $^1/_2$) and that, as it follows from (5.3), dim L is an even number, we have

$$\dim L \geqslant 2, \tag{5.4}$$

from which the simple relation

$$\ell \geq N \tag{5.5}$$

follows. Eq. (5.5) means that the standard geometrical invariants like (5.2) give contribution beginning from the 3rd loop for the N = 3 supergravity, from the 4th loop for N = 4 and from the 8th loop for N = 8.

Thus it follows that in the N = 8 theory the standard geometrical invariants begin from the 8th loop level

$$\begin{aligned} S^{\ell=8}_{N=8} &= k^{14} \int d^4x\, d^{32}\theta \text{ Ber } E\, (W^{a}_{ijk}\, \bar{W}^{ijk}_{\dot{a}})^2 = \\ &= k^{14} \int d^4x\, (R_{abcd}\, D_{u_1} \ldots D_{u_5}\, \bar{R}_{\dot{a}\dot{b}\dot{c}\dot{d}})^2 + \ldots \end{aligned} \tag{5.6}$$

The first geometrical invariant in the N = 4 is the 4-loop counterterm

$$S^{\ell=4}_{N=4} = k^6 \int d^4x\, d^{16}\theta \text{ Ber } E(W^{\ell}_{a}\, \bar{W}^{\dot{a}}_{\ell})^2 , \tag{5.7}$$

where

$$\bar{W}^{\dot{a}}_{\ell} = \varepsilon_{ijk\ell}\, T^{ijk\,\dot{a}}_{bc}\, \varepsilon^{bc} . \tag{5.8}$$

The first standard geometrical invariant in the N = 3 supergravity is the three-loop invariant

$$S^{\ell=3}_{N=3} = k^4 \int d^4x\, d^{12}\theta \text{ Ber } E\, (W_a\, \bar{W}_{\dot{a}})^2 , \tag{5.9}$$

where

$$\bar{W}^{\dot{a}} = \varepsilon_{ijk}\, \varepsilon^{ab}\, T^{\dot{a}\,ijk}_{ab} . \tag{5.10}$$

Thus if we would not be aware of superinvariants which contain the integration over the sub-supermanifold, the role of which was stressed by Zumino (preprint), we could conclude that, e.g., the N = 8 supergravity is finite up to the 8-th loop order and the N = 4 supergravity is finite up to the forth loop order. But this conclusion is wrong, at least until it is proved that the linearized 3-loop invariants in the theories N = 4, ..., 8 have no non-linear generalizations.

Let us consider the relation between linearized and exact

superinvariants. From the absence of linearized counterterms follows the absence of exact non-linear ones. However the existence of the linearized counterterms does not guarantee the existence of the corresponding non-linear ones. Though there was no examples of the last situation, we stress that one should be careful with the new linearized superinvariants, especially for high N. In the N = 1 supergravity the linearized invariants constructed from chiral fields were known long ago. But only after Ogievetsky and Socatchev (1978) have shown that in the exact non-linear theory with prepotentials there exist the invariant volumes in the left and right chiral spaces $\{X_L, \theta_L\}$, $\{X_R, \bar{\theta}_R\}$ and there exist chiral superfields, which depend only on X_L, θ_L or $X_R, \bar{\theta}_R$, it becomes clear that the full non-linear chiral superinvariants exist in the theory.

Since at present none of the supergravities $N \geq 2$ is constructed geometrically on the base of unconstrained prepotentials the problem of existence in the theories of invariant (or covariant) subvolumes is far from being solved. Nevertheless it is interesting in this connection to discuss the 3-loop invariants in the theories N = 4 (4.9) and N = 8 (4.10), (4.11). The expression (4.9) could be written down in geometrical terms by means of differential forms (Akulov, Volkov & Soroka 1975, Wess & Zumino 1977)

$$E^M = dz^{\mathcal{N}} E_{\mathcal{N}}^M = dx^\mu E_\mu{}^M + d\theta^{\mathcal{A}} E_{\mathcal{A}}^M , \tag{5.11}$$

where dx^μ ($d\theta^{\mathcal{A}}$) are anticommuting (commuting) variables. We shall use also the "dual" differential forms introduced by Berezin (1979)

$$\tilde{E}_M = E_M^\mu \, d\tilde{x}_\mu + E_M^{\mathcal{A}} \, d\tilde{\theta}_{\mathcal{A}} , \tag{5.12}$$

where $d\tilde{x}_\mu$ ($d\tilde{\theta}_{\mathcal{A}}$) are commuting (anticommuting) variables. In this case we could write the superinvariant (4.9) in the form

$$S_{N=4}^{\ell=3} = \int \prod_{u=1}^{4} E^u \, \mathcal{D}_{ii_1}^{jj_1} \left(W_a^i W_a^{i_1} \bar{W}_{j\dot{a}} \bar{W}_{j_1\dot{a}} \right)_{tr=0} , \tag{5.13}$$

where $\mathcal{D}_{ii_1}^{jj_1}$ is defined in (4.5), (4.7) and the new definition of $Q_{i\,abc}$ is

$$Q_{i\,abc} \equiv \varepsilon_{ijk\ell} \, \tilde{E}_a^j \, \tilde{E}_b^k \, \tilde{E}_c^\ell , \tag{5.14}$$

E_u, $\tilde{E}^j_a$ being defined in (5.11), (5.12) correspondingly. The expression (5.13) contains now only geometrical objects, at linear level it is a superinvariant and seems to be a good starting point for the investigations of the non-linear theory.

We consider now the N = 8 supergravity. The three-loop invariant in (4.10) can be written in the form containing only integral over d^4x, i.e., all θ-integrations can be performed. After that procedure the expression contains only the geometrical spinor superfield W^a_{mnp} and its derivatives (the scalar superfield W_{mnpq} is not a geometrical object). The corresponding equation has the form

$$S^{\ell=3}_{N=8} = k^4 \int d^4x\, e\, [D^m_{(a} D^n_b D^p_c W_{mnp\, d)} \cdot \bar{D}_{m_1(\dot{a}} \bar{D}_{n_1\dot{b}} \bar{D}_{p_1\dot{c}} \times$$

$$\times W^{m_1 n_1 p_1}_{\dot{d})}]^2 + \ldots + [D_{a\dot{a}} P_{b\dot{b}mnpq} D_{a_1\dot{a}_1} P_{b_1\dot{b}_1}{}^{mnpq}]^2, \tag{5.15}$$

where at linear level

$$\partial_{a\dot{a}} W_{ijk\ell} = P_{a\dot{a}\, ijk\ell}\ , \tag{5.16}$$

and there exists the exact relation (Brink & Howe 1978),

$$P_{a\dot{a}\, ijk\ell} = \bar{D}_{\dot{a}i} W_{jk\ell\, a} = \frac{1}{24} \varepsilon_{ijk\ell mnpq} \bar{P}_{a\dot{a}}{}^{mnpq}\ . \tag{5.17}$$

It is not clear whether the exact non-linear generalizations of the 3-loop invariants in the N = 4, ... 8 supergravities exist. But if we assume that they exist, the form of linearized subvolume could be indicative, as it was in the N = 1 supergravity.

Now we shall consider the self-duality in supergravity mainly in the N = 1 theory. Note that in this theory all invariants on-shell ($G_{a\dot{b}} = R = \bar{R} = 0$) are constructed from the superfields W_{abc}, $\bar{W}_{\dot{a}\dot{b}\dot{c}}$,

$$W_{abc} = \varepsilon^{\dot{a}\dot{b}}\, T_{a\dot{a}\, b\dot{b}}{}^{c}\ , \tag{5.18}$$

where $T_{a\dot{a}\, b\dot{b}}{}^{c}$ is one of the components of the torsion tensor. In the theory N = 1 there exist an infinite number of higher order invariants. But they possess remarkable properties in the self-dual field:

$$W_{abc} = 0, \qquad \bar{W}_{\dot{a}\dot{b}\dot{c}} \neq 0. \tag{5.19}$$

Eqs. (5.19) mean that the curvature tensor and the field strength of the spin $^3/_2$ field are self-dual

$$R_{abcd} \Rightarrow R_{\mu\nu\lambda\delta} - \varepsilon_{\mu\nu\lambda'\delta'} R^{\lambda'\delta'}{}_{\lambda\delta} = 0, \tag{5.20}$$

$$\Psi_{abc} \Rightarrow \psi^a_{\mu\nu} - \varepsilon_{\mu\nu}{}^{\zeta\eta} \psi^a_{\zeta\eta} = 0. \tag{5.21}$$

Taking into account the chiral invariance (Ogievetsky & Sokatchev 1978) of the N = 1 supergravity it is possible to prove (Kallosh 1979, Christensen & Duff 1979) that all ℓ-loop counterterms vanish in the self-dual field (5.19) with the exception of the topological invariants $\int d^4x\, d^2\theta\, W^2_{abc} \pm \text{h.c.}$

A simple definition of self-duality is given (Kallosh 1979) in terms of the differential terms of torsion

$$T^a = 0, \qquad T^{\dot{a}} \neq 0, \tag{5.22}$$

where

$$T^a = dz^M dz^N T_{MN}{}^a \quad . \tag{5.23}$$

It is natural to call the corresponding superspace the half-flat superspace. Note that in the purely gravitational self-dual background (5.20) with $\psi_{\mu\nu}{}^a = \psi_{\mu\nu}{}^{\dot{a}} = 0$, as well as in the purely spin $^3/_2$ self-dual background 5.21 with $R_{\mu\nu\lambda\sigma} = 0$, it is possible to prove that all ℓ-loop counterterms vanish even without the use of the γ_5-invariance (Christensen & Duff 1979).

It seems extremely interesting that in the extended supergravities in the gravitational self-dual background also all ℓ-loop counterterms vanish (Christensen & Duff 1979). However the straightforward supergeneralization of self-duality can be done only at $N \leq 4$ where there exist chiral superfields. In these theories super-self-duality means that

$$W_{a_1 \ldots a_{4-N}} = 0, \qquad \bar{W}_{\dot{a}_1 \ldots \dot{a}_{4-N}} \neq 0. \tag{5.24}$$

However in the case of the most interesting theory N = 8 instead of the pair of chiral superfields $W...$, $\bar{W}...$ there is only one superfield $W_{ijk\ell}$, which contains among its components both left R_{abcd} and right $\bar{R}_{\dot{a}\dot{b}\dot{c}\dot{d}}$ curvature spinors!

Therefore it seems probable that the analysis of self-duality in the N = 8 supergravity may give us some new and unexpected information about the structure of divergences in this theory.

Thus the hope to have the finite N = 8 supergravity simply because of the absence of the corresponding invariants on-shell (as it was the case in the N = 1 supergravity in the first two loops) is not realized. At least starting from the 8-th loop order there exist super-invariants respecting all the necessary symmetries of the theory such as global E_7 and local SU(8). But now it becomes especially important to make a progress in the second point of the renormalizability program, i.e., to calculate the coefficients in front of the counterterms or to find the hidden symmetry which forbids the ultraviolet divergences in extended supergravities.

6 SUPERSPACE GEOMETRY, CONFORMAL ANOMALIES AND BACKGROUND FORMULATION OF QUANTUM SUPERGRAVITY

The usual analysis of conformal anomalies in supergravity is based on the investigations of the one-loop approximation in the gravitational background. In some cases conformal anomaly is equal to zero (Duff lectures). In the N = 3 supergravity this fact was known long ago. In the N = 8 supergravity conformal anomaly is equal to zero, as pointed out by Siegel, if one takes into account that the theory contains a definite number of antisymmetric tensor fields, which follows from the 11-dimensional theory. In this analysis it was essential that conformal anomalies of the classically equivalent fields described by different Lorentz group representations are not equivalent, as was shown by Duff and van Nieuwenhuizen (1980). For the N = 4 supergravity there also exists a version (Nicolai & Townsend 1981) where conformal anomaly is equal to zero. For the theories $5 \leq N \leq 7$ there exists a set of fields for which the anomaly vanishes (Duff lectures).

When conformal anomalies are calculated in the gravitational background, in particular, when the zero modes in the self-dual gravitational background are calculated, the vanishing of conformal anomalies looks like an accident which takes place at $3 \leq N \leq 8$ (for special representations). The question arises whether it is possible to understand the vanishing of conformal anomalies from the point of view of the superspace geometry.

We shall represent a few observations which may help us to understand why conformal anomalies in the $N \geq 3$ supergravities are equal to zero. We shall use the results of the paper of Brink & Howe (1978) where it is shown which superfields describe the N = 8 supergravity on-shell. The basic superfield

$$W_{ijk}{}^{a} = T^{a}_{\dot{a}\dot{b}\; ijk}\, \varepsilon^{\dot{a}\dot{b}} \tag{6.1}$$

has in the first component the spin $^1/_2$ field $\chi^{\dot{a}}_{[ijk]}$. This component of the torsion tensor remains not equal to zero up to N = 3 where

$$W^{a}_{ijk} = \varepsilon_{ijk}\, \chi^{a} = T_{\dot{a}\dot{b}}{}^{a}{}_{ijk}\, \varepsilon^{\dot{a}\dot{b}} \quad . \tag{6.2}$$

In the theories N = 1, N = 2 $T^{a}_{\dot{a}\dot{b}}$ is equal to zero which is connected with the absence of spin $^1/_2$ fields in these theories. Therefore not every superfield could exist in the high N supergravity background and the greater is N the less constrained superfields are allowed.

We present a few examples.

1. In the N = 3, ..., 8 superbackground no chiral superfields could exist (no scalars, spinors etc.). The proof is based on the fact that $T^{\dot{a}}_{ab}{}^{ijk}$ in these theories is non-trivial. The chirality condition has the form

$$D^{i}_{a}\, \phi_{a_1 \ldots a_{A/2},\, \dot{a}_1 \ldots \dot{a}_{B/2}} = 0, \quad i = 1, \ldots, N, \tag{6.3}$$

where A, B are arbitrary numbers. In eq. (6.3) the covariant derivative contains the vielbeins and connections from the external background geometry. From eq. (6.3) it follows that

$$\{D^{i}_{a}, D^{j}_{b}\}\, \phi_{a_1 \ldots a_{A/2},\, \dot{a}_1 \ldots \dot{a}_{B/2}} = R^{ij}_{ab,\, a_1}{}^{\xi}\, \phi_{\xi\, a_2 \ldots, \ldots \dot{a}_{B/2}}$$

$$+ \ldots + T^{\dot{a}}_{ab}{}^{ijk}\, D_{\dot{a}k}\, \phi_{a_1 \ldots, \ldots \dot{a}_{B/2}} \quad . \tag{6.4}$$

Therefore we come to the conclusion that for the consistency of eqs. (6.4), (6.3) for $N \geq 3$ the external torsion $T_{ab}^{\dot{a}}$, $T_{\dot{a}\dot{b}}^{a}$ must be equal to zero or the superfield $\phi_{\ldots,\ldots}$ with the chirality condition (6.3) must be equal to zero. When we consider the field ϕ which is equal, e.g., to W_a^2 in the $N = 3$ theory, where $W_a = T^{a}_{\dot{a}\dot{b}\, ijk}\, \varepsilon^{ijk}\, \varepsilon^{\dot{a}\dot{b}}$, then eq. (6.4) is consistent with eq. (6.3) even for nontrivial $T_{ab}^{\dot{a}}$ and $\Phi = W_a^2$. However in analysing conformal anomalies in the background method we are interesting only in the quantum superfield ϕ which is independent of the background curvature and torsion. Note that all other superfields of the $N = 3, \ldots, 8$ supergravities are the derivatives of $T_{ab}^{\dot{c}}$, $T_{\dot{a}\dot{b}}^{c}$, which means that the existence of chiral superfields (6.3) (and conjugated ones) is possible only in the flat $N = 3, \ldots, 8$ superspace.

2. In the $N = 2$ supergravity scalar chiral superfields may exist, but spinor chiral superfields cannot exist since from

$$D_{\dot{a}\, i}\, \Psi_b = 0 \tag{6.5}$$

follows that

$$\{D_{\dot{a}\, i}, D_{\dot{b}\, j}\}\, \Psi_c = R_{\dot{a}\dot{b},c\; ij}{}^{d}\, \Psi_d = 0. \tag{6.6}$$

In the $N = 2$ supergravity $R_{\dot{a}\dot{b},c\; ij}{}^{d}$ is not equal to zero but is connected with the vector field strength

$$R_{\dot{a}\dot{b},c\; ij}{}^{d} = M_{cd}^{ij}\, \varepsilon_{\dot{a}\dot{b}} \;. \tag{6.7}$$

This analysis could be developed also in the case of self-dual background when $M_{cd}^{ij} = 0$, $M_{\dot{c}\dot{d}}^{ij} \neq 0$. In this case the right chiral spinors can exist since $D_a \Psi_{\dot{b}} = 0$ and the left chiral spinors cannot exist.

3. In the $N = 8$ supergravity we consider the linearized superfield $W_{ijk\ell}$ with the condition

$$D^{\dot{a}}_{(j}\, W_{k)\ell mn} = 0, \tag{6.8}$$

where the derivatives are defined as before in the external background. From (6.8) we have

$$\sum_{\substack{\text{sym (im)}\\ \text{(jn)}}} \{D^{\dot a}_{j}, D^{\dot b}_{i}\} W_{mnpq} = R^{\dot a \dot b}_{ji\;m}{}^{m_1} W_{m_1 npq} + \ldots + \tag{6.9}$$
$$+ T^{a}_{\dot a \dot b}\; j(i\ell\, D^{\ell}_{a} W_{m)npq} = 0.$$

It can be verified that eq. (6.9) is inconsistent when both $T^{c}_{\dot a \dot b}$ and $W_{ijk\ell}$ are different from zero. Either the external superspace must be flat, or the superfield $W_{ijk\ell}$ with the condition (6.8) could not exist in the nontrivial background.

4. Let us consider the N = 4 super Yang-Mills theory. The superfield $\phi_{ij} = \frac{1}{4!}\, \varepsilon_{ijk\ell}\, \bar\phi^{k\ell}$ with the condition

$$D^{\dot a}_{(m}\, \phi_{i)j} = 0 \tag{6.10}$$

cannot exist in the N = 4 superbackground. The proof is analogous to the case considered above since

$$\sum_{\substack{\text{sym (mi)}\\ \text{(nj)}}} \{D^{\dot a}_{m}, D^{\dot b}_{n}\} \phi_{ij} = \varepsilon^{\dot a \dot b} F_{m(n} \times \phi_{i)j} \;. \tag{6.11}$$

The left hand side of eq. (6.11) is equal to zero due to eq. (6.10), and the right hand side is equal to zero only when $F_{mn} = \phi_{mn}$, which is not the case when F_{mn} is the external curvature and the field ϕ_{ij} is the quantum field.

5. In the N = 2 super Yang-Mills theory eqs. (6.10), (6.11) take the form

$$D^{\dot a}_{i}\, \phi = 0,$$
$$\{D^{\dot a}_{i}, D^{\dot b}_{j}\}\, \phi = \varepsilon^{\dot a \dot b} F_{ij}\, \phi\;, \tag{6.12}$$

where $\phi_{ij} = \varepsilon_{ij}\, \phi$, $\bar\phi^{ij} = \varepsilon^{ij}\, \bar\phi$. Eqs. (6.12) mean that the chiral superfield ϕ, $\bar\phi$ cannot live in the N = 2 super Yang-Mills background. But this theory instead of chiral superfields can be described with the help of the hypermultiplet (Fayet 1976) ϕ_i

$$D^{\dot a}_{(i}\, \phi_{j)} = 0, \tag{6.13}$$

$$\sum_{(ijk)} \{D^{\dot a}_{i}, D^{\dot b}_{j}\}\, \phi_k = \sum_{(ijk)} \varepsilon^{\dot a \dot b} F_{ij}\, \phi_k = 0. \tag{6.14}$$

Eq. (6.14) is not inconsistent when $F_{(ij)} = 0$ which is the case since F_{ij} is antisymmetric in internal indices.

Thus the main conclusion which could be drawn from the above-mentioned observations is the following. When analysing the conformal anomalies in extended supergravities it is important i) to work not in the gravitational but in the supergravitational background ii) to find which superfields could live in the extended superspace and could describe the corresponding quantum theory (instead of the usual analysis of the Lorentz representations which could exist in the Riemannian geometry).

It is probable to reach the understanding of conformal anomalies in this way. We stress thereby that there is a correlation between the absence of chiral extended superfields in N = 3, ..., 8 supergravities, absence of conformal anomalies in these theories and the fact (Grisaru lectures) that in the N = 1 supergravity only chiral superfields give a contribution to conformal anomalies.

We hope that the more deep investigation of the renormalizability problem than merely listing all possible counterterms will bring a new and unexpected information about ultraviolet properties of the maximally extended supersymmetric theories.

7 ADDITIONAL COORDINATES FOR DESCRIBING EXTENDED SUPERGRAVITIES WITH HIDDEN SYMMETRY

The possibility and necessity of introducing central charges into supersymmetry were stressed in many papers. The purpose of the following is to show that when N is growing we also get the possibility (and probably the necessity) of the introduction of additional coordinates in superspace. The reasons for that are the following:

i) to have the possibility to describe the hidden symmetry (E_7 in the N = 8 theory) by means of the natural geometrical terms

ii) to try to find the possibility for closure of the left and right differential forms in the superspace with additional coordinates (which in the N = 1 supergravity is the base for the unconstrained prepotential geometry).

Similar extensions of the superspace were proposed also in the papers (Siegel 1981, Howe & Lindström 1981) and in the talks of J. Taylor and J. Luckiersky at the Trieste supergravity workshop 1981.

The superspace which we propose has the form

$$z = \{x^{\mu}, \theta^{A}, \bar{\theta}^{A}, t_{[IJ]}, \bar{t}^{[IJ]}\}, \qquad I,J = 1,\ldots,8 \quad (7.1)$$

where the new bosonic coordinates are connected originally with Cartan's antisymmetric tensors x^{ij}, y_{ij}, $i = 1,\ldots,8$ which gives the explicit form of the E_7 infinitesimal generators

$$\delta x^{ij} = \Lambda^{i}_{k} x^{kj} + \Lambda^{j}_{k} x^{ik} + \frac{1}{24}\varepsilon^{ijk\ell mnop}\Sigma_{mnop}\, y_{k\ell},$$

$$\delta y_{ij} = \Lambda^{k}_{i} y_{kj} + \Lambda^{k}_{j} y_{ik} + \Sigma_{ijk\ell} x^{k\ell}, \qquad (7.2)$$

with $\Lambda^{i}{}_{k} = -\Lambda_{k}{}^{i}$, $\Lambda^{i}{}_{i} = 0$, $\Sigma_{[ijk\ell]}$. Then t_{IJ} corresponds to $x^{ij} + iy_{ij}$ and $\bar{t}^{IJ}$ corresponds to $x^{ij} - iy_{ij}$. In the superspace (7.1) there are new vielbein forms $dz^{M} E^{M}_{M}$, in addition to the old ones we have E_{ij}, $\bar{E}^{ij}$. In the flat superspace we now have new nontrivial vielbein components $E_{ai}^{t_{k\ell}} = \theta^{j}_{a}\, \delta^{[k}_{j}\, \delta^{\ell]}_{i}$, $E^{ij}_{IJ} = \delta^{[i}_{I}\, \delta^{j]}_{J}$, etc. In the curved superspace at $\theta = \bar{\theta} = 0$ we have the new part of the vielbein matrix connected with the scalar field

$$V = \begin{vmatrix} U^{IJ}{}_{ij} & \bar{V}^{IJij} \\ V_{IJij} & \bar{U}_{IJ}{}^{ij} \end{vmatrix}. \qquad (7.3)$$

As was shown in (Gates & Siegel 1979), in the N = 1 theory in the superspace $(x,\theta,\bar{\theta})$ for $E_{a} = E^{M}_{a}\partial_{M}$ we have

$$\{E_{a}, E_{b}\} = C^{c}_{ab} E_{c}; \quad C^{c}_{ab} = T^{c}_{ab} - \phi^{c}_{ab}, \qquad (7.4)$$

where ϕ^{c}_{ab} is the connection. Eq. (7.4.) shows that the left and right forms are closed separately. In the theories N = 3,...,8 instead of (7.4) in the superspace $(x, \theta_{i}, \bar{\theta}^{i})$, i = 3,...,N we have

$$\{E^{i}_{a}, E^{j}_{b}\} = C^{c}_{ab}{}^{ij}{}_{k} E^{k}_{c} + T^{\dot{c}}_{ab}{}^{ijk} E_{\dot{c}k}, \qquad (7.5)$$

where the component $T^{\dot{c}}_{ab}{}^{ijk}$ of the supertorsion is connected with the presence of spin 1/2 field. Eq. (7.5) shows that the left and right forms in the usual superspace are mixed and we cannot expect the simple formulation of the minimal group in extended theories. In the space with additional coordinates (7.1) we could try (probably by introducing some nontrivial dependence on the new coordinates) to reach the closure of the left and right forms (e.g., of E_a, E_t and separately of $E_{\dot{a}}$, $E^{\bar{t}}$). We hope that the new description of extended theories may allow one to define the theory off-shell and also to perform a more deep analysis of its ultraviolet behaviour.

REFERENCES

Akulov, V.P. Volkov, D.V. & Soroka, V.A. (1975), JETPh Lett. 22 187.
Berezin, F.A. (1979), Yad. Fiz. 30, 1168.
Brink, L. & Howe, P.(1978), Phys. Lett. 79B 222.
Christensen, S.M. & Duff, M.J., (1979), Nucl. Phys. B154, 301.
Deser, S, & Kay, J.H. (1980), Phys. Lett. 76B 400.
Deser, S. Kay, J.H. & Stelle, K. (1977), Phys. Rev. Lett. 38 527.
de Wit, B. private communication; Bergshoeff, E, de Roo, M. de Wit, B. preprint NIKHEF-H/80-07, to be published in Nucl. Phys.
Duff, M. lectures in this volume.
Duff, M.J. & van Nieuwenhuizen, P. (1980), Phys. Lett B94, 179.
Fayet, P. (1976), Nucl. Phys. B113, 135.
Ferrara, S. Scherk, J, Zumino, B (1977), Nucl. Phys. B121, 393.
Ferrara, S. & Zumino, B. (1978), Nucl. Phys. B134, 301.
Gates, S.J. & Siegel, W. (June 1979), Harvard preprint HUTP/99/A034.
Grisaru, M.T. (1977), Phys. Lett. 66B, 75.
Grisaru, M.T. lectures in this volume.
Grisaru, M.T. van Nieuwenhuizen, P., Vermaseren, J.A.M. (1976), Phys. Rev. Lett. 37, 1662.
Howe, P.S. & Lindström, U., (1981) Nucl. Phys. B181, 487.
Howe, P.S., Stelle, K.S. & Townsend, P.K. Preprint Ref. TH3065-CERN.
Kallosh, R. (1979), J.E.T.Ph. Lett. 29, 172; Nucl. Phys. B165, 119.
Kallosh, R. (1981), Phys. Lett. 99B, 122.
Kallosh, R. (1981), J.E.T.Ph. Lett. 33, 292.
Nicolai, H. & Townsend, P.R. (1981), Phys. Lett. B98, 257.
Ogievetsky, V. & Sokatchev, E. (1978), Phys. Lett. 79B, 222.
Siegel, W. (1981), Nucl. Phys. B177, 325.
van Nieuwenhuizen, P, Vermaseren, J.A.M. (1976), Phys. Lett. 65B, 263.
Wess, J. & Zumino, B. (1977), Phys. Lett. 66B, 361.
Zumino, B. Preprint Ref. TH.2852-CERN.

GROUP MANIFOLD APPROACH TO GRAVITY AND SUPERGRAVITY THEORIES

Riccardo d'Auria, Pietro Fré and Tullio Regge

Istituto di Fisica Teorica, Università di Torino, Italy

and

I.N.F.N. - Sezione di Torino

It is very difficult to summarize the present status of gravity theories because of the variety of different approaches which are being tried and developed at the moment in this rapidly evolving field.

We have chosen to present gravity theories from the point of view of group manifold formulation. This has some distinct advantages. First of all one utilizes consistently the language of differential forms which turns out to be considerably more compact than the usual tensor calculus with an essentially equivalent physical content.

Secondly, the invariance properties of the theory are exhibited in a more natural and straightforward way from the Lagrangian. This point is also relevant in the discussion of auxiliary fields for supersymmetric theories.

Finally, by writing these theories on the group manifold in the first order formalism we hope to arrive at a consistent classification of the rheonomic lagrangians parallelling that of ordinary lagrangians invariant under group action. Here by rheonomic we intend a special class of actions to be precisely defined in the sequel, which admit a non-trivial class of solutions, besides the vacuum, related by supersymmetric transformations.

The lectures are organized as follows. In Section I we discuss the differential geometry of groups and supergroups introducing the notion of connection and related Yang Mills potentials. In Section II we discuss ordinary Einstein gravity in the Cartan formulation. This provides a first example which will then be generalized to more complicated theories, in particular supergravity. At this point we shall discuss the distinction between "pure" and "impure" theories. In Section III we develop an axiomatic approach to rheonomic theories related to the concept of Chevalley cohomology on group manifolds. In Section IV we apply these principles to N=1 supergravity. In Section V we present the panorama of so-far-constructed pure and impure group manifold supergravities. In particular we discuss in some detail the pure d = 5 N = 2 case and as examples of the impure theories N = 2 and N = 3 in d = 4. The way a pure theory becomes impure after dimensional reduction is illustrated. In Section VI we discuss the role of kinematical superspace constraints as a subset of the group-manifold equations of motion demonstrating how the auxiliary fields are obtained in this approach.

Section VII deals with the application of the group manifold method to supersymmetric Super Yang-Mills theories.

Sec. I - Differential Geometry of Groups and Supergroups

Let G be a generic Lie group, that is a differentiable manifold endowed with a differentiable group structure. In particular let U be a local chart containing the identity e of G. In U we shall denote with $y_1 \ldots y_n$, $u_1 \ldots u_n$, $v_1 \ldots v_n$ the coordinates of elements (y,u,v) G. We suppose e = 0. By dy we mean the 1 x n matrix given by $dy_1 \ldots dy_n$. Since G is a Lie group there is a differentiable map $G \times G \xrightarrow{f} G$ which in the local chart can be written as:

$$y^M = f^M(u_1 \ldots u_n, v_1 \ldots v_n) \tag{1.1}$$

where y = uv. By differentiation we obtain:

$$dy^M = du^N \frac{\partial f}{\partial u^N} + dv^N \frac{\partial f}{\partial v^N} \tag{1.2}$$

which can be rewritten briefly as:

$$dy = d(uv) = du\,T(u,v) + dv\,U(u,v) \tag{1.3}$$

Associativity of the group product implies: (tu)v = t(uv).

By differentiating this relation we obtain:

$$\left.\begin{aligned} &T(t,uv) = T(t,u)\,T(tu,v) \\ &U(t,u)\,T(tu,v) = T(u,v)\,U(t,uv) \\ &U(tu,v) = U(u,v)\,U(t,uv) \end{aligned}\right\} \tag{1.4}$$

From (1.4) we see that the forms:

$$\sigma^A = dy^M\, U(y^{-1}y)^{A}_{\cdot M} \tag{1.5}$$

are invariant under translation to the left (right) on G. Namely consider the map L(b) : L(b)y = by and R(b) : R(b)y = yb. We have immediately from (1.4) that:

$$\begin{aligned} L^*(b)\sigma &= d(by)\,U(y^{-1}b^{-1}, by) = \\ &= dy\,U(b,y)\,U^{-1}(y^{-1}b^{-1}, by) = dy\,U(y^{-1}, y) = \sigma \end{aligned} \tag{1.6}$$

where $L^*(b)\sigma$ is called the pull-back of σ under L(b).

The adjoint automorphism defined by:

$$ad(b)y \longrightarrow byb^{-1} \tag{1.7}$$

acts as:

$$ad^*(b)\sigma = \sigma\, Ad(b^{-1}) \tag{1.8}$$

The matrix Ad(b) yields the adjoint representation of G. The differentiation and multiplication of forms are natural under differentiable

maps of manifold, they commute with the pull-back. Therefore also $d\sigma^A$ is invariant under L(b).
Expanding $d\sigma^A$ into the σ^B one finds a relation:

$$d\sigma^A + \frac{1}{2} C^A_{\cdot BC}\, \sigma^B \wedge \sigma^C = 0 \qquad (1.9)$$

where the scalars $C^A_{\cdot BC}$ must be invariant under left translations. It follows that they must be constants on G; they are named the structure constants of G. The (1.9) are called the Maurer Cartan equations of G. The set of forms σ^A correspond to a frame in cotangent space: in the physicist's language they are a complete set of covariant vectors on G. We may introduce the dual frame of tangent (contravariant) vectors on G defined by:

$$\vec{T} = \mathcal{U}(y\cdot e)\frac{\vec{\partial}}{\partial y} \qquad (1.10)$$

We may consider the $\vec{T}_A$ as differential operators on G. Their commutator can be computed easily from the Maurer Cartan equations and turns out to be:

$$[\vec{T}_A, \vec{T}_B] = C^L_{\cdot AB}\, \vec{T}_L \qquad (1.11)$$

The Jacobi identity is then:

$$C^M_{\cdot LA} C^L_{\cdot BC} + C^M_{\cdot LB} C^L_{\cdot CA} + C^M_{\cdot LC} C^L_{\cdot AB} = 0 \qquad (1.12)$$

A particular example, of interest in the following, is provided by the Poincaré group. We define P as the set of 5 x 5 matrices

$$g = \begin{pmatrix} \Lambda, & \xi \\ 0, & 1 \end{pmatrix} \qquad (1.13)$$

where Λ is a Lorentz matrix. The inverse of the above matrix (1.13) is given by:

$$g^{-1} = \begin{pmatrix} \Lambda^{-1}, & -\Lambda^{-1}\xi \\ 0, & 1 \end{pmatrix} \tag{1.14}$$

The definition of left invariant forms given in (1.8) can be bypassed by simply noticing that:

$$g^{-1}dg = \begin{pmatrix} \Lambda^{-1}, & -\Lambda^{-1}\xi \\ 0, & 1 \end{pmatrix}\begin{pmatrix} d\Lambda, & d\xi \\ 0, & 0 \end{pmatrix} \tag{1.15}$$

is a matrix valued 1-form which is left invariant. Its matrix elements are the left invariant forms. They are better organized into the forms:

$$\omega = \Lambda^{-1}d\Lambda \qquad ; \quad \Lambda^{-1}d\xi = V \tag{1.16}$$

which are 4 x 4 matrix valued and vector valued respectively. By differentiation we see immediately that:

$$d\omega = \Lambda^{-1}d\Lambda \wedge \Lambda^{-1}d\Lambda = \omega\wedge\omega \, ; \; dV = \Lambda^{-1}d\Lambda\wedge\Lambda^{-1}d\xi = \omega\wedge V \tag{1.17}$$

These are the Maurer Cartan equations. The Lie algebra is given by the tangent vectors P_r, J_{rs}

$$V^a(\vec{P}_r) = \delta^a_r \, ; \; \omega^{ab}(i\vec{J}_{rs}) = \frac{1}{2}\left(\delta^a_r\delta^b_s - \delta^a_s\delta^b_r\right) = \delta^{ab}_{rs} \tag{1.18}$$

with the commutation relations:

$$\begin{aligned} &[\vec{P}_r, \vec{P}_s] = 0 \qquad [\vec{J}_{rs}, \vec{P}_\ell] = -i\left(\eta_{r\ell}\vec{P}_s - \eta_{s\ell}\vec{P}_r\right) \\ &[\vec{J}_{rs}, \vec{J}_{\ell m}] = -i\left(\eta_{r\ell}\vec{J}_{sm} + \eta_{sm}\vec{J}_{r\ell} - \eta_{rm}\vec{J}_{s\ell} - \eta_{s\ell}\vec{J}_{rm}\right) \end{aligned} \tag{1.19}$$

Our theory will be formalized on the group manifold G. Objects of the theory are forms μ^A forming a non-singular frame in every point of G. The σ^A are understood as particular cases of the μ^A corresponding to the "vacuum" configuration. In general the μ^A do not satisfy the Maurer Cartan equation and we have instead the definition of curvature:

$$R^A[\mu] = d\mu^A + \frac{1}{2} C^A_{\cdot BG}\, \mu^B \wedge \mu^G = R^A_{\cdot FG}\, \mu^F \wedge \mu^G \tag{1.20}$$

which is a 2-form obeying the Bianchi identities:

$$\nabla R^A = dR^A + C^A_{\cdot BG}\, \mu^B \wedge R^G = 0 \tag{1.21}$$

The dual tangent frames to the μ^A are tangent vectors defined by:

$$\mu^A(\tilde{T}_B) = \delta^A_B \tag{1.22}$$

and satisfying the commutation relations:

$$[\tilde{T}_A, \tilde{T}_B] = \left(C^L_{\cdot AB} - 2R^L_{\cdot AB}\right)\tilde{T}_L \tag{1.23}$$

which, however, do not close as in the vacuum case because of the intrinsic components of the curvature $R^A_{\cdot FG}$. It is also possible to base a gravity theory entirely in terms of tangent vectors which carry the same information as the forms μ^A.

We shall refer to the μ^A or the $\tilde{T}_A$ as the "soft" forms and tangent vectors on G. Suitable modifications must be introduced in the case of supergroups. The symmetry of the $C^L_{\cdot AB}$ is now given by:

$$C^L_{\cdot AB} = -(-)^{AB}\, C^L_{\cdot BA} \tag{1.24}$$

and the Jacobi identities by:

$$C^{M}_{\cdot AL} C^{L}_{\cdot BC} + (-)^{A(B+C)} C^{M}_{\cdot BL} C^{L}_{\cdot CA} + (-)^{B(C+A)} C^{M}_{\cdot CL} C^{L}_{\cdot AB} = 0 \quad (1.25)$$

where the quantities A,B appearing in the exponent are understood as 0,1 if the corresponding index labels a bose, fermi direction respectively in the supergroup.

Of great interest are the following formal devices:

1. Covariant differentiation. It applies to any p form with values in a representation of G and yields a set of p + 1 forms with values in the same multiplet. We suppose given a matrix representation D of the Lie algebra G satisfying the commutation relations:

$$D(T_A)^{i}_{\cdot k} D(T_B)^{k}_{\cdot j} - (-)^{AB} D(T_B)^{i}_{\cdot k} D(T_A)^{k}_{\cdot j} = C^{L}_{\cdot AB} D(T_L)^{i}_{\cdot j} \quad (1.26)$$

We have then as covariant derivative of the generic form ω^i the quantity:

$$\nabla \omega^i = d\omega^i + \mu^A \wedge D(T_A)^{i}_{\cdot j} \omega^j \quad (1.27)$$

From which follows that the second derivative is:

$$\nabla\nabla \omega^i = R^A \wedge D(T_A)^{i}_{\cdot j} \omega^j \quad (1.28)$$

The curvature can be considered as a 2 form with values in the adjoint representation of G. Its covariant derivative then vanishes according to (1.21).

2. Contraction. To any tangent vector $\tilde{T}_A$ we associate a map from the p-forms to the p-i forms. A generic p-form is specified by expanding it as a polynomial in the μ^A:

$$\omega^i = \omega^{i}_{\cdot A_1 \ldots A_p} \mu^{A_1} \wedge \ldots \wedge \mu^{A_p} \quad (1.29)$$

We define:

$$\vec{t}\rfloor \omega^i(\tilde{T}_{B_1},\ldots,\tilde{T}_{B_{p-1}}) = p\,\omega^i(\vec{t},\tilde{T}_{B_1},\ldots,\tilde{T}_{B_{p-1}}) \tag{1.30}$$

The form $_A\rfloor\omega^i$ is called the contraction of ω^i with $\tilde{T}_A$. Obviously the contraction $_A\rfloor\mu^B$ is given by:

$$_A\rfloor\mu^B = \delta_A^B \tag{1.31}$$

3. Differentiation raises the order of the form, contraction lowers it. By combining the two we obtain an operator which leaves invariant the order of the form. The Lie derivative along the tangent vector $\vec{t}$ is defined by:

$$\ell_t\omega^i = {}_t\rfloor d\omega^i + d({}_t\rfloor\omega^i) \tag{1.32}$$

The Lie derivative along $\tilde{T}_A$ is then given by:

$$\ell_A\omega^i = {}_A\rfloor d\omega^i + d({}_A\rfloor\omega^i) \tag{1.33}$$

The Lie derivative has the following geometrical interpretation. In ref. [1] the anholonomized coordinate transformation (AGCT) was introduced and a formula was derived detailing the change of the μ^A under infinitesimal coordinate transformations:

$$y^\Lambda \longrightarrow y^\Lambda + \epsilon^\Lambda \; ; \quad \vec{\epsilon} = \epsilon^\Lambda \partial_\Lambda \tag{1.34}$$

in (1.34) the $\vec{\epsilon}$ form a tangent vector. We define $\epsilon^A = \mu^A_{.\Lambda}\epsilon^\Lambda$ as the (intrinsic) components of $\vec{\epsilon}$. They constitute a multiplet of 0-forms (ordinary functions). The infinitesimal change brought in by the ϵ^A is now:

$$\delta\mu^A = dy^{\Sigma}\left[\partial_{\Sigma}\epsilon^A + \epsilon^{\Lambda}\left(\partial_{\Lambda}\mu^A_{\Sigma} - (-)^{\Lambda\Sigma}\partial_{\Sigma}\mu^A_{\Lambda}\right)\right] \tag{1.35}$$

By rearranging the differentials we see that the change can be rewritten as:

$$\delta\mu^A = \nabla\epsilon^A + 2\epsilon^F\mu^G R^A_{\cdot FG} = \nabla\epsilon^A + \underline{\epsilon}\rfloor R^A \tag{1.36}$$

and this in turn is the Lie derivative along the tangent vector $\vec{\epsilon}$. The result can be generalized to any form of any degree. Replacing the ordinary derivative with the covariant derivative yields the covariant Lie derivative of a multiplet of forms. Explicitly one has:

$$L_{\epsilon}\omega^i = \underline{\epsilon}\rfloor\nabla\omega^i + \nabla\left(\underline{\epsilon}\rfloor\omega^i\right) \tag{1.37}$$

Of interest are the Leibnitz rules for differentiation and contraction:

$$\underline{t}\rfloor\left(\omega^i_{(p)}\wedge\omega^j_{(q)}\right) = \underline{t}\rfloor\omega^i_{(p)}\wedge\omega^j_{(q)} + (-)^p\omega^i_{(p)}\wedge\underline{t}\rfloor\omega^j_{(q)} \tag{1.38a}$$

$$d\left(\omega^i_{(p)}\wedge\omega^j_{(q)}\right) = d\omega^i_{(p)}\wedge\omega^j_{(q)} + (-)^p\omega^i_{(p)}\wedge d\omega^j_{(q)} \tag{1.38b}$$

In this way the covariant Lie derivative can be rewritten as:

$$L_t\omega^i = \ell_t\omega^i + t^A D(T_A)^i_{\cdot j}\,\omega^j \tag{1.39}$$

Finally, the AGCT variation of a multiplet of p forms is given by:

$$\delta_{\epsilon}\omega^i = -\epsilon^A D(T_A)^i_{\cdot j}\,\omega^j + L_{\epsilon}\omega^i \tag{1.40}$$

A form is equivariant if:

$$L_{\epsilon}\omega^{i}=0 \tag{1.41}$$

In this case a coordinate transformation simply induces a linear transformation on the multiplet. A weaker property is the H-equivariance in which one requires only:

$$L_{h}\omega^{i}=0 \tag{1.42}$$

where h is now an element of the Lie algebra of a subgroup $H \subset G$.

Finally, we have a Killing vector if the ϵ^{A} corresponds to a coordinate transformation which leaves the μ^{A} invariant. This notion corresponds to the classical one in general relativity and is of great importance in carrying out dimensional reduction. Accordingly we have:

$$\delta_{\epsilon}\mu^{A}=\nabla\epsilon^{A}+\underline{\epsilon}\rfloor R^{A}=0 \tag{1.43}$$

as a condition for a Killing vector.

Sec. II - Ordinary Gravity

General relativity as a theory on a soft manifold was essentially introduced by Cartan and later by Sciama, Kibble and Utiyama through the formalism of tetrads or vierbeins. Since this is now a well-known procedure, we recall briefly the essential points. The metric tensor is parametrized as:

$$g_{\mu\nu}(x)=V^{a}_{\mu}(x)V^{b}_{\nu}(x)\eta_{ab}\ ;\quad \eta_{ab}=\begin{pmatrix}1&0&0&0\\0&-1&0&0\\0&0&-1&0\\0&0&0&-1\end{pmatrix} \tag{2.1}$$

and we introduce correspondingly the 1-forms (vierbeins):

$$V^a = V^a_\mu(x)\, dx^\mu \tag{2.2}$$

The Cartan connection is the ω^{ab} defined through the equation:

$$R^a = dV^a - \omega^{ab} \wedge V_b = \mathcal{D}V^a = 0 \tag{2.3}$$

where the operator D denotes the covariant differentiation with respect to the Lorentz group SO(1,3). The curvature tensor is then defined from the connection as:

$$R^{ab} = \mathcal{R}^{ab} = d\omega^{ab} + \omega^{ac} \wedge \omega^{b}_{\cdot c} \tag{2.4}$$

In this way gravity appears as a theory in the V^a and ω^{ab}, the latter appearing in the second order formulation as the first derivatives of the vierbeins and not as an independent propagating field. The action is then written as:

$$A = \frac{1}{k^2} \int R^{ab} \wedge V^c \wedge V^d \epsilon_{abcd} \tag{2.5}$$

Upon variation in the ω^{ab} this action yields the constraints (2.3) in the form:

$$\epsilon_{abcd} R^c \wedge V^d = 0 \tag{2.6}$$

We refer to this equation as the "torsion" equation. Variation in the V^a instead gives the usual Einstein field equations in the vacuum:

$$\epsilon_{abcd} R^{bc} \wedge V^d = 0 \Longrightarrow R^{am}_{\cdot bm} - \frac{1}{2} \delta^a_b R^{mn}_{\cdot mn} = 0 \tag{2.7}$$

The following observations are helpful in understanding the group manifold approach. The V^a form a basis in cotangent space and the ω^{ab} seem to serve no useful purpose except covariant differentiation. However, any solution V^a, ω^{ab} can be extended to a basis on a 10 dimensional space by considering instead of a locally Minkowski space, a fibre bundle with fibre SO(3,1). For simplicity we work locally in some open set U such that the restriction of the bundle on U is trivial and can be taken as $U \otimes G$. In $U \otimes G$ we have coordinates $x^\mu \in R^4$ and $\eta^{\mu\nu} \in SO(3,1)$ along the fibre. On $U \otimes G$ we define the forms:

$$\hat{V}^a = \left(\Lambda^{-1}(\eta)V\right)^a \; ; \; \hat{\omega}^{ab} = \left(\Lambda^{-1} d\Lambda\right)^{ab} + \left(\Lambda^{-1}\omega\Lambda\right)^{ab} \tag{2.8}$$

(where Λ is the Lorentz matrix of parameters $\eta^{\mu\nu}$), which turn out to give a non singular basis in the cotangent space. The forms $\hat{V}, \hat{\omega}$ differ from V, ω through an undefined gauge transformation in SO(3,1). Any choice of a section $\Lambda(x)$ in the bundle specifies the transformation. For these reasons it is easy to prove that all forms $\hat{V}, \hat{\omega}$ still satisfy equations (2.6)-(2.7) if the original forms did so. Therefore, the Einstein equations can be extended to a 10 dimensional manifold which is (at least locally) diffeomorphic to the Poincaré group. Therefore equations (2.6)-(2.7) have the (2.8)as solutions if interpreted on the Poincaré group. We are aware that there are some nontrivial questions to be asked globally but in what follows we speak of the Poincaré group by abuse of language.

This possibility of extending equations (2.6)-(2.7) to a manifold of larger dimension rests on the fact that they use the exterior algebra only, without, for instance, the concept of Hodge dual. Moreover all these equations are trivially covariant under arbitrary changes of coordinates even in the extended manifold.

This raises the interesting possibility of considering the equations (2.6)-(2.7) extended on the whole Poincaré group as the primary field equations of the theory. Clearly any solution of the conventional theory is also a solution of the extended one. But also the converse is true. The field equations (2.6)-(2.7), if interpreted on a space of larger dimensionality, imply many more conditions on the R^a, R^{ab} than we had before, even considering the fact that the number of independent components of the combined vierbeins+ connections is now 100 instead of 40. Heuristically, one sees that the extended equations admit the solution (2.8)as the only one. We refer to a structure of the kind (2.8) where the Lorentz group is factorized in a trivial manner as an undefined gauge transformation as a "factorized" solution. In general we expect that a gravity-like theory, if written on a soft group manifold,will admit factorization under all $H \subset G$ subgroups which are exact gauge symmetries of the theory. This is true of SO(3,1) in gravity. Yet it is also true that translations are not included in the gauge transformations of gravity. What is considered as translational or amended "translational" gauge invariance is simply a relabelling coming from the usual coordinate transformations of the theory. Upon use of the language of differential forms this last invariance becomes trivial and need not be reinstated. In particular it does not yield any new conditions on the theory.

A second point to be raised is the fact that all field equations permit the vacuum $R^a = R^{ab} = 0$ as a solution. This depends on the particular structure of the action (2.5). In general we can think of an action of the form:

$$A = \int \left(\Lambda + R^A \wedge \nu_A + R^A \wedge R^B \wedge \nu_{AB} + \cdots \right) \qquad (2.9)$$

where the Λ, ν_A, ν_{AB},... are polynomials in the μ^A. Upon variation we would obtain the field equations:

$$\underline{A}\rfloor \Lambda + \nabla \nu_A + (-)^{AB} R^B \wedge \underline{A}\rfloor \nu_B + 2 R^B \wedge \nabla \nu_{AB} + (-)^{A(B+C)} R^B \wedge R^C \wedge \underline{A}\rfloor \nu_{BC} + \cdots = 0 \qquad (2.10)$$

which are compatible with vacuum solutions only if

$$\underline{A}\rfloor \Lambda + \nabla \nu_A = 0 \quad \text{at } R^A = 0 \qquad (2.11)$$

which is of course true for gravity. In the case of gravity one has

$\Lambda = 0 \quad \nu_A = \begin{cases} \nu_{ab} = \epsilon_{abcd} V^c \wedge V^d \\ \nu_a = 0 \end{cases}$ and since ν_A is a multiplet in the co-adjoint representation of the Poincaré group:

$$\nabla \nu_A = \begin{cases} \nabla \nu_{ab} = \mathcal{D}\nu_{ab} + \frac{1}{2} V_a \wedge \nu_b - V_b \wedge \nu_a = 2\epsilon_{abcd} \mathcal{D}V^c \wedge V^d = \\ \qquad = 2\epsilon_{abcd} R^c \wedge V^d = 0 \quad \text{at } R^A = 0 \\ \nabla \nu_a = \mathcal{D}\nu_a = 0 \end{cases}$$

This condition suitably generalized can be used to narrow the choices of lagrangians, it is therefore a building principle for generalized gravity theories. A second, less obvious condition, is that the field equations admit solutions other than the vacuum. We all know that this is not true for gravity in 3 dimensions. Combining these requirements plus that of parity we see that gravity is practically unique. Summarizing, we started from a theory which had all the 10 potentials for a Yang-Mills theory of the Poincaré group but then we decided to forego the whole group invariance while retaining only SO(3,1) invariance. In general an incomplete invariance destroys the uniqueness of the action. Here instead the simple requirement of the existence of the vacuum and other non-trivial solution results in a straightforward selection of the action. It is interesting to notice that the scheme maintains its usefulness in more complicated cases.

Gravity is the prototype of a "pure" theory, namely one in which the only field variables are the μ^A and there is no Hodge duality. The scheme can be generalized to impure theories in which we allow the use of other forms, not necessarily of degree 1, and therefore not necessarily interpretable as Yang-Mills potentials. We insist, however, on the absence of a dual in order to make the theory extendable into higher dimensional spaces and obtain factorization. An example of an impure theory is gravity with electromagnetic field:

$$A = \iint \left\{ \frac{1}{K^2} R^{ab}{}_\wedge V^c{}_\wedge V^d - \frac{1}{12} F^{rs} F_{rs} V^a{}_\wedge V^b{}_\wedge V^c{}_\wedge V^d + F^{ab} dA_\wedge V^c{}_\wedge V^d \right\} \epsilon_{abcd} \tag{2.12}$$

which has the 0-forms F^{ab} as a multiplet besides the other fields. Another example is the scalar field coupled with gravity. We can achieve this in many impure ways. The most intriguing is the following:

$$A = \iint \left\{ \frac{1}{K^2} R^{ab}{}_\wedge V^c{}_\wedge V^d \epsilon_{abcd} + \cos\Phi \, R^{ab}{}_\wedge V_a{}_\wedge V_b \right\} \tag{2.13}$$

Upon elimination of the ω^{ab} in the second order formalism one obtains an action of the kind:

$$\mathcal{A} = \int \left(R\sqrt{-g} + \partial_\mu \Phi \partial^\mu \Phi \right) d^4x \tag{2.14}$$

which implies propagation of the massless scalar field Φ .

A number of interesting formal points can be raised for impure theories. Some of them can be obtained from pure theories through the device of dimensional reduction; this is indeed the case of the conventional E.M. field which has been long known to follow from ordinary gravity in 5 dimensions, as exemplified by the paper of Klein and Kaluza. Other theories may look impure because we simply choose to look only at some subgroup of G. Gravity is an impure SO(3,1) theory because then the vierbeins are not considered as Yang-Mills potentials. In this example the game is obvious, it may not be so in more complicated settings. It would be nice if all interesting physical theories would be pure or at least derivable from pure theories through dimensional reduction. This makes the search for the pure theories worthwhile. Some non-trivial examples are known in 5 dimensions, almost certainly others exist in higher dimensions as well but are beyond our present computational capability.

Before we leave ordinary gravity let us see briefly how factorization follows from the field equations. We expand the curvature as a generic polynomial in the ω^{ab} and V^{a}, considered as independent forms since we work in 10 dimensions: in formulas:

$$R^{ab} = R^{ab}_{\cdot mn} V^m \wedge V^n + R^{ab}_{\cdot (kl)m}\, \omega^{kl} \wedge V^m + R^{ab}_{\cdot (kl)(rs)}\, \omega^{kl} \wedge \omega^{rs}$$

$$R^{a} = R^{a}_{\cdot mn} V^m \wedge V^n + R^{a}_{\cdot (kl)m}\, \omega^{kl} \wedge V^m + R^{a}_{\cdot (kl)(rs)}\, \omega^{kl} \wedge \omega^{rs} \tag{2.15}$$

Next we introduce (2.15) into (2.6)-(2.7). After some simple algebra we see that the all intrinsic components of the curvature containing at least one (kl) vanish. The remaining equations have exactly the same content as the old theory. In general the "normal" components along forms other than the vierbein either vanish or they can be expressed in terms of the tangential components. The last weaker property is called rheonomy and is equivalent to supersymmetry in supergravity. In higher dimensions a generic theory has so many equations as to leave room for the vacuum only.

In order to avoid this trap one has to impose very stringent conditions on the theory which then turn out to be practically unique. The search for a suitable action, once the group G is given, is a purely algebraic problem which is similar to the search for cohomology classes on G. The definition of a cohomology class is intrinsic to the group, but the form representing the class is not in general invariant under group operations. Similarly, the search for the analog of the polynomials $\{\Lambda, \nu_A, \nu_{AB}, \dots\}$ is intrinsic to the group G although the polynomial may break the G gauge invariance. In this way it is the group itself which tells us how to break it.

It is this economy in axioms which makes the search for pure theories interesting, theories of this sort give information and are related to the topology of the group G. They are intrinsic in character. This may explain their uniqueness in spite of their non-invariance.

Sec. III - General Features of the Action on Group Manifolds

Maxwell theory and chromodynamics are based on the complete invariance of the action under the gauge group. Here we take a different attitude. Normally invariance of the theory under a particular group, to be taken as large as possible, is required in order to fix the action as narrowly as possible, furthermore invariance leads to conservation laws.

However, already in ordinary gravity one sees that, although the theory is built from the potentials of the full Poincaré group, the corresponding action is invariant only under gauge transformations of the Lorentz subgroup. We stress here that gauge transformations are understood in the narrow and strict sense of the definition:

$$\delta\mu^A = \nabla\epsilon^A = d\epsilon^A + C^A_{\cdot BC}\,\mu^B\epsilon^C \qquad (3.1)$$

not containing derivatives of the potentials. Actually, if this condition is relaxed one can obtain a sort of extension of the gauge invariance under the full Poincaré group. In fact the translational gauge transformation contained in the Poincaré group reads as:

$$\delta V^a = \mathcal{D}\epsilon^a \quad ; \quad \delta\omega^{ab} = 0 \qquad (3.2)$$

As one can easily check, this transformation does not leave the action (2.5) invariant. However, one can amend the transformation by allowing changes in the connection ω^{ab} which contain derivatives, that is the curvatures:

$$\delta V^a = \mathcal{D}\epsilon^a \quad ; \quad \delta\omega^{ab} = \underline{\epsilon}\rfloor R^{ab} \qquad (3.3)$$

This transformation is often quoted as the "true" gauge transformation of gravity corresponding to translations. Quite clearly it is just an infinitesimal coordinate transformation in the AGCT form as one can check from (1.36). Therefore its physical content is trivial insofar as the theory being written in the exterior algebra of forms is quite transparently in covariant form. Invariance under coordinates transformation need not be proved and says nothing new about the action.

We need instead criteria which select the action as uniquely as possible. We do not have yet a clear cut set of axioms but the following principles seem to insure that only a very narrow class of actions will survive on the group manifold. In a sense our procedure is identical in content to the search for supersymmetric lagrangians in a completely different algebraic setting. We ask that:

1) The world is a Lie group manifold. This manifold should include either the Poincaré or De Sitter group as a subgroup.

2) A pure theory should contain as fields the potentials μ^A only, that is a 1-form with values in the Lie algebra of G. The action is then a p-form formed from the potentials by using the exterior algebra with the operations d and $\wedge$ only, excluding the Hodge dual.

3) The theory should admit the G-vacuum as a solution. The G-vacuum is the field configuration where the μ^A are the left-invariant Cartan forms, the curvature defined by (1.20) is then vanishing and the potentials are gauge null.

4) The rheonomy principle. The theory should admit other solutions besides the G-vacuum.

Comment. The degree p of the action does not coincide with the dimension of the group and is lower than that of the quotient G/H where H is a subgroup of G under which the action is now exactly invariant.

How do we select p once the group G is given? We have no generic mathematical procedure which would yield the existing theories without working very hard. In practice the principles which we just stated lead immediately to the conclusion that in the groups so far explored there is only one dimension in which the theory produces an acceptable action. Most choices of dimension lead to "rigid" theories in which the vacuum is the only solution. It is quite plausible that most groups do not have any acceptable action at all. For instance in some groups it is not possible to balance exactly the fermion degrees of freedom with the bose ones on-shell and one has to introduce extra fields making the theory impure. In this case it is better to consider the theory as derived from a pure one through dimensional reduction. The condition for the existence of the vacuum is however easy to implement. If the action is of the form:

$$A = \int \left\{ \Lambda + R^A \wedge \nu_A + R^A \wedge R^B \wedge \nu_{AB} + \cdots \right\} \tag{3.4}$$

then the existence of the vacuum implies

$$\underline{{}_A|}\Lambda + \nabla \nu_A = 0 \quad at \ R^A = 0 \tag{3.5}$$

(A set of a p-form Λ and p-2 form ν_A which satisfy eq. (3.5) and which is build up with G/H potentials only is called an H-orthogonal cosmococycle the reasons for which we are going to discuss).

Eq. (3.5) is responsible already for the fixing of many parameters in the action of interesting theories. Examples are:

1) The De Sitter gravity. Here it fixes the coefficient of the cosmological term.

2) Supergravity. Here it fixes the value of the coefficient of the Rarita Schwinger term relative to the conventional term.

3) In d = 5 supergravity it reduces the number of free parameters of the theory from 10 to six only. It is obvious at this stage that similar reductions will take place in other theories.

Once the theory has been written according to 1) and 2) the stage is set for the rheonomy principle to be applied. Its strength is quite evident from the simple example of supergravity, as we shall see in the next section. Here the most general theory satisfying 1) and 2) only is given by:

$$A = \int \left\{ R^{ab} \wedge V^c \wedge V^d \epsilon_{abcd} + 4 \bar{\psi} \wedge \gamma_5 \gamma_a \rho \wedge V^a + a R^{ab} \wedge \bar{\psi} \wedge \vec{Z}_1^{cd} \psi \epsilon_{abcd} \right\} \tag{3.6}$$

However the extra term is seen to yield rigid theories no matter how small the parameter a. We must set a = 0 and conventional supergravity is recovered as the only possibility. Here we see that the parameter a is not dimensionless and that this is most probably the reason why it must be discarded. In more complicated cases, discarding of the parameters with the wrong dimension is necessary but not sufficient condition for rheonomy. Finally, we recall that conventional gravity in dimension 3 is a rigid theory, quite possibly the simplest case of rigid theory that one can build out of the principles 1) and 2).

Considering theories in which there are only linear terms in the curvature we see that the corresponding condition 2) appears now as:

$$\nabla \nu_A = 0 \quad \text{at } R^A = 0 \tag{3.7}$$

which is the condition for the form multiplet given by ν_A to be covariantly closed. In this case it represents an element of the Chevalley cohomology on G. Quite appropriately there is one such element for the Poincaré group and it yields ordinary gravity; there is no such element for the De Sitter unless one generalizes the closure condition by admitting the cosmological term Λ. There are arguments based on the relative exact sequence of cohomology groups which state that the absolute cohomology of a group is isomorphic to the one relative to the biggest semi-simple subgroup, in our case the Lorentz group. If we translate this statement into the more familiar one which affirms that the form must be orthogonal with respect to the Lorentz group, that is the theory is gauge invariant under the Lorentz group. This gauge invariance is therefore a consequence of the postulates at least in the simple cases considered. In general, it is not possible to reduce the corresponding problem to the search for a cohomology class on G. We stress, however, that the problem is nevertheless intrinsic in character. A cohomology group has a definition which is intrinsic to the group G, it states in fact a topological property of G. The cohomology class need not be represented by an invariant form. In some sense the group G is telling us how it can be broken preferentially while preserving some subgroup intact. Rheonomy, on the other hand, is a much more difficult and technically involved condition to implement, but finally it fixes all remaining parameters of the theory. If satisfied, the deduction of the supersymmetry transformations is an immediate matter. It is quite likely that there are no physically acceptable actions of degree higher than quadratic in the curvature; once the connections are expressed in terms of the other fields, the equation of motion would be of order higher than second. Quite likely the known 11 dimensional theory of Julia Cremmer and Scherk can be derived from a pure theory of the form described here.

Sec. IV - N = 1 Supergravity

Supergravity in 4 dimensions is the first non-trivial pure theory constructed besides gravity. It originates from the graded De Sitter group which includes the ordinary De Sitter group as a subgroup. Besides the 10 parameters of SO(2,3), all Bose in character, it contains 4 spinor components which are fermi. Correspondingly the set of forms consists of a connection ω^{ab}, of a vierbein V^a and of a spinorbein ψ. The curvatures are defined by:

$$\begin{aligned} R^{ab} &= \mathcal{R}^{ab} + V^a \wedge V^b - i\bar{\psi} \wedge \sigma^{ab} \psi \\ R^a &= \mathcal{D}V^a - \frac{i}{2} \bar{\psi} \wedge \gamma^a \psi \\ \rho &= \mathcal{D}\psi + \frac{i}{2} V^a \wedge \gamma_a \psi \end{aligned} \qquad (4.1)$$

These objects are related to the physical quantities via the relations

$$\omega^{ab} = \omega^{ab}_{phys.} \qquad V^a = \frac{e}{K} V^a_{phys.} \qquad \psi = \sqrt{eK}\, \psi_{phys.} \qquad (4.2)$$

here $K = \sqrt{4\pi G}$ is the Planck length, e is a parameter which is dimensionless and will be fixed in extended supergravity.

The Bianchi identities can be written as:

$$\begin{aligned}
\nabla R^{ab} &= \mathcal{D}R^{ab} + R^{[a}\wedge V^{b]} + 2i\,\bar{\psi}\wedge\Sigma^{ab}\rho = 0 \\
\nabla R^{a} &= \mathcal{D}R^{a} + R^{ab}\wedge V_b - i\,\bar{\psi}\wedge\gamma^a\rho = 0 \\
\nabla\rho &= \mathcal{D}\rho - \frac{i}{2}\gamma_a\psi\wedge R^a - \frac{i}{2}\Sigma_{ab}\psi\wedge R^{ab} = 0
\end{aligned} \tag{4.3}$$

In the limit of $e \longrightarrow 0$ we obtain the graded Poincaré group with the curvatures:

$$\begin{aligned}
R^{ab} &= \mathcal{R}^{ab} \\
R^{a} &= \mathcal{D}V^a - \frac{i}{2}\bar{\psi}\wedge\gamma^a\psi \; ; \quad \rho = \mathcal{D}\psi
\end{aligned} \tag{4.4}$$

and Bianchi identities:

$$\begin{aligned}
\nabla R^{ab} &= \mathcal{D}R^{ab} = 0 \\
\nabla R^{a} &= \mathcal{D}R^{a} + R^{ab}\wedge V_b - i\,\bar{\psi}\wedge\gamma^a\rho = 0 \\
\nabla\rho &= \mathcal{D}\rho - \frac{i}{2}\Sigma_{ab}\psi\wedge R^{ab} = 0
\end{aligned} \tag{4.5}$$

The problem is now that of constructing an action of the kind:

$$A = \int \mathcal{L}[V^a, \omega^{ab}, \psi]$$

$$\begin{aligned}
\mathcal{L} = \Lambda &+ R^{ab}\wedge\nu_{ab} + R^a\wedge\nu_a + 2\bar{\rho}\wedge m + R^{ab}\wedge R^{cd}\,\nu_{ab|cd} \\
&+ R^{ab}\wedge R^c\,\nu_{ab|c} + R^{ab}\wedge\rho^\alpha\,\nu_{ab|\alpha} + R^a\wedge R^b\,\nu_{a|b} + R^a\wedge\rho^\alpha\,\nu_{a|\alpha} + \\
&+ \rho^\alpha\wedge\rho^\beta\,\nu_{\alpha|\beta}
\end{aligned} \tag{4.6}$$

It is easily seen that all quadratic invariants in the curvature are either divergencies or can be reduced to first order terms modulo a divergence. They will be omitted.

We require now SO(3,1) gauge invariance of the theory. This

implies, besides other things, that the ω^{ab} cannot enter in Λ, ν_A for we cannot construct a set of ν_A transforming homogeneously by using algebraic operations only on the μ^A. Therefore, Λ, ν_A are some polynomials in the V^a and ψ only, having the proper tensorial character. Moreover we must select terms having the proper parity properties. These terms appear as pseudoscalars in tangent space since the volume element d^4x is already included in the definition. Taking this into account we see that the action must be of the form:

$$\begin{aligned} \Lambda &= \alpha_1 \epsilon_{abcd} V^a \wedge V^b \wedge V^c \wedge V^d + \alpha_2 \epsilon_{abcd} \bar{\psi} \wedge \Xi^{ab} \psi \wedge V^c \wedge V^d ; \\ \nu_a &= 0 \; ; \quad n = \delta_1 \gamma_5 \gamma_a \psi \wedge V^a \\ \nu_{ab} &= \beta_1 \epsilon_{abcd} V^c \wedge V^d + a \, \epsilon_{abcd} \bar{\psi} \wedge \Xi^{cd} \psi \end{aligned} \tag{4.7}$$

At this stage we use eq. (2.11) and recall that the covariant derivative of a coadjoint supermultiplet is given by:

$$\begin{aligned} \nabla \nu_a &= \mathcal{D} \nu_a - V^b \wedge \nu_{ba} - \frac{i}{2} \left(\bar{\psi} \wedge \gamma^a n - \bar{n} \wedge \gamma^a \psi \right) \\ \nabla \nu_{ab} &= \mathcal{D} \nu_{ab} + V_{[a} \wedge \nu_{b]} - \frac{i}{2} \left(\bar{\psi} \wedge \Xi_{ab} n - \bar{n} \wedge \Xi_{ab} \psi \right) \\ \nabla n &= \mathcal{D} n + \frac{i}{2} V^a \wedge \gamma_a n - \frac{i}{2} \gamma^a \psi \wedge \nu_a - i \Xi^{ab} \psi \wedge \nu_{ab} \end{aligned} \tag{4.8}$$

The equations (2.11):

$$\underline{A} \rfloor \Lambda + \nabla \nu_A = 0 \qquad \text{at } R^A = 0 \tag{4.9}$$

determine the remaining parameters with one exception of notable interest. The resulting action is of the form:

$$A = \int \left\{ R^{ab} \wedge V^c \wedge V^d \epsilon_{abcd} + 4 \bar{\psi} \wedge \gamma_5 \gamma_a \rho \wedge V^a + a R^{ab} \wedge \bar{\psi} \wedge \Xi^{cd} \psi \epsilon_{abcd} \right\} \tag{4.10}$$

The parameter a appears as multiplying a Pauli-like term. Rheonomy forces a = 0, as we shall see.

The variational equations are now given by:

Torsion equation: (variation in ω^{ab})

$$2\epsilon_{abcd} R^c \wedge V^d - 2a\bar{\psi}\wedge \Xi^{cd} \rho\, \epsilon_{abcd} \tag{4.11}$$

Einstein equation: (variation in V^a)

$$2\epsilon_{abcd} R^{bc} \wedge V^d + 4\bar{\psi}\wedge\gamma_5\gamma_a \rho = 0 \tag{4.12}$$

Gravitino "Rarita Schwinger" equation: (variation in ψ)

$$8\gamma_5\gamma_a \rho\wedge V^a - 4\gamma_5\gamma_a\psi\wedge R^a + 2a\epsilon_{abcd}\Xi^{cd}\psi\wedge R^{ab} = 0 \tag{4.13}$$

If a = 0 by projecting on ψVV we obtain the equation:

$$4\epsilon_{abcd} R^c_{.\alpha[m}\delta^d_{n]} - 2a\epsilon_{abcd}\left(\Xi^{cd}\rho_{mn}\right)_\alpha = 0 \tag{4.14}$$

which implies $\rho_{.mn} = 0$ if a = 0. But $\rho_{.mn} = 0$ in turn implies that all curvatures vanish and that vacuum is the only solution. Therefore we select a = 0. Now the remaining equations give:

$$\begin{aligned} &R^a = 0 \\ &R^{ab} = R^{ab}_{.cd} V^c \wedge V^d + 2\bar{\psi}^\alpha R^{ab}_{\alpha m} V^m \\ &\rho = \rho_{mn} V^m \wedge V^n \end{aligned} \tag{4.15}$$

The inner components $\rho_{.mn}$ and $R^{ab}_{.mn}$ obey the Rarita Schwinger and Einstein equations:

$$\begin{aligned} &\epsilon^{rsmn}\gamma_5\gamma_s\rho_{mn} = 0 \\ &R^{am}_{.bm} - \frac{1}{2}\delta^a_b R^{mn}_{.mn} = 0 \end{aligned} \tag{4.16}$$

The outer components are determined by the following conditions:

$$R^{ab}_{\ \cdot\alpha m} = \mathrm{const}\left(B^{ab}_{\ \cdot\alpha m} + \delta^{[a}_{m} B^{b]n}_{\alpha n}\right)$$
$$B^{ab}_{\alpha m} = \epsilon^{ablt}(\gamma_5\gamma_m)_{\alpha\beta}\,\rho^{\beta}_{\ \cdot lt} \tag{4.17}$$

This is the moment to elaborate the concept of rheonomy which was briefly introduced in Section II.

The idea is that a solution of the field equations is given on a standard 4 dimensional hypersurface M_4 contained in G. Using the normal components of the field equations we see that, by knowing the tangential components of the curvature we can retrieve the normal ones. Heuristically, this means that once a solution is fixed on M_4 it can be propagated through the whole G in a unique way. This statement is possibly just of local character but for the moment we want to draw qualitative conclusions.

Consider now the injection map $M \xrightarrow{i} G$ and a generic diffeomorphism $G \xrightarrow{\lambda} G$. Given the μ^A on G the pull-back defines $\mu^A_{|M} = i^*\mu^A$. Inversely, the field equations yield a unique extension of the input $\mu^A_{|M}$ to a solution $\chi\mu^A_{|M}$ on G such that $i^*\chi$ = identity. Therefore there is a map:

$$Rh(\lambda) = i^*\lambda^*\chi \tag{4.18}$$

acting on the forms $\mu^A_{|M}$ on M and yielding a new solution $Rh(\lambda)\mu^A_{|M}$ of the field equations on M. This solution, depending on the choice of λ, will follow from $\mu^A_{|M}$ through generic coordinate, gauge and supersymmetry transformations.

The infinitesimal form of these transformations can be easily obtained from the AGCT in eq. (1.36). Here there appear intrinsic components of the curvatures along the μ^A. Generally these components are called the supercovariant curvature components. In conventional supergravity, it pays off to distinguish between inner components (those along V) and outer (along ψ). Rheonomy implies that all components can be expressed uniquely in terms of those which are completely inner. If M is orthogonal to the ψ this means that if we know the tangential components of the curvature R^A along M then we can reconstruct all the others. Therefore the AGCT can be locally reconstructed on M and the infinitesimal rheonomy transformations recovered. For instance using the rheonomic conditions (4.15) and (4.17) in the AGCT formula (1.36) we find the first order supersymmetry transformations of Deser-Zumino:

$$\delta_\epsilon V^a = \nabla\epsilon^a + {}_\epsilon\rfloor R^a = \nabla\epsilon^a = i\bar{\epsilon}\gamma^a\psi \tag{4.19a}$$

$$\delta_\epsilon\psi = \nabla\epsilon + {}_\epsilon\rfloor\rho = \nabla\epsilon = \mathcal{D}\epsilon \tag{4.19b}$$

$$\begin{aligned}\delta_\epsilon\omega^{ab} &= \nabla\epsilon^{ab} + {}_\epsilon\rfloor R^{ab} = {}_\epsilon\rfloor R^{ab} = \\ &= \mathrm{const}\left(\bar{\epsilon}B^{ab}_m + \delta^{[a}_m\bar{\epsilon}B^{b]n}{}_n\right)V^m\end{aligned} \tag{4.19c}$$

In general one splits the Lie algebra $\mathbb{G}$ of G into a $\mathbb{H}$ part (which corresponds to the Lorentz+ internal subgroup) and a $\mathbb{K}$ part (the vierbeins and the ψ). $\mathbb{K}$ must be further split into an inner part $\mathbb{I}$ (of the same dimensionality as M) and an outer part $\mathbb{O}$ (orthogonal to M). Rheonomy implies that we can always express any outer part through the inner one. From the following examples it will be clear how this works.

Sec. V - Extended Supergravities as Impure Theories in d=4 and Other Pure Theories in d > 4

In previous sections we have considered the pure theory of N = 1 Supergravity and we have explicitly shown that it is the local theory of the supergroup Osp(4/1) or of its contraction Osp(4/1), namely the graded Poincaré group. Coming now to extended N > 1 Supergravities, it is quite evident that they should be the local theories of the corresponding orthosymplectic groups Osp(4/N) (or continuation thereof). In this respect, however, all the authors have been aware that, at least for $N \geq 3$, these theories could not be regarded as pure in that they involve more fundamental fields than the components of the Yang-Mills potential μ^A of the corresponding supergroup. Indeed, in the $N = 3$ case[10], for example, besides the graviton V^a_μ, the three gravitino ψ^A_μ and the three spin 1-fields A^A_μ, which constitute the Osp(4/3) potential, the supermultiplet contains a spin 1/2 Majorana particle λ, whose geometrical nature is that of a 0-form. Therefore the action of $N = 3$ Supergravity and, similarly, of all the $N > 3$ theories has got to be impure, in the sense elucidated above.

The case $N = 2$[11,20], however, was thought to be comparable to the pure $N = 1$ case, since the $N = 2$ supermultiplet contains only the graviton V^a_μ, the 2-gravitinos ψ^A_μ and the spin 1 A_μ, which complete the Osp(4/2) potentials. Accordingly in[25] Townsend and van Nieuwenhuizen wrote the action of $N = 2$ theory using only the Yang-Mills potentials and curvatures of the Osp(4/2) supergroup. Their formulation, however, is not yet "geometrical" in the strict sense we defined in previous sections. Indeed their Lagrangian makes essential use of the Hodge duality operation on the space-time manifold, so that the equations of motion cannot be naturally extended to the whole superspace, as we did for the $N = 1$ theory. The price you pay for such a limitation to the x-space surface is the inability to determine the form of the supersymmetry transformation laws directly from the equations of motion; actually you do not even know whether such transformations exist which leave the x-space action invariant. Indeed the authors of[25] had to construct the supersymmetry transformation laws starting from some explicit ansatz and found that, besides the Osp(4/2) covariant derivative of the infinitesimal parameters, the gravitino law had an additional $F_{\mu\nu}(x)$ dependent term which they named "non-geometrical". From the point of view of the group-manifold approach this term is an indication that the ρ^A curvature of the gravitino ψ^A, must have a $V^a \wedge \psi^B$ non zero component which, by means of some rheonomic condition, is given in terms of the inner $V_a \wedge V_b$ components of the photon curvature (namely F_{ab}). Such a rheonomic condition will be provided by the outer projections of the equations of motion in a true geometrical formulation which avoids the Hodge duality. The first order group-manifold formulation of the N = 2 and N = 3 theories was given in [18,19]. It turns out that also the N = 2 case is "impure" as much as the $N \geq 3$ ones. In fact, although the particle content is that of the Osp(4/2) potential, the physical propagation of the spin 1 field and the requirement of non-trivial solutions on the group manifold (rheonomy) demands the introduction of the 0-form $F_{ab}(x)$ as an independent field.

Quite differently, as we shall see, the N = 2 theory in 5 space-time dimensions, which is based on the SU(2,2/1) supergroup, is completely pure. Indeed in that case we also have a graviton $V^a_\mu(x)$, a U(1)-potential $B_\mu(x)$ and two gravitinos $\psi^A_\mu(x)$, which can be embedded in the SU(2,2/1) pseudoconnection, yet differently from the d = 4 case the propagator of the spin 1 B_μ is obtained from a totally geometric Lagrangian which avoids both 0-forms and the Hodge duality and which is rheonomic in the sense discussed above. The N = 2 theory in d = 4 is impure because it can be obtained as a fragment of the dimensional reduction of the pure N = 2 theory in d = 5. The other fragment is the N = 2 super Yang-Mills theory to be discussed in later sections.

At this point we try to make the above discussion on extended Supergravities explicit. Let us consider the supergroups Osp(4/N). They are described by the following curvatures[18]:

$$R^{ab} = \mathcal{R}^{ab} + \frac{e^2}{K^2} V^a \wedge V^b - i e K \bar{\psi}_A \wedge \Sigma^{ab} \psi_A \tag{5.1a}$$

$$R^a = \mathcal{D} V^a - \frac{i}{2} K^2 \bar{\psi}_A \wedge \gamma^a \psi_A \tag{5.1b}$$

$$\rho_A = \mathcal{D}\psi_A + \frac{ie}{2K} V^a_\wedge \gamma_a \psi_A + e A_{AB} \wedge \psi_B \tag{5.1c}$$

$$R^{AB} = dA^{AB} + e A^{AC}{}_\wedge A^{CB} + \frac{K}{2} \bar{\psi}^A{}_\wedge \psi^B \tag{5.1d}$$

where V^a is the Vierbein 1-form, ψ_A the N-gravitinos (Majorana spinor 1-forms),

$$A_{AB} = - A_{BA} \tag{5.2}$$

the $\frac{N(N-1)}{2}$ SO(N) gauge potentials and ω^{ab}, contained in R^{ab} and D, the SO(1,3) Lorentz connection. In eqs. (5.1) $V^a, \psi_A, A_{AB}, \omega^{ab}$ have already their physical dimensions and K and e are respectively the Planck length and a dimensionless gauge coupling constant. In the limit $e \to 0$ we obtain the contracted $\overline{Osp(4/N)}$ supergroup which, instead of the SO(N) internal group has N central charges. In the same limit the De Sitter subgroup $SO(2,3) \subset Osp(4/N)$ contracts to the Poincaré group.

Following the scheme of the previous sections we try to construct a purely geometrical action of the type:

$$A_{MIN}(N) = \int \left\{ R^{ab}{}_\wedge \nu_{ab} + R^a{}_\wedge \nu_a + \bar{\rho}_A \wedge n_A + R^{AB}{}_\wedge \nu_{AB} + \Lambda \right\} \tag{5.3}$$

where Λ, ν_{ab}, ν_a, n_A is an $SO(1,3) \otimes SO(N)$ orthogonal cosmococycle of parity P = -1:

$$\left.\begin{array}{l} a \rfloor \Lambda + \nabla \nu_a = 0 \\ A\alpha \rfloor \Lambda + \nabla n_{A\alpha} = 0 \\ \nabla \nu_{ab} = \nabla \nu_{AB} = 0 \end{array}\right\} \text{ at } R^{ab} = R^{AB} = R^a = \rho^A = 0 \tag{5.4}$$

and ∇ denotes the Osp(4/N) covariant derivative (in this case in the coadjoint representation). Excluding the Pauli-like terms which have been shown to trivialize the theory already in the N = 1 case, we get the answer

$$A_{MIN}(N) = \int \Big\{ \frac{1}{K^2} R^{ab} \wedge V^c \wedge V^d \epsilon_{abcd} + 4 \bar{\rho}_A \wedge \gamma_5 \gamma_a \psi_A \wedge V^a + 4Ki R^{AB} \wedge \bar{\psi}_A \wedge \gamma_5 \psi_B + 2iK^2 \bar{\psi}_A \wedge \psi_B \wedge \bar{\psi}_A \wedge \gamma_5 \psi_B - \frac{e^2}{2K^4} V^a \wedge V^b \wedge V^c \wedge V^d \epsilon_{abcd} + \frac{ie}{2K} \bar{\psi}_A \wedge \Sigma^{ab} \psi_A \wedge V^c \wedge V^d \epsilon_{abcd} \Big\} \tag{5.5}$$

which does not contain any further free parameter besides the contraction parameter e.

The theory described by the action principle (5.5) is trivial for all N $\geqslant$ 2. In fact the variation of (5.5) in A^{AB}, instead of giving the Maxwell inhomogeneous equation gives:

$$8iK \bar{\rho}_A \wedge \gamma_5 \psi_B = 0 \tag{5.6}$$

which implies

$$\rho^A_{\cdot ab} = 0 \tag{5.7}$$

where $\rho^A_{\cdot ab}$ are the space-time components of the gravitino curvature. Eq. (5.7) rules out any propagation of the spin 3/2 particle and, used in the other equations of motion, leads to the conclusion that the vacuum

$$R^{ab} = R^{AB} = \rho^A = R^a = 0 \tag{5.8}$$

is the only solution of (5.5). The case N = 1 is exceptional just because, in this limit, there is no A^{AB} field and eq. (5.6) does not exist.

In order to detrivialize the theory (5.5) we are forced to add to it a new piece, called the detrivializing action, which contains additional 0-form fields. The structure of the detrivializing action and the number and type of 0-forms depends on N.

N = 2 Theory. The only component of the connection A^{AB} is A^{12}. If we set

$$A^{12} = A^{\otimes} \quad ; \quad R^{12} = R^{\otimes} \tag{5.9}$$

and we introduce a 0-form field $F^{ab} = -F_{ba}$, we obtain the complete action of the N = 2 theory by setting:

$$A(N=2) = A_{MIN}(N=2) + A_{DETR.}(N=2) \tag{5.10}$$

where

$$A_{DETR}(N=2) = \int \left\{ -4F^{ab} V^c_\wedge V^d_\wedge R^{\otimes} \epsilon_{abcd} + \frac{1}{3} F_{ab} F^{ab} V^i_\wedge V^j_\wedge V^k_\wedge V^l \epsilon_{ijkl} \right\} \tag{5.11}$$

is essentially the Maxwell action of the electromagnetic field written in first order formalism.

The equations of motion, obtained from the independent variation of the fields ω^{ab}, V^a, ψ_A, A and F^{ab} are now written using only the exterior algebra operations d and $\wedge$ and hold on the entire group manifold. They imply the following rheonomic conditions:

$$R^a = 0 \tag{5.12a}$$

$$R^{\otimes} = F_{ab} V^a_\wedge V^b \tag{5.12b}$$

$$\rho^A = \rho^A_{\cdot ab} V^a_\wedge V^b + K \epsilon^{AB} \left(\gamma_a \psi_B \wedge V_b F^{ab} + \frac{1}{2} \gamma_5 \gamma_a \psi_B \wedge V_b F_{cd} \epsilon^{abcd} \right) \tag{5.12c}$$

$$R^{ab} = R^{ab}_{\cdot cd} V^c \wedge V^d - \frac{1}{4} \epsilon^{abcd} \bar{\rho}_{A|cd} \gamma_5 \gamma_m \psi_A \wedge V^m -$$
$$- \frac{1}{8} \epsilon^{amcd} \bar{\rho}_{A|cd} \gamma_5 \gamma_m \psi_A \wedge V^b + \frac{1}{8} \epsilon^{bmcd} \bar{\rho}_{A|cd} \gamma_5 \gamma_m \psi_A \wedge V^a \tag{5.12d}$$
$$- K^3 \left(F^{ab} \bar{\psi}_A \wedge \psi_B + \frac{1}{2} \epsilon^{abcd} \bar{\psi}_A \wedge \gamma_5 \psi_B F_{cd} \right) \epsilon_{AB}$$

where the space-time components F_{ab}, $\rho^A_{\cdot ab}$, $R^{ab}_{\cdot cd}$ are subject to the following propagation equations:

$$\mathcal{D}_m F^{ma} = 0 \tag{5.13a}$$

$$\gamma_5 \gamma_a \rho^A_{\cdot bc} \epsilon^{abcd} = 0 \tag{5.13b}$$

$$R^{ab}_{\cdot mb} - \frac{1}{2} \delta^a_m R^{bc}_{\cdot bc} = 0 \tag{5.13c}$$

In eq. (5.12c) one can recognize the "<u>non-geometrical term</u>" of the gravitino supersymmetry transformation law. Indeed using eq. (5.12c) in the formula for the AGCT of infinitesimal fermionic parameter ϵ^B we obtain

$$\begin{aligned} \delta_\epsilon \psi^A &= \nabla \epsilon^A + \underline{\epsilon} | \rho^A = \\ &= \mathcal{D} \epsilon^A + \frac{ie}{2K} V^a \gamma_a \epsilon^A + e \epsilon^{AB} A \epsilon^B \\ &\quad + K \epsilon^{AB} \left(i \gamma_a \epsilon^B V_b F^{ab} + \frac{1}{2} \gamma_5 \gamma_a \epsilon^B V_b F_{cd} \epsilon^{abcd} \right) \end{aligned} \tag{5.14}$$

which is the standard formula in ref.[25].

<u>N = 3 Theory</u>. We set:

$$A^{AB} = \epsilon^{ABC} B^C ; \quad R^{AB} = \epsilon^{ABC} R^C ; \quad C = 1,2,3 \tag{5.15}$$

The detrivializing action contains, besides a 0-form F^A_{ab} which, in second order, will represent the space-time components of R^C, another 0-form Majorana spinor λ which stands for the spin 1/2 particle of the N = 3 supermultiplet.

Explicitly[18] one finds:

$$A(\mathcal{N}=3) = A_{MIN}(\mathcal{N}=3) + A_{DETR.}(\mathcal{N}=3) \tag{5.16}$$

where, using the substitution (5.15) in (5.5) one has

$$\begin{aligned} A_{MIN}(\mathcal{N}=3) = \int \Big\{ & \frac{1}{K^2} R^{ab}\wedge V^c \wedge V^d \epsilon_{abcd} + 4\bar{\rho}_A \wedge \gamma_5 \gamma_a \psi_A \wedge V^a \\ & + 4Ki\, R^A \wedge \bar{\psi}^B \wedge \gamma^5 \psi^C \epsilon_{ABC} + 2iK^2 \bar{\psi}_A \wedge \psi_B \wedge \bar{\psi}_A \wedge \gamma_5 \psi_B \\ & - \frac{e^2}{2K^4} V^a \wedge V^b \wedge V^c \wedge V^d \epsilon_{abcd} + \frac{ie}{2K} \bar{\psi}_A \wedge \Sigma^{ab} \psi_A \wedge V^c \wedge V^d \epsilon_{abcd} \Big\} \end{aligned} \tag{5.17}$$

and

$$\begin{aligned} A_{DETR}(\mathcal{N}=3) = \int \Big\{ & 4R^A \wedge V^a \wedge V^b F^{cd}_A \epsilon^{abcd} - 4K\bar{\lambda}\gamma_5\gamma_m \psi_A \wedge V^m \wedge R^A \\ & + \frac{1}{3} F^{ab}_A F^A_{ab} V^i \wedge V^j \wedge V^k \wedge V^\ell \epsilon_{ijk\ell} - \frac{i}{3}\bar{\lambda}\gamma^a \nabla\lambda \wedge V^b \wedge V^c \wedge V^d \epsilon_{abcd} \\ & + 2iK F^A_{ab} V_c \wedge V_d \wedge \bar{\lambda}\gamma_m \psi_A \wedge V^m \epsilon^{abcd} + \\ & + \frac{K^2}{2}\bar{\lambda}\gamma_5 \lambda \bar{\psi}_A \wedge \Sigma_{ab} \psi_A \wedge V^a \wedge V^b + i\frac{K^2}{2}\bar{\lambda}\lambda \bar{\psi}_A \wedge \Sigma_{ab}\psi_A \wedge V_c \wedge V_d \epsilon^{abcd} \Big\} \end{aligned} \tag{5.18}$$

The variation of the action (5.16) in all the independent fields (ω^{ab}, V^a, ψ_A, B_A, F^{ab}_A, λ) gives equations of motion which imply the following rheonomic conditions:

$$R^a = \frac{K^2}{8} \epsilon^{abcd} \bar{\lambda}\gamma_5\gamma_b \lambda V_c \wedge V_d \tag{5.19a}$$

$$R^A = F^A_{ab} V^a \wedge V^b - \frac{K}{2} \bar{\lambda} \gamma_m \psi^A \wedge V^m \tag{5.19b}$$

$$\begin{aligned} \rho^A = {} & \rho^A_{\cdot ab} V^a \wedge V^b + K \epsilon^{ABC} \left\{ i F^B_{ab} \gamma_a \psi^C \wedge V_b + \frac{1}{2} F^B_{ab} \gamma_5 \gamma_c \psi^C \wedge V_d \, \epsilon^{abcd} \right\} \\ & + \frac{i}{8} K^2 \left\{ \epsilon^{abcd} \Sigma_{ab} \psi^A \wedge \bar{\lambda} \gamma_5 \gamma_c \lambda V_d - \gamma_5 \psi^A \wedge \bar{\lambda} \gamma_5 \gamma_m \lambda V^m \right\} \\ & + \frac{K^2}{4} \left\{ \lambda \bar{\psi}_B \wedge \psi_C - \gamma_5 \lambda \bar{\psi}_B \wedge \gamma_5 \psi_C \right\} \epsilon_{ABC} \end{aligned} \tag{5.19c}$$

$$\nabla \lambda = \Lambda_m V^m - \frac{iK}{8} F^{ab}_A \Sigma_{ab} \psi_A \tag{5.19d}$$

From equations (5.19) one can read off the transformation law, under supersymmetry of the physical fields V^a, B^A, ψ^A, λ.

The derivation of the impure theories of extended supergravity for $N > 3$ is feasible with the same procedure used in the first two cases but it is increasingly complicated and painful. Essentially one has to introduce a 0-form for all the spins less than 1 appearing in the corresponding supermultiplet and, in the case of bosons, for each of them, also a 0-form representing their space-time derivatives (i.e. for a scalar field one has φ and φ_a, later to be identified with $\partial_a \varphi$). Then all possible $SO(N) \otimes SO(1,3)$ gauge invariant terms of the correct parity $P = -1$ are written down and summed to the minimal action (5.5) with undetermined coefficients. Checking one by one the equations of motion these coefficients are now fixed in such a way as to allow non-vacuum solutions having the rheonomic property.

Apart from the labour involved, this procedure lacks in aesthetic satisfaction. The extra 0-forms are not part of the potential μ^A, rather they live in the background described by it, the same way as matter fields, in General Relativity, are superimposed on the background metric. Moreover the action is not completely determined by the group structure, the same way as it happens in pure theories.

In this sense the impure theories of extended supergravity seem to be no more unified than coupled Matter + Einstein systems are. The essential difference is that in General Relativity matter is optional, here instead it is mandatory in order to avoid trivialization.

Quite possibly impure theories can be derived through dimensional reduction from pure theories in higher dimension. The 0-forms are components of the Yang-Mills potentials associated with group generators which are inoperative after dimensional reduction.

A simple example will clarify the situation. Let us consider the pure theory of gravity in D = 5. It is based on the Poincaré group ISO(1,4). The action is:

$$A(d=5) = \int \mathbb{R}^{\hat{a}\hat{b}} \wedge \mathbb{V}^{\hat{c}} \wedge \mathbb{V}^{\hat{d}} \wedge \mathbb{V}^{\hat{e}} \, \epsilon_{\hat{a}\hat{b}\hat{c}\hat{d}\hat{e}} \tag{5.20}$$

where

$$\mathbb{R}^{\hat{a}\hat{b}} = d\omega^{\hat{a}\hat{b}} + \omega^{\hat{a}\hat{c}} \wedge \omega^{\hat{b}\cdot}{}_{\hat{c}} \tag{5.21a}$$

$$\mathbb{R}^{\hat{a}} = d\mathbb{V}^{\hat{a}} - \omega^{\hat{a}\hat{b}} \wedge \mathbb{V}_{\hat{b}} \tag{5.21b}$$

are the curvatures of the ISO(1,4) group, $\omega^{\hat{a}\hat{b}}$ being the SO(1,4) connection and $\mathbb{V}^{\hat{a}}$ the fünfbein. Our convention is that the indices $\hat{A}, \hat{b}$, .. run from 1 to 5 while a,b,c .. run from 1 to 4. The equations of motion of the theory (5.20) imply, the same as in d = 4, the factorization of the Lorentz subgroup SO(1,4) and the vanishing of the torsion $\mathbb{R}^{\hat{a}}$. We get:

$$\mathbb{R}^{\hat{a}} = 0 \tag{5.22a}$$

$$\mathbb{R}^{\hat{a}\hat{b}} \wedge \mathbb{V}^{\hat{c}} \wedge \mathbb{V}^{\hat{d}} \, \epsilon_{\hat{a}\hat{b}\hat{c}\hat{d}\hat{e}} = 0 \tag{5.22b}$$

The dimensional reduction of the theory (5.20) restricts the manifold of solutions of eqs. (5.22a) and (5.22b) to those admitting a constant

Killing vector. We recall that, under a AGCT transformation of parameters $\epsilon = (\epsilon^{\hat{a}}, \epsilon^{\hat{a}\hat{b}})$ we have

$$\delta_\epsilon V^{\hat{a}} = \nabla \epsilon^{\hat{a}} + \epsilon \rfloor R^{\hat{a}} = \mathcal{D}\epsilon^{\hat{a}} - V^{\hat{b}} \epsilon^{\hat{a}}{}_{.\hat{b}} \tag{5.23a}$$

$$\delta_\epsilon \omega^{\hat{a}\hat{b}} = \nabla \epsilon^{\hat{a}\hat{b}} + \epsilon \rfloor R^{\hat{a}\hat{b}} = \mathcal{D}\epsilon^{\hat{a}\hat{b}} + \epsilon^{\hat{m}} \,{}_{\hat{m}}\rfloor R^{\hat{a}\hat{b}} \tag{5.23b}$$

A Killing vector $\mathbb{K}$ is a set of parameters $(\epsilon^{\hat{a}}, \epsilon^{\hat{a}\hat{b}})$ such that the corresponding variation of the fünfbein and connection is identically zero. We assume that $\mathbb{K}$ has a constant space-like $\epsilon^{\hat{a}}$ component, for instance:

$$\epsilon^{a} = 0 \qquad \epsilon^{5} = 1 \tag{5.24}$$

From $\delta_{\mathbb{K}} V^{\hat{a}} = 0$ using (5.24) in (5.23a) we obtain:

$$\epsilon^{a5} = 0 \tag{5.25a}$$

$$\omega^{a5} = \epsilon^{ab} V_b \tag{5.25b}$$

The constraint (5.25b) inserted into eq. (5.22a) yields:

$$R^{a} = dV^{a} - \omega^{ab} \wedge V_b - \epsilon^{ab} V_b \wedge V_5 = 0 \tag{5.26a}$$

$$R^{5} = dV^{5} + \epsilon^{ab} V_b \wedge V_a = 0 \tag{5.26b}$$

Therefore, if we make the following redefinitions:

$$V^{a} = V^{a} \tag{5.27a}$$

$$V^{5} = A \tag{5.27b}$$

$$\mathbb{E}^{ab} = \mathbb{F}^{ab} \tag{5.27c}$$

$$\boldsymbol{\omega}^{ab} - \mathbb{E}^{ab} \mathbb{V}_5 = \omega^{ab} \tag{5.27d}$$

we obtain

$$dV^a - \omega^{ab} \wedge V_b = R^a = 0 \tag{5.28a}$$

$$dA - \mathbb{F}^{ab} V_a \wedge V_b = R^{\otimes} - \mathbb{F}^{ab} V_a \wedge V_b = 0 \tag{5.28b}$$

where $R^a(\omega, V)$ and $R^{\otimes} = dA$ can now be reinterpreted as the torsion of the 4-dimensional space-time spanned by V^a and the field strength of an electromagnetic potential respectively.

In this way it appears that the torsion equation in d = 5 contains, after dimensional reduction, the analogue of eq. (5.12b) discussed in the context of N = 2 Supergravity.

Using now (5.25b) and the redefinitions (5.27) to compute the $\mathbb{R}^{ab}$ and $\mathbb{R}^{a5}$ curvatures we obtain:

$$\mathbb{R}^{ab} = R^{ab}(\omega) + \mathcal{D}\mathbb{F}^{ab} \wedge A + \mathbb{F}^{ab} R^{\otimes} + \mathbb{F}^{am} \mathbb{F}^{bn} V_m \wedge V_n \tag{5.29a}$$

$$\mathbb{R}^{a5} = \mathcal{D}\mathbb{F}^{am} \wedge V_m - \mathbb{F}^{am} R_m + \mathbb{F}^{ab} \mathbb{F}_b^{\cdot m} V_m \wedge A \tag{5.29b}$$

If we insert eq. (5.29) into eq. (5.22b) and implement eq. (5.28) we get:

$$\mathcal{D}_m \mathbb{F}^{ma} = {}_{\underline{m}}\rfloor \mathcal{D}\mathbb{F}^{ma} = 0 \tag{5.30a}$$

$$R^{am}_{\cdot bm} - \frac{1}{2}\delta^a_b R^{mn}_{\cdot mn} = \text{const}\left(\mathbb{F}^{am}\mathbb{F}_{bm} - \frac{1}{4}\delta^a_b \mathbb{F}^{mn}\mathbb{F}_{mn}\right) \tag{5.30b}$$

which are the dynamical equations of the Maxwell-Einstein system in 4 space-time dimensions.

We note that starting from eq. (5.27) as an ansatz and substituting eq. (5.29) into the action (5.20) one obtains:

$$A = \int \Big\{ 3 \Big(\mathcal{R}^{ab} \wedge V^c \wedge V^d \epsilon_{abcd} + F^{ab} R^{\otimes} V^c \wedge V^d \epsilon_{abcd} - \\ - \frac{1}{12} F_{rs} F^{rs} V^i \wedge V^j \wedge V^k \wedge V^l \epsilon_{ijkl} \Big) \wedge A - 6 F^{am} V_m \wedge R^b \wedge V^c \wedge V^d \epsilon_{abcd} \Big\} \quad (5.31)$$

One easily realizes that in (5.31) the coefficient (in the exterior algebra sense) of the 1-form A is the "impure" Lagrangian of the Maxwell-Einstein system as it appears in the bosonic sector of N = 2 Supergravity (see eq. (5.11)).

We can therefore conclude that dimensional reduced theories can be obtained in two steps, first assuming equations similar to (5.29) which correspond to the existence of a Killing vector (in the standard language this means trivial dependence on some coordinates), secondly taking as Lagrangian the coefficient of the volume element of the suppressed dimensions.

The important point is that performing this exercise in the first order formalism we naturally obtain the appearance of the 0-forms. In this particular example we see that F_{ab} is the component of ω^{a5} namely the gauge field of the J_{a5} generator of SO(1,4). In the reduction this generator is broken because we are no longer allowed to mix spatial and internal dimensions. This confirms the statement we started with: 0-forms are remnants of broken generators.

<u>N = 2 Theory in d = 5</u>. The minimal supergravity theory in 5 space-time dimensions can be formulated as a pure theory although it contains a propagating spin 1 particle[5]. We need a suitable supergroup G. From a physical point of view we may guess what the structure of G will be by looking at the field content of the theory. In d = 5 the spin 2 graviton has 5 and the spin 3/2 complex gravitino field has 8 on-shell

degrees of freedom (we remind the reader that in d = 5 no Majorana spinor can be defined). The mismatch 8 - 5 = 3 between the fermionic and bosonic degrees of freedom may be compensated by the presence of a spin 1 vector field (Maxwell field). Thus we expect the minimal supergroup G to contain, besides the non-propagating Lorentz connection ω_μ^{ab}, a fünfbein $V_\mu^a(x)$, a Maxwell field B_μ and a complex gravitino field ξ_μ. This is exactly the field content of the Yang-Mills potential for the superalgebra SU(2,2/1) and its contractions. The μ can be written as follows:

$$\mu = \mu^A T_A = \frac{i}{2}\omega^{ab} J_{ab} + \frac{i}{2} V^a P_a + B Z_\otimes + \bar{\xi} Q + \bar{Q} \xi \tag{5.32}$$

where J_{ab} are the SO(1,4) Lorentz generators, P_a the (in general) non-commuting translations, $Z_\otimes$ the U(1) generator and $Q, (\bar{Q})$ the complex supersymmetries.

The structure of SU(2,2/1) is described writing the curvatures:

$$R^{ab} = \mathcal{R}^{ab} + \frac{e^2}{K^2} V^a \wedge V^b - i e K^4 \bar{\xi} \wedge \Sigma^{ab} \xi \tag{5.33a}$$

$$R^a = \mathcal{D} V^a - \frac{i}{2} K^5 \bar{\xi} \wedge \Gamma^a \xi \tag{5.33b}$$

$$R^\otimes = dB - i K^4 \bar{\xi} \wedge \xi \tag{5.33c}$$

$$\rho = \mathcal{D}\xi + \frac{ie}{2K} V^a \wedge \Gamma_a \xi - \frac{3i}{4} e B \wedge \xi \tag{5.33d}$$

where K is a length parameter and e is a dimensionless gauge coupling constant. In this way the fields V^a, ξ, B and ω^{ab} have their canonical scale dimension (in d = 5):

$$[V^a] = [L]; \quad [\omega] = [B] = [L^0]; \quad [\xi] = [L^{-2}] \tag{5.34}$$

The conventions used for the Γ-matrix algebra in d = 5 are the following ones:

$$\{\Gamma_a, \Gamma_b\} = 2\eta_{ab} \quad ; \quad \eta_{ab} = \mathrm{diag}(+,-,-,-,-) \tag{5.35a}$$

$$\Gamma_0^\dagger = \Gamma_0 \quad ; \quad \vec{\Gamma}^\dagger = -\vec{\Gamma} \tag{5.35b}$$

$$\Sigma_{ab} = \frac{i}{4}[\Gamma_a, \Gamma_b] \tag{5.35c}$$

$$C^2 = -\mathbb{1} \; ; \; C^T = -C \; ; \; C\Gamma_a^T C^{-1} = \Gamma_a \tag{5.35d}$$

If we perform the limit e $\rightarrow$ 0 in eq. (5.33) we get the contracted supergroup SU(2,2/1). In this limit the U(1) charge of the gravitino is zero and $Z_\otimes$ becomes a central charge.

In reference[5] we determined the action of D = 5 supergravity through the following steps:

i) First we assumed that the theory is pure and we looked for an action of the type:

$$A = \int_{M_5} \left\{ \Lambda + R^A \wedge \nu_A + R^A \wedge R^B \wedge \nu_{AB} \right\} \tag{5.36}$$

where Λ, ν_A, ν_{AB} are Lorentz orthogonal cochains of order 5, 3 and 1 transforming respectively in the identity, coadjoint and adjoint representation of the SU(2,2/1) supergroup. Lorentz orthogonality means that the cochains Λ, ν_A, ν_{AB} are constructed using only V^a, B and ξ but not ω^{ab}. The belief that the theory would be pure was based on the previous results of paper[3] which demonstrated how a spin 1 propagating field can be described, in 5 space-time dimensions by a Lagrangian free from 0-forms and Hodge duality.

ii) Secondly we solved the cosmococycle condition

$$\underline{A}\rfloor \Lambda + \nabla \nu_A = 0 \qquad \text{at } R^A = 0 \tag{5.37}$$

which guarantees the existence of the vacuum solution and we obtained a purely geometrical action with 6 free parameters.

iii) Finally, we examined the equations of motion and we determined the parameter conditions to be satisfied in order for the equations of motion to admit non-vacuum solutions. The result is the following unique action where η is a discrete parameter which can be either 1 or -1 ($\eta = \pm 1$):

$$\begin{aligned} A = \int \Big\{ & -\frac{2}{15}\frac{e^2}{K^5}\epsilon_{abijk} V^a \wedge V^b \wedge V^i \wedge V^j \wedge V^k + \frac{i}{3} eK \bar{\xi} \wedge \Xi^{ab} \xi \wedge V^i \wedge V^j \wedge V^k \epsilon_{abijk} \\ & + K^7 \left(\bar{\xi} \wedge \xi \wedge \bar{\xi} \wedge \Gamma^a \xi \wedge V_a - \frac{1}{2} K \bar{\xi} \wedge \xi \wedge \bar{\xi} \wedge \xi \wedge B \right) + iK^3 R^a \wedge \bar{\xi} \wedge \Gamma_a \xi \wedge B \\ & + R^{ab} \wedge \left(\frac{1}{3K^3} \epsilon_{abijk} V^i \wedge V^j \wedge V^k + \frac{1}{K^2} V_a \wedge V_b \wedge B \right) \\ & + R^{\otimes} \wedge \left(\frac{i}{2} K^4 \bar{\xi} \wedge \xi \wedge B - 2iK^3 \bar{\xi} \wedge \Gamma_a \xi \wedge V^a \right) \\ & + 2K^2 \left(\bar{\rho} \wedge \Xi_{ab} \xi + \bar{\xi} \wedge \Xi_{ab} \rho \right) \wedge V^a \wedge V^b + \frac{1}{4} R^{\otimes} \wedge R^{\otimes} \wedge B \\ & + \frac{1}{K^2} R^a \wedge R_a \wedge B - \frac{1}{K^2}(1-\eta) R^a \wedge V_a \wedge R^{\otimes} \Big\} \end{aligned} \tag{5.38}$$

The vacuum of the above theory corresponds to $R^A = 0$ which in particular means

$$R^{ab} = \mathcal{R}^{ab} + \frac{e^2}{K^2} V^a \wedge V^b - ieK^4 \bar{\xi} \wedge \Xi^{ab} \xi = 0 \tag{5.39}$$

Since the Riemannian curvature of the space-time manifold is R^{ab}, the vacuum of the non-contracted SU(2,2/1) theory is a 5-dimensional De Sitter space with a De Sitter radius of order

$$r_{\text{de Sitter}} \sim \frac{K}{e} \tag{5.40}$$

In the contraction limit $e \to 0$ we have $r_{\text{De Sitter}} \to \infty$ and the vacuum becomes a standard Minkowski space. We see that, in the same limit the first two terms disappear from eq. (5.38). They are the cosmological terms.

The equations of motion of (5.38) yield the following rheonomic conditions:

$$R^{\otimes} = F_{ab} V^a \wedge V^b \tag{5.41a}$$

$$R^a = -\frac{\eta K}{4} \epsilon^{abcde} F_{bc} V_d \wedge V_e \tag{5.41b}$$

$$\rho = \rho_{ab} V^a \wedge V^b + K\left(\frac{1}{2}\Gamma_a \xi F^{ab} - \frac{i}{8}(1-\eta)\epsilon^{bacde} F_{ac} \Sigma_{de} \xi\right) \wedge V_b \tag{5.41c}$$

$$\begin{aligned} R^{ab} = {} & R^{ab}{}_{\cdot cd} V^c \wedge V^d + K^5 \epsilon^{abijk} \left\{ \bar{\rho}_{ij} \left(\Sigma_{mk} - i(1-\eta)\delta_{km} \mathbb{1} \right) \xi \right. \\ & \left. + \text{c.c.} \right\} \wedge V^m + K^6 \left\{ \frac{i}{2} F^{ab} \bar{\xi} \wedge \xi - \frac{i}{8}(1+\eta) \epsilon^{abijk} F_{ij} \bar{\xi} \wedge \Gamma_k \xi \right\} \end{aligned} \tag{5.41d}$$

As usual from eq. (5.41) we can read off the supersymmetry transformation laws of the various fields. In particular from eq. (5.41c) we see that the gravitino transformation law has a non-geometrical term similar to the one appearing in $N = 2$ $d = 4$ supergravity (see eq. (5.14)).

The important and novel feature of the $d = 5$ theory is given by eq. (5.41b) and (5.41a) which are just a yield of the torsion equation, namely the variational equation of the ω^{ab} 1-form. Combining (5.41a) and (5.41b) we obtain an algebraic equation for ω_μ^{ab} which can be solved. The result gives ω_μ^{ab} as a function of ξ_μ, V_μ^a, B_μ and the spatial derivatives of the last two.

If we restrict the original group-manifold action to the space-time surface, taking only the coefficient of $dx^{\mu_1} \dots dx^{\mu_5}$ in the expansion of the Lagrangian 5-form and if we replace ω_μ by its explicit solution, we obtain the second order formulation of the theory. Setting for simplicity K = 1, $e = 0$ the second order Lagrangian turns out to be:

$$\mathcal{L}^{N=2,D=5}_{2^{nd}\,order} = \left(4R(\overline{\omega}) - \frac{3}{2}\mathcal{F}_{\mu\nu}\mathcal{F}^{\mu\nu} + 3i\,\mathcal{F}_{\mu\nu}\overline{\xi}^\mu\xi^\nu + \frac{3}{2}\overline{\xi}_\mu\xi_\nu\overline{\xi}^\mu\xi^\nu\right)\times$$
$$\times(\det V) + \left(\frac{1}{4}\mathcal{F}_{\mu\nu}\mathcal{F}_{\rho\sigma}B_\tau + \frac{3}{4}i\,\mathcal{F}_{\mu\nu}\overline{\xi}_\rho\Gamma_\sigma\xi_\tau + \right. \tag{5.42}$$
$$\left. + \overline{\xi}_\mu\xi_\nu\overline{\xi}_\rho\Gamma_\sigma\xi_\tau + 2\mathcal{D}_\mu(\overline{\omega})\overline{\xi}_\nu\Xi_{\sigma\tau}\xi_\rho + \overline{\xi}_\rho\Xi_{\sigma\tau}\mathcal{D}_\mu(\overline{\omega})\xi_\nu\cdot 2\right)\epsilon^{\mu\nu\rho\sigma\tau}$$

where

$$\mathcal{F}_{\mu\nu} = \partial_{[\mu}B_{\nu]} \tag{5.43a}$$

$$\overline{\omega}_{\mu|\lambda\nu} = c_{\lambda|\mu\nu} + c_{\nu|\lambda\mu} - c_{\mu|\nu\lambda} + h_{\lambda\nu|\mu} \tag{5.43b}$$

$$c_{\lambda|\mu\nu} = \frac{1}{2}\left(V_\lambda^a\left[\partial_\mu V_\nu^b - \partial_\nu V_\mu^b\right]\eta_{ab}\right) \tag{5.43c}$$

$$h_{\lambda\nu|\mu} = -\frac{i}{4}\left(\overline{\xi}_\mu\Gamma_\lambda\xi_\nu + \overline{\xi}_\lambda\Gamma_\nu\xi_\mu + \overline{\xi}_\lambda\Gamma_\mu\xi_\nu - [\lambda\leftrightarrow\nu]\right) \tag{5.43d}$$

It is evident from eq. (5.42) that, in the second order formalism, the η-dependence has disappeared: moreover the Maxwell kinetic term of the B_μ field ($F_{\mu\nu}F^{\mu\nu}$) has been generated. The propagation equation of B_μ is already present in the group manifold first order equations although somewhat hidden. Actually one of the projections of the group manifold equations yields:

$$\epsilon_{abcde}\left(\eta R^{ab|cd} + \frac{1}{4}(3-\eta) F^{ab} F^{cd}\right) = 0 \tag{5.44}$$

which upon substitution of the explicit value of ω_{μ}^{ab} as a function of V_{μ}^{a}, B_{μ}, ξ_{μ} gives:

$$\Box B_{\mu} + \text{other terms} = 0 \tag{5.45}$$

Before leaving the present section it is worthwhile mentioning that the same group SU(2,2/1) can be used in d = 4 to construct the group-manifold formulation of conformal supergravity recently obtained by Castellani, Fré and van Nieuwenhuizen(8). Such a theory is somewhat exceptional in the present framework because besides being impure it is written only on a submanifold of the SU(2,2/1) group. Space limitations prevent us from a full-fledged discussion of this theory and we refer the reader to the original paper.

Sec. VI - Kinematical Constraints and Auxiliary Field in the Group Manifold Framework.

In the investigation of supergravity and Super Yang-Mills theories a central place is occupied by the problem of finding for each theory the convenient set of auxiliary fields. Having the auxiliary fields means dealing with a closed supersymmetry algebra which is extremely important for the quantization programme. In the standard approach the auxiliary fields are also necessary in order to couple matter multiplets to supergravity (tensor calculus(12)).

As we shall presently see the auxiliary fields are nothing else than "outer components" of the group curvatures R^{A} which are set to zero by the equations of motion of the matter-free theory, but develop finite values in the presence of matter sources. From this point of view the search of auxiliary fields and the related tensor calculus is irrelevant in our approach. Indeed the problem of coupling matter to supergravity is transformed into the problem of writing a consistent group manifold action for the matter multiplet. Once this is obtained the coupling is simply given by the sum of the two actions of supergravity + matter. The equations of motion of the coupled system will provide new rheonomic symmetry conditions where formerly zero outer components of the group curvatures are now non-zero and expressed in terms of the matter fields. In this way we automatically obtain the correct supersymmetry laws avoiding the use of tensor calculus.

We shall explicitly illustrate this mechanism by coupling N = 1 Supergravity to the Wess-Zumino multiplet[26] as it was done in[4]. However, it is in view of the quantization programme that the search of auxiliary fields is still of interest in our approach. The reason is the following. As we have seen in previous sections the group-manifold variational equations provide us with a set of rheonomic conditions on group curvatures corresponding to a given set of super symmetry transformation laws. The algebra of these transformations is usually named "on-shell" because they satisfy Jacobi identities only upon use of the propagation equations of physical fields. This result can be understood in a different way: if we take the rheonomic conditions and we use them in the Bianchi identities:

$$\nabla R^A = dR^A + C^A_{\cdot BC}\, \mu^B \wedge R^C = 0 \tag{6.1}$$

we see that certain constraints on the inner curvature components $R^A_{\cdot mn}$ are implied. It goes without saying that these constraints are the propagation equations.

In N = 1 d = 4 Supergravity, for instance, if we insert in the Bianchi's identities:

$$\mathcal{D} R^{ab} = 0 \tag{6.2a}$$

$$\mathcal{D}R^a + R^{ab}\wedge V_b - i\overline{\Psi}\wedge\gamma_a\rho = 0 \tag{6.2b}$$

$$\mathcal{D}\rho - \frac{i}{2}\Sigma^{ab}\psi\wedge R_{ab} = 0 \tag{6.2c}$$

the rheonomic conditions given in (4.15), (4.17), namely:

$$R^a = 0 \tag{6.3a}$$

$$\rho = \rho_{ab} V^a \wedge V^b \tag{6.3b}$$

$$R^{ab} = R^{ab}_{\cdot\, cd} V^c \wedge V^d + \text{const}\left(\overline{\Psi}B^{ab}_m + \delta^{[a}_m \overline{\Psi}B^{b]n}_n\right)\wedge V^m \tag{6.3c}$$

we get the propagation equations (4.16):

$$\epsilon^{rsmn}\gamma_5\gamma_s\rho_{mn} = 0\ ;\ R^{am}_{\cdot\, bm} - \frac{1}{2}\delta^a_b R^{mn}_{\cdot\, mn} = 0 \tag{6.4}$$

In order to construct a supersymmetric quantum theory we need a representation of the supersymmetry algebra which closes (satisfies Jacobi identities) without use of the propagation equations (in the present case eqs. (6.4)). Such a representation is already provided by the AGCT transformations (interpreted on x-space) when no rheonomic condition is imposed on the curvatures and all outer components of the group curvatures are treated as extra independent fields. In this sense we can say that unrestricted AGCT transformations correspond to the maximal set of auxiliary fields.

Of interest to physics is the minimal set of auxiliary fields. This is the one which originates from the minimal parametrization of outer curvature components satisfying Bianchi identities without use of propagation equations. Therefore, the problem is that of writing down the maximal number of constraints on outer curvature components which can be implemented, still maintaining the inner ones unrestricted. These constraints are usually named "kinematical". The difference between standard superspace approach and ours is that, in the first there is no a priori suggestion for the choice of these kinematical constraints and you have to invent them out of ingenuity;in the second, instead, you already have a set of constraints (which is certainly consistent because it originates from a variational principle) but which is too restrictive for an off-shell theory. The problem is, therefore, that of finding the minimal number of rheonomic conditions to be relaxed. This relaxation gives birth to the minimal set of auxiliary fields. To say it in plain words, in the rheonomy scenario one trades the problem of understanding why certain constraints are to be imposed for the question of why some of them remain while others are to be removed in the off-shell formulation.

This point of view has some distinct advantages. Indeed, as we shall presently see the kinematical constraints, at least in the few theories where they have been derived, are just two. The first is the statement that the torsion is, so to say, "on-shell", namely, satisfies its own equation of motion obtained through the variation of the non-propagating 1-form ω^{ab} (or in impure theories the non-propagating 0-forms); the second one, as far as we know, cannot be written as yet, like the first, as an exterior algebra statement, but it corresponds to the elimination in the $\psi\wedge\psi$ component of the curvature of certain representations of the factorized subgroup H = Lorentz $\otimes$ SO(N). We do not yet have a full understanding of why this further constraint can be maintained, however, we feel that it resembles very much a representation selection rule whose final geometrical meaning will soon be retrieved. The first constraint on the other hand is crystal clear: it can be maintained because it comes down from the equation of a non-propagating field. Hence it cannot imply propagation equations.

In order to briefly illustrate how these ideas work we consider a few cases. A more detailed discussion of these examples, together with the explanation of an algebraic technique which allows a very quick and compact analysis of the Bianchi identities will be presented in a forthcoming paper(6).

<u>N = 1 d = 4 Supergravity</u>. In this case the equation of motion of the non-propagating ω^{ab} Lorentz connection is:

$$\epsilon_{abcd} R^c \wedge V^d = 0 \Longrightarrow \mathcal{R}^a = 0 \tag{6.5}$$

If we insert eq. (6.5) into the Bianchi identity (6.2b) we obtain:

$$\rho = \rho_{ab} V^a \wedge V^b + i A_m \gamma_5 \psi \wedge V^m + P \gamma_5 \gamma_m \psi \wedge V^m + i S \gamma_m \psi \wedge V^m + 2 \gamma_5 \Sigma^{ab} \psi \wedge A'_a V_b + \gamma_5 \psi \wedge \bar{\psi} \wedge \gamma_5 \chi \tag{6.6a}$$

$$R^{ab} = R^{ab}{}_{cd} V^c \wedge V^d + 2i \bar{\rho}^{m[a} \gamma^{b]} \psi \wedge V_m + i \bar{\rho}^{ab} \gamma^m \psi \wedge V_m - 2P \bar{\psi} \wedge \gamma_5 \Sigma^{ab} \psi + 2iS \bar{\psi} \wedge \Sigma^{ab} \psi - i \epsilon^{abcd} A'_c \bar{\psi} \wedge \gamma_d \psi \tag{6.6b}$$

where A_m, A'_a are two axial vectors, P is a pseudoscalar, S a scalar and χ a Majorana spinor. They are auxiliary fields appearing in the gravitino transformations law. It is evident that they are more than necessary since the mismatch between bosons and fermions was originally 6 and we just needed 6 extra bosonic auxiliary degrees of freedom rather than 10. This suggests that we can still implement a constraint. Indeed if we set $\chi = 0$, namely if we demand

$$\underline{\psi|\psi|}\,\rho = 0 \tag{6.7}$$

and we insert (6.6a), (6.6b) in the remaining Bianchis (6.2c) and (6.2a) we get:

$$\mathcal{A}'_a = \frac{1}{2}\mathcal{A}_a \tag{6.8}$$

together with equations which express the spinor derivative $\mathcal{D}_\alpha = \underline{\alpha}|\mathcal{D}$ of $\mathcal{A}_a, \mathcal{S}, \mathcal{P}$ in terms of $j^a = \epsilon^{abcd}\gamma_5\gamma_b\rho_{cd}$

It is therefore evident that the minimal set of auxiliary fields of Ferrara and van Nieuwenhuizen(13) and Stelle and West(24) are obtained simply by erasing the constraint

$$\underline{V}|\underline{\psi}|\rho = 0 \tag{6.9}$$

that is, implementing the torsion equation plus the spin 1/2 exclusion constraint (6.7).

We shall show in(6) that the minimal set of auxiliary fields for N = 2 supergravity both in d = 4(27) and d = 5(6) can be retrieved by an analogous procedure. Now, we discuss the coupling of the Wess-Zumino multiplet to N = 1 supergravity. This will enable us to show that contrary to what is usually said the minimal set of Supergravity auxiliary fields is not sufficient. Indeed in this coupling the constraint (6.7) is explicitly violated. This should not be surprising because the final on-shell transformation rules obtained from our approach coincide with those obtained from tensor calculus after elimination of the auxiliary fields and show a term proportional to ψ_μ in the $\delta\psi_\mu$ supersymmetry variation. This fact can happen only if:

$$\rho^\gamma_{\alpha\beta} \neq 0 \tag{6.10}$$

The reason why, in tensor calculus people insist that the minimal set (A_a, S, P) is sufficient is that they perform such field redefinitions so to have some of the minimal set auxiliary fields (S, P, A^a) containing ψ_μ terms.

<u>Wess-Zumino supermultiplet</u>. For matter multiplets, as for impure su-

pergravity theories, there is no "cohomology" framework to construct the action. However, we have the non-triviality condition which is still severely restrictive. The logical steps are the following[4].

STEP 1. Suppose that we start with the physical fields of some supermultiplet. In the present case they are a scalar $A(x)$ a pseudoscalar $B(x)$ and a Majorana spinor $\lambda(x)$. Let us now assume that these fields, originally defined on x-space, are extended to functions on the quotient space G/H where in this case $G = Osp(4/1)$ and $H = SO(1,3)$. This means that A, B and λ are now regarded as superfields on the supergravitational background described by the Vielbein $E = (V^a, \psi)$.

STEP 2. The exterior derivative of A, B and λ can be expanded in the (V^a, ψ) basis:

$$dA = \Phi_a V^a + \bar{\xi}\psi \tag{6.11a}$$

$$dB = \Pi_a V^a + i\bar{\tau}\gamma_5\psi \tag{6.11b}$$

$$\mathcal{D}\lambda = d\lambda + \frac{i}{2}\omega^{ab}\Sigma_{ab}\lambda = \Lambda_a V^a + M\psi \tag{6.11c}$$

where we have introduced the coefficients $\Phi_a, \Pi_a, \xi, \Lambda^a, \tau$ and M which are all superfields and have the following transformation properties under the Lorentz group:

Φ_a = vector

Π_a = axial vector

ξ, τ = Majorana spinor

M = Matrix in spinor space transforming in Majorana spinors

If we are in a flat supergravitational background (= rigid superspace), namely if:

$$R^a = \mathcal{D}V^a - \frac{i}{2}\bar{\psi}\wedge\gamma^a\psi = 0$$

$$\left.\begin{array}{l} R^{ab} = 0 \\ \rho = \mathcal{D}\psi = 0 \end{array}\right\} \underset{\text{fixing a Lorentz gauge}}{\Longleftrightarrow} \left\{\begin{array}{l} \omega^{ab} = 0 \\ V^a = dx^a + \frac{i}{2}\bar{\theta}\gamma^a d\theta \\ \psi = d\theta \end{array}\right. \tag{6.12}$$

the fields Φ_a, Π_a, ξ, τ, and M can be written as follows:

$$\Phi_a = \vec{P}_a A = \partial_a A \tag{6.13a}$$

$$\Pi_a = \vec{P}_a B = \partial_a B ; \Lambda_a = \vec{P}_a \lambda = \partial_a \lambda \tag{6.13b}$$

$$\xi = \vec{Q} A = \left(\frac{\partial}{\partial\bar{\theta}} - \frac{i}{2}\gamma^a \theta \partial_a\right) A \tag{6.13c}$$

$$\tau = -i\gamma_5 \vec{Q} B = -i\gamma_5\left(\frac{\partial}{\partial\bar{\theta}} - \frac{i}{2}\gamma^a \theta \partial_a\right) B \tag{6.13d}$$

$$M = (C\vec{Q}\lambda)^T = \left[C\left(\frac{\partial}{\partial\bar{\theta}} - \frac{i}{2}\gamma^a \theta \partial_a\right)\lambda\right]^T \tag{6.13e}$$

where $\vec{P}_a = \partial_a$ and $Q = \frac{\partial}{\partial\bar{\theta}} - \frac{i}{2}\gamma^a \theta \partial_a$ are the tangent vectors dual to the 1-forms V^a and ψ given in (6.12):

$$V^a(\vec{P}_b) = \delta^a_b \; ; \; V^a(\vec{Q}) = 0 \tag{6.14a}$$

$$\psi(\vec{P}_a) = 0 \; ; \; \psi(\vec{Q}) = \mathbb{1} \tag{6.14b}$$

STEP 3. In order to be supersymmetric our theory must be rheonomic. This means that, upon use of the equations of motion, the outer components ξ, τ, M of the exterior derivative, which is the analogue for matter fields of the group curvature, must be expressible in terms of the inner ones, namely Φ_a, Π_a, Λ_a or in terms of the A,B and λ fields themselves.

In full generality we can write

$$dA = \Phi_a V^a + \alpha_1 \bar{\lambda}\psi + i a_1 \bar{\Lambda}_a \gamma^a \psi \tag{6.15a}$$

$$dB = \Pi_a V^a + i\alpha_2 \bar{\lambda}\gamma_5\psi + a_2 \bar{\Lambda}_a \gamma^a \gamma_5 \psi \tag{6.15b}$$

$$\mathcal{D}\lambda = \Lambda_a V^a + i\beta_1 \gamma_m \psi \Phi^m + \beta_2 \gamma_5 \gamma_m \psi \Pi^m + b_1 A\psi + i b_2 B\psi \tag{6.15c}$$

We observe that while $\alpha_1, \alpha_2, \beta_1, \beta_2$ are dimensionless constants since:

$$[\psi] = [d\theta] = [L^{1/2}]$$
$$[B] = [A] = [L^{-1}] \quad ; \quad [\lambda] = [L^{-3/2}] \tag{6.16}$$

a_1, a_2, b_1, b_2 have instead the dimensions of a length. Therefore if we want a supersymmetric theory without dimensional parameters in the transformation laws we have to set ($a_1 = a_2 = b_1 = b_2 = 0$).

STEP 4. The remaining parameters $\alpha_1, \alpha_2, \beta_1, \beta_2$ are fixed by the implementation of the "Bianchi Identities" (= integrability conditions):

$$ddA = ddB = \mathcal{D}\mathcal{D}\lambda = 0 \tag{6.17}$$

The result is:

$$dA = \Phi_a V^a - \bar{\lambda}\psi \tag{6.18a}$$

$$dB = \Pi_a V^a - i\bar{\lambda}\gamma_5\psi \tag{6.18b}$$

$$\mathcal{D}\lambda = \Lambda_a V^a - \frac{1}{2}\left(i\gamma^n \psi \Phi_m + \gamma_5 \gamma_m \psi \Pi^m\right) \tag{6.18c}$$

The Bianchi consistent rheonomic conditions (6.18) correspond to the standard on-shell supersymmetry transformation rules of the W-Z multiplet:

$$\delta_\epsilon A = \ell_\epsilon A = \epsilon\rfloor dA = \bar{\epsilon}\lambda \tag{6.19a}$$

$$\delta_\epsilon B = \ell_\epsilon B = \epsilon\rfloor dB = i\bar{\epsilon}\gamma_5\lambda \tag{6.19b}$$

$$\delta_\epsilon\lambda = \ell_\epsilon\lambda = \epsilon\rfloor \mathcal{D}\lambda = -\frac{1}{2}\left(i\gamma^m\epsilon\partial_m A + \gamma_5\gamma^m\epsilon\partial_m B\right) \tag{6.19c}$$

The above transformations are on-shell because the Bianchi identities (6.17) + the conditions (6.18) imply the Dirac equation:

$$\gamma^m\Lambda_m = 0 \tag{6.20}$$

This result can be further illustrated by the following remarks. Assuming a flat supergravity background and comparing equations (6.18) with (6.13) we see that the following identities hold:

$$\lambda = -\vec{Q}A = i\gamma_5\vec{Q}B \tag{6.21a}$$

$$-\frac{1}{2}\left(i\gamma^n\partial_m A + \gamma^5\gamma^n\partial_m B\right) = -\left(C\vec{Q}\vec{Q}\right)^T A = i\left(C\vec{Q}\gamma_5\vec{Q}\right)^T B \tag{6.21b}$$

Eq. (6.21a) has a two-fold meaning. On one hand it defines λ as a derivative of A (or B); on the other it imposes a constraint on the complex superfield

$$\varphi(x,\theta) = A(x,\theta) + i\,B(x,\theta) \tag{6.22}$$

It is not difficult to see that (6.21a) is the familiar chirality condition which makes $\varphi(x,\theta)$ irreducible. Indeed a simple manipulation shows that (6.21a) is equivalent to:

$$\frac{1+\gamma_5}{2}\vec{Q}\varphi = 0 \; ; \quad \frac{1-\gamma_5}{2}\vec{Q}\varphi^* = 0 \tag{6.23}$$

The constraint (6.21b) on the other hand is quadratic in the derivatives and implies the equations of motion. It is therefore suggested that assuming only eqs. (6.18a) and (6.18b), while relaxing eq.(6.18c), should correspond to dealing with a single irreducible chiral superfield. As it is well known this latter contains, besides A,B and λ, two auxiliary fields F and G. These are immediately retrieved from Bianchi identities. Indeed relaxing (6.18c), while retaining (6.18a) and (6.18b), we find that the most general solution of the Bianchi identities is the following:

$$\begin{aligned}
dA &= \Phi_a V^a - \bar{\lambda}\psi \\
dB &= \Pi_a V^a - i\bar{\lambda}\gamma_5\psi \\
\mathcal{D}\lambda &= \Lambda_a V^a - \frac{1}{2}\left(i\gamma^m\psi\Phi_m + \gamma_5\gamma^n\psi\Pi_m\right) + \left(F + i\gamma_5 G\right)\psi \\
dF &= F_m V^m - \frac{i}{2}\bar{\psi}\gamma^m\Lambda_m \\
dG &= G_m V^m - \frac{1}{2}\bar{\psi}\gamma^5\gamma^m\Lambda_m
\end{aligned} \tag{6.24}$$

which gives the off-shell algebra of the W-Z system. We note that, as it is well known, the fields F and G do not have canonical dimensions. This is just the reason why they are auxiliary fields.

STEP 5. Knowing the rheonomic conditions consistent with Bianchi identities the action of the multiplet can now be constructed. We want a first order action and therefore we treat Φ_a and Π_a as independent dynamical variables.

Moreover we want the variation of these fields to yield the rheonomic conditions (6.18a) and (6.18b) which have been shown to be equivalent to chirality condition. All the remaining terms in the Lagrangian are then uniquely fixed by the non-triviality requirement. For a massive multiplet the action turns out to be:

$$A_{WZ} = \int\Big\{ -\big[\Phi^a(dA+\bar\lambda\psi)+\Pi^a(dB+i\bar\lambda\gamma_5\psi)-im\bar\lambda(A-iB\gamma_5)\gamma^a\psi - i\bar\lambda\gamma^a\mathcal{D}\lambda\big]\wedge V^b\wedge V^c\wedge V^d\epsilon_{abcd}+\frac{1}{4}\big[\Phi^a\Phi_a+\Pi_a\Pi^a+\frac{m}{2}(mA^2+mB^2+2\bar\lambda\lambda)\big]V^i\wedge V^j\wedge V^k\wedge V^\ell\epsilon_{ijk\ell}+\big[6(dA\wedge\bar\lambda\gamma_5\Sigma_{ab}\psi+idB\wedge\bar\lambda\Sigma_{ab}\psi - \frac{3i}{2}\bar\lambda\gamma_5\gamma_a\lambda\bar\psi\wedge\gamma_b\psi-3m(A\bar\psi\gamma_5\Sigma_{ab}\psi-iB\bar\psi\wedge\Sigma_{ab}\psi\big]\wedge V^a\wedge V^b + \frac{3i}{2}(AdB-BdA)\wedge\bar\psi\wedge\gamma^a\psi\wedge V_a\Big\} \tag{6.25}$$

The structure of this action is determined by the non-triviality requirement in the flat superspace. This implies the existence of all the various ψ terms. In flat superspace $\psi = d\theta$ and therefore these terms do not represent physical interactions. However, as soon as the theory is established in flat superspace it can be generalized to any supergravitational curved background replacing the left-invariant (V^a, ψ) with generic soft forms (V^a, ψ). In this case the ψ terms of the Lagrangian (6.25) become physical interactions between the gravitino and the W-Z fields. These interactions are exactly those found in the early papers on the coupling of matter to supergravity based on the Noether method(14). We stress that the coupling is already determined by the flat symmetry structure exactly as it happens with the coupling of matter to ordinary gauge theories or Einstein gravity.

The variational equations associated with (6.25) besides (6.18a) and (6.18b) yield:

$$\mathcal{D}\lambda = \Lambda_m V^m - \frac{1}{2}\left(i\gamma^n\psi\Phi_m+\gamma_5\gamma_m\psi\Pi^m\right)-\frac{m}{2}(A-iB\gamma_5)\psi \tag{6.26}$$

where Λ_m, Φ^n and Π^n satisfy the following propagation equations

$$i\gamma^n\Lambda_n+m\lambda=0 \qquad \partial_m\Phi^m+m^2A=0$$
$$\partial_m\Pi^m+m^2B=0 \tag{6.27}$$

STEP 6. As we have already emphasized the coupled action of supergravity and W-Z system is simply given by the sum of eq. (4.10) (with a = 0) and eq. (6.25). The supergravity eqs. (4.11), (4.12) and (4.13) are modified by the presence of source terms:

$$2\epsilon_{abcd} R^c \wedge V^d = J_{ab} \tag{6.28a}$$

$$2\epsilon_{abcd} R^{bc} \wedge V^d - 4\bar{\psi}\wedge\gamma_5\gamma_a\rho = J_a \tag{6.28b}$$

$$8\gamma_5\gamma_a\rho\wedge V^a - 4\gamma_5\gamma_a\psi\wedge R^a = 2j \tag{6.28c}$$

where J_{ab}, J_a, and 2j are defined by the matter Lagrangian as follows:

$$\underline{ab}\rfloor \mathcal{L}_{matter} = -J_{ab} \; ; \; \underline{a}\rfloor \mathcal{L}_{matter} = -J_a \; ; \; \underline{\alpha}\rfloor \mathcal{L}_{matter} = -2j_\alpha \tag{6.29}$$

Explicitly one finds:

$$J_{ab} = \frac{1}{2} t_m V^m \wedge V_a \wedge V_b \tag{6.30a}$$

$$\begin{aligned} 2j = {} & j^a V^b \wedge V^c \wedge V^d \epsilon_{abcd} - \frac{1}{2} t_a \gamma_5 \gamma_b \psi \wedge V_c \wedge V_d \epsilon^{abcd} \\ & - 8i \mathcal{A}_a \gamma_b \psi \wedge V^a \wedge V^b - 16 f \Sigma_{ab} \psi \wedge V^a \wedge V^b \\ & - 16 i \rho \Sigma_{ab} \psi \wedge V^a \wedge V^b - 8 \gamma_a \psi \wedge \bar{\chi} \gamma_5 \psi \wedge V^a \end{aligned} \tag{6.30b}$$

where

$$t_m = 3(\bar{\lambda}\gamma_5\gamma_m\lambda) \tag{6.31a}$$

$$\mathcal{A}_m = \frac{3}{8}(A\Pi_m - B\Phi_m) \tag{6.31b}$$

$$\mathcal{S} = -\frac{3}{8} m A \tag{6.31c}$$

$$\mathcal{P} = \frac{3}{8} m B \tag{6.31d}$$

$$\chi = \frac{3}{8}(A + iB\gamma_5)\lambda \tag{6.31e}$$

$$j^a = (\Phi^a + i\gamma_5 \Pi^a)\lambda + (\gamma_5 \vec{\Sigma}_{bc}\Phi_d + \vec{\Sigma}_{bc}\Pi_d)\epsilon^{abcd} + im(A - iB\gamma_5)\gamma^a\lambda \tag{6.31f}$$

As it is evident from eq.s (6.30) and (6.31) the supergravity sources besides VVV terms which modify the propagation equations of the graviton and the gravitino (and also the torsion) have ψVV and $\phi\psi$V terms which work as sources for outer components of the curvatures, namely, supergravity auxiliary fields. Indeed the new rheonomic conditions obtained from eq. (6.28) are the following:

$$\begin{aligned} R^a &= -\frac{1}{8}\epsilon^{abcd} t_b V_c \wedge V_d \\ \rho &= \rho_{ab} V^a \wedge V^b + i\mathcal{A}_a \gamma_5 \psi \wedge V^a + i\mathcal{S}\gamma_a \psi \wedge V^a \\ &\quad + \mathcal{P}\gamma_5\gamma_a \psi \wedge V^a + \gamma_5 \psi \wedge \bar{\chi}\gamma_5 \psi \end{aligned} \tag{6.32}$$

Comparing eq. (6.32) with eq. (6.6a) we see that the W-Z multiplet excites all the auxiliary fields except A'_a. The coupling is manifestly non-minimal because $A_a = A'_a$ and $\chi \neq 0$. The fact that χ is different from zero is unavoidable because, whatever approach we rely on, we must end up with an on-shell transformation rule of the ψ_μ field which contains a term of the following type:

$$\delta\psi_\mu = \text{other terms} + \frac{3}{8}\gamma_5\left(\bar{\lambda}(A - iB\gamma_5)\gamma_5\psi - \psi_\mu\bar{\lambda}(A - iB\gamma_5)\gamma_5\epsilon\right) \tag{6.33}$$

already featured in eq. (20) of ref.[15]. The appearance of ψ_μ in $\delta\psi_\mu$ means that $\underline{\psi}|\underline{\psi}|\rho \neq 0$ which in turn means a non-minimal set of auxiliary fields.

Sec. VII - Supersymmetric Yang-Mills Theories on the Supergroup Manifold.

Supersymmetric Yang-Mills theories have been formulated on the group-manifold in[16] and [17] in the N = 1 and N = 2 case. A thorough discussion of the N = 1, D = 10 theory[22] and of its N = 4 descendant in D = 4 will be given in a forthcoming paper[7]. These theories are matter theories and are constructed with the same procedure used for the W-Z multiplet.

As group manifold G one chooses (in d = 4)

$$G = \mathcal{G}_{Y.M.} \otimes \overline{Osp(4/N)} \tag{7.1}$$

where $\mathcal{G}_{Y.M.}$ is the local Yang-Mills group. The action will then be constructed in such a way as to produce factorization of the subgroup

$$H = \mathcal{G}_{Y.M.} \otimes SO(1,3) \otimes SO(N) \tag{7.2}$$

We briefly discuss the simplest example of N = 1 d = 4 theory and we give a few anticipations on the intriguing results obtained for the D = 10 theory by R. d'Auria, P. Fré and da Silva and to be published in the forthcoming paper[7].

N = 1 d = 4 Theory. We have a potential 1-form B for the Yang-Mills group Y.M. and a Majorana spinor 0-form λ. Both B and λ are matrices generating the adjoint representation of $\mathcal{G}_{Y.M.}$. The action was determined in[17] using the requirement of non triviality in flat superspace exactly the same way as we constructed the W-Z action. It is as follows:

$$A_{Y.M.}(N=1) = \int \mathcal{L}_{Y.M.}(N=1) = \int \mathrm{Tr} \Big\{ \mathcal{F} F^{ab} \wedge V^c \wedge V^d \epsilon_{abcd} - \\ - \frac{1}{12} F^{ab} F_{ab} V^i \wedge V^j \wedge V^k \wedge V^\ell \epsilon_{ijk\ell} + \frac{1}{2} (\bar{\lambda}\gamma_5\lambda) \bar{\psi} \wedge \Xi^{ab} \psi \wedge V_a \wedge V_b \\ + \frac{1}{4} (\bar{\lambda}\lambda) \bar{\psi} \wedge \Xi^{ab} \psi \wedge V^c \wedge V^d \epsilon_{abcd} + 2 \bar{\lambda}\gamma_5\gamma_m \psi \wedge V^m \wedge \mathcal{F} \\ - i F_{ab} V_c \wedge V_d \wedge \bar{\lambda}\gamma_m \psi \wedge V^m \epsilon_{abcd} - \frac{1}{3} \bar{\lambda}\gamma^a \nabla\lambda \wedge V^b \wedge V^c \wedge V^d \epsilon_{abcd} \Big\} \qquad (7.3)$$

(Trace in the Yang-Mills space).

In eq. (7.3) F^{ab} is a 0-form, also belonging to the adjoint presentation of $\mathcal{G}_{Y.M.}$, which is treated as an independent field. F is the Yang-Mills curvature:

$$\mathcal{F} = dB + B \wedge B \quad ; \quad \mathcal{F} = F_{ab} V^a \wedge V^b + i \bar{\lambda}\gamma_m \psi \wedge V^m \ \left(\text{from eq. } F^{ab}\right) \qquad (7.5)$$

The variation in the remaining fields λ and B gives

$$\nabla\lambda = \Lambda_m V^m + i F_{ab} \Xi^{ab} \psi \qquad (7.6)$$

together with the propagation equations:

$$\gamma^m \Lambda_m = 0 \qquad (7.7a)$$

$$\underline{m} \rfloor \mathcal{D} F^{ma} = 0 \qquad (7.7b)$$

We must note that if we insert the whole stock of rheonomic conditions (7.5) and (7.6) into the Bianchi identities

$$\nabla \mathcal{F} = 0 \quad ; \quad \nabla\nabla\lambda = 0 \qquad (7.8)$$

we obtain the propagation equations (7.7a) and (7.7b). Therefore the supersymmetry transformation laws associated with eqs. (7.5), (7.6) constitute the on-shell algebra. On the other hand if we insert only eq. (7.5) into (7.8) relaxing (7.6) we get:

$$\mathcal{F} = F_{ab} V^a \wedge V^b + i\bar{\lambda}\gamma_m \psi \wedge V^m$$
$$\nabla\lambda = \Lambda_m V^m + i F_{ab}\Sigma^{ab}\psi + i P\gamma_5 \psi \qquad (7.9)$$
$$\nabla P = P_m V^m - \frac{1}{2}\bar{\psi}\gamma_5\gamma^m \Lambda_m$$

where P is a pseudoscalar auxiliary field in the adjoint representation of $\mathcal{G}_{Y.M.}$. Eq. (7.9) gives rise to the off-shell supersymmetry algebra.

We conclude therefore that the variation of the F_{ab} field, similarly to what happens in the W-Z system for Φ_a and Π_a, plays a double role. On one hand it eliminates F_{ab} setting:

$$F_{ab} = \frac{1}{2}\left\{\partial_a B_b - \partial_b B_a + [B_a, B_b]\right\} \qquad (7.10)$$

and on the other it places an irreducibility condition on the superfield $B_\Lambda(x, \theta)$ which is the counterpart of the chirality condition (6.23). The irreducible superfield contains the off-shell multiplet (B_μ, λ, P).

We note in passing that the action (7.3) is determined by the requirement of non-triviality in flat space. When we go over to curved superspace the terms in ψ of eq. (7.3) become the physical interactions of the Yang-Mills fields with the gravitino. They are just the same as they were determined in the early Noether approach of paper[(14)].

Special Features of N = 1 D = 10. It is well know that the efforts to find a set of auxiliary fields for the N = 1 D = 10 super Yang-Mills theory have so far been frustrated. In paper[(7)] we shall give a full account of the investigation of this theory which two of us (R. d'Auria P. Fré) have carried on together with A. Da Silva. Since the computations involved are quite massive and based on a group-theoretical algebraic technique which requires a lot of explanations we refer the reader to our forthcoming paper for all the details. We just want to mention the new property which emerges from this analysis. For the N = 1 D = 10 theory we have a group-manifold action very similar in structure to eq. (7.3). The variation in F_{ab} gives similarly to (7.5):

$$\mathcal{F} = F_{ab} V^a \wedge V^b + i \bar{\lambda} \Gamma_m \psi \wedge V^m \qquad (7.11)$$

where now Γ_m are the SO(1,9) gamma matrices. It was shown in[17] that, while in d = 4 eq. (7.5) is an irreducibility condition on the $B_\Lambda(x, \theta)$ superfield which leaves the theory off-shell in D = 10 eq. (7.11), once inserted into the Bianchis, already implies the Dirac equation

$$\gamma^m \Lambda_m = 0 \qquad (7.12)$$

We have therefore examined in an extensive way the Bianchi identities of this exceptional theory and we have come to the conclusion that whatsoever linear or differential constraint of order 1, 2 or three on the components of the curvature $F = dB + B \wedge B$ places the theory on-shell. All the indications we have collected from our computations point toward the conclusion that in this maximally extended supersymmetric theory there are no kinematical constraints. This implies that all the components of the unrestrained superfield $B_\Lambda(x, \theta)$ are necessary auxiliary fields.

In particular we can immediately prove that the off-shell multiplet must contain spins larger than 1 notably a spin 3/2 representation is unavoidable.

A sketch of our derivation of N = 1 Super Yang-Mills theory in 10-space time dimensions[7] is the following. First we examine the structure of D = 10 Superspace which is spanned by a zehnbein 1-form V^a and a gravitino 1-form ψ, being both Majorana and Weyl:

$$C(\overline{\psi})^T = \psi \quad ; \quad \Gamma_{11}\psi = \psi \tag{7.13}$$

As we emphasize in (6), the key to understanding all properties of superspace and therefore all on-shell and off-shell representations of supersymmetry is given by the systematics of Fierz identities, namely the decomposition of wedge products $\psi\wedge\psi, \psi\wedge\psi\wedge\psi, \psi\wedge\psi\wedge\psi\wedge\psi$ into irreducible representations of the relevant Lorent group. In (7) we give a full account of both the Bosonic and Fermionic SO(1,9) representations, up to the spin $[2,2,2,1,1]$, and applying extensively the techniques developed in (6), we derive the following decomposition for D = 10 superspace:

$$(\psi\wedge\overline{\psi})_{\alpha\beta} = \frac{1}{16}(\Gamma^a)_{\alpha\beta} X^{(10)}_a + \frac{1}{5!\,32}(\Gamma^{a_1-a_5})_{\alpha\beta} X^{(126)}_{a_1-a_5} \tag{7.14a}$$

$$\psi\wedge\overline{\psi}\wedge\Gamma_a\psi = \frac{1}{336}\,\Xi^{(144)}_a \tag{7.14b}$$

$$\psi\wedge\overline{\psi}\wedge\Gamma_{a_1-a_5}\psi = \Xi^{(672)}_{a_1-a_5} + \frac{1}{336}\,\Gamma_{[a_1-a_4}\,\Xi_{a_5]} \tag{7.14c}$$

In eqs. (7.14) $X^{(10)}_a$ is the representation $[1,0,0,0,0]$ with 10 components (the vector) $X^{(126)}_{a_1-a_5}$ is the representation $[1,1,1,1,1]$ with 126 components (the antisymmetric antiself dual tensor), $\Xi^{(144)}_a \left(\Gamma^a\,\Xi^{(144)}_a = 0\right)$ is the representation $[3/2,\ 1/2,\ 1/2,\ 1/2, 1/2]$ and $\Xi^{(672)}_{a_1-a_5} \left(\Gamma^m\,\Xi^{(672)}_{a_1-a_4 m} = 0\right)$ is the representation $[3/2,\ 3/2,\ 3/2,\ 3/2,\ 3/2]$. The last two have respectively 144 and 672 components which sum up to 816, that is the number of independent components of the wedge product

$$\psi^\alpha\wedge\psi^\beta\wedge\psi^\gamma \quad .$$

Considering now the flat superspace or supergravity vacuum described by zero curvature and torsion:

$$R^{ab} = \mathcal{R}^{ab} = d\omega^{ab} + \omega^{ac} \wedge \omega^{bc} = 0 \tag{7.15a}$$

$$R^{a} = \mathcal{D}V^{a} - \frac{i}{2}\bar{\psi} \wedge \Gamma^{a}\psi = 0 \tag{7.15b}$$

$$\rho = \mathcal{D}\psi \tag{7.15c}$$

and given an arbitrary Yang-Mills group $\mathcal{G}$ whose Lie Algebra is:

$$[h_\alpha, h_\beta] = K^{\gamma}_{\cdot\alpha\beta} h_\gamma \tag{7.16}$$

we introduce a $\mathcal{G}$-Lie Algebra valued 1-form A in D=10 superspace:

$$A = A^{\alpha} h_\alpha = \left(A^{\alpha}_{a}(x,\theta)V^{a} + \bar{\psi} W^{\alpha}(x,\theta)\right) h_\alpha \tag{7.17}$$

and two $\mathcal{G}$-Lie Algebra valued 0-forms $\lambda = \lambda^{\alpha} h_\alpha$; $F_{ab} = F^{\alpha}_{ab} h_\alpha$ being a Majorana-Weyl spinor and an antisymmetric tensor respectively. In terms of these variables and of the Yang-Mills curvature:

$$\mathcal{F} = dA + A \wedge A \tag{7.18}$$

and using the already discussed superspace decomposition (7.14) in[7] we determine the most general action satisfying the following principles:

i) The action is the integral of a 10-form $\mathcal{L}_{Y.M.}$ performed on an arbitrary 10-dimensional surface immersed in superspace

$$\mathcal{A} = \int_{M_{10}} \mathcal{L}_{Y.M.}(D=10) \tag{7.19}$$

ii) The action is required to be stationary both with respect to the variations in the fields and in the surface M_{10}. This simply means that the differential form equations obtained by variation of $\mathcal{L}_{Y.M.}$ in the fields hold in all superspace.

iii) The 10-form $\mathcal{L}_{Y.H.}$ is SO(1,9) $\otimes\mathcal{G}$ gauge invariant and it is constructed out of A, λ, F_{ab}, V^a and ψ using the d and $\wedge$ operations only (No Hodge duality)

iv) $\mathcal{L}$ is at most cubic in the Yang-Mills fields A, λ, F_{ab} and at most quadratic in ψ

v) $\mathcal{L}$ is non-trivial. This means that in the flat superspace background (7.15) the variational equations associated to $\mathcal{L}$ admit non-trivial solutions, namely they do not force $F_{ab} = \lambda = \mathcal{F} = 0$

vi) Restricted on space-time the variational equations associated to $\mathcal{L}$ yield the Maxwell and Dirac wave-equations of the spin 1 and spin 1/2 field respectively.

The action which satisfies all the above criteria is unique up to a 1-parameter (a) freedom. It reads as follows:

$$\begin{aligned}\mathcal{L}_{Y.H.} = \mathrm{Tr}\Big\{ & \mathcal{F}\wedge F^{a_1a_2} V^{a_3}\wedge\ldots\wedge V^{a_{10}}\,\epsilon_{a_1-a_{10}} + \\ & + 2i\, F^{a_1a_2}\bar{\lambda}\Gamma_m\psi\wedge V^m\wedge V^{a_3}\wedge\ldots\wedge V^{a_{10}}\,\epsilon_{a_1-a_{10}} \\ & + \frac{4}{9} i\,\bar{\lambda}\Gamma^{a_1}\mathcal{D}\lambda\wedge V^{a_2}\wedge\ldots\wedge V^{a_{10}}\,\epsilon_{a_1-a_{10}} \\ & - \frac{1}{90} F_{ab}F^{ab} V^{a_1}\wedge\ldots\wedge V^{a_{10}}\,\epsilon_{a_1-a_{10}} \\ & + \frac{8}{3} i\,\bar{\lambda}\Gamma^{a_1\ldots a_3}\psi\wedge\mathcal{F}\wedge V^{a_4}\wedge\ldots\wedge V^{a_{10}}\,\epsilon_{a_1-a_{10}}\end{aligned} \tag{7.20}$$

$$- 84\, i\, \mathcal{F} \wedge A \wedge \bar{\psi} \wedge \Gamma^{a_1 - a_5} \psi \wedge V_{a_1} \wedge \ldots \wedge V_{a_5}$$

$$+ \left(1 + \frac{8}{3} a\right) \bar{\lambda} \Gamma^{a_1 - a_3} \lambda\, \bar{\psi} \wedge \Gamma_m \psi \wedge V^m \wedge V^{a_4} \wedge \ldots \wedge V^{a_{10}} \epsilon_{a_1 - a_{10}}$$

$$+ a\, \bar{\lambda} \Gamma^{a_1 a_2 m} \lambda\, \bar{\psi} \wedge \Gamma_m \psi \wedge V^{a_3} \wedge \ldots \wedge V^{a_{10}} \epsilon_{a_1 - a_{10}}$$

$$- \frac{84}{3} i\, A \wedge A \wedge A \wedge \bar{\psi} \wedge \Gamma^{a_1 - a_5} \psi \wedge V_{a_1} \wedge - \wedge V_{a_5} \Big\}$$

The action (7.20) is a first-order action entirely written in terms of superfields $\lambda(x,\theta)$, $F_{ab}(x,\theta)$, $A_a(x,\theta)$, $W(x,\theta)$ and, being geometric, it is coordinate invariant in superspace. This means that it is invariant under supersymmetry transformation closing an off-shell algebra, namely:

$$\delta_\epsilon A = \epsilon \rfloor \mathcal{F} \tag{7.21a}$$

$$\delta_\epsilon \lambda = \epsilon \rfloor \nabla \lambda \tag{7.21b}$$

$$\delta_\epsilon \mathcal{F}_{ab} = \epsilon \rfloor \nabla \mathcal{F}_{ab} \tag{7.21c}$$

The question whether there exists a second-order off-shell formulation with a limited amount of auxiliary fields can be answered by a careful study of Bianchi identities:

$$\nabla\mathcal{F} = d\mathcal{F} + A\wedge\mathcal{F} = \nabla\nabla\lambda = \nabla\nabla\mathcal{F}_{ab} = 0 \tag{7.22}$$

in order to discover whether some of the components of the curvature $\mathcal{F}$ or of the covariant derivative $\nabla\lambda$ can be set to zero without implying the Maxwell or Dirac wave equations. In[7] we wrote down the most general parametrization of $\mathcal{F}$ and $\nabla\lambda$ in terms of irreducible representations:

$$\mathcal{F} = F'_{ab} V^a \wedge V^b - 2\overline{\psi}\left(\xi^{(144)}_a - i\Gamma_a \lambda'^{(16)}\right)\wedge V^a + i\mathcal{B}^{(10)}_a \overline{\psi}\wedge\Gamma^a\psi + i\mathcal{B}^{(126)}_{a_1 - a_5}\overline{\psi}\wedge\Gamma^{a_1 - a_5}\psi \tag{7.23a}$$

$$\mathcal{D}\lambda = \mathcal{D}_m\lambda' V^m + \left(U^{(1)}\mathbb{1} + U^{(45)}_{a_1a_2}\Gamma^{a_1a_2} + U^{(210)}_{a_1 - a_4}\Gamma^{a_1 - a_4}\right)\psi \tag{7.23b}$$

and similarly for all the derivatives of the irreducible representations appearing in (7.23), for instance:

$$\mathcal{D}\mathcal{B}^{(10)}_a = \mathcal{D}_m\mathcal{B}^{(10)}_a V^m + \overline{\psi}\left(\chi^{(144)}_a + \frac{1}{10}\Gamma_a\chi^{(16)}\right) \tag{7.24}$$

Imposing the Bianchi identities and working out their information content by means of the superspace decomposition (7.14) we obtain a very long string of relations among the irreducible representations ξ_a, $\mathcal{B}_a$, $\mathcal{B}_{a_1 - a_5}$, $U^{(1)}$, $U^{(45)}_{a_1a_2}$, $\ldots$. Once this relations are used and a parametrization of the curvature and covariant derivatives in terms of an independent set of irreducible representations is obtained we reach the following conclusion: elimination of whatsoever of the independent irreps implies the wave-equations:

$$\mathcal{D}_m F^{ma'} = 0 \quad ; \quad \Gamma^m\mathcal{D}_m\lambda' = 0 \tag{7.25}$$

Hence no constraint can be imposed without going on-shell.
On the other hand the variational equations associated to the action (7.20) yield eq. (7.11) and

$$\mathcal{D}\lambda = \mathcal{D}_m\lambda - \frac{1}{4}\Gamma_{ab}\psi F^{ab}$$

together with eqs. (7.25). Therefore there is an off-shell first order formulation which propagates only the physical fields but there is no second order off-shell formulation of D = 10 super Yang-Mills Theory. In the second part of[7] we also perform the dimensional reduction $D=10 \longrightarrow D=4$ and we obtain the off-shell first order action of $\mathcal{N}=4$ super Yang-Mills; since the computations involved are very massive and the result is very long to write we refer the reader to the original paper.

Aknowledgements.

The authors are indebted to their graduate students and collaborators F. Giani, E. Maina and to their friend A.J. Da Silva for essential assistence in the preparation of the present course of lectures. It is also a pleasure for them to express their gratitude to Prof. A. Salam for hospitality at the International Center of Theoretical Physics in Trieste where these lectures were first issued in preprint form.

REFERENCES

(1) - A. D'Adda, R. d'Auria, P. Fré, T. Regge - Rivista del Nuovo Cimento N° 6 (1980)

(2) - R. d'Auria, P. Fré, T. Regge - Rivista del Nuovo Cimento 3, N 12 (1980).

(3) - R. d'Auria, P. Fré - Nucl. Phys. B173 (1980) 456.

(4) - R. d'Auria, P. Fré, T. Regge - The Supergroup manifold formulation of the Wess-Zumino multiplet ... - Princeton preprint Nov. 1980 to appear in Nucl. Phys.

(5) - R. d'Auria, P. Fré, E. Maina, T. Regge - Geometrical First Order Supergravity in 5 space-time dimensions - IFTT 404 (1981) to appear in Annals of Physics.

(6) - R. d'Auria, P. Fré, E. Maina, T. Regge - IFTT 410 (1981)

(7) - R. d'Auria, P. Fré, A. Da Silva - IFTT 412 (1981)

(8) - L. Castellani, P. Fré, P. van Nieuwenhuizen - ITP-SB-81-4 to appear in Annals of Physics.

(9) - C. Chevalley, S. Eilenberg - Trans. Am. Math. Soc., 63, 85(1948).

(10)- S. Ferrara, J. Scherk, B. Zumino - Phys. Lett. B66, 35 (1977).

(11)- S. Ferrara, P. van Nieuwenhuizen - Phys. Rev. Lett. 37, 1669 (1976).

(12)- S. Ferrara, P. van Nieuwenhuizen - Phys. Lett. 78B 573 (1978).

(13)- S. Ferrara, P. van Nieuwenhuizen - Phys. Lett. 74B (1978) 333.

(14)- S. Ferrara, F. Gliozzi, J. Scherk, P. van Nieuwenhuizen - Nucl. Phys. B117 (1976) 333.

(15)- S. Ferrara, D.Z. Freedman, P. van Nieuwenhuizen, P. Breitenlohner F. Gliozzi, J. Scherk - Phys. Rev D15, 1013, (1977)

(16)- P. Fré - "N=1 Supersymmetric Yang-Mills Theory on the Supergroup Manifold and the Supercurrent" CALT-68-806 - Lettere al Nuovo Cimento 30 (1981) 507

(17)- P. Fré - Remarks on the Supergroup Manifold Approach to supersymmetric Yang-Mills Theories and the explicit construction of the N = 2 case - ITP-SB-80-79 - Nucl. Phys. B187 (1981) 376

(18)- P. Fré - Nucl. Phys. B 186 (1981), 44.

(19)- P. Fré - Nucl. Phys. B179 (1981) 417

(20) - D.Z. Freedman, A. Das - Nucl. Phys. B120, 221 (1977).

(21) - D.Z. Freedman, P. van Nieuwenhuizen, S. Ferrara - Phys. Rev. D13, 3214 (1976) - S. Deser, B. Zumino - Phys. Lett. B, 62, 335 (1976).

(22) - F. Gliozzi, J. Scherk, D. Olive - Nucl. Phys. B133 (1978) 253. L. Brink, J. Schwarz, J. Scherk - Nucl. Phys. B121 (1977) 77.

(23) - Y. Ne'eman, T. Regge - Rivista del Nuovo Cimento 1, 5, 1 (1978).

(24) - K. Stelle, P.C. West - Phys. Lett. 74B (1978) 330.

(25) - P.K. Townsend, P. van Nieuwenhuizen - Phys. Lett. 67B, 439 (1977).

(26) - J. Wess, B. Zumino - Nucl. Phys. B70 (1974) 39.

(27) - B. de Wit, Van Holten - J.W. Nucl. Phys. B155 (1979) 530 E.S. Fradkin, M.A. Vasiliev - Lettere al Nuovo Cimento 25 (1979) 79